Progress in Brain Research
Volume 266

Brain Protection Strategies and Nanomedicine

Serial Editor

Vincent Walsh
Institute of Cognitive Neuroscience
University College London
17 Queen Square
London WC1N 3AR UK

Editorial Board

Mark Bear, *Cambridge, USA.*
Medicine & Translational Neuroscience

Hamed Ekhtiari, *Tehran, Iran.*
Addiction

Hajime Hirase, *Wako, Japan.*
Neuronal Microcircuitry

Freda Miller, *Toronto, Canada.*
Developmental Neurobiology

Shane O'Mara, *Dublin, Ireland.*
Systems Neuroscience

Susan Rossell, *Swinburne, Australia.*
Clinical Psychology & Neuropsychiatry

Nathalie Rouach, *Paris, France.*
Neuroglia

Barbara Sahakian, *Cambridge, UK.*
Cognition & Neuroethics

Bettina Studer, *Dusseldorf, Germany.*
Neurorehabilitation

Xiao-Jing Wang, *New York, USA.*
Computational Neuroscience

Progress in Brain Research
Volume 266

Brain Protection Strategies and Nanomedicine

Edited by

Hari Shanker Sharma
International Experimental Central Nervous System Injury & Repair (IECNSIR), Department of Surgical Sciences, Anesthesiology & Intensive Care Medicine, Uppsala University Hospital, Uppsala University, Uppsala, Sweden

Aruna Sharma
International Experimental Central Nervous System Injury & Repair (IECNSIR), Department of Surgical Sciences, Anesthesiology & Intensive Care Medicine, Uppsala University Hospital, Uppsala University, Uppsala, Sweden

Elsevier
Radarweg 29, PO Box 211, 1000 AE Amsterdam, Netherlands
The Boulevard, Langford Lane, Kidlington, Oxford, OX5 1GB, United Kingdom
50 Hampshire Street, 5th Floor, Cambridge, MA 02139, United States

First edition 2021

Copyright © 2021 Elsevier B.V. All rights reserved.

No part of this publication may be reproduced or transmitted in any form or by any means, electronic or mechanical, including photocopying, recording, or any information storage and retrieval system, without permission in writing from the publisher. Details on how to seek permission, further information about the Publisher's permissions policies and our arrangements with organizations such as the Copyright Clearance Center and the Copyright Licensing Agency, can be found at our website: www.elsevier.com/permissions.

This book and the individual contributions contained in it are protected under copyright by the Publisher (other than as may be noted herein).

Notices

Knowledge and best practice in this field are constantly changing. As new research and experience broaden our understanding, changes in research methods, professional practices, or medical treatment may become necessary.

Practitioners and researchers must always rely on their own experience and knowledge in evaluating and using any information, methods, compounds, or experiments described herein. In using such information or methods they should be mindful of their own safety and the safety of others, including parties for whom they have a professional responsibility.

To the fullest extent of the law, neither the Publisher nor the authors, contributors, or editors, assume any liability for any injury and/or damage to persons or property as a matter of products liability, negligence or otherwise, or from any use or operation of any methods, products, instructions, or ideas contained in the material herein.

ISBN: 978-0-323-98927-5
ISSN: 0079-6123

For information on all Elsevier publications
visit our website at https://www.elsevier.com/books-and-journals

Publisher: Zoe Kruze
Acquisitions Editor: Sam Mahfoudh
Developmental Editor: Federico Paulo Mendoza
Production Project Manager: Abdulla Sait
Cover Designer: Alan Studholme

Typeset by STRAIVE, India

Contributors

Artur Biktimirov
Department of Fundamental Medicine, School of Biomedicine, Far Eastern Federal University, Vladivostok, Russia

Andrey S. Bryukhovetskiy
ZAO NeuroVita Clinic of the Interventional Restorative Neurology and Therapy, Moscow, Russia

Igor Bryukhovetskiy
Department of Fundamental Medicine, School of Biomedicine, Far Eastern Federal University; Laboratory of Pharmacology, National Scientific Center of Marine Biology, Far East Branch of the Russian Academy of Sciences; School of Biomedicine, Medical Center, Far Eastern Federal University, Vladivostok, Russia

Anca D. Buzoianu
Department of Clinical Pharmacology and Toxicology, "Iuliu Hatieganu" University of Medicine and Pharmacy, Cluj-Napoca, Romania

Rudy J. Castellani
Department of Pathology, University of Maryland, Baltimore, MD, United States

Lianyuan Feng
Department of Neurology, Bethune International Peace Hospital, Shijiazhuang, Hebei Province, China

Lyudmila Y. Grivtsova
A. Tsyb Medical Radiological Center of the FGBU National Medical Research Radiological Center of the Ministry of Health of the Russian Federation, Obninsk, Russia

Chao He
Department of Neurosurgery, Zhuji People' Hospital of Zhejiang Province, Zhuji Affiliated Hospital of Shaoxing University, Zhuji, Zhejiang, China

Hongyun Huang
Institute of Neurorestoratology, Third Medical Center of General Hospital of PLA, Beijing, China

José Vicente Lafuente
LaNCE, Department of Neuroscience, University of the Basque Country (UPV/EHU), Leioa, Bizkaia, Spain

Cong Li
Department of Neurosurgery, Chinese Medicine Hospital of Guangdong Province; The Second Affiliated Hospital, Guangzhou University of Chinese Medicine, Yuexiu District, Guangzhou, China

Xu Longbiao
Department of Neurosurgery, Zhuji People' Hospital of Zhejiang Province, Zhuji Affiliated Hospital of Shaoxing University, Zhuji, Zhejiang, China

Igor Manzhulo
Laboratory of Pharmacology, National Scientific Center of Marine Biology, Far East Branch of the Russian Academy of Sciences, Vladivostok, Russia

Preeti K. Menon
Department of Biochemistry and Biophysics, Stockholm University, Stockholm, Sweden

Dafin F. Muresanu
Department of Clinical Neurosciences, University of Medicine & Pharmacy; "RoNeuro" Institute for Neurological Research and Diagnostic, Cluj-Napoca, Romania

Ala Nozari
Anesthesiology & Intensive Care, Massachusetts General Hospital, Boston, MA, United States

Asya Ozkizilcik
Department of Biomedical Engineering, University of Arkansas, Fayetteville, AR, United States

Oleg Pak
School of Biomedicine; Department of Neurosurgery, Medical Center, Far Eastern Federal University, Vladivostok, Russia

Ranjana Patnaik
Department of Biomaterials, School of Biomedical Engineering, Indian Institute of Technology, Banaras Hindu University, Varanasi, India

Seaab Sahib
Department of Chemistry & Biochemistry, University of Arkansas, Fayetteville, AR, United States

Aruna Sharma
International Experimental Central Nervous System Injury & Repair (IECNSIR), Department of Surgical Sciences, Anesthesiology & Intensive Care Medicine, Uppsala University Hospital, Uppsala University, Uppsala, Sweden

Hari Shanker Sharma
International Experimental Central Nervous System Injury & Repair (IECNSIR), Department of Surgical Sciences, Anesthesiology & Intensive Care Medicine, Uppsala University Hospital, Uppsala University, Uppsala, Sweden

Valery Shevchenko
Institute of Carcinogenesis, N.N. Blokhin National Medical Research Center of Oncology, Moscow, Russia

Per-Ove Sjöquist
Division of Cardiology, Department of Medicine, Karolinska Institutet, Karolinska University Hospital, Stockholm, Sweden

Stephen D. Skaper
Anesthesiology & Intensive Care, Department of Pharmacology, University of Padua, Padova, Italy

Z. Ryan Tian
Department of Chemistry & Biochemistry, University of Arkansas, Fayetteville, AR, United States

Lin Wang
Department of Neurosurgery, Second Affiliated Hospital of Zhejiang University School of Medicine, Hangzhou, Zhejiang, China

Lars Wiklund
International Experimental Central Nervous System Injury & Repair (IECNSIR), Department of Surgical Sciences, Anesthesiology & Intensive Care Medicine, Uppsala University Hospital, Uppsala University, Uppsala, Sweden

Qilin Wu
Department of Neurosurgery, Zhuji People' Hospital of Zhejiang Province, Zhuji Affiliated Hospital of Shaoxing University, Zhuji, Zhejiang, China

Gang Yang
Department of Neurosurgery, Zhuji People' Hospital of Zhejiang Province, Zhuji Affiliated Hospital of Shaoxing University, Zhuji, Zhejiang, China

Sergei Zaitsev
School of Biomedicine, Medical Center, Far Eastern Federal University, Vladivostok, Russia

Zhiquiang Zhang
Department of Neurosurgery, Chinese Medicine Hospital of Guangdong Province; The Second Affiliated Hospital, Guangzhou University of Chinese Medicine, Yuexiu District, Guangzhou, China

Ming Zhao
Department of Neurosurgery, Zhuji People' Hospital of Zhejiang Province, Zhuji Affiliated Hospital of Shaoxing University, Zhuji, Zhejiang, China

Contents

Acknowledgments

We are highly indebted to Sam Mahfoudh, Acquisitions Editor at Elsevier Science, Oxford, United Kingdom, for his excellent and constant help, encouragement, and support since the inception of this volume and throughout various stages of compilation, communication, and support leading to its realization. Without his support, the volume in its current form would not have been possible. We thank Federico Paulo S. Mendoza, Editorial Project Manager, Elsevier, Manila, Philippines, for helping us with the compilation and publication procedure. We are grateful to Abdulla Sait, Project Manager, Elsevier, Chennai, India, for his excellent help during the production of the volume. We also thank Suraj Sharma, Blekinge Institute of Technology, Karlskrona, Sweden, and Dr. Saja Alshafeay, University of Arkansas, Fayetteville, AR, United States, for providing computer and graphics support.

Hari Shanker Sharma
Aruna Sharma
Uppsala University, Uppsala, Sweden

Preface

"todo hombre puede ser, si se lo propone, escultor de su propio cerebro"
[Every man can, if he so desires, become the sculptor of his own brain]

Ramón y Cajal (1909)

Brain protection or neuroprotection refers to the mechanisms and/or strategies that protect the central nervous system (CNS) from adverse cellular or system events causing trauma, stroke, ischemia, or strenuous stress leading to inflammation, apoptosis, or degeneration (Lipton et al., 1998). Neurodegenerative diseases including Alzheimer's disease, Parkinson's disease, and multiple sclerosis as well as acute insults to the CNS cause disruption of the blood–brain barrier (BBB), induce cerebral edema, and neuroinflammation, leading to molecular and cellular brain pathology (Leist and Nicotera, 1998; Montal, 1998; Sharma, 2009; Sharma et al., 1998a). There are reasons to believe that the BBB is the gateway to various neurological diseases (Rosenberg, 2012; Sharma, 2009; Sharma et al., 1998a; Sweeney et al., 2019). This is evident from the fact that the BBB, the regulatory physiological barrier between the blood and the brain and vice versa, maintains the homeostasis of the CNS in normal conditions (Weiss et al., 2009). Another barrier property lies between the blood and the cerebrospinal fluid (CSF), known as the blood-CSF barrier (BCSFB) responsible for maintaining the composition of the CSF, in which the brain and the spinal cord covered with meninges are suspended within the cranial cavity and the spinal column (Bradbury, 2000; Johanson et al., 2011; Pardridge, 2016; Sharma, 1982; Sharma and Johanson, 2007). These regulatory barriers, the BBB and the BCSFB, are responsible for the exchange of nutrients to the brain and spinal cord and for the removal of the metabolic waste products (Engelhardt and Sorokin, 2009). Any perturbation among the BBB or BCSFB allows accumulation of unwanted substances into the brain and spinal cord, leading to disease processes that in turn precipitate into neurodegeneration (Sharma, 1982, 1999; Sharma and Westman, 2004).

The BBB and BCSFB are disturbed in pathological conditions due to alterations in the cellular, molecular, or biochemical dysfunction in association with other immunological or synaptic abnormalities (Sharma, 2009, 2004a,b; Sharma and Westman, 2004). This is well reflected in changes in neuromodulators and neurotransmitters and their receptors (Sharma, 2009). Thus, the restoration of the BBB and BCSFB is one of the important factors in strategies for brain protection in neurological diseases (Sharma and Sharma, 2010; Sharma and Westman, 2004).

The brain and spinal cord have intrinsic properties to adapt to various conditions of adverse circumstances by changing the synaptic interaction and/or neuronal network (Bain, 1868; Carpenter, 1874; Dumont, 1876; Ramon y Cajal, 1999; Zilles, 1992). These adaptive properties of the CNS were first called "plasticity" in relation to behavioral changes by James (1890) in *The Principles of Psychology* (James, 1890). The term "neuroplasticity" was coined by a famous Polish neuroscientist Jerzy Konorski in 1948. According to Konorski, neurons in close contact with the active neuronal network may incorporate themselves in adapting the new situations (Konorski, 1948). This adaptation in the CNS may be structural, synaptic, or biochemical at cellular or molecular levels in response to new situations that occur throughout the life (Fuchs and Flügge, 2014; Griesbach and Hovda, 2015; Gulyaeva, 2017; Stahnisch and Nitsch, 2002; Taub et al., 1999). Later, Hilmar Bading used the term "acquired neuroprotection" to refer to the altered physiological activity of the synaptic functions that are needed for neurons to survive under the changed situations (Bading and Greenberg, 1991). These changed synaptic activities related to neuroprotection induced by calcium entry into the cell nucleus through synaptic NMDA receptors (Bading and Greenberg, 1991; Hardingham and Bading, 2003; Zhang et al., 2009).

The term "neuroprotection" or "brain protection" is used invariably to denote any kind of insults that induces brain pathology, and this is partially restored using various pharmacological, cellular, and/or other biobehavioral approaches (Sharma and Sharma, 2010; Sharma and Westman, 2014). Neuroprotection includes the rescue of all components of the CNS including glial cells, endothelial cells, and the neurons (Sharma, 2007a,b; Sharma and Sharma, 2012a,b; Sharma et al., 2007). Likewise, brain protection is not limited to the rescue of brain components alone as protection of the spinal cord structure is also included (Sharma, 2005a,b; Sharma and Olsson, 1990; Sharma et al., 2007). Anatomically, the spinal cord is the extension of brain components and according to some scientists brain is the outgrowth of the spinal cord (Windle, 1980; Sharma et al., 1998; Sharma and Westman, 2004). Thus, physiological properties of the brain and spinal cord in relation to trauma and alterations in the BBB or blood-spinal cord barrier (BSCB), edema formation, and cell injury are largely similar in nature (Sharma, 2004a,b, 2005a, 2007a,b).

Several clinical and preclinical studies suggest that the maintenance of a healthy BBB or BSCB is the key to neuroprotection (Griffiths, 1978; Griffiths et al., 1978; Sharma, 2009). This is further evident from the findings that almost all clinical or preclinical disease stages exhibit leakage across the BBB or BSCB (Sharma and Westman, 2004). In several early reports, the leakage of the BSCB under pathological conditions are referred to as the BBB leakage (Beggs and Waggener, 1975; Fabis et al., 2007; Sharma and Westman, 2004). Thus, the term BBB also refers to BSCB breakdown (Griffiths, 1978; Griffiths et al., 1978; Sharma and Westman, 2014). Leakage of protein across the BBB indicates vasogenic edema formation that is instrumental in CNS cellular pathology (Sharma and Olsson, 1990; Sharma et al., 1991a,b,c). Breakdown of the BBB apart from trauma is also seen following several stress situation including depression (Sharma and Sharma, 2021; Sharma et al., 1991a,b,c, 1995), immobilization (Sharma and Dey, 1986; Sharma et al., 2013), heat stress or heat stroke (Sharma and Dey, 1987; Sharma and Hoopes, 2003; Sharma, 2005b,c, 2006a,b; Sharma et al., 2011, 2015a,b,c), substance abuse, and sleep deprivation (Sharma et al., 2015b, 2019). These stressful situations and traumatic injuries to the CNS are associated with cellular and molecular pathologies. Apart from acute or chronic stressful situations, neurodegenerative changes following Alzheimer's or Parkinson's disease are equally associated with pathological changes in the CNS in relation to the BBB alterations (Sharma et al., 2012, 2016, 2020, 1990).

Thus, therapeutic strategies to reduce cellular and molecular pathologies require restoration of the BBB in brain disease. To further enhance therapeutic strategies in brain diseases to induce superior neuroprotection, the use of nanomedicine or nanodelivery of effective agents is recommended (Santos et al., 2012; Sharma and Sharma, 2007, 2012a,b, 2013; Thangudu et al., 2020). Available evidences suggest that the nanoformulation of neurotrophic factors could be the possible therapeutic agents for enhancing superior neuroprotection. Thus, the use of nanomedicine or nanodelivery of effective compounds is the need of the hour to treat neurological diseases. However, there are principal differences between experimental and clinical brain diseases regarding treatment approaches. Clinical neurological diseases are often accompanied by several comorbidity factors, and thus treatments alone based on reclinical research will not working effectively. To counteract these problems, preclinical research in neurological diseases associated with comorbidity factors are needed to explore further. In such situations, the drug therapy should be modified to take care of associated comorbidity factors during the treatment procedures. One of the key factors in modifying suitable therapeutic measures associated with comorbidity factors is to use nanomedicine. There are evidences that nanodelivery may enhance drug penetration across the BBB and could act longer in the biological system for enhanced therapeutic effects. Thus, the future clinical therapy associated with comorbidity factors requires nanomedicine for the enhanced quality of life of patients.

This volume deals with the use of nanomedicine or nanodelivery of effective compounds in situations of neurodegenerative disease models or trauma and stressful models in inducing neuroprotection. The salient feature of this volume is to explore nanomedicine in brain diseases associated with several common comorbidity factors. There are reasons to believe that this volume will open up new avenues of treatments in brain diseases associated with several types of comorbidity factors and encourage further research in this key area of neurological disorders.

The volume comprises 11 invited peer-reviewed chapters from the world's leading experts in neurosciences in the cutting-edge clinical and preclinical research areas across the globe. This excellent collection of chapters may enhance further research and development in the treatment strategies of neurological disorders and nanomedicine.

In Chapter 1, Aruna Sharma (Uppsala, Sweden) and colleagues discuss the evidences showing the involvement of histamine receptors in Parkinson's disease. Their work suggests that the nanowired delivery of histamine H3 and H4 modulators together with histamine antibodies induces profound neuroprotection in PD.

In Chapter 2, Andrey S. Bryukhovetskiy (Moscow, Russian Federation) and colleagues describe novel biological markers for the diagnosis of malignant and neurodegenerative diseases using immunologic profiles of autologous hematopoietic stem cells.

In Chapter 3, Hari Shanker Sharma (Uppsala, Sweden) and colleagues discuss the neuroprotective effects of insulin-like growth factor-1 (IGF-1) in spinal cord injury after intoxication with Ag-, Cu-, and Al-engineered nanoparticles-induced exacerbation of cord pathology. Their work suggests that IGF-1 is able to attenuate nitric oxide synthase (NOS) upregulation following nanoparticle-induced exacerbation in spinal cord injury.

Drugs of abuse including methamphetamine induce neurotoxicity. Military personnel are prone to high-altitude hazard where brain trauma could easily occur during combat operations. In some people, methamphetamine abuse is quite common at high altitude. Hari Shanker Sharma (Uppsala, Sweden) and colleagues in Chapter 4 explore the effect of high altitude on concussive head injury and its effect on methamphetamine intoxication. Their novel data suggest that high altitude exacerbates head injury-induced brain pathology and methamphetamine further aggravates the situation. The therapeutic potential of antioxidants and nanomedicine is discussed in great detail.

In Chapter 5, Oleg Pak (Vladivostok, Russian Federation) and colleagues present novel data regarding the effectiveness of bortezomib and temozolomide in recurrent human glioblastoma cells that are resistant to radiation.

Excitotoxicity plays a key role in concussive head injury (CHI)-induced brain pathology. In Chapter 6, Hari Shanker Sharma (Uppsala, Sweden) and colleagues discuss the potential role of excitatory amino acids and inhibitory amino acids in CHI. Their data indicate that cerebrolysin—a balanced composition of several neurotrophic factors and active peptide fragments—restores the balance between excitatory and inhibitory amino acids in CHI.

In Chapter 7, Artur Biktimirov (Vladivostok, Russian Federation) and colleagues indicate that neuromodulation could be one of the important components of neuroprotection in spinal cord injury.

Neurotrophic factors are one of the key agents that induce powerful neuroprotection. In Chapter 8, Hari Shanker Sharma (Uppsala, Sweden) and colleagues discuss the antioxidant and antiischemic properties of cerebrolysin in heat stroke in comparison to several stroke therapies. The data suggest that cerebrolysin exerts superior neuroprotective ability in heat stroke as compared to other strike treatments.

Hypertension is known to induce blood–brain barrier (BBB) disruption and ventricular hemorrhage causing brain pathology. In Chapter 9, He Chao (Zhuji, Zhejiang, China) and colleagues indicate that urokinase treatment is beneficial for hypertension-induced ventricular hemorrhage.

Brain-derived neurotrophic factor (BDNF) and insulin-like growth factor-1 (IGF-1) are capable of thwarting upregulation of nitric oxide synthase (NOS) or hemeoxygenase (HO) enzymes after spinal cord injury and inducing neuroprotection. In Chapter 10, Aruna Sharma (Uppsala, Sweden) and colleagues indicate that a suitable combination of BDNF, ciliary neurotrophic factor (CNTF), and glial cell-derived neurotrophic factor (GDNF) reduces HO-2 upregulation and induces neuroprotection in spinal cord injury.

The pathophysiology of the spinal cord is associated with spinal stenosis. In Chapter 11, He Chao (Zhuji, Zhejiang, China) and colleagues discuss the diagnostic experience and literature review of patients with cervical, thoracic, and lumbar multisegment spinal stenosis.

The new knowledge presented in this volume for the novel treatment strategies for brain protection will enhance the pharmacotherapy of brain tumor, brain and spinal cord injury, heat stroke, concussive head injury associated with PD, high-altitude brain pathology in combination with head injury, and substance abuse as well as novel diagnostic parameters in neurodegeneration, brain tumors, hypertensive encephalopathy, and spinal stenosis.

We sincerely hope that the new knowledge emanating from this volume will further expand our understanding about these brain diseases and shed light on better therapeutic efficacy in reducing brain pathology.

This volume is indispensable for educators, researchers, policy makers, clinicians, and the drug development and healthcare industries. Teachers, researchers, and students in several neuroscience disciplines including neuropathology, neuropharmacology, neuroradiology, neurology, neurorehabilitation, neuropsychiatry, and neurophysiology will benefit from the research and state of the art in scientific literature presented in the volume. We sincerely hope that the volume will encourage further research in neuroscience and clinical practices with nanomedicine for the betterment of the quality of life of patients with brain diseases and associated comorbidity factors.

Hari Shanker Sharma
Aruna Sharma
Uppsala University, Uppsala, Sweden

References

Bading, H., Greenberg, M.E., 1991. Stimulation of protein tyrosine phosphorylation by NMDA receptor activation. Science 253, 912–914.

Bain, A., 1868. Mental and Moral Science: A Compendium of Psychology and Ethics. Longmans, New York, NY.

Beggs, J.L., Waggener, J.D., 1975. Vasogenic edema in the injured spinal cord: a method of evaluating the extent of blood-brain barrier alteration to horseradish peroxidase. Exp. Neurol. 49 (1 Pt. 1), 86–96.

Bradbury, M.W., 2000. Hugh Davson- -his contribution to the physiology of the cerebrospinal fluid and blood-brain barrier. Cell Mol. Neurobiol. 20 (1), 7–11. https://doi.org/10.1023/a:1006987709018.

Carpenter, W.B., 1874. Principles of Mental Physiology. Appleton, New York, NY.

Dumont, L., 1876. De l'habitude. In: Revue Philosophique de la France et de l'Étranger. PUF, Paris, pp. 321–366.

Engelhardt, B., Sorokin, L., 2009. The blood-brain and the blood-cerebrospinal fluid barriers: function and dysfunction. Semin. Immunopathol. 31 (4), 497–511. https://doi.org/10.1007/s00281-009-0177-0. Epub 2009 Sep 25.

Fabis, M.J., Scott, G.S., Kean, R.B., Koprowski, H., Hooper, D.C., 2007. Loss of blood-brain barrier integrity in the spinal cord is common to experimental allergic encephalomyelitis in knockout mouse models. Proc. Natl. Acad. Sci. U. S. A. 104 (13), 5656–5661. https://doi.org/10.1073/pnas.0701252104. Epub 2007 Mar 19.

Fuchs, E., Flügge, G., 2014. Adult neuroplasticity: more than 40 years of research. Neural Plast. 541870. P. 1-10. https://doi.org/10.1155/2014/541870.

Griesbach, G.S., Hovda, D.A., 2015. Cellular and molecular neuronal plasticity. Handb. Clin. Neurol. 128, 681–690. https://doi.org/10.1016/B978-0-444-63521-1.00042-X.

Griffiths, I.R., 1978. Spinal cord injuries: a pathological study of naturally occurring lesions in the dog and cat. J. Comp. Pathol. 88 (2), 303–315.

Griffiths, I.R., Burns, N., Crawford, A.R., 1978. Early vascular changes in the spinal grey matter following impact injury. Acta Neuropathol. (Berl.) 41 (1), 33–39.

Gulyaeva, N.V., 2017. Molecular mechanisms of neuroplasticity: an expanding universe. Biochemistry (Moscow) 82 (3), 237–242. https://doi.org/10.1134/S0006297917030014.

Hardingham, G.E., Bading, H., 2003. The Yin and Yang of NMDA receptor signalling. Trends Neurosci. 26, 81–89.

James, W., 1890. The Principles of Psychology. Henry Holt and Company, New York, NY, p. 1393. Pages xviii.

Johanson, C., Stopa, E., Baird, A., Sharma, H., 2011. Traumatic brain injury and recovery mechanisms: peptide modulation of periventricular neurogenic regions by the choroid plexus-CSF nexus. J. Neural Transm. (Vienna) 118 (1), 115–133. https://doi.org/10.1007/s00702-010-0498-0. Epub 2010 Oct 10.

Konorski, J., 1948. Conditioned Reflexes and Neuron Organization. Cambridge University Press, New York, p. 267. First published: January 1950. https://doi.org/10.1002/1097-4679(195001)6:1 < 107::AID-JCLP2270060132 > 3.0.CO;2-0.

Leist, M., Nicotera, P., 1998. Apoptosis, excitotoxicity and neuropathology. Exp. Cell Res. 239, 183–201.

Lipton, S.A., Choi, Y.B., Sucher, N.J., Chen, H.S., 1998. Neuroprotective versus neurodestructive effects of NO-related species. Biofactors 8, 33–40.

Montal, M., 1998. Mitochondria, glutamate neurotoxicity and the death cascade. Biochim. Biophys. Acta 1366, 113–126.

Pardridge, W.M., 2016. CSF, blood-brain barrier, and brain drug delivery. Expert Opin. Drug Deliv. 13 (7), 963–975. https://doi.org/10.1517/17425247.2016.1171315. Epub 2016 Apr 11.

Ramón y Cajal, S., 1909. Histologie du système nerveux de l'homme & des vertébrés. 1852–1934. Paris: Maloine (French).

Ramon y Cajal, S., 1999. In: Pasik, P., Pasik, T. (Eds.), Texture of the Nervous System of Man and the Vertebrates: Volume I (Texture of the Nervous System of Man & the Vertebrates), 1999th ed. Springer, p. 675. ISBN-13: 978-3211830574; ISBN-10: 321183057X.

Rosenberg, G.A., 2012. Neurological diseases in relation to the blood-brain barrier. J. Cereb. Blood Flow Metab. 32 (7), 1139–1151. https://doi.org/10.1038/jcbfm.2011.197. Epub 2012 Jan 18.

Santos, T., Maia, J., Agasse, F., Xapelli, S., Ferreira, L., Bernardino, L., 2012. Nanomedicine boosts neurogenesis: new strategies for brain repair. Integr. Biol. (Camb) 4 (9), 973–981. https://doi.org/10.1039/c2ib20129a. Epub 2012 Jul 17.

Sharma, H.S., 1982. Blood-Brain Barrier in Stress. Ph D Thesis, Banaras Hindu University Press, Varanasi, India.

Sharma, H.S., 1999. Pathophysiology of blood-brain barrier, brain edema and cell injury following hyperthermia: new role of heat shock protein, nitric oxide and carbon monoxide. an experimental study in the rat using light and electron microscopy. Acta Universitatis Upsaliensis 830, 1–94.

Sharma, H.S., 2004a. Blood-brain and spinal cord barriers in stress. In: Sharma, H.S., Westman, J. (Eds.), The Blood-Spinal Cord and Brain Barriers in Health and Disease. Elsevier Academic Press, San Diego, pp. 231–298.

Sharma, H.S., 2004b. Pathophysiology of the blood-spinal cord barrier in traumatic injury. In: Sharma, H.S., Westman, J. (Eds.), The Blood-Spinal Cord and Brain Barriers in Health and Disease. Elsevier Academic Press, San Diego, pp. 437–518.

Sharma, H.S., 2005a. Pathophysiology of blood-spinal cord barrier in traumatic injury and repair. Curr. Pharm. Des. 11 (11), 1353–1389. https://doi.org/10.2174/1381612053507837.

Sharma, H.S., 2005b. Neuroprotective effects of neurotrophins and melanocortins in spinal cord injury: an experimental study in the rat using pharmacological and morphological approaches. Ann. N. Y. Acad. Sci. 1053, 407–421. https://doi.org/10.1111/j.1749-6632.2005.tb00050.x.

Sharma, H.S., 2005c. Heat-related deaths are largely due to brain damage. Indian J. Med. Res. 121 (5), 621–623.

Sharma, H.S., 2006a. Hyperthermia induced brain oedema: current status and future perspectives. Indian J. Med. Res. 123 (5), 629–652.

Sharma, H.S., 2006b. Hyperthermia influences excitatory and inhibitory amino acid neurotransmitters in the central nervous system. An experimental study in the rat using behavioural, biochemical, pharmacological, and morphological approaches. J. Neural Transm. (Vienna) 113 (4), 497–519. https://doi.org/10.1007/s00702-005-0406-1.

Sharma, H.S., 2007a. Nanoneuroscience: emerging concepts on nanoneurotoxicity and nanoneuroprotection. Nanomedicine (Lond) 2 (6), 753–758. https://doi.org/10.2217/17435889.2.6.753.

Sharma, H.S., 2007b. Neurotrophic factors in combination: a possible new therapeutic strategy to influence pathophysiology of spinal cord injury and repair mechanisms. Curr. Pharm. Des. 13 (18), 1841–1874. https://doi.org/10.2174/138161207780858410.

Sharma, H.S., 2009. Blood–central nervous system barriers: the gateway to neurodegeneration, neuroprotection and neuroregeneration. In: Lajtha, A., Banik, N., Ray, S.K. (Eds.), Handbook of Neurochemistry and Molecular Neurobiology: Brain and Spinal Cord Trauma. Springer Verlag, Berlin, Heidelberg, New York, pp. 363–457.

Sharma, H.S., Castellani, R.J., Smith, M.A., Sharma, A., 2012. The blood-brain barrier in Alzheimer's disease: novel therapeutic targets and nanodrug delivery. Int. Rev. Neurobiol. 102, 47–90. https://doi.org/10.1016/B978-0-12-386986-9.00003-X.

Sharma, H.S., Dey, P.K., 1986. Influence of long-term immobilization stress on regional blood-brain barrier permeability, cerebral blood flow and 5-HT level in conscious normotensive young rats. J. Neurol. Sci. 72 (1), 61–76. https://doi.org/10.1016/0022-510x(86)90036-5.

Sharma, H.S., Dey, P.K., 1987. Influence of long-term acute heat exposure on regional blood-brain barrier permeability, cerebral blood flow and 5-HT level in conscious normotensive young rats. Brain Res. 424 (1), 153–162. https://doi.org/10.1016/0006-8993(87)91205-4.

Sharma, H.S., Hoopes, P.J., 2003. Hyperthermia induced pathophysiology of the central nervous system. Int. J. Hyperthermia 19 (3), 325–354. https://doi.org/10.1080/0265673021000054621.

Sharma, H.S., Johanson, C.E., 2007. Blood-cerebrospinal fluid barrier in hyperthermia. Prog. Brain Res. 162, 459–478. https://doi.org/10.1016/S0079-6123(06)62023-2.

Sharma, H.S., Olsson, Y., 1990. Edema formation and cellular alterations following spinal cord injury in the rat and their modification with p-chlorophenylalanine. Acta Neuropathol. 79 (6), 604–610. https://doi.org/10.1007/BF00294237.

Sharma, H.S., Olsson, Y., Dey, P.K., 1990. Changes in blood-brain barrier and cerebral blood flow following elevation of circulating serotonin level in anesthetized rats. Brain Res. 517 (1-2), 215–223. https://doi.org/10.1016/0006-8993(90)91029-g.

Sharma, H.S., Sharma, A., 2007. Nanoparticles aggravate heat stress induced cognitive deficits, blood-brain barrier disruption, edema formation and brain pathology. Prog. Brain Res. 162, 245–273. https://doi.org/10.1016/S0079-6123(06)62013-X.

Sharma, H.S., Sharma, A., 2010. Breakdown of the blood-brain barrier in stress alters cognitive dysfunction and induces brain pathology: new perspectives for neuroprotective strategies. In: Ritsner, M. (Ed.), Brain Protection in Schizophrenia, Mood and Cognitive Disorders. Springer, Dordrecht. https://doi.org/10.1007/978-90-481-8553-5_9.

Sharma, A., Sharma, H.S., 2012a. Monoclonal antibodies as novel neurotherapeutic agents in CNS injury and repair. Int. Rev. Neurobiol. 102, 23–45. https://doi.org/10.1016/B978-0-12-386986-9.00002-8.

Sharma, H.S., Sharma, A., 2012b. Nanowired drug delivery for neuroprotection in central nervous system injuries: modulation by environmental temperature, intoxication of nanoparticles, and comorbidity factors. Wiley Interdiscip. Rev. Nanomed. Nanobiotechnol. 4 (2), 184–203. https://doi.org/10.1002/wnan.172. Epub 2011 Dec 8.

Sharma, H.S., Sharma, A., 2013. New perspectives of nanoneuroprotection, nanoneuropharmacology and nanoneurotoxicity: modulatory role of amino acid neurotransmitters, stress, trauma, and co-morbidity factors in nanomedicine. Amino Acids 45 (5), 1055–1071. https://doi.org/10.1007/s00726-013-1584-z.

Sharma, H.S., Sharma, A., 2021. Amine precursors in depressive disorders and the blood-brain barrier. In: Riederer, P., Laux, G., Nagatsu, T., Le, W., Riederer, C. (Eds.), NeuroPsychopharmacotherapy. Springer, Cham. https://doi.org/10.1007/978-3-319-56015-1_423-1.

Sharma, H.S., Winkler, T., Stålberg, E., Olsson, Y., Dey, P.K., 1991a. Evaluation of traumatic spinal cord edema using evoked potentials recorded from the spinal epidural space. An experimental study in the rat. J. Neurol. Sci. 102 (2), 150–162. https://doi.org/10.1016/0022-510x(91)90063-d.

Sharma, H.S., Cervós-Navarro, J., Dey, P.K., 1991b. Acute heat exposure causes cellular alteration in cerebral cortex of young rats. Neuroreport 2 (3), 155–158. https://doi.org/10.1097/00001756-199103000-00012.

Sharma, H.S., Cervós-Navarro, J., Dey, P.K., 1991c. Increased blood-brain barrier permeability following acute short-term swimming exercise in conscious normotensive young rats. Neurosci. Res. 10 (3), 211–221. https://doi.org/10.1016/0168-0102(91)90058-7.

Sharma, H.S., Westman, J., Navarro, J.C., Dey, P.K., Nyberg, F., 1995. Probable involvement of serotonin in the increased permeability of the blood-brain barrier by forced swimming. An experimental study using Evans blue and 131I-sodium tracers in the rat. Behav. Brain Res. 72 (1-2), 189–196. https://doi.org/10.1016/0166-4328(96)00170-2.

Sharma, H.S., Nyberg, F., Gordh, T., Alm, P., Westman, J., 1998. Neurotrophic factors attenuate neuronal nitric oxide synthase upregulation, microvascular permeability disturbances, edema formation and cell injury in the spinal cord following trauma. In: Stålberg, E., Sharma, H.S., Olsson, Y. (Eds.), Spinal Cord Monitoring. Basic Principles, Regeneration, Pathophysiology and Clinical Aspects. Springer, Wien New York, pp. 118–148.

Sharma, H.S., Westman, J., Cervós-Navarro, J., Dey, P.K., Nyberg, F., 1998a. Blood-brain barrier in stress: a gateway to various brain diseases. In: Levy, A., Grauer, E., Ben-Nathan, D., de Kloet, E.R. (Eds.), New Frontiers of Stress Research: Modulation of Brain Function. Harwood Academic Publishers Inc, Amsterdam, pp. 259–276.

Sharma, H.S., Lundstedt, T., Flärdh, M., Skottner, A., Wiklund, L., 2007. Neuroprotective effects of melanocortins in CNS injury. Curr. Pharm. Des. 13 (19), 1929–1941. https://doi.org/10.2174/138161207781039797.

Sharma, H.S., Muresanu, D.F., Patnaik, R., Stan, A.D., Vacaras, V., Perju-Dumbrav, L., Alexandru, B., Buzoianu, A., Opincariu, I., Menon, P.K., Sharma, A., 2011. Superior neuroprotective effects of cerebrolysin in heat stroke following chronic intoxication of Cu or Ag engineered nanoparticles. A comparative study with other neuroprotective agents using biochemical and morphological approaches in the rat. J. Nanosci. Nanotechnol. 11 (9), 7549–7569. https://doi.org/10.1166/jnn.2011.5114.

Sharma, H.S., Muresanu, D.F., Patnaik, R., Sharma, A., 2013. Exacerbation of brain pathology after partial restraint in hypertensive rats following SiO_2 nanoparticles exposure at high ambient temperature. Mol. Neurobiol. 48 (2), 368–379. https://doi.org/10.1007/s12035-013-8502-y. Epub 2013 Jul 6.

Sharma, H.S., Feng, L., Lafuente, J.V., Muresanu, D.F., Tian, Z.R., Patnaik, R., Sharma, A., 2015a. TiO_2-nanowired delivery of mesenchymal stem cells thwarts diabetes- induced exacerbation of brain pathology in heat stroke: an experimental study in the rat using morphological and biochemical approaches. CNS Neurol. Disord. Drug Targets 14 (3), 386–399. https://doi.org/10.2174/1871527314666150318114335.

Sharma, H.S., Kiyatkin, E.A., Patnaik, R., Lafuente, J.V., Muresanu, D.F., Sjöquist, P.O., Sharma, A., 2015b. Exacerbation of methamphetamine neurotoxicity in cold and hot environments: neuroprotective effects of an antioxidant compound H-290/51. Mol. Neurobiol. 52 (2), 1023–1033. https://doi.org/10.1007/s12035-015-9252-9. Epub 2015 Jun 26.

Sharma, A., Muresanu, D.F., Lafuente, J.V., Patnaik, R., Tian, Z.R., Buzoianu, A.D., Sharma, H.S., 2015c. Sleep deprivation-induced blood-brain barrier breakdown and brain dysfunction are exacerbated by size-related exposure to Ag and Cu nanoparticles. Neuroprotective effects of a 5-HT3 receptor antagonist ondansetron. Mol. Neurobiol. 52 (2), 867–881. https://doi.org/10.1007/s12035-015-9236-9. Epub 2015 Jul 2.

Sharma, H.S., Muresanu, D.F., Sharma, A., 2016. Alzheimer's disease: cerebrolysin and nanotechnology as a therapeutic strategy. Neurodegener. Dis. Manag. 6 (6), 453–456. https://doi.org/10.2217/nmt-2016-0037. Epub 2016 Nov 9.

Sharma, A., Muresanu, D.F., Ozkizilcik, A., Tian, Z.R., Lafuente, J.V., Manzhulo, I., Mössler, H., Sharma, H.S., 2019. Sleep deprivation exacerbates concussive head injury induced brain pathology: neuroprotective effects of nanowired delivery of cerebrolysin with α-melanocyte-stimulating hormone. Prog. Brain Res. 245, 1–55. https://doi.org/10.1016/bs.pbr.2019.03.002. Epub 2019 Apr 2.

Sharma, A., Muresanu, D.F., Castellani, R.J., Nozari, A., Lafuente, J.V., Sahib, S., Tian, Z.R., Buzoianu, A.D., Patnaik, R., Wiklund, L., Sharma, H.S., 2020. Mild traumatic brain injury exacerbates Parkinson's disease induced hemeoxygenase-2 expression and brain pathology: Neuroprotective effects of co-administration of TiO(2) nanowired mesenchymal stem cells and cerebrolysin. Prog. Brain Res. 258, 157–231. https://doi.org/10.1016/bs.pbr.2020.09.010. Epub 2020 Nov 9.

Sharma, H.S., Westman, J., 2004. The Blood-Spinal Cord and Brain Barriers in Health and Disease, Academic Press, San Diego, pp. 1–617. (Release date: Nov. 9, 2003).

Stahnisch, F.W., Nitsch, R., 2002. Santiago Ram´on y Cajal's concept of neuronal plasticity: the ambiguity lives on. Trends Neurosci. 25 (11), 589–591.

Sweeney, M.D., Zhao, Z., Montagne, A., Nelson, A.R., Zlokovic, B.V., 2019. Blood-brain barrier: from physiology to disease and back. Physiol. Rev. 99 (1), 21–78. https://doi.org/10.1152/physrev.00050.2017.

Taub, E., Uswatte, G., Pidikiti, R., 1999. Constraint-induced movement therapy: a new family of techniques with broad application to physical rehabilitation- -a clinical review. J. Rehabil. Res. Dev. 36 (3), 237–251.

Thangudu, S., Cheng, F.Y., Su, C.H., 2020. Advancements in the blood-brain barrier penetrating nanoplatforms for brain related disease diagnostics and therapeutic applications. Polymers (Basel) 12 (12), 3055. https://doi.org/10.3390/polym12123055.

Weiss, N., Miller, F., Cazaubon, S., Couraud, P.O., 2009. The blood-brain barrier in brain homeostasis and neurological diseases. Biochim. Biophys. Acta 1788 (4), 842–857. https://doi.org/10.1016/j.bbamem.2008.10.022. Epub 2008 Nov 11.

Windle, W.F., 1980. The spinal cord and its reaction to traumatic injury. Anatomy-physiology-pharmacology-therapeutics. In: Modern Pharmacology-Toxicology. vol. 18. Marcel Dekker, New York, NY, USA, p. 384. ISBN-10:0824766881; ISBN-13:978-0824766887.

Zhang, S.J., Zou, M., Lu, L., Lau, D., Ditzel, D.A., Delucinge-Vivier, C., Aso, Y., Descombes, P., Bading, H., 2009. Nuclear calcium signaling controls expression of a large gene pool: identification of a gene program for acquired neuroprotection induced by synaptic activity. PLoS Genet. 5 (8), e1000604. https://doi.org/10.1371/journal.pgen.1000604. Epub 2009 Aug 14.

Zilles, K., 1992. Neuronal plasticity as an adaptive property of the central nervous system. Ann. Anat. 174 (5), 383–391.

CHAPTER

1

Histamine H3 and H4 receptors modulate Parkinson's disease induced brain pathology. Neuroprotective effects of nanowired BF-2649 and clobenpropit with anti-histamine-antibody therapy

Aruna Sharma[a,*], Dafin F. Muresanu[b,c], Ranjana Patnaik[d], Preeti K. Menon[k], Z. Ryan Tian[e], Seaab Sahib[e], Rudy J. Castellani[f], Ala Nozari[g], José Vicente Lafuente[h], Anca D. Buzoianu[i], Stephen D. Skaper[j], Igor Bryukhovetskiy[l,m], Igor Manzhulo[m], Lars Wiklund[a], and Hari Shanker Sharma[a,*]

[a]*International Experimental Central Nervous System Injury & Repair (IECNSIR), Department of Surgical Sciences, Anesthesiology & Intensive Care Medicine, Uppsala University Hospital, Uppsala University, Uppsala, Sweden*

[b]*Department of Clinical Neurosciences, University of Medicine & Pharmacy, Cluj-Napoca, Romania*

[c]*"RoNeuro" Institute for Neurological Research and Diagnostic, Cluj-Napoca, Romania*

[d]*Department of Biomaterials, School of Biomedical Engineering, Indian Institute of Technology, Banaras Hindu University, Varanasi, India*

[e]*Department of Chemistry & Biochemistry, University of Arkansas, Fayetteville, AR, United States*

[f]*Department of Pathology, University of Maryland, Baltimore, MD, United States*

[g]*Anesthesiology & Intensive Care, Massachusetts General Hospital, Boston, MA, United States*

[h]*LaNCE, Department of Neuroscience, University of the Basque Country (UPV/EHU), Leioa, Bizkaia, Spain*

[i]*Department of Clinical Pharmacology and Toxicology, "Iuliu Hatieganu" University of Medicine and Pharmacy, Cluj-Napoca, Romania*

[j]*Anesthesiology & Intensive Care, Department of Pharmacology, University of Padua, Padova, Italy*

[k]*Department of Biochemistry and Biophysics, Stockholm University, Stockholm, Sweden*

Progress in Brain Research, Volume 266, ISSN 0079-6123, https://doi.org/10.1016/bs.pbr.2021.06.003
Copyright © 2021 Elsevier B.V. All rights reserved.

[l]*Department of Fundamental Medicine, School of Biomedicine, Far Eastern Federal University, Vladivostok, Russia*

[m]*Laboratory of Pharmacology, National Scientific Center of Marine Biology, Far East Branch of the Russian Academy of Sciences, Vladivostok, Russia*

**Corresponding authors: Tel.: +46-70-21-95-963; Fax: +46-18-24-38-99 (Aruna Sharma); Tel.: +46-70-2011-801; Fax: +46-18-24-38-99 (Hari Shanker Sharma), e-mail address: aruna.sharma@surgsci.uu.se; sharma@surgsci.uu.se; harishanker_sharma55@icloud.com; hssharma@aol.com*

Abstract

Military personnel deployed in combat operations are highly prone to develop Parkinson's disease (PD) in later lives. PD largely involves dopaminergic pathways with hallmarks of increased alpha synuclein (ASNC), and phosphorylated tau (p-tau) in the cerebrospinal fluid (CSF) precipitating brain pathology. However, increased histaminergic nerve fibers in substantia nigra pars Compacta (SNpc), striatum (STr) and caudate putamen (CP) associated with upregulation of Histamine H3 receptors and downregulation of H4 receptors in human cases of PD is observed in postmortem cases. These findings indicate that modulation of histamine H3 and H4 receptors and/or histaminergic transmission may induce neuroprotection in PD induced brain pathology. In this review effects of a potent histaminergic H3 receptor inverse agonist BF-2549 or clobenpropit (CLBPT) partial histamine H4 agonist with H3 receptor antagonist, in association with monoclonal anti-histamine antibodies (AHmAb) in PD brain pathology is discussed based on our own observations. Our investigation shows that chronic administration of conventional or TiO_2 nanowired BF 2649 (1 mg/kg, i.p.) or CLBPT (1 mg/kg, i.p.) once daily for 1 week together with nanowired delivery of HAmAb (25 μL) significantly thwarted ASNC and p-tau levels in the SNpC and STr and reduced PD induced brain pathology. These observations are the first to show the involvement of histamine receptors in PD and opens new avenues for the development of novel drug strategies in clinical strategies for PD, not reported earlier.

Keywords

Parkinson's disease, Histamine, Alpha synuclein, Phosphorylated tau, BF-2549, Clobenpropit, Anti-histamine antibodies, Dopamine, Brain pathology, Neuroprotection

1 Introduction

Parkinson's disease (PD) is one of the most prominent neurodegenerative movement disorders affecting mankind across the globe (Cuenca et al., 2019; Kalia and Lang, 2015; Samii et al., 2004; Tolosa et al., 2006). PD burden on mankind is associated with either due to environmental, toxicological or genetic defects (Axelsen and

Woldbye, 2018; Balestrino and Schapira, 2020; Bellou et al., 2016; Jankovic and Tan, 2020; Hatcher et al., 2008; Kim and Alcalay, 2017; Lang and Espay, 2018). About 10 million people are suffering from PD World-wide out of which about 1.8 million are living in the United States (Collier et al., 2017; Tysnes and Storstein, 2017). Epidemiological data suggest that about 60k people are diagnosed with PD every year with mortality rate of 23k each year in the United States (Ascherio and Schwarzschild, 2016). In Europe the prevalence of PD are estimated at about 110–258 cases per 100k populations (von Campenhausen et al., 2005). In terms of PD each year about 10–20 cases per 100k populations are diagnosed in Europe (Canevelli et al., 2019).

The cardinal symptoms of PD include tremor, rigidity, bradykinesia or akinesia together with postural instability but several non-motor symptoms are also identifies in clinical cases (Armstrong and Okun, 2020; Balestrino and Schapira, 2020; Jankovic, 2008; Marino et al., 2020; Peball et al., 2020). The neuropathological symptoms of PD are the loss of dopaminergic neurons in the substantia nigra pars compacta (SNpC) and striatum (STr) (Lopiano et al., 2019; Marino et al., 2020; Santos-García et al., 2020). In addition, the hallmark of PD includes accumulation of misfolded protein alpha synuclein (ASNC) and deposits of Lewy bodies (LBs) (Chatterjee et al., 2020; Guo et al., 2020; Kyle and Bronstein, 2020; Pennington et al., 2020; Shahnawaz et al., 2020; Sonninen et al., 2020). With advancement of time the neurodegenerative changes spread all over the brain leading to serious consequences and death. Treatment strategies for PD cases are so far symptomatic and largely focused on the dopaminergic pathways (Qian et al., 2020; Ruppert et al., 2020). However, no satisfactory treatment regimen so far has emerged. Thus, this is highly warranted to expand our knowledge on the PD pathophysiology in order to explore novel clinical therapy using nanomedicine.

PD is characterized by the progressive loss of dopaminergic neurons in the SNpC and STr leading to dopamine deficiency (Drew et al., 2020; Schweitzer et al., 2020; Yang et al., 2020). However, other neurotransmitters are also present in the brain affecting motor functions and related to PD pathology (Iarkov et al., 2020; De Deurwaerdère and Di Giovanni, 2020). Histamine is one of the key regulatory neurotransmitters that are widely distributed in the brain (Esbenshade et al., 2008; Provensi et al., 2020; Prell and Green, 1986; Schwartz et al., 1980). The cell bodies of approximately 64k histaminergic neurons are located in the posterior hypothalamus of the tuberomamillary nucleus (TMN) and send its projections to almost all areas of the cerebral and cerebellar cortex in the brain (Giannoni et al., 2009; Khedkar et al., 2012; Moriwaki et al., 2015; Umehara et al., 2012). From TMN the exclusive site of neuronal histamine production SNpC receives dense histaminergic innervation (Anichtchik et al., 2000a,b; Maisonnette et al., 1998; Panula et al., 1989; Shan et al., 2012). Apart from dense histaminergic innervation in SNpC enlarged axonal varicosities and increased histamine levels were found in SNpC of PD patients (Rinne et al., 2002; Shan et al., 2012, 2013, 2015). Evidences suggest that increased endogenous histamine levels induce degeneration of dopaminergic neurons in PD (Anichtchik et al., 2001; Liu et al., 2007). The blood levels of histamine and histamine metabolite in the cerebrospinal fluid (CSF) are increased in PD (Anichtchik et al., 2000a,b; Kempuraj et al., 2016; Prell and Green, 1986). In addition, blockade of histamine H2 receptors in PD reduces bradyphrenia and improve motor functions in some patients of PD (Molinari et al., 1995). Histaminergic innervation also increased in the areas of dopaminergic innervation in SNpC and STr in PD (Auvinen and Panula, 1988; Panula et al., 1989, 1990).

Histamine has several receptors in the brain and central nervous system that regulate or modulate several functions and effects of other neurotransmitters in health and disease (Annamalai et al., 2020; del Rio et al., 2012; Moreno-Delgado et al., 2020; Parsons and Ganellin, 2006; Panula et al., 2015; Tiligada et al., 2011).

Recent studies demonstrate that drug targeting of histamine H3 and H4 receptors could be beneficial in a large number of neurological diseases (del Rio et al., 2012; Naddafi and Mirshafiey, 2013; Patnaik et al., 2018; Saligrama et al., 2012; Teuscher et al., 2007). Histamine H3 and H4 receptors are altered in the brain of PD patients (Shan et al., 2012). Thus, a possibility exists that Histamine H3 and H4 receptors could modulate brain pathology in PD (Chen et al., 2020; Koski et al., 2020,b; Nuutinen and Panula 2010; Panula and Nuutinen, 2013). It would be interesting to compare nanodelivery of select histamine H3 and H4 receptor modulator drugs with nanowired delivery in PD to induce superior neuroprotective effects (Patnaik et al., 2000, 2018; Sharma et al., 2017a,b, 2019a,b). Histamine appears to be involved in the pathophysiology of PD. Thus, it would be interesting to see whether administration of histamine antibodies could exert additional beneficial effects in PD together with histamine H3 or H4 receptor modulator agents in inducing superior neuroprotection (Sharma and Cervós-Navarro, 1991; Sharma et al., 1992a,b, 2006a,b, 2016a,b,c, 2017a,b).

In this chapter we used nanowired delivery of histamine H3 and H4 rector modulator agents together with histamine antibodies to explore potential therapeutic roles of histaminergic neurotransmission in PD based on our own investigations. The functional significance of our findings in the light of current histamine research is discussed.

1.1 Histaminergic system in the brain

Histamine [2-(1H-imidazol-4-yl)ethanamine] is a biogenic amine distributed ubiquitously throughout the brain and several tissues in the body (Haas et al., 2008; Yoshikawa et al., 2019). Histamine was first discovered by Sir Henry Dale in 1910 in mammalian tissues that dilates blood vessels and contracts smooth muscles and could be involved in anaphylaxis reaction (Dale and Laidlaw, 1910). Histamine is synthesized by amino acid L-histidine by catalytic activity of histidine decarboxylase (HDC) (Holeček, 2020; Moriguchi and Takai, 2020; Watanabe and Ohtsu, 2002). Histamine is catabolized by the cytosolic enzyme histamine *N*-methyltransferase to *N*-methylhistamine or by extracellular enzyme diamine oxidase to imidazole acidic acid (Kitanaka et al., 2016; Shan et al., 2017; Weinshilboum et al., 1999).

Histamine in the brain largely in the gray matter was for the first time described by Kwiatkowski (1941) and White demonstrated its formation and catabolism in the brain in 1959 (White, 1959, 1961). In 1974 histaminergic projections in the brain was shown by Garbarg using lesion of the median forebrain bundle (Garbarg et al., 1974). However, fluorescent immunohistochemical demonstration of histaminergic neurons in brain was first identified in 1983 by Watanabe in Osaka, Japan (Watanabe et al., 1983) and later immunohistochemical distribution of histaminergic nerve cells and fibers was shown in all most all parts of the brain extensively in 1984 by Panula in Washington DC, USA (Panula et al., 1984). Later, Panula, Steinbusch and others from 1988 to 1998 showed extensive details of histaminergic neurons and its projections throughout the rat brain and spinal cord and in human brain (Smits et al., 1990; Steinbusch 1991; Steinbusch et al., 1986; Wouterlood et al., 1986, 1987, 1988). These immunocytochemical mapping of histaminergic neurons established its role as a neurotransmitter in the central nervous system (CNS) (Brandes et al., 1990; Haas et al., 2008; Schwartz et al., 1980).

1.2 Histaminergic receptors in the brain

With recent advancements in drugs targeting histamine receptors for various diseases, four metabotropic receptor types are identified as histamine H1, H2, H3 and H4 (Chazot, 2013; Panula et al., 2015; Roeder, 2003; Tiligada and Ennis, 2020). The histamine H1, H2 and H3 are expressed in abundance in the brain whereas histamine H4 receptors are largely found in peripheral tissues but recently their functional expression in brain is also described (Connelly et al., 2009; Fang et al., 2021; Martinez-Mir et al., 1990; Parsons and Ganellin, 2006; Schneider et al., 2015; Zhou et al., 2006). Drugs targeting H4 receptors in brain showed Histamine H4 receptors were prominently expressed in neurons of human cerebral cortex in layer VI (Feliszek et al., 2015). In mouse histamine H4 expression is seen in nerve cells located in the thalamus, hippocampal CA4, stratum lucidum and cerebral cortex layer IV (Feliszek et al., 2015. Based on these studies several histamine H3 agonists and H4 antagonists are developed that are used to target CNS disorders (Nieto-Alamilla et al., 2016; Patnaik et al., 2018; Sharma et al., 2017a,b; Zhou et al., 2019. All metabotropic histamine receptors belong to class A rhodopsin like family of G-protein coupled receptors (GPCRs) (Conrad et al., 2020; Hishinuma et al., 2016; Strasser et al., 2013). The histamine H1 and H2 receptors have low-affinity for histamine as compared to histamine H3 and H4 receptors that are high affinity histamine receptors (Čarman-Kržan and Lipnik-Štangelj, 2000; Fossati et al., 2001.

1.3 Histamine in brain diseases

Histamine in the brain is implicated in a variety of normal brain functions as well as in several disease processes leading to neurodegeneration (Barata-Antunes et al., 2017). These effects are mediated through various histamine receptors (McClain et al., 2020).

1.3.1 Cognitive dysfunction and Alzheimer's disease

Behavioral studies indicate role of histaminergic system in cognitive functions through histamine H1, H2 and H3 receptors (Santangelo et al., 2017). Thus is further supported by high degree of expression of these receptors in brain regions including cerebral cortex, thalamus, hypothalamus, hippocampus and amygdala involved in cognition (Alvarez, 2009). Intracerebroventricular (i.c.v.) administration of histamine facilitates step-down inhibitory avoidance behavior that is mediated through histamine H1 and H2 receptors (Mayer et al., 2018). Furthermore mice lacking histamine H1 and H2 receptors exhibited impaired object recognition and acquisition of spatial memory (Dai et al., 2007; da Silveira et al., 2013; Passani et al., 2017).

Pharmacological data suggests that histamine H3 receptor antagonists improve cognitive functions in animal models and alleviates cognitive deficits in transgenic amyloid precursor protein (APP) mutant mouse model of Alzheimer's disease (AD) (Brioni et al., 2011; Medhurst et al., 2009; Witkin and Nelson 2004). In human cases of AD histamine levels are decreased in the hippocampus, temporal cerebral cortex and hypothalamus (Cacabelos et al., 1989; Jaarsma et al., 1994). This suggests that histaminergic neurons degeneration is likely to contribute cognitive decline in AD. Several histamine H3 receptor antagonists are being tried in clinical trials Phase I and II to improve cognitive functions in AD, schizophrenia and attention-deficit hyperactivity disorder (ADHD) (Ellenbroek and Ghiabi, 2015; Kubo et al., 2015; Minzenberg and Carter, 2008; Rapanelli and Pittenger, 2016; Sadek et al., 2016; Weisler et al., 2012). However, the results are still inconclusive.

1.3.2 Motor disorders and Parkinson's disease

Increased levels of histamine are found in the substantia nigra, putamen and globus pallidus in Parkinson's disease patients (Rinne et al., 2002). Postmortem finding show a decrease in TMN neurons in patients of multiple brain atrophy, but not seen in PD cases (Shan et al., 2012; Yamada et al., 2020). In animal models of PD induced with 6-hydroxydopamine (6-OHDA) lesioned animals histamine levels are increased in the brain similar to human cases in PD (Nowak et al., 2009). This suggests that increased histamine in brain is associated with PD associated with insomnia. Interestingly histamine H3 and H4 receptor antagonist reduces apomorphine-induced stereotyped behavior in 6-OHDA lesioned animals (Farzin and Attarzadeh, 2000; Liu et al., 2008).

1.3.3 Schizophrenia

As mentioned above histamine H1, H2 and H3 receptors are expressed in high densities within the brain areas in the cerebral cortex, hippocampus, dorsal thalamus (Arrang, 2007; Ellenbroek and Ghiabi, 2015; Iwabuchi et al., 2005). These brain regions are involved in cognitive functions and are severely disrupted in schizophrenia (Besteher et al., 2020; Guo et al., 2019). In addition, brains of schizophrenic patients histaminergic system are severely perturbed. Increased levels of histamine metabolite tele-methylhistamine reflecting histamine release are found in the CSF of schizophrenic patients (Brioni et al., 2011; Nakai et al., 1991). The histamine H1 receptor binding is reduced in prefrontal, frontal and cingulate cortices in brains of patients with schizophrenia (Ito 2004; Mahmood et al., 2012; Orange et al., 1996). On the other hand, histamine H3 receptor biding is increased in dorsolateral prefrontal cortex in schizophrenic patients. Treatment with histamine H3 receptor antagonist induced beneficial therapeutic effects in schizophrenia patients (Brioni et al., 2011; Browman et al., 2004; Mahmood et al., 2012).

1.3.4 Multiple sclerosis

The autoimmune disease multiple sclerosis (MS) is associated with demyelination, neuroinflammation, axonal degeneration and neuronal loss (Absinta et al., 2020; Filippi et al., 2020; Hauser and Cree, 2020). In animals models experimental autoimmune encephalomyelitis (EAE) clearly indicated involvement of histamine in the pathological processes (Jadidi-Niaragh and Mirshafiey, 2010; Rafiee Zadeh et al., 2018). In remitting or progressive patients of MS histamine levels in the CSF are more than 60% higher than healthy cases (Kallweit et al., 2013; Tuomisto et al., 1983). Also the histamine H1 receptor mRNA is upregulated in MS lesions (Lu et al., 2010). Treatment with histamine H1 receptor antagonist improves neurological benefits in MS patients and the onset of the disease (Alonso et al., 2006; Logothetis et al., 2005).

1.3.5 Sleep disorders

Involvement of histamine is amply shown in human cases and in animal models in sleep dysfunction (Diez-Garcia and Garzon, 2017; Scammell et al., 2019; Thakkar, 2011). Available data show that histamine levels are decreased in the cerebrospinal fluid (CSF) in humans with idiopathic hypersomnia or narcolepsy (Dauvilliers et al., 2013; Szakacs et al., 2017). This suggests that histamine is involved in alertness and wakefulness. Treatment with histamine H2 receptor antagonists improves sleep quality in narcolepsy patients (Thorpy, 2020). Based on these data it appears that low histamine levels in brain or CSF is related with increased sleepiness (Bassetti et al., 2010; Franco et al., 2019; Kanbayashi et al., 2009; Nishino et al., 2009).

1.3.6 Addictive disorders

Studies showing lesion of TMN neurons or treatment with histamine H1 receptor antagonist induces an inhibitory effect on reward and reinforcement as well as addictive behavior (Brabant et al., 2010; Halpert et al., 2002; Pallardo-Fernández et al., 2020). Histamine H3 receptor antagonists potentiate self-administration of morphine or alcohol (Mobarakeh et al., 2009; Owen et al., 1994; Panula, 2020; Panula and Nuutinen, 2011; Popiolek-Barczyk et al., 2018; Vanhanen et al., 2013). This treatment also reinforces methamphetamine or cocaine preference (Kitanaka et al., 2011; Okuda et al., 2009; Munzar et al., 2004; Moreno et al., 2014; Pallardo-Fernández et al., 2020). This potentiation of methamphetamine and cocaine are also modulated by histamine H3 and H4 receptors (Corrêa and dos Santos Fernandes, 2015; Schneider et al., 2014; Sharma and Ali, 2006, 2008; Sharma et al., 2014,

2009a,b; Thurmond, 2015). Experiments in our laboratory show that morphine dependence and withdrawal symptoms are lessened in rats with histamine H3 and H4 receptor modulation (Sharma HS unpublished observation).

2 Histaminergic system alterations in Parkinson's disease

Several postmortem studies in Human PD brain show significant alterations in histaminergic system as compared to normal brain (Anichtchik et al., 2000a,b; Rinne et al., 2002; Shan et al., 2012, 2013). These observations indicate that histaminergic system play key roles in PD.

2.1 Histamine concentration in PD

Measurement of histamine in different brain areas of PD patient and multiple system atrophy (MSA) showed significant increase from the control groups (Shan et al., 2012). In PD postmortem brain histamine level increased in caudate nucleus by 128%, putamen by 159%, pars compacta of substantia nigra (SNpC) by 201%, pars reticularis of substantia nigra (SNpR) by 164%, internal globus pallidus by 234%, external globus pallidus by 200% from control group. Other brain areas such as hippocampus 116%, hypothalamus 147%, prefrontal cortex 117%, temporal cortex 95%, and occipital cortex 88% increase in histamine concentration was seen as compared to the controls (Anichtchik et al., 2000a,b; Rinne et al., 2002; Shan et al., 2013). The other brain areas did not show any significant differences in histamine content between PD and control brains. Interestingly, no significant changes in histamine levels in any brain areas were observed in multiple system atrophy [MSA] brain with control group (Rinne et al., 2002). These observations suggest that histamine levels are selectively increased in PD brain in the regions involved in motor function.

2.2 Histaminergic nerve fibers in PD

Immunohistochemical studies on histaminergic fibers in human brain of PD cases did not differ at the level of superior colliculus in orientation as compared to the normal brain (Anichtchik et al., 2000a,b). However, histaminergic nerve fibers are distinctly different in densities and thinner in SNpC and SNpR in PD than the normal brain. The mean diameter of histaminergic fibers in PD is 0.71 μm as compared to 1.15 μm in normal brains (Anichtchik et al., 2000a,b). In PD brain many varicosities in histaminergic fibers ranging from 1 to 6 μm were also present (see Anichtchik et al., 2000a,b). These observations suggest that altered histaminergic innervation in SNpC and SNpR reflects neurodegeneration in these areas in PD.

2.3 Histaminergic gene and receptors in PD

In human postmortem brain histamine gene expression for histamine methyltransferase (HMT) the main enzyme for catabolizing histamine and histidine decarboxylase (HDC) the enzyme that catabolize histidine to form histamine was examined in PD brains in substantia nigra, caudate nucleus and putamen. In addition histamine H3 and H4 receptors and mRNA expression was also investigated in identical PD brain (Anichtchik et al., 2000a,b).

2.3.1 HMT and HDC gene expression are altered in PD brain

Histamine methyltransferase (HMT) mRNA increased significantly by 51% in substantia nigra of PD brain as compared to control group. Interestingly there was a negative correlation between HMT mRNA level and PD duration from diagnosis to death. In putamen HMT mRNA expression showed more than 50% increase in PD brain. There was a significant correlation of increase in HMT mRNA between substantia nigra and putamen in controls but not in PD group (Garbarg et al., 1983; Koski et al., 2020).

In caudate nucleus and putamen a very low level of histamine decarboxylase (HDC) mRNA was seen in PD cases along with substantia nigra (Anichtchik et al., 2000a,b; Palada et al., 2012).

2.3.2 Histamine H3 and H4 receptor expression are altered in PD brain

In PD brain histamine H3 receptor mRNA showed a significant decrease by 40% in the substantia nigra whereas histamine H4 receptor mRNA was unaltered (Anichtchik et al., 2001). On the other hand histamine H4 receptor mRNA was increased by more than 6.3-fold in caudate nucleus but histamine H3 receptor mRNA was unaltered in PD brain (Anichtchik et al., 2000a,b, 2001; Zhou et al., 2006). Similarly in putamen histamine H4 receptor mRNA in PD brain increased by 4.2-fold from control but histamine H3 receptor mRNA was not affected (Aquino-Miranda et al., 2012; Goodchild et al., 1999).

These observations show that a decrease in histamine H3 receptor mRNA with significant increase in HMT mRNA in the substantia nigra of PD brains (Rodríguez-Ruiz et al., 2017). Interestingly, HMT mRNA was higher in putamen. In this study histamine H4 receptor mRNA was first time shown in the brain in substantia nigra, caudate nucleus and in putamen areas in the brain (Anichtchik et al., 2000a,b). The expression of histamine H4 receptor mRNA was also seen significantly higher in putamen and in caudate nucleus in PD brain (Corrêa et al., 2021). Taken together these findings indicate that histaminergic system is significantly altered in nigrostriatal network in PD.

3 Our own investigations on histamine modulation of Parkinson's disease

We have initiated a series of laboratory investigation factors affecting in PD induced bran pathology. Furthermore, we are exploring several pathways to induce neuroprotection in PD using a variety of agents and nanodelivery of novel compounds. In this investigation we used histamine H3 and H4 receptor modulation with antibodies of histamine using nanowired delivery in PD to achieve neuroprotection. The salient features of our investigation are summarized below.

3.1 Methodological consideration

3.1.1 Animals

Experiments were carried out on in bred C57BL/6 mice (weighing 28–33 g, age 12–14 weeks) housed at controlled ambient temperature (21 ± 1 °C) with 12 h dark and 12 h light schedule. Food and tap water were provided ad libitum.

All experiments were conducted according to the National Institute of Health (NIH) Guidelines of Handling and Care of Experimental Animals and approved by the Local Institutional Ethics Committee (National Research Council 2011; https://www.nap.edu).

3.2 Animal model of PD

For induction of PD like symptoms 1-methyl-4-phenyl-1,2,3,6-tetrahydropyridine (MPTP, 20 mg/kg, i.p.) twice daily was administered within 2-h intervals for 5 days as described earlier (Ozkizilcik et al., 2018a, 2019; Niu et al., 2020; Sharma et al., 2020).

On the 8th day, PD like symptoms are confirmed by the marked decrease in the number of tyrosine hydroxylase (TH) positive cells in the substantia nigra pars compact (SNpc) and striatum (STr). Also, biochemical analysis exhibited profound decrease in dopamine (DA) and its metabolites 3,4-Dihydroxyphenylacetic acid (DOPAC) and homovanillic acid (HVA) in these brain regions (Niu et al., 2020; Ozkizilcik et al., 2018a, 2019; Sharma et al., 2020).

Further evidence of the success of PD model was strengthened by measuring the cerebrospinal fluid (CSF) levels of alpha-synuclein (ASNC) and phosphorylated tau (p-tau) that exhibited profound increase in these mice (Niu et al., 2020; Ozkizilcik et al., 2018a, 2019; Sharma et al., 2020).

Control group of animals received 0.9% saline instead of MPTP in identical manner (See Sharma et al., 2020).

3.3 Modulation of histaminergic agents

In order to evaluate the influence of histaminergic modulation of PD the following agents were administered in mice model of PD using conventional and/or their TiO_2 nanowired delivery as follows.

3.3.1 Histamine H3 receptor inverse agonist BF2649

The potent and selective histamine H3 receptor inverse agonist BF2649 hydrochloride (1-[3-[3-(4-chlorophenyl)propoxy]propyl]-piperidine hydrochloride, Cat. Nr. 3743, CAS Nr. 903576-44-3, Batch Nr. 2, Tocris, Bristol, UK) soluble in water was administered in a dose of 1 mg/kg, i.p. once daily for 7 days after the onset of MPTP administration on the 1st day (Patnaik et al., 2018). On the 8th day other parameters were measured as described earlier (Patnaik et al., 2018; Sharma et al., 2017a,b, 2020).

3.3.2 Histamine H3 receptor antagonist and partial H4 receptor agonist clobenpropit

The extremely potent histamine H3 receptor antagonist with partial H4 receptor agonist clobenpropit dihydrobromide (*N*-(4-Chlorobenzyl)-*S*-[3-(4(5)-imidazolyl) propyl]isothiourea dihydrobromide Cat. Nr. 0752, CAS Nr. 145231-35-2, Batch Nr. 3, Tocris, Bristol, UK) soluble in water was administered in a dose of 1 mg/kg, i.p.) once daily for 7 days after the onset of MPTP administration on the 1st day (Patnaik et al., 2018: Sharma et al., 2017a,b). All the parameters were measured on the 8th day after completion of the clobenpropit (CLBPT) administration.

3.3.3 Anti-histamine mouse monoclonal antibody

Monoclonal mouse anti-histamine antibody (MBS358003; MyBioSource, San Diego, CA, USA) was administered into the left lateral cerebral ventricle using chronic cannula implanted 1 week before under aseptic conditions. About 25 or 50 μL anti-histamine monoclonal antibody (HAmAb) was administered once daily for 7 days in PD or control group with BF2649 or CLBPT injections given intraperitoneally (Sharma et al., 2017a,b).

3.3.4 TiO_2 nanodelivery of pharmacological agents

In separate groups, TiO_2-nanowired delivery of BF2649, CLBPT or AHmAB was nanowired using standard procedures as described earlier (Sharma et al., 2007a,b, 2009a,b; Sharma and Sharma, 2008; Ozkizilcik et al., 2018b; see Sharma et al., 2017a,b, 2018, 2019a,b, 2020; Tian et al., 2012). These nanowired drugs were delivered using identical protocol and groups for 7 days. On the 8th day, various parameters were measured for comparison between conventional and nanodelivered drugs induced neuroprotection in PD.

3.4 Parameters measured

The following physiological, behavioral, biochemical and histopathological parameters were measured using standard protocol in PD induced brain pathology and their modifications with the conventional or nanowired delivery of histaminergic drugs and/or antibodies.

3.4.1 Physiological parameters

The body weight, rectal temperature, pain perception is measured in mice intoxicated with MPTP on the 3rd day and on the 8th day using standard protocol. The rectal temperature was measured with thermistor probes (Yellow Springfield, NJ, USA) connected to a 12-channel telethermometer (Harvard Apparatus, Holliston, MA, USA) (Sharma and Dey 1986a,b, 1987). The body weight was measured in rodent weighing machine (Mettler Toledo, Columbus, OH, USA, sensitivity 1 g) (Sharma 1987). The pain sensitivity in MPTP treated mice was assessed using thermal and mechanical nociception apparatuses (Harvard Apparatus, Holliston, MA; USA) as describe earlier (Sharma et al., 2020).

3.4.2 Physiological variables

Mean arterial blood pressure (MABP), arterial pH and arterial blood gasses $PaCO_2$ and PaO_2 were measured using Radiometer apparatus (Copenhagen, Denmark). For recording of MABP the left common carotid artery was cannulated using a PE10 polythene cannula implanted retrogradely toward heart under Equithesin anesthesia aseptically 1 week before the start of the experiments. At the time of MABP recording, the carotid artery cannula was connected to a Strain Gauge Pressure Transducer (Statham P23, USA) and the output was connected to a chart recorder (Electromed, UK). Immediately before the carotid artery cannula was connected to the pressure transducer, about 1 mL of blood was withdrawn for the later measurement of arterial pH and blood gasses as described earlier (Sharma, 1987). Recording of heart rate (beats per min) and respiration cycle (cycles per min) was also recorded on the chart recorder using appropriate leads (Sharma and Dey, 1986a,b).

3.4.3 Blood-brain barrier permeability

The blood brain barrier (BBB) permeability was examined using two exogenous tracers namely Evans blue albumin (EBA) and radioiodine ([131]-Iodine) and endogenous serum albumin extravasation using albumin immunohistochemistry as describe earlier (Gordh et al., 2006; Sharma, 2011a,b; Sharma et al., 1990, 2011a,b, 2013).

In brief, EBA (3 m/kg of 2% solution in physiological saline, pH 7.4) and radioiodine (100 μCi/kg) were administered intravenously into the right jugular vein through a polythene cannula PE10 implanted aseptically 1 week before the experiments (Sharma, 1987) and the tracers were allowed to circulate for 5–10 min. After that the mice were deeply anesthetized with Equithesin (3 mL/kg, i.p.) and chest was rapidly opened and the right auricle was incised. A butterfly needle (21G) was inserted into the left ventricle connected to cold physiological saline (0.9% NaCl) through a peristaltic pump (Harvard Apparatus, Holliston, MA; USA). About 50 mL saline was perfused at 90 Torr perfusion pressure to remove intravascular tracers. Immediately before perfusion, about 0.5 mL blood was withdrawn or later determination of whole blood radioactivity or EBA content as described earlier (Sharma, 1987; Sharma et al., 1990).

After saline perfusion, the brains were removed and inspected for EBA staining on the dorsal and ventral surfaces of the brain followed by mid sagittal section to see dye penetration within the cerebral ventricles and underlying brain areas (Sharma and Dey, 1986a,b, 1987, 1988). After recording EBA dye staining into the brain small tissue pieces were dissected out, weighed immediately and the radioactivity counted in a 3-in Gamma Counter (Packard, USA). Extravasation of the radioactivity in the brain was expressed as percentage increase over the while blood radioactivity as describe earlier (Sharma and Dey, 1986a,b).

After counting the radioactivity the in brain the tissue samples were homogenized in a mixture of analytical grade of Acetone (Extra pure, BDH, UK) and 0.5% sodium sulfate (pH 7.1) to extract Evans blue dye entered into the brain. The samples were then centrifuged at $900 \times g$ for 10min and the supernatant was measured for blue dye concentration in the Spectrophotometer at 620nm. The dye concentration was calculated from the standard curve prepare earlier (See Sharma, 1987).

3.4.3.1 Albumin immunohistochemistry

For endogenous albumin penetration into the brain standard immunohistochemistry was used for albumin using anti-albumin antibody (mouse monoclonal Anti-albumin antibody ALB/2144 (ab 236492, abcam, San Diego, CA, USA) according to commercial protocol (Gordh et al., 2006; Sharma et al., 2011a,b). For this purpose, after saline wash, mice were perfused with about 100mL of cold 4% buffered paraformaldehyde in situ and wrapped in an aluminum foil and kept overnight in a refrigerator (4°C). On the next day, the brains were dissected out and kept in the same fixative for 3–4 days. Coronals sections were cut and processed for paraffin embedding using Sakura FineTek automatic tissue processer (Torrance, CA, USA).

About 3-μm thick paraffin sections were cut and processed for albumin immunostaining after incubating the section in albumin-antibody (1:200) and the reaction was visualized using avidin-biotin-complex (Burlingame, CA, USA) as standard procedures (See Gordh et al., 2006; Sharma et al., 2011a,b). The sections were analyzed in an inverted microscope (Carl Zeiss, Germany) and the images were captured using the attached digital camera. The images were stored in an Apple Macintosh Power Book (Mac OS system El Capital 10.11.6) and processed through commercial software Adobe Photoshop 12.4 using identical brightness and color filters for control and experimental groups (Sharma and Sjöquist, 2002).

3.4.4 Cerebral blood flow

Cerebral blood flow (CBF) was measured using radiolabeled carbonized microspheres (o.d. 15 ± 0.6 μm) and peripheral reference blood flow (RBF) techniques as described earlier (Sharma, 1987; Sharma and Dey, 1986a,b, 1987, 1988; Sharma et al., 1990, 1992a,b). In brief, about 90k microspheres were administered as bolus through an indwelling carotid artery cannula (PE 10) implanted retrogradely toward heart 1 week before within 45s (Sharma, 1987). Peripheral blood samples from the right femoral arterial cannula was withdrawn at the rate of 0.1 mL/min starting from 30s before microsphere injection and continued until 90min after completion of microsphere administration at every 30s. After that the mice were decapitated and the brains were removed and placed on filter paper wetted with cold physiological saline. The large superficial blood vessels and blood clots, if any were removed immediately and the brains were dissected into small pieces of the desired area, weighed immediately and the radioactivity counted in a 3-in well type Gamma counter (Packard, Ramsey, MN, USA). Serial samples of blood collected during the microsphere infusion were also counted for the accumulation of radioactivity. The CBF (mL/g/min) was calculated from the amount of radioactivity in brain samples with reference to the peripheral blood flow as described earlier (Sharma, 1987; Sharma et al., 1990).

3.4.5 Brain edema formation

Brain edema formation was evaluated using increase in brain water content and volume swelling (%*f*) as described earlier (Sharma and Cervós-Navarro, 1990, 1991; Sharma and Olsson, 1990; Sharma et al., 1991a,b, 1994). In brief, after the end of the experiments, mice were decapitated and the brains are removed and placed on filter paper wetted with cold 0.9% NaCl and the large superficial blood vessels and/or blood clots were removed immediately. After that the brains were dissected and tissue pieces were placed on preweighed filter paper (Whatman Nr. 1, USA) and weighed immediately in a sensitive Toledo Mettler electronic balance (sensitivity 0.1 mg, Columbus, OH, USA). The tissue pieces were then placed in the laboratory incubator (Thomas Scientific, Swedesboro, NJ, USA) maintained at 90 °C for 72 h to obtain their dry weight. When the dry weight measurements were became constant in three consecutive readings the brain water content was calculated from the differences between wet and dry weight as describe earlier (Sharma and Cervós-Navarro, 1990).

The volume swelling was calculated from the differences between water content of control and experimental group using the formula of Elliott and Jasper, (1949) as described earlier (Sharma and Olsson, 1990). In general a difference of 1% in water content will be approximately 4% increase in volume swelling (Sharma et al., 1991a,b, 1994).

3.5 Behavioral functions

The following behavioral functions were analyzed in PD group and their modification with drugs and antibodies.

3.5.1 Rota Rod treadmill

The Rota-Rod treadmill assesses the overall cognitive and motor functions (Sharma, 2006). We measured Rota-Rod function for mice at the speed of 16 rpm and counted their ability to stay on the treadmill. The cut off time was 120 s. 1 week before the experiments all mice were trained to stay on Rota-Rod treadmill for 120 s (Sharma, 2006, 2010).

3.5.2 Immobility during forced swimming

For this purpose a cylindrical glass jar was used which is filled with water at 24 °C (depth 18 cm). Mice were placed in the jar for 60 s and the period of immobility was recorded manually. About 1 week before mice were trained to forced swimming for 60 s daily to get accustomed with this procedure (Lafuente et al., 2018; Sharma et al., 1991b, 1995a).

3.6 Biochemical measurements

The following biochemical measurements were done in the SNpC and STr in mice using commercial ELISA kit as described earlier (Ozkizilcik et al., 2018a, 2019; Sharma et al., 2020).

3.6.1 α-synuclein

Alpha-synuclein (ASNC) was measured in SNpC and STr using mouse alpha synuclein ELISA kit (MBS725030, My BioSource, San Diego, CA, USA, sensitivity 1.0 pg/mL) according to commercial protocol (Ozkizilcik et al., 2018a; Sharma et al., 2020).

3.6.2 p-tau

p-tau was measured in SNpC and STr using mouse phospho tau ELISA kit (MBS02 6489, My BioSource, San Diego, CA, USA, sensitivity 0.1 ng/mL) using commercial protocol (Niu et al., 2020; Sharma et al., 2017a,b, 2019a,b, 2020).

3.6.3 Tyrosine hydroxylase

The tyrosine hydroxylase (TH) content was evaluated in SNpC and STr using mouse tyrosine hydroxylase ELISA Kit (MBS2019774, My BioSource, San Diego, CA, USA, sensitivity 0.56 ng/mL) according to instructions and commercial protocol (Fauss et al., 2013; Reinhard and O'Callaghan, 1991).

3.6.4 Dopamine and its metabolites

Dopamine (DA), 3,4-dihydroxyphenylacetic acid (DOPAC), homovanillic acid (HVA) are measured in tissue samples from SNpC and ST4 using HPLC with electrochemical detection as described earlier (Niu et al., 2020; Sharma et al., 2020).

3.7 Neuropathology of Parkinson's disease

Pathological changes in the PD brain were evaluated using standard protocol at light and electron microscope as describe earlier (Sharma et al., 1991a,b, 1995a,b, 1998a,b, 2006a,b, 2011a,b).

3.7.1 Light microscopy

For light microscopy, after the experiments, the animals were anesthetized with Equithesin (3 mL/kg, i.p.) and the brains were perfused in situ with 4% buffered paraformaldehyde as described earlier (Sharma and Cervós-Navarro, 1990). In brief, after anesthesia, the chest was rapidly opened and the heart was exposed and right auricle was incised. A butterfly needle 21G was inserted into the left cardiac ventricle and the other end was connected to a peristaltic pump (Harvard apparatus, Holliston, MA, USA) attached to a bottle of 0.1 M phosphate buffer sodium (PBS, pH 7.0). About 50 mL of PBS was perfused at 90 Torr to remove blood from the vessels. This was followed by perfusion of about 100 mL of 4% buffered paraformaldehyde either alone or containing 0.25% picric acid at 90 Torr (Sharma and Cervós-Navarro, 1990).

After paraformaldehyde perfusion, the animals were wrapped in aluminum foil and kept overnight at 4 °C in a refrigerator. On the next day the brains were dissected, coronal sections are cut and the tissues were processed for histological preparation and paraffin embedding in Sakura FineTek Tissue processor (USA Inc., Torrance, CA, USA).

About 3-μm thick paraffin sections were cut and stained with standard histological procedures for Haematoxylin & Eosin (H&E), Nissl, Toluidine Blue and Luxol Fast Blue (LFB) staining for neuronal and myelin evaluation (Sharma et al., 2018, 2019a,b, 2020).

3.7.1.1 Immunohistochemistry

Some sections of the paraffin were processed for immunohistochemistry according to commercial protocol as described below (Sharma et al., 2018, 2019a,b, 2020).

3.7.2 Tyrosine hydroxylase

Using mouse monoclonal anti-tyrosine hydroxylase antibody (MAB7566 R&D System, Minneapolis, MN, USA) immunohistochemistry on paraffin sections were processed according to the commercial protocol (Sharma et al., 2020). In brief, after deparaffinization and blocking of background activity the sections were exposed at 4 °C to tyrosine hydroxylase monoclonal antibodies (mAb) for 6 h udder agitation (Sharma et al., 2020). The immunoreaction was developed using avidin-biotin-complex (Burlingame, CA, USA) according to commercial protocol and examined under an Inverted Microscope (Carl-Zeiss, Oberkochen, Germany). The immunoreactive cells were counted manually at high magnification in the SNpC and STr in a blinded fashion by two independent workers three times and the median values were used for statistical analyses between control, PD and drug treated groups as described earlier (Niu et al., 2020; Sharma et al., 2009a,b).

3.7.3 Glial fibrillary acidic protein

Using mouse monoclonal anti-GFAP antibody (Anti-GFAP antibody [2A5] (ab4648), abcam, Cambridge, MA, USA) was used to develop GFAP immunoreactivity on paraffin sections as described earlier (Sharma et al., 1992a,b). In brief, the paraffin sections were incubated with GFAP primary antibody (1:500) at room temperature for 8 h and immunoreaction was developed using avidin-biotin-complex (Burlingame, CA, USA) according to commercial protocol (Cervós-Navarro et al., 1998; Sharma et al., 1992a,b). The images were examined under Carl-Zeiss (Oberkochen, Germany) microscope and nr of reactive astrocyte was counted in a blinded fashion by two independent observers. The median value was used for statistical analyses (Sharma et al., 1992a,b, 1993a,b, 1995a,b, 1997a,b,c).

3.7.4 Transmission electron microscopy

For ultrastructural investigation of neuropil in PD and its modification by histaminergic drugs, transmission electron microscopy (TEM) was used (Sharma and Olsson, 1990; Sharma et al., 1991a,b, 1992a,b). In brief, mice brains perfused with paraformaldehyde containing picric acid were used for TEM investigation. For this purpose, small tissue pieces were post fixed in osmium tetraoxide (OsO_4) and embedded in plastic (Epon 812). About 1 μm thick semithin sections were cut and stained with Toluidine blue to identify areas for ultrastructural investigation. The Toluidine blue stained sections were examined under a bench microscope and the desired areas of the neuropil was identified. Then the Epon blocks were trimmed and ultrathin sections were cut on an ultramicrotome (LKB, Stockholm, Sweden) using diamond knife (Diatome, Bern-Nidau, Switzerland). About 50nm thin serial sections were cut and placed on one hole copper TEM grid. Some sections were examined unstained while some sections were counterstained with uranyl acetate and lead citrate (Sharma and Olsson, 1990). These TEM grids were examined under Phillips Electron microscope (Eindhoven, The Netherlands) and images were captured on an attached digital camera (Gatan Image, USA) at 4–12k magnifications (Muresanu et al., 2020; Sharma et al., 2010a,b,c,d, 2011a, b, 2015, 2018). The images were stored in an Apple Macintosh Power Book (System El Capitan, Mc Os 12.11.6) and processed using identical filters in commercial Adobe Photoshop Extension 12.4 software (Sharma and Sjöquist, 2002).

3.7.4.1 Statistical analysis of data

Statistical analysis of data obtained were performed using commercial software StatView 5 (Abacus Concepts, USA) in Macintosh Classical Environment employing ANOVA followed by Dunnett's test for multiple group comparison using one control for the quantitative data obtained. Semiquantitative data were analyzed using non-parametric Chi Square test. A p-value less than 0.05 was considered significant.

4 Results

The following parameters are examined in mouse model of PD induced by MPTP as reported earlier (Ozkizilcik et al., 2018a, 2019). A brief description of our findings is presented below.

4.1 MPTP induced PD mouse model

Administration of MPTP (20mg/kg, i.p.) per day two times within the interval of 5h (1st injection 8:30AM and 2nd injection 1:30PM) results in Parkinson's disease like syndrome in mice on the 8th day (Ozkizilcik et al., 2018a, 2019). MPTP is neurotoxic and this dose and frequency degenerate dopaminergic neurons in mice and leads to symptoms of PD (Jackson-Lewis and Przedborski 2007; Jiang et al., 2019; Liu et al., 2020). The usefulness and success of MPTP model of mouse PD is further characterized by the following parameters (see below).

4.1.1 Physiological parameters following MPTP administration

We examined physiological parameters including body weight; body temperature and pain perception following MPTP induced neurotoxicity in mice throughout the exposure period.

4.1.1.1 Body weight

The body weights of mice in saline treated group (30 ± 3 g) were slightly reduced after 3rd day of MPTP administration (27 ± 2 g, $P < 0.05$). On 7th day the body weight showed continued decline mildly (26 ± 6 g, $P < 0.05$ from control). However, on the 8th day the body weight is slightly recovered (28 ± 4 g, $P < 0.05$ from control) (Fig. 1).

4.1.1.2 Rectal temperature

Measurement of rectal temperature showed significant elevation (38.34 ± 0.08 °C) on the 3rd day of MPTP treatment as compared to saline treated group (36.81 ± 0.25 °C, $P < 0.05$). On the 7th day of MPTP treatment the rectal temperature is continued to be elevated (38.56 ± 0.05 °C, $P < 0.05$) from control group. On the 8th day the rectal temperature was almost returned to normal value of 37.04 ± 0.04 °C in MPTP treated group (Fig. 1).

4.1.1.3 Pain perception

Using thermal pain stimulus, MPTP treated mice a latency of tail flick response of 10 ± 2 s was seen on the 3rd day of treatment as compared to saline control latency of mice (5 ± 2 s, $P < 0.05$). On the 7th day the tail flick latency to thermal stimulus in MPTP mice was 8 ± 2 s ($P < 0.05$) that almost returned to the normal value in MPTP treated group on the 8th day (6 ± 2 s) (Fig. 1). This suggests that MPTP induced neurotoxicity does not alter pain perception to thermal noxious stimuli (Park et al., 2015; Rosland et al., 1992).

Using mechanical noxious stimuli, the latency of tail flick response in saline treated animals was 6 ± 3 s. On the 3rd day of MPTP treatment the mechanical pain stimulus latency was see significantly enhanced (12 ± 4 s, $P < 0.05$). On the 7th day after MPTP treated mice exhibited tail flick latency after mechanical noxious stimulus to 9 ± 3 s ($P < 0.05$). However, on the 8th day after MPTP administration the tail flick latency after mechanical noxious stimulation returned to near normal values (7 ± 3 s) (Fig. 1). These observations suggest that MPTP neurotoxicity on the 8th day did note alter pain perception to mechanical noxious stimuli (Biagioni et al., 2021).

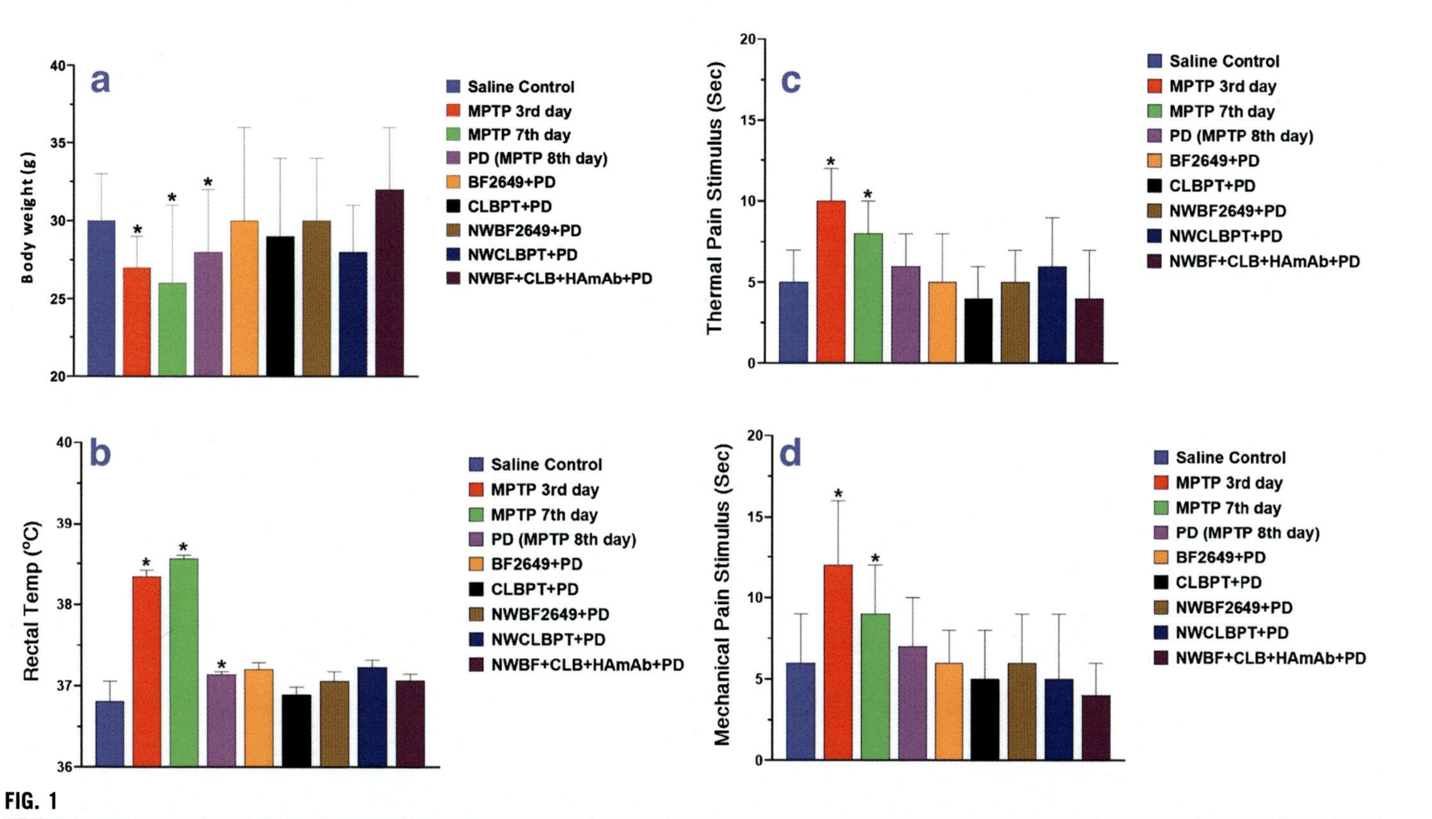

FIG. 1

Physiological parameters body weight (A) rectal temperature (B), thermal nociception (C) and mechanical nociception (D) following MPTP administration in mice and their modification with histaminergic agents. Values are Mean ± SD of 6–8 mice at each point. * $P < 0.05$ from saline control, ANOVA followed by Dunnett's test for multiple group comparison from one control. MPTP was administered 20 mg/kg, i.p. twice daily with interval of 5 h for 5 days. BF2649 or NWBF2649 were administered once daily in a dose of 1 mg/kg, i.p. for 1 week. CLBPT or NWCLBPT were administered separately (1 mg/kg, i.p.) for 1 week one injection daily. Histamine mAb was administered with NWBF and NWCLB daily once for week. HAmAb was administered into the i.c.v. route 50 μL once daily for 1 week. Other drugs were administered 1 mg/kg, i.p. Daily once for 1 week. For details see text. CLBPT, clobenpropit; mAb, monoclonal antibody; HA, histamine; NW, TiO_2 nanowired.

4.1.2 Physiological variables in MPTP model of PD

Measurement of cardiovascular physiological parameters on MABP, arterial pH, and blood gases along with heart and respiration rate showed significant alterations throughout the exposure period of mice to MPTP.

4.1.2.1 Mean arterial blood pressure

The MABP significantly declined on the 3rd day of MPTP intoxication (108 ± 4 Torr) as compared to the saline treated control mice (112 ± 6 Torr, $P < 0.05$). On the 7th day after MPTP administration the MABP was significantly increased (116 ± 6 Torr, $P < 0.05$) from the saline treated control group. On the 8th day MPTP treated mice exhibited almost restored MABP of 110 ± 4 Torr (Fig. 2). These observations suggest that MPTP neurotoxicity affects MABP during dopaminergic degeneration (Liu et al., 2020).

4.1.2.2 Arterial pH

The arterial pH in MPTP intoxicated mice showed significant elevation on the 7th day (7.39 ± 0.06) as compared to saline treated group (7.36 ± 0.04, $P < 0.05$). On the 8th day after MPTP administration the arterial pH was slightly decreased from the control group to 7.34 ± 0.04 that is very similar after MPTP injection on the 3rd day (7.35 ± 0.07) (Fig. 2).

4.1.2.3 Arterial $PaCO_2$

The arterial $PaCO_2$ in MPTP administered group showed a gradual significant elevation on 3rd day (35.18 ± 0.09 Torr, $P < 0.05$), 7th day (35.56 ± 0.12 Torr, $P < 0.05$) and on the 8th day (35.78 ± 0.08 Torr, $P < 0.05$) as compared to the saline treated control group (34.23 ± 0.08 Torr) (Fig. 2).

4.1.2.4 Arterial PaO_2

The arterial PaO_2 also showed significant variation in MPTP intoxicated mice as compared to the saline treated group. Thus, a significant increase in PaO_2 was observed in MPTP treated mice on the 3rd day (80.32 ± 0.06 Torr, $P < 0.05$) as compared to the saline control group (79.32 ± 0.07 Torr). However, on the 7th day the PaO_2 declined significantly (78.34 ± 0.09 Torr, $P < 0.05$) and remained almost steady on the 8th day after MPTP administration (78.38 ± 0.12 Torr, $P < 0.05$) as compared to saline treated group (Fig. 2).

4.1.2.5 Heart rate

The heart rate significantly varied during the period of MPTP exposure in mine. Thus, on the 3rd day of MPTP administration the heart rate decreased significantly (370 ± 6 bpm) as compared to the saline treated control group (380 ± 12 bpm, $P < 0.05$). On the 7th day the heart rate was 376 ± 8 bpm that further elevates to 385 ± 10 bpm on the 8th day after MPTP administration ($P < 0.05$ from 7th day) (Fig. 2).

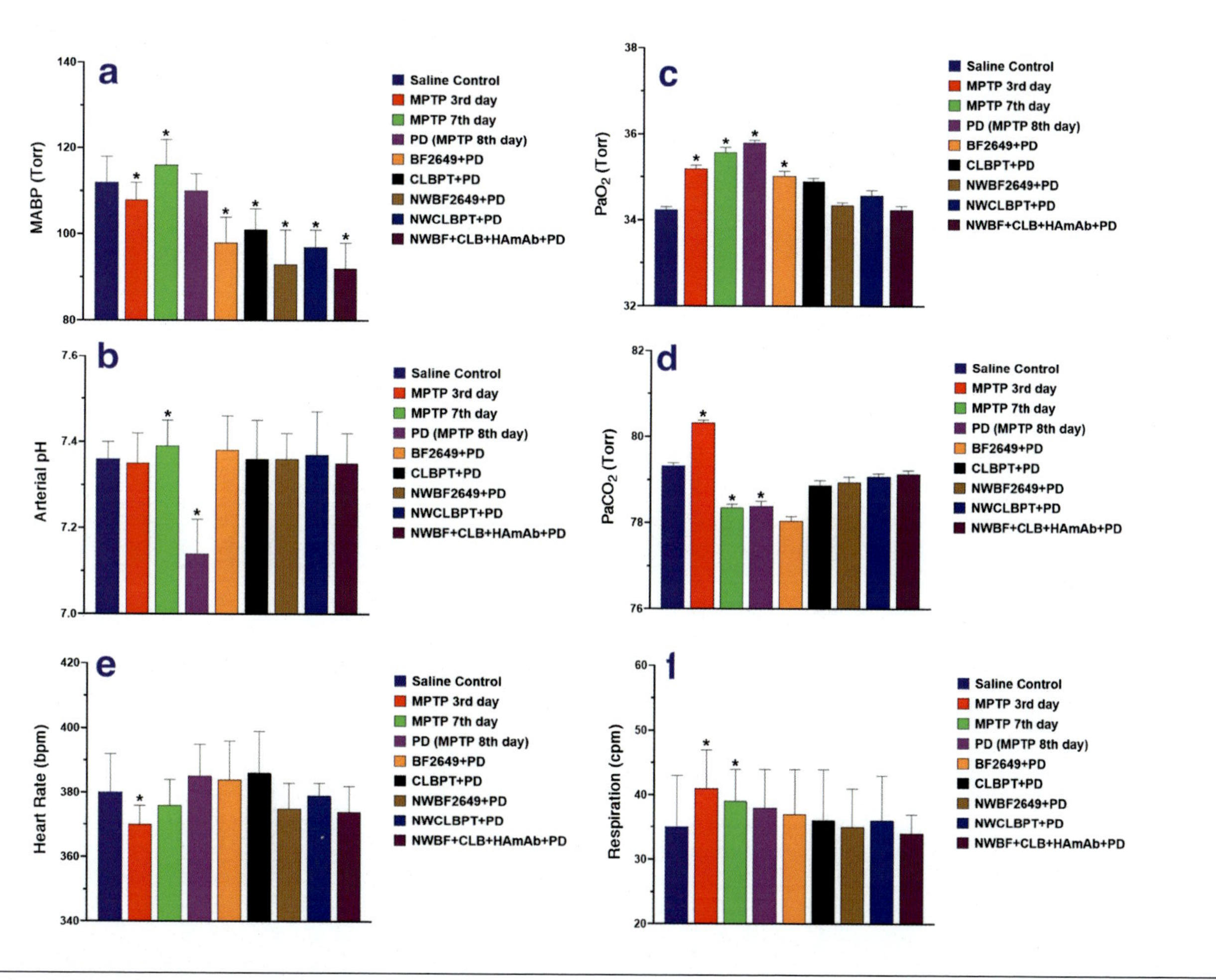

FIG. 2

Physiological variables mean arterial blood pressure (MABP) (A) arterial pH (B), PaO_2 (C) $PaCO_2$ (D) heart rate (E) and respiration rate (F) following MPTP administration in mice and their modification with histaminergic agents. Values are Mean ± SD of 6–8 mice at each point. * $P < 0.05$ from saline control, ANOVA followed by Dunnett's test for multiple group comparison from one control. MPTP was administered 20 mg/kg, i.p. twice daily with interval of 5 h for 5 days. BF2649 or NWBF2649 were administered once daily in a dose of 1 mg/kg, i.p. for 1 week. CLBPT or NWCLBPT were administered separately (1 mg/kg, i.p.) for 1 week one injection daily. Histamine mAb was administered with NWBF and NWCLB daily once for week. HAmAb was administered into the i.c.v. route 50 μL once daily for 1 week. Other drugs were administered 1 mg/kg, i.p. Daily once for 1 week. For details see text. bpm, beats per min; CLBPT, clobenpropit; cpm, cycle per min; HA, histamine; mAb, monoclonal antibody; NW, TiO_2 nanowired.

4.1.2.6 Respiration rate

The rate of respiration also varied during MPTP exposure in mice. Thus, a significant increase in respiration rate (41 ± 6 cpm) was seen in MPTP intoxicated mice on the 3rd day from saline treated group (35 ± 8 cpm, $P < 0.05$). The respiration rate was further declined gradually on the 7th day (39 ± 5 cpm) and 8th day (38 ± 6 cpm) after MPTP administration in mice (Fig. 2).

These observations support the idea that MPTP induced neurotoxicity of dopaminergic neurons affects cardiovascular response effectively (Liu et al., 2020).

4.1.3 Behavioral parameters in MPTP model of PD

MPTP administration significantly affects cognitive and sensory motor performances and results in depression like symptoms (Lesemann et al., 2012).

4.1.3.1 Rota-Rod performance

Thus, a significant decrease on Rota Rod treadmill performance latency was observed in MPTP intoxicated mice on the 3rd day (92 ± 8 s) as compared to saline treated control mice (118 ± 4, $P < 0.05$). This decrease in Rota Rod performance latency was continued to decline in MPTP treated mice on the 7th day (80 ± 6 s, $P < 0.05$) and on the 8th day after MPTP intoxication (61 ± 6 s, $P < 0.05$) (Fig. 3). This suggests that stay on Rota-Rod treadmill that requires both cognitive and sensory motor performances are declined in MPTP induced dopaminergic neurotoxicity (Sharma et al., 2010a,b,c,d).

4.1.3.2 Forced swimming immobility

Immobility during forced swimming is a sign of depression (Sharma et al., 1991b,c). Thus, when MPTP intoxicated mice were exposed to forced swimming the latency of immobility response was increased significantly on the 3rd day (8 ± 4 s) as compared to saline treated control group (6 ± 2 s, $P < 0.05$). This increase in immobility latency was further enhanced following 7th day (12 ± 5 s, $P < 0.05$) and 8th day (28 ± 6 s, $P < 0.05$) as compared to the control group (Fig. 3). This observation indicates that MPTP induced neurotoxicity of dopaminergic neurons results in depressive like behavior (Gorton et al., 2010).

4.1.4 Biochemical parameters of MPTP model of PD

In PD MPTP induces loss of tyrosine hydroxylase enzyme, dopamine and its metabolite 3,4-Dihydroxyphenylacetic acid (DOPAC) and Homovanillic acid (HVA) in SNpC and STr (Daubner et al., 2011; Soares-da-Silva and Garrett 1990). Thus we measured these enzymes biochemically in MPTP treated mouse (Ozkizilcik et al., 2018a, 2019; Sharma et al., 2020).

4.1.4.1 Tyrosine hydroxylase enzyme

Measurement of tyrosine hydroxylase enzyme (TH) in SNpC and STr in MPTP treated mice exhibited significant progressive decrease after MPTP intoxication as compared to the saline treated mice. Thus, on the 3rd day of MPTP administration, significant decrease in TH was observed in SNpC (0.53 ± 0.14 ng/μg protein) and in STr (0.49 ± 0.12 ng/μg protein) significantly from saline treated SNpC

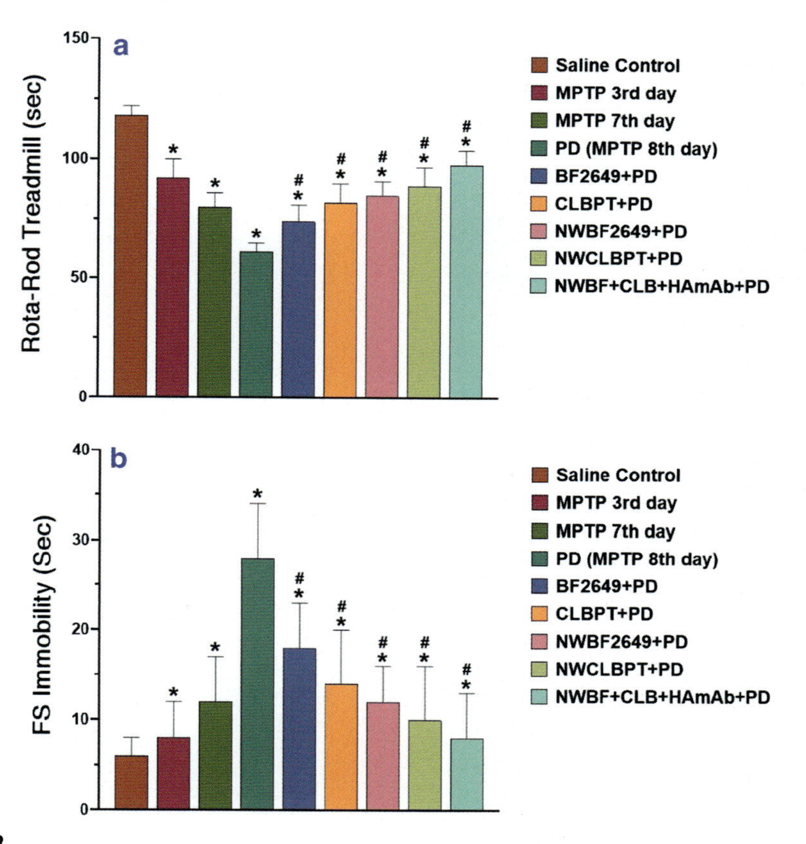

FIG. 3

Behavioral analysis following MPTP administration induced PD in mice on (A) Rota-Rod performance and Immobility in Forced Swimming (B) and their modification with histaminergic agents. Values are Mean ± SD of 6–8 mice at each point. * $P < 0.05$ from saline control, # $P < 0.05$ from MPTP-PD cases; ANOVA followed by Dunnett's test for multiple group comparison from one control. MPTP was administered 20 mg/kg, i.p. twice daily with interval of 5 h for 5 days. BF2649 or NWBF2649 were administered once daily in a dose of 1 mg/kg, i.p. for 1 week. CLBPT or NWCLBPT were administered separately (1 mg/kg, i.p.) for 1 week one injection daily. Histamine mAb was administered with NWBF and NWCLB daily once for week. HAmAb was administered into the i.c.v. route 50 μL once daily for 1 week. Other drugs were administered 1 mg/kg, i.p. Daily once for 1 week. For details see text.

(0.86 ± 0.12 ng/μg protein, $P < 0.05$) and STr (0.74 ± 0.14 ng/μg protein, $P < 0.05$). On the 7th day after MPTP treatment the TH values decrease further in SNpC (0.35 ± 0.10 ng/μg protein, $P < 0.05$) and in STr (0.30 ± 0.09 ng/μg protein, $P < 0.05$). On the 8th day the TH values declined further in SNpC (0.24 ± 0.08 ng/μg protein, $P < 0.05$) and in STr (0.18 ± 0.06 ng/μg protein, $P < 0.05$) as compared to saline treated control group (Fig. 4). This suggests that MPTP treatment progressively decreases the TH levels in SNpC and STr.

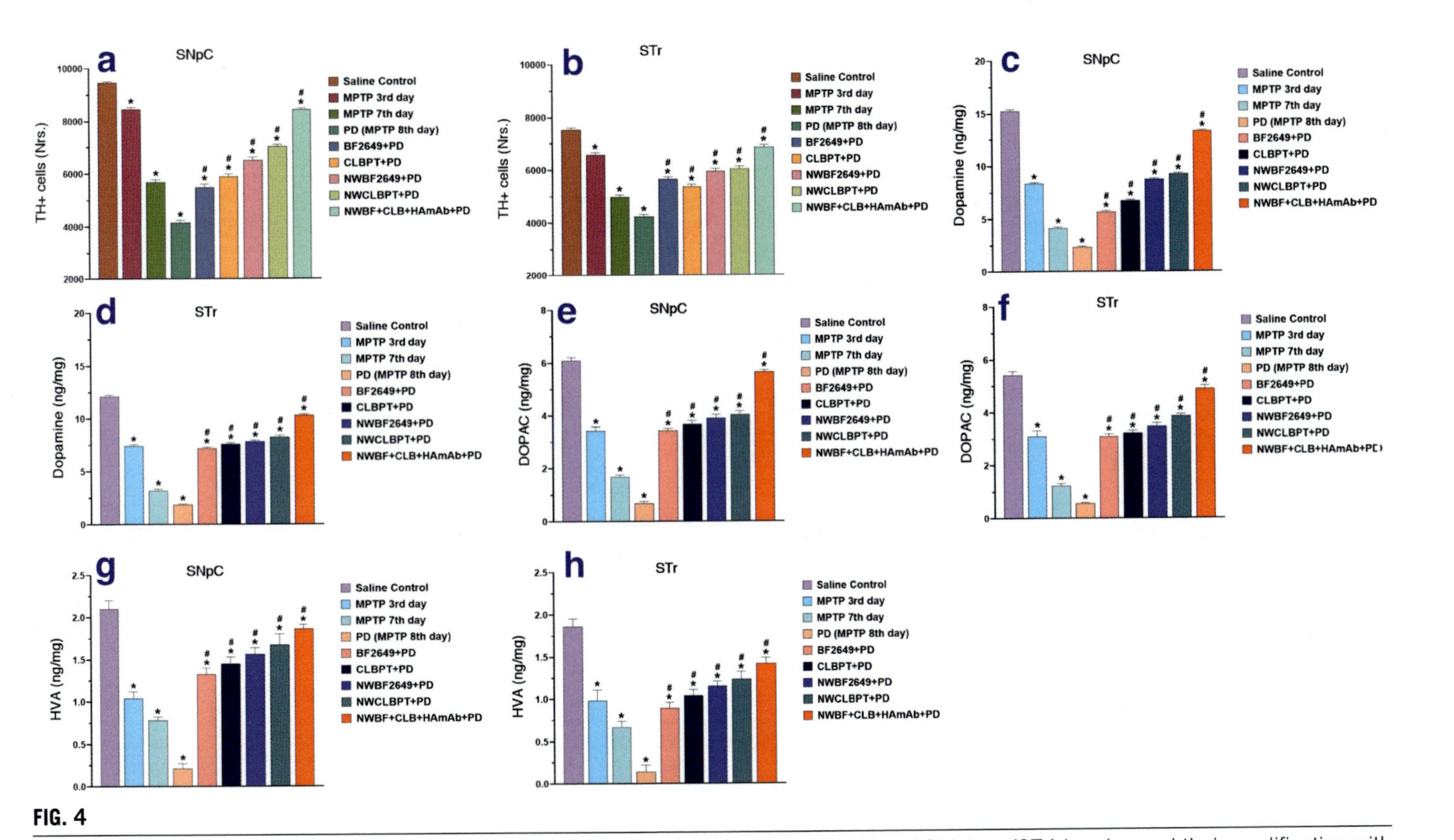

FIG. 4

MPTP administration induced PD biomarkers in Substantia nigra pars compacta (SNpC) and Striatum (STr) in mice and their modification with histaminergic agents. TH positive cells (A and B), dopamine (C and D), DOPAC (E and F), and HVA (G and H) was measured in SNpC and STr biochemically using standard procedures in PD. Values are Mean ± SD of 6–8 mice at each point. * $P < 0.05$ from saline control, # $P < 0.05$ from MPTP-PD cases; ANOVA followed by Dunnett's test for multiple group comparison from one control. MPTP was administered 20 mg/kg, i.p. twice daily with interval of 5 h for 5 days. BF2649 or NWBF2649 were administered once daily in a dose of 1 mg/kg, i.p. for 1 week. CLBPT or NWCLBPT were administered separately (1 mg/kg, i.p.) for 1 week one injection daily. Histamine mAb was administered with NWBF and NWCLB daily once for week. HAmAb was administered into the i.c.v. route 50 μL once daily for 1 week. Other drugs were administered 1 mg/kg, i.p. Daily once for 1 week. For details see text. CLBPT, clobenpropit; DOPAC, 3,4-dihydroxyphenylacetic acid; HVA, homovanillic acid; NW, TiO_2 nanowired; SNpC, Substantia nigra pars compacta; STr, Striatum; TH, tyrosine hydroxylase.

4.1.4.2 Dopamine

Measurement of dopamine (DA) in MPTP intoxicated mice shows a progressive decline on 3rd, 7th and 8th day in SNpC and STr. Thus, the DA level was significantly decreased in SNpC (08.34 ± 0.12 ng/mg) and in STr (07.43 ± 0.012 ng/mg) as compared to saline treated SNpC (15.24 ± 0.18 ng/mg, $P < 0.05$) and in STr (12.13 ± 0.15 ng/mg, $P < 0.05$). This decrease in DA was further continued to 7th day after MPTP treatment in SNpC (04.12 ± 0.14 ng/mg, $P < 0.05$) and in STr (03.21 ± 0.13 ng/mg, $P < 0.05$). On the 8th day the values of DA is decreased further in SNpC (02.32 ± 0.12 ng/mg, $P < 0.05$) and in STr (01.86 ± 0.08 ng/mg, $P < 0.05$) (Fig. 4). This indicates that MPTP neurotoxicity decreased DA content in SNpC and STr progressively up to 8th day (Ozkizilcik et al., 2018a, 2019; Sharma et al., 2020).

4.1.4.3 DOPAC

The DA metabolite also exhibited progressive decrease in MPTP treated ice up to 8th day period. Thus, on the 3rd day of MPTP treatment the DOPAC decreased in SNpC (3.42 ± 0.14 ng/mg, $P < 0.05$) and in STr (3.10 ± 0.21 ng/mg, $P < 0.05$) and continue to decline on the 7th day in SNpC (1.68 ± 0.08 ng/mg, $P < 0.05$) and in STr (1.23 ± 0.09 ng/mg, $P < 005$) as compared to the saline treated mice. On the 8th day the DOPAC values were further reduced in SNpC (0.68 ≈ 0.07 ng/mg, $P < 0.05$) and in STr (0.56 ± 0.04 ng/mg, $P < 0.05$) as compared to the saline control group (Fig. 4). These observations indicate that MPTP intoxicated mice reduces DOPAC levels progressively up to 8th period (Ozkizilcik et al., 2018a, 2019).

The HVA level also showed a gradual decrease in MPTP intoxicated mice in the SNpC and STr over 8 days. Thus, the HVA level in SNpC (1.04 ± 0.08 ng/mg) and in STr (0.98 ± 0.13 ng/mg) was observed on the 3rd day of MPTP treatment in mice as compared to the saline treated SNpC (2.10 ± 0.16 ng/mg, $P < 0.05$) and in STr (1.80 ± 0.09 ng/mg, $P < 0.05$). This decrease in HVA was further emphasized on 7th day in SNpC (0.78 ± 0.04 ng/mg, $P < 0.05$) and in STr (0.67 ± 0.07 ng/mg, $P < 0.05$) and on the 8th day in SNpC (0.21 ± 0.06 ng/mg, $P < 0.05$) and in STr (0.14 ± 0.08 ng/mg, $P < 0.05$) after MPTP exposure in mice as compared to the saline control group (Fig. 4). These studies showed that HVA level declined progressively after MPTP treatment in mice up to 8th day period (Ozkizilcik et al., 2018a, 2019).

4.1.5 Hallmarks of biomarkers in MPTP model of PD

The classical hallmarks of PD are ASNC and p-tau in various brain regions and in CSF (Ozkizilcik et al., 2018a, 2019; Sharma et al., 2020). We measured ASNC and p-tau in MPTP intoxicated mice in the SNpC and STr that showed progressive increase over the 8th day period. This indicates that MPTP induced neurotoxicity in mice mimic PD symptoms in clinical cases (Goris et al., 2007; Hadi et al., 2020; Hu et al., 2020; Sharma et al., 2020).

4.1.5.1 ASNC

Measurement of ASNC in SNpC and STr showed a progressive increase in MPTP intoxicated mice over the 8th day period. Thus, a significant increase in ASNC was seen in SNpC (6.76 ± 0.12 ng/μg, $P < 0.05$) and in STr (7.89 ± 0.21 ng/μg,

$P < 0.05$) as compared to the saline control mice in SNpC (3–48 ± 0.08 ng/μg) and in STr (4.38 ± 0.06 ng/μg). The ASNC values increased further in SNpC (10.34 ± 0.21 ng/μg, $P < 0.05$) and in STr (15.39 ± 0.34 ng/μg, $P < 0.05$) on the 7th day after MPTP treatment in mice as compared to saline treated group. On the 8th after MPTP treatment the ASNC values are further increased in SNpC (12.34 ± 0.12 ng/μg, $P < 0.05$) and in STr (18.53 ± 0.15 ng/μg, $P < 0.05$) as compared to the saline control group (Fig. 5).

4.1.5.2 p-tau

The hallmark of PD p-tau also showed a progressive increase in MPTP intoxicated mice over the 8th day period. Thus, on the 3rd day of MPTP intoxication the p-tau was significantly elevated in SNpC (26.31 ± 0.32 ng/μg, $P < 0.05$) and in STr (40.32 ± 0.23 ng/μg, $P < 0.05$) as compared to the saline treated mice SNpC (22.34 ± 0.08 ng/μg) and in STr (38.28 ± 0.07 ng/μg). On the 7th day the p-tau vales were further increased in SNpC (28.43 ± 0.10 ng/μg, $P < 0.05$) and in STr (43.56 ± 0.28 ng/μg, $P < 0.05$) as compared to the control group. On the 8th day p-tau level in SNpC (42.56 ± 0.13 ng/μg, $P < 0.05$) and in STr (45.43 ± 0.21 ng/μg, $P < 0.05$) was seen that was significantly higher that the saline control group (Fig. 5). This suggests that MPTP model of PD could be used to evaluate drug treatment for neuroprotection (Ozkizilcik et al., 2018a, 2019; Sharma et al., 2020).

4.1.6 Pathological parameters of MPTP model of PD

In PD degeneration of dopaminergic neurons in SNpC and STr is the main cause of motor dysfunction (Chinta and Andersen, 2005; Foffani and Obeso, 2018). The biomarker of dopaminergic neuron is tyrosine hydroxylase (An et al., 2013; Pasinetti et al., 1992; Weihe et al., 2006). Our immunohistochemical studies on tyrosine hydroxylase in SNpC and STr showed significant degeneration indicating loss of dopaminergic neurons in MPTP intoxication in our mouse model of PD (Ozkizilcik et al., 2018a, 2019; Sharma et al., 2020).

We observed a gradual decrease in the number of tyrosine hydroxylase positive cells in the SNpC and STr after MPTP administration in mice. Thus, on the 3rd day of MPTP administration the number of tyrosine hydroxylase positive neurons was significantly decreased in SNpC (8456 ± 78 cells) and in STr (6574 ± 90 cells) as compared to the saline treated controls in SNpC (9456 ± 49 cells, $P < 0.05$) and in STr (7534 ± 87 cells, $P < 0.05$). This decrease in tyrosine hydroxylase motive neurons was further decreased on the 7th day after MPTP administration in SNpC (5688 ± 101 cells, $P < 0.05$) and in STr (4987 ± 79 cells, $P < 0.05$) as compared to the saline treated control group. On the 8th day after MPTP injection the number of tyrosine hydroxylase positive neurons were further decreased in SNpC (157 ± 93 cells, $P < 0.05$) and in STr (4236 ± 79 cells, $P < 0.05$) as compared to the saline treated animals (Fig. 5). This observation suggests that 5 days MPTP administration 2 times daily with an interval of 5 h significantly induced dopaminergic neurons degeneration as seen in clinical PD cases (Marino et al., 2020; Ozkizilcik et al., 2018a; Sharma et al., 2020).

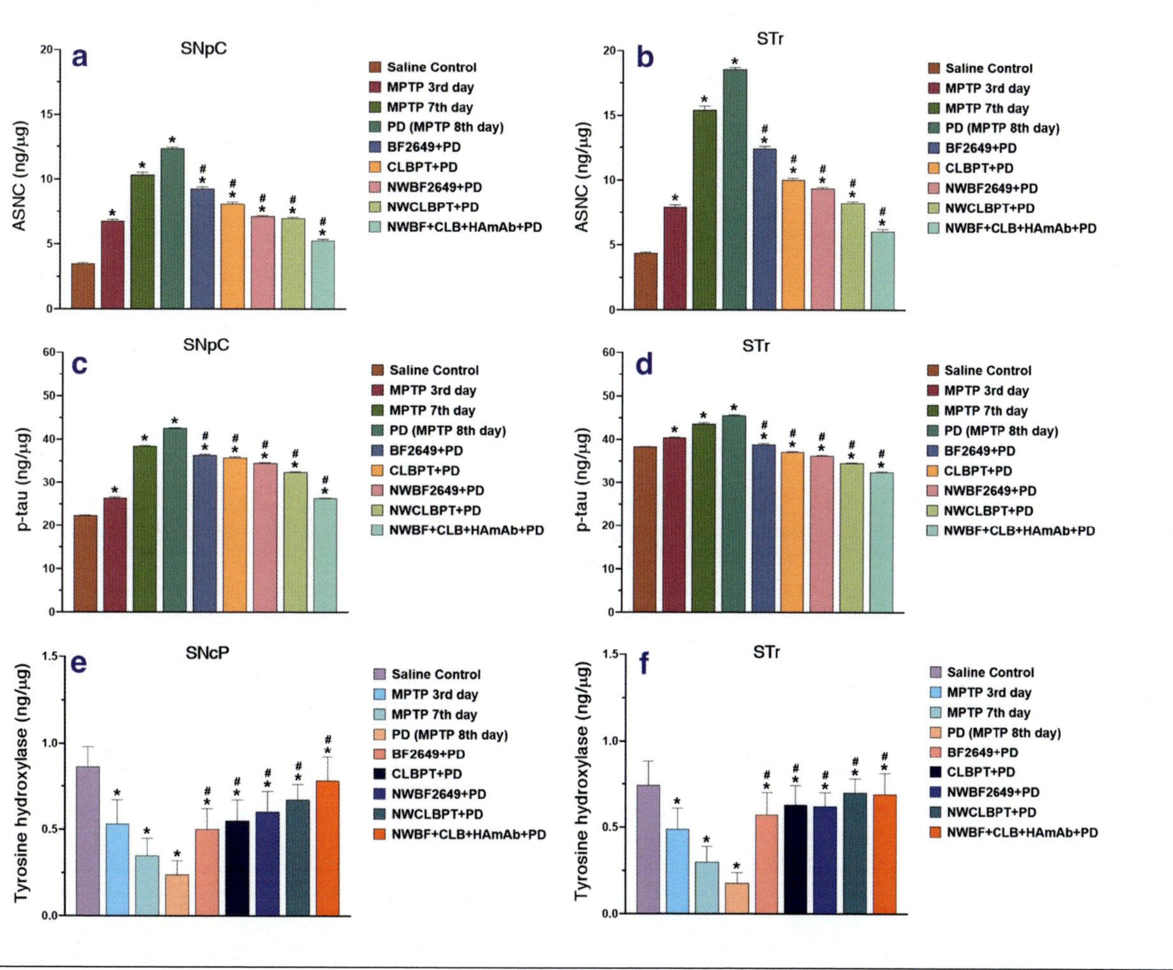

FIG. 5

MPTP administration induced PD biomarkers in Substantia nigra pars compacta (SNpC) and Striatum (STr) in mice and their modification with histaminergic agents. Alpha synuclein (ASNC) (A and B), phospho tau (C and D), and tyrosine hydroxylase enzyme (E and F) was measured in SNpC and STr biochemically using standard procedures in PD. Values are Mean ± SD of 6–8 mice at each point. * $P < 0.05$ from saline control, # $P < 0.05$ from MPTP-PD cases; ANOVA followed by Dunnett's test for multiple group comparison from one control. MPTP was administered 20 mg/kg, i.p. Twice daily with interval of 5 h for 5 days. BF2649 or NWBF2649 were administered once daily in a dose of 1 mg/kg, i.p. for 1 week. CLBPT or NWCLBPT were administered separately (1 mg/kg, i.p.) for 1 week one injection daily. Histamine mAb was administered with NWBF and NWCLB daily once for week. HAmAb was administered into the i.c.v. route 50 µL once daily for 1 week. Other drugs were administered 1 mg/kg, i.p. Daily once for 1 week. For details see text. ASNC, alpha synuclein; CLBPT, clobenpropit; NW, TiO_2 nanowired; p-tau, phospho tau; SNpC, Substantia nigra pars compacta; STr, Striatum; TH, Tyrosine hydroxylase.

4.2 Blood-brain barrier disturbances in PD

Several lines of evidences suggests significant breakdown of the BBB in animal models and in Human cases of PD (Ahlskog et al., 1989; Al-Bachari et al., 2020; Elabi et al., 2021; Gray and Woulfe, 2015).

We observed a significant increase in the BBB breakdown to EBA and radioiodine during the time course of MPTP intoxication in mice model of PD (Ozkizilcik et al., 2018a, 2019). Thus, there was a significant increase in the BBB breakdown to EBA (0.34 ± 0.08 mg%) and radioiodine (0.48 ± 0.06%) on the 3rd day of MPTP administration as compared to the saline treated control group for EBA (0.18 ± 0.02 mg %, $P < 0.05$) and radioiodine (0.28 ± 0.04%, $P < 0.05$) (Fig. 6). The extravasation of EBA (0.98 ± 0.06mg%, $P < 0.05$) and radioiodine (1.10 ± 0.05%, $P < 0.05$) was further enhanced on the 7th day after MPTP administration as compared to the saline control. On the 8th day the BBB leakage continued to enhance for EBA (1.58 ± 0.09 mg%, $P < 005$) and radioiodine (2.04 ± 0.10%, $P < 0.05$) as compared to the saline treated group (Fig. 6). This indicates that MPTP model of PD is very similar to clinical cases of BBB leakage (Al-Bachari et al., 2020; Niu et al., 2020; Ozkizilcik et al., 2018a, 2019; Sharma et al., 2020).

Albumin immunohistochemistry showed marked albumin positive cells in the neuropil of PD mice. The albumin immunoreactivity was present around microvessels and also several neurons are albumin positive in the neuropil (results not shown). This indicates leakage of serum albumin in the brain of PD group on the 8th day after MPTP administration.

4.3 Cerebral blood flow in PD

Alterations in the cerebral blood flow (CBF) leading to ischemia and stroke are frequent findings in PD (Kim and Vemuganti, 2017; Perdomo-Lampignano et al., 2020; Song et al., 2017). We measured CBF using microsphere technique in MPTP induced mouse model of PD. Our observations show a progressive decline in CBF during the course of MPTP administration and development of PD like symptoms.

There was a significant decline in CBF (1.04 ± 0.07 mL/g/min) on the 3rd day of MPTP intoxication as compared to the saline treated control mice (1.78 ± 0.04 mL/g/min, $P < 0.05$). This decrease in the CBF was further enhanced on the 7th day after MPTP administration to 0.98 ± 0.05 mL/g/min ($P < 0.05$) and further declined on the 8th day (0.76 ± 0.08 mL/g/min, $P < 0.05$) as compared to the saline treated group (Fig. 6). These observations suggest that MPTP neurotoxicity is associated with severe CBF decline.

4.4 Brain edema and volume swelling in PD

Breakdown of the BBB and reduction in the CBF lead to disturbances in fluid microenvironment of the brain in PD (Niu et al., 2020; Sharma et al., 2020). These pathophysiological events lead to brain edema and volume swelling in pre-clinical

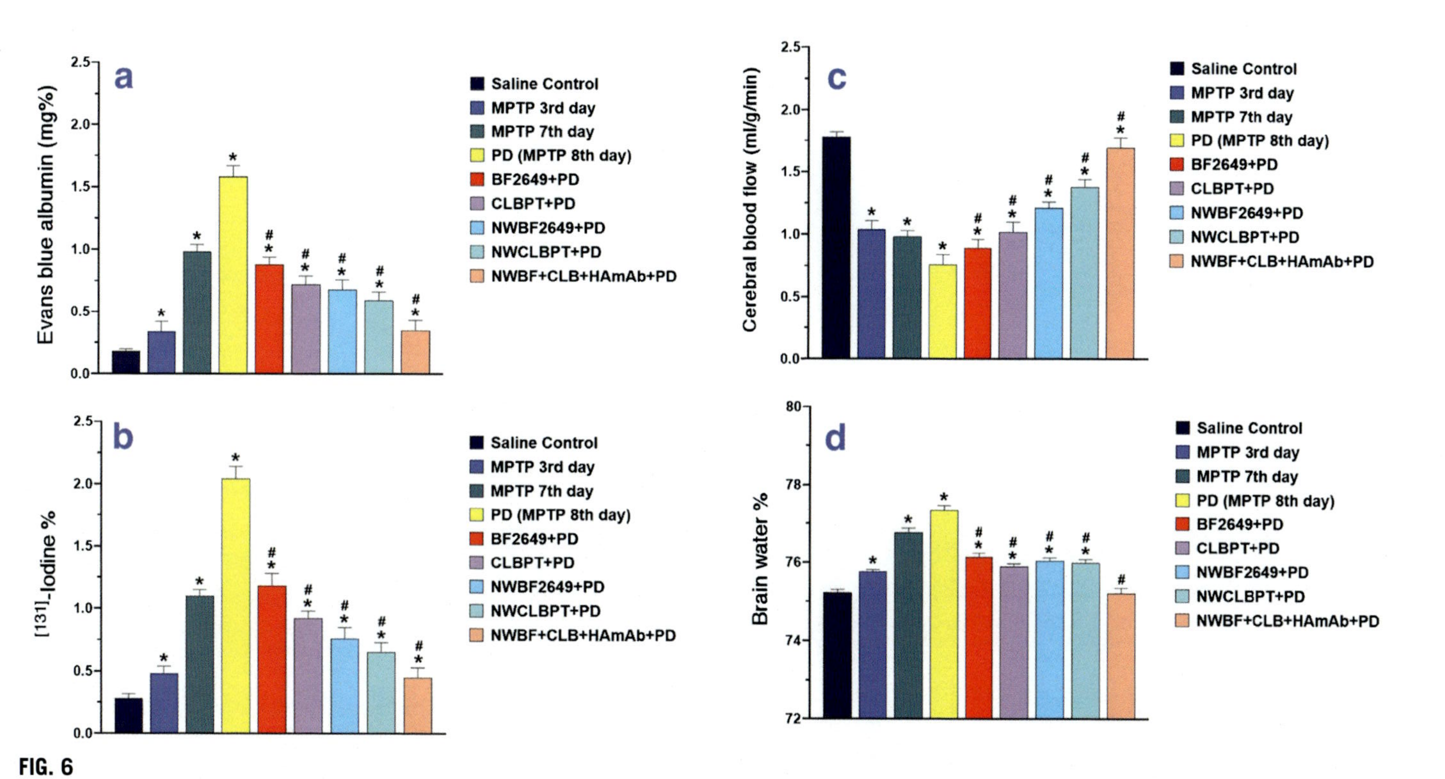

FIG. 6

Blood-brain barrier permeability to Evans blue albumin (A), radioiodine (b), cerebral blood flow (C) and brain edema (D) following MPTP administration induced PD in mice and their modification with histaminergic agents. Values are Mean ± SD of 6–8 mice at each point. * $P < 0.05$ from saline control, # $P < 0.05$ from PD; ANOVA followed by Dunnett's test for multiple group comparison from one control. MPTP was administered 20 mg/kg, i.p. twice daily with interval of 5 h for 5 days. BF2649 or NWBF2649 were administered once daily in a dose of 1 mg/kg, i.p. for 1 week. CLBPT or NWCLBPT were administered separately (1 mg/kg, i.p.) for 1 week one injection daily. Histamine mAb was administered with NWBF and NWCLB daily once for week. HAmAb was administered into the i.c.v. route 50 μL once daily for 1 week. Other drugs were administered 1 mg/kg, i.p. Daily once for 1 week. For details see text. CLBPT, clobenpropit; EBA, Evans blue albumin; NW, TiO_2 nanowired.

(Niu et al., 2020; Ozkizilcik et al., 2018a, 2019; Sharma et al., 2020) and in clinical cases of PD (Borellini et al., 2019; Nazzaro et al., 2017; Trezza et al., 2018).

We measured brain edema and volume swelling during the course of MPTP administration in the mouse model of PD. Our observations show that a gradual and significant increase in the brain water content and volume swelling occurred in MPTP model of PD in mice.

There was a significant increase in the brain water content in MPTP intoxicated mice on the 3rd day ($75.78 \pm 0.065\%$) equivalent to 2% volume swelling as compared to the saline treated control mice ($75.23 \pm 0.08\%$, $P < 0.05$). On the 7th day after MPTP intoxication the brain water content increased to $76.76 \pm 0.12\%$ ($P < 0.05$) equal to 6% volume swelling and was further increased to $77.34 \pm 0.13\%$ brain water content ($P < 0.05$) leading to about 8% of volume swelling on the 8th day after MPTP administration (Figs. 6 and 7). These observations clearly indicate that MPTP induced neurotoxicity is associated with brain edema formation and volume swelling in mouse model of PD (Ozkizilcik et al., 2018a, 2019).

4.5 Brain pathology in PD

PD is associated with brain pathology involving neuronal (Braak and Braak, 2000; Braak et al., 2000, 2003; Rietdijk et al., 2017), glial (Morales et al., 2016; Oeckl et al., 2019; Teismann and Schulz, 2004) and axonal (Bellucci et al., 2016; Caminiti et al., 2017; Tagliaferro and Burke, 2016) damages in various parts of the brain.

Using neuropathological investigations in MPTP intoxicated mice during the course of treatment and development of PD like symptoms, profound pathological changes in neuronal, glial and axonal morphology are seen (Fig. 7).

Using Nissl or H&E staining on paraffin sections significant neuronal distortion with either swelling or shrinkage was observed on the 3rd day of MPTP administration as compared to the saline treated group. There was about 12 ± 4 neurons appear distorted in MPTP group as compared to the saline treated mice (2 ± 1, $P < 0.05$). These neuronal damages further advanced to 123 ± 21 ($P < 0.05$) on the 7th day while on the 8th day after MPTP administration about 186 ± 14 ($P < 0.05$) neurons were damaged or distorted as compared to saline treated group.

The damaged neurons include dark and distorted nerve cells with perineuronal edema (Fig. 8). Both cytoplasm and karyoplasm appear dark and dense and most of the cells are devoid of a centrally located nucleus. A general sponginess and expansion of neuropil is clearly evident in MPTP intoxicated mice on the 8th day as compared to the saline treated mice. The neuronal damages were frequently seen in the cerebral cortex, cerebellum, hippocampus, SNpC and STr (Figs. 8–10; other results not shown).

Immunohistochemical staining with glial fibrillary acidic protein (GFAP) showed activated astrocytes located in the several brain areas in MPTP intoxicated mice. Thus, about 18 ± 8 GFAP positive cells are distributed with the brain on the 3rd day of MPTP treatment as compared to the saline group where only 4 ± 2 astrocyte was GFAP reactive ($P < 0.05$). On the 7th day the GFAP positive cells were enhanced to

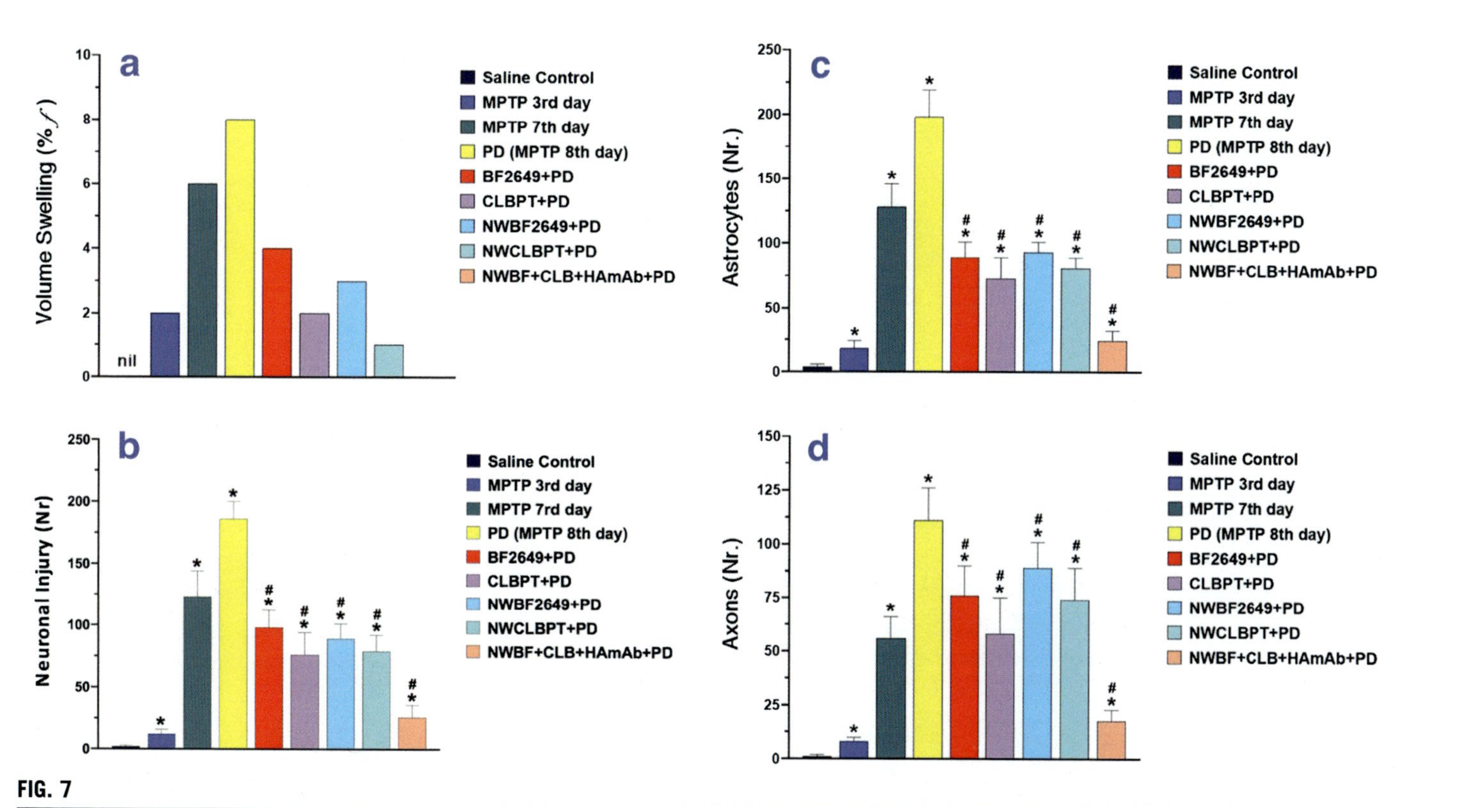

FIG. 7

Volume swelling (% *f*) calculated from changes in brain water content from control group (semiquantitative analysis of neuronal injury (C) glial fibrillary acidic protein (GFAP) positive cells at light microscopy and axonal injury or myelin vesiculation at transmission electron microscope (D) following MPTP administration induced PD in mice and their modification with histaminergic agents. Values are Mean ± SD of 6–8 mice at each point. * $P < 0.05$ from saline control, # $P < 0.05$ from PD; ANOVA followed by Dunnett's test for multiple group comparison from one control. MPTP was administered 20 mg/kg, i.p. twice daily with interval of 5 h for 5 days. BF2649 or NWBF2649 were administered once daily in a dose of 1 mg/kg, i.p. for 1 week. CLBPT or NWCLBPT were administered separately (1 mg/kg, i.p.) for 1 week one injection daily. Histamine mAb was administered with NWBF and NWCLB daily once for week. HAmAb was administered into the i.c.v. route 50 μL once daily for 1 week. Other drugs were administered 1 mg/kg, i.p. Daily once for 1 week. For details see text. CLBPT, clobenpropit; EBA, Evans blue albumin; NW, TiO_2 nanowired.

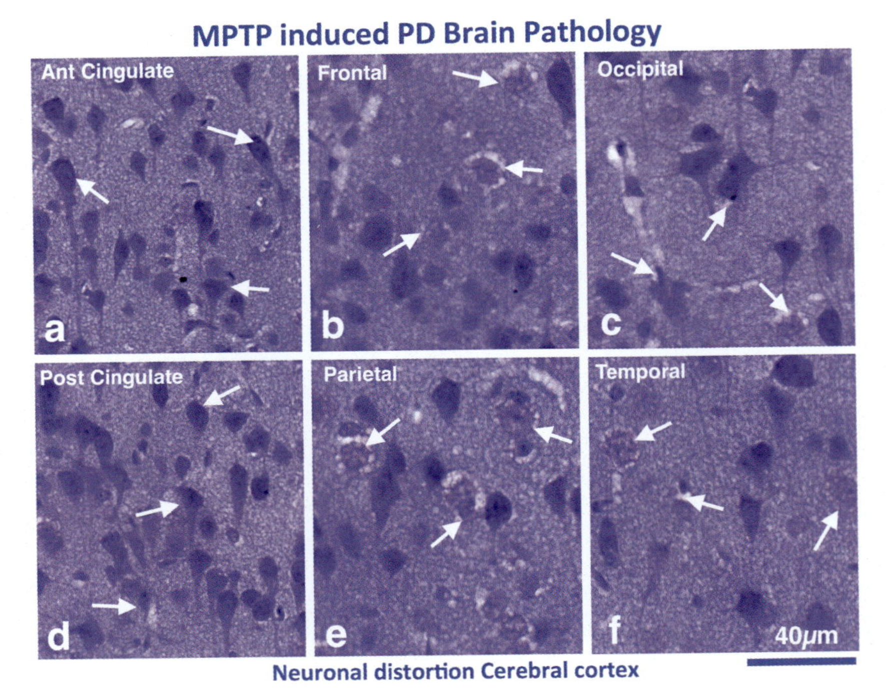

FIG. 8

Light Microscopy of Cresyl violet staining of nerve cells on 3-μm thick paraffin sections of cerebral cortex from (A) anterior cingulate cortex, (B) frontal cortex, (C) occipital cortex, (D) posterior cingulate cortex, (E) parietal cortex, and (F) temporal cortex of MPTP administration induced Parkinson's Disease (PD) like syndrome. Several neurons were showing distortion, dark neurons (arrows, A and D), degeneration and perineuronal edema (arrows, B, C, E and F) in PD. Expansion of the neuropil and edema is clearly seen in PD cases (B, C, E and F). Bar=40 μm. Nissl Stain.

128 ± 18 cells ($P < 0.05$) and that was further enhanced to 198 ± 21 GFAP positive astrocytes ($P < 0.05$) on the 8th day of MPTP administered mice (Fig. 7). These GFAP positive astrocytes are located in the cerebral cortex, SNpC, STr and hippocampus areas (results not shown).

Axonal damages seen at the ultrastructural level exhibited profound myelin vesiculation, distorted axonal morphology with periaxonal edema, vacuolation and membrane damage in MPTP group of mice (Fig. 11). These axonal changes were seen as early as on the 3rd day of MPTP administration (8 ± 2 axons) as compared to the saline treated group (1 ± 1, $P < 0.05$). On the 7th day of MPTP injection 56 ± 10 axons were severely distorted ($P < 005$) as compared to saline treated group (Fig. 7). These axonal damages were further enhanced to 111 ± 15 ($P < 0.05$) on the 8th day of

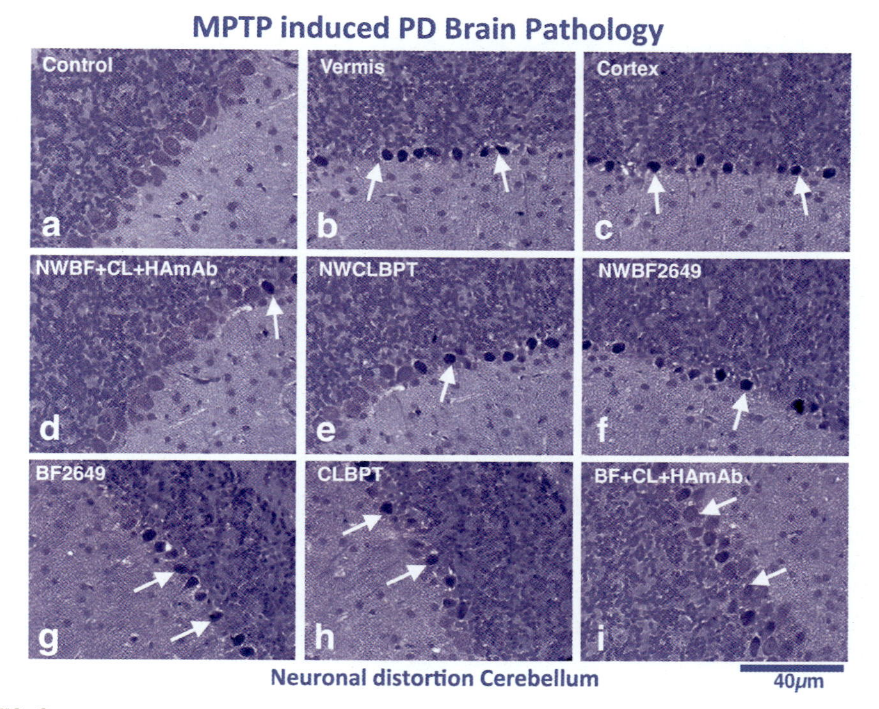

FIG. 9

Light Microscopy of Cresyl violet of Purkinje Cells and granule cells in the cerebellum in (a) control, and PD mice in (B) vermis, (cerebellar cortex) on 3-μm thick paraffin sections in MPTP administration induced Parkinson's Disease (PD) and its modification with nanowired BF2649+ clobenpropit+ Histamine mAb (D), nanowired clobenpropit (E), nanowired BF2649 (F) and conventional delivery of BF2649 (G), clobenpropit (H) and BF22649+ clobenpropit+ Histamine mAb (I). Several dark and distorted Purkinje Cells were seen in cerebellum of PD mice (B and C) as compared to saline control (A). Treatment with nanowired combination of BF2649+ clobenpropit and histamine monoclonal antibody has the most superior neuroprotective effects on Purkinje cells and granule cells (D) whereas nanowired BF2649 and NW clobenpropit alone has limited effects on neuroprotection in cerebellum in PD cases. Conventional treatment of drugs in combination with histamine monoclonal antibodies (I) still induced superior neuroprotection as compared the BF2649 or clobenpropit alone in PD cases (G and H). Bar=40μm, Nissl Stain.

MPTP treatment as compared to the saline treated mice. These axonal damages are present in several brain areas in the cerebral cortex, hippocampus, SNpC and in STr (results not shown).

These morphological observations clearly suggest that MPTP administration in mice induces PD symptoms very similar to the clinical cases.

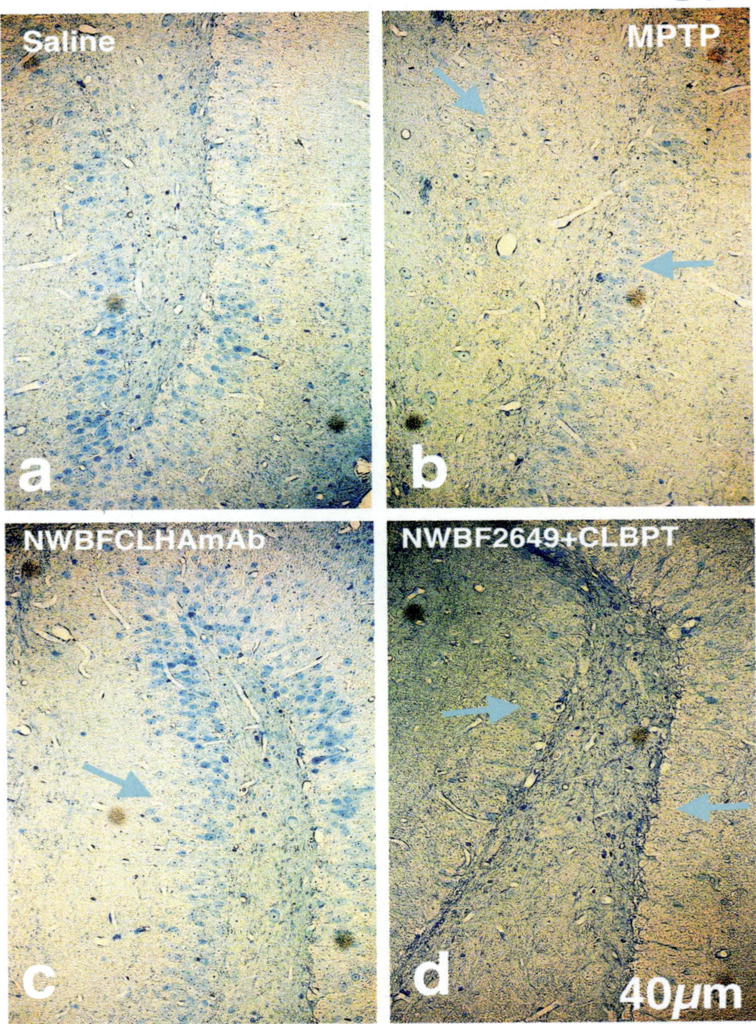

FIG. 10

Light Microscopy of Luxol Fast Blue (LFB) stain of hippocampus on 3-μm thick paraffin sections in MPTP administration induced Parkinson's disease (PD) (B) and its modification with nanowired BF2649+ clobenpropit+ Histamine mAb (C), nanowired BF264+ clobenpropit (D), as compared to saline control (A). Treatment with nanowired combination of BF2649+ clobenpropit and histamine monoclonal antibody has the most superior neuroprotective effects on dentate gyrus area of the hippocampus in PD mice (C), whereas nanowired BF2649 and NW clobenpropit alone has limited effects on neuroprotection in dentate gyrus area of hippocampus (D) as compared to untreated PD (B). Bar = 40 μm, Luxol Fast Blue Stain.

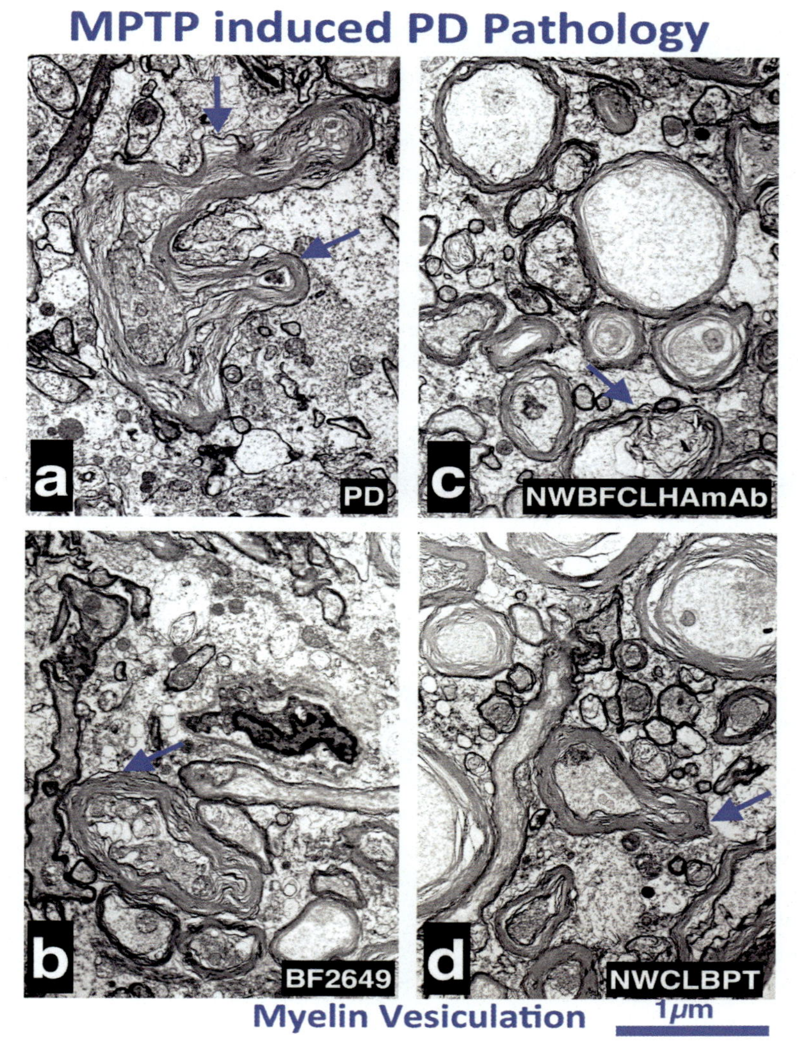

FIG. 11

Transmission electron micrograph (TEM) of substantia nigra pars compacta (SNpC) showing myelin vesiculation (A) in MPTP administration induced Parkinson's disease (PD) and its modification with conventional BF2649 (B) nanowired BF2649+ clobenpropit+ Histamine mAb (C), nanowired clobenpropit (D). Treatment with nanowired combination of BF2649+ clobenpropit and histamine monoclonal antibody has the most superior neuroprotective effects on reducing myelin vesiculation and ultrastructural changes in SNpC (C) as compared to other treatments in PD mice (B and D). Bar = 1 μm.

5 Histamine modulating agent's for neuroprotection in PD

MPTP model of PD in rodents induced profound pathophysiological response as described earlier (Niu et al., 2020; Ozkizilcik et al., 2018a, 2019; Sharma et al., 2020). Several reports suggest that histamine is actively involved in PD pathophysiology (Anichtchik et al., 2000a,b, 2001; Rinne et al., 2002). In PD histamine H3 and H4 receptors are particularly involved in the pathophysiology of PD (Gomez-Ramirez et al., 2006; Schneider 2019). Keeping these views in consideration we have used histamine H3 receptor inverse agonist BF2649 and histamine H3 receptor agonist with histamine H4 receptor antagonist clobenpropit (CLBPT) was used to influence PD pathophysiology in our MPTP mouse model of PD. In addition, to counteract histamine activity in PD, monoclonal histamine antibody (HAmAb) was used to induce neuroprotection. A brief description of results obtained is given below.

5.1 Histamine H3 receptor inverse agonist BF2649

Treatment with BF2649, a potent H3 receptor inverse agonist (Dudek et al., 2016; Ligneau et al., 2007; Uguen et al., 2013) significantly reduced the pathophysiology of MPTP induced PD in mice model.

5.1.1 Physiological parameters

BF2649 did not affect body weight changes or the rectal temperature in MPTP administered mice (Fig. 1) as compared to saline control. The thermal or mechanical pain perception was also remaining unaltered in PD group as compared to the untreated PD mice (Fig. 1). The MABP was significantly reduced (98 ± 6 Torr, $P < 0.05$, untreated PD group 110 ± 4 Torr) in BF2649 treated PD mice group as compared to the untreated PD mice. However, in BF2649 treated PD group the arterial pH, blood gases (Fig. 2) and the heart rate or respiration were not affected as compared to the untreated PD mice (Fig. 2).

5.1.2 Behavioral parameters

Treatment with BF2649 significantly improved behavioral functions on Rota Rod performances (74 ± 7 s) as compared to the untreated PD group (61 ± 4 s, $P < 0.05$). This treatment also reduced the time of immobility during forced swimming (18 ± 5 s) as compared to untreated PD mice (28 ± 6 s, $P < 0.05$) (see Fig. 3).

5.1.3 Blood-brain barrier permeability

BF2649 was significantly reduced the BBB breakdown to EBA and radioiodine in MPTP treated PD mice as compared to the untreated group. Thus, BF2649 resulted in a significant reduction in EBA (0.88 ± 0.06 mg %, $P < 0.05$) and radioiodine (1.18 ± 0.10, $P < 0.05$) extravasation in the brain as compared to the untreated PD group. There was also a significant reduction in albumin positive cells in the neuropil (Fig. 6).

5.1.4 Cerebral blood flow

The CBF in BF2649 treated PD group was significantly elevated (0.89 ± 0.07 mL/g/min) as compared to the untreated PD group (0.76 ± 0.08 mL/g/min, $P < 0.05$) seen on the 8th day (Fig. 6).

5.1.5 Brain edema and volume swelling

Treatment with BE2649 significantly attenuated brain water content (76.14 ± 0.10%, $P < 0.05$) as compared to the untreated PD group (77.34 ± 0.13%). This reduction in brain edema was equal to a 4% decrease in volume swelling in treated group as compared to the untreated PD mice (Figs. 6 and 7).

5.1.6 Brain pathology

Treatment with the potent histamine H3 inverse agonist BF2649 induced marked brain protection in neuronal, glial and axonal pathology in PD as compared to the untreated PD mice. Thus, the number of distorted neurons significantly reduced in BF2649 treated PD mice to 98 ± 14 cells ($P < 0.05$) while the GFAP positive astrocytes were reduced to 89 ± 12 cells ($P < 0.05$) as compared to the untreated PD mice (186 ± 14 neurons and 198 ± 21 astrocytes), respectively (Fig. 7). The number of axonal distortion was also reduced in BF2649 treated PD group to 76 ± 14 as compared to the untreated PD mice (111 ± 15 axons) (Fig. 7).

5.1.7 Biochemical and immunohistochemical markers

In BF2649 treated PD mice significantly attenuated the biochemical and immunohistological marker in SNpC and STr as compared to the untreated PD group (Fig. 4). Thus, tyrosine hydroxylase immunohistochemistry showed significantly increase in cell numbers in BF2649 treated PD group in SNpC (6489 ± 123 cells, $P < 0.05$, untreated 4157 ± 93 cells) and in STr (5640 ± 89 cells, $P < 0.05$, untreated 4236 ± 79 cells) as compared to the untreated PD mice.

This effect is also seen in biochemical measurement of tyrosine hydroxylase in BD2649 treated PD mice. Thus, in treated group tyrosine hydroxylase enzyme content in SNpC (0.50 ± 0.12 ng/μg protein, $P < 0.05$, untreated 0.24 ± 0.08 ng/μg protein) and in STr (0.57 ± 0.13 ng/μg protein, $P < 0.05$, untreated 0.18 ± 0.06 ng/μg protein) was significantly enhanced as compared to the untreated PD mice (Fig. 4).

The dopamine level was also enhanced in BF2649 treated PD group in SNpC (5.68 ± 0.12 mg/mg, $P < 0.05$, untreated 2.32 ± 0.12 ng/mg) and in STr (7.17 ± 0.12 ng/mg, $P < 0.05$, untreated 1.86 ± 0.08 ng/mg) as compared to the untreated PD mice. The DOPAC level also increased in SNpC (3.42 ± 0.08 ng/mg, $P < 0.05$, untreated 0.68 ± 0.07 ng/mg) and in STr (3.08 ± 0.10, $P < 0.05$, untreated 0.56 ± 0.04 ng/mg) of PD mice in BF 2549 treatment as compared to the untreated PD group (Fig. 4). The HVA level in BF2649 treated PD group was significantly higher in SNpC (1.32 ± 0.08 ng/mg, $P < 0.05$, untreated 0.21 ± 0.06 ng/mg) and in STr (0.89 ± 0.07 ng/mg, $P < 0.05$, untreated 0.14 ± 0.08 ng/mg) as compared to the untreated PD mice (Fig. 4).

5.1.8 PD hallmarks

Treatment with BF2649 significantly reduced the PD hallmarks in MPTP administered mice on the 8th day (Fig. 5). Thus, the ASNC in BF2649 treated PD mice showed significant decrease in SNpC (9.38 ± 0.15 ng/μg, $P < 0.05$, untreated 12.34 ± 0.12 ng/μg) and in STr (12.28 ± 0.38 ng/μg, $P < 0.05$, untreated 18.54 ± 0.15 ng/μg) as compared to the untreated PD mice (Fig. 5).

The p-tau was also greatly reduced in BF2649 treated PD mice in SNpC (36.37 ± 0.21 ng/μg, $P < 0.05$, untreated 42.56 ± 0.13 ng/μg) and in STr (38.78 ± 0.31 ng/μg, $P < 0.05$, untreated 45.43 ± 0.21 ng/μg) as compared to the untreated PD mice (Fig. 5).

5.2 Histamine H3 receptor antagonist and partial H4 receptor agonist clobenpropit

We used clobenpropit a potent histamine H3 antagonist with partial H4 agonist properties (Buckland et al., 2003; Liu et al., 2001; Wanot et al., 2018; Yokoyama et al., 1994) to evaluate its neuroprotective effects in MPTP induced PD in mice.

5.2.1 Physiological parameters

Clobenpropit treatment in PD mice did not alter body weight, rectal temperature and pain perception after thermal or mechanical noxious stimulation as compared to the untreated PD group (Fig. 1). However, the MABP (101 ± 5 Torr, $P < 0.05$, untreated 110 ± 4 Torr) and the arterial $PaPaCO_2$ (34.89 ± 0.08 Torr, $P < 0.05$, untreated 35.78 ± 0.08 Torr) were significantly lowered in clobenpropit treated PD mice as compared to the MPTP intoxicated untreated PD group (Fig. 2). Other parameters such as the arterial pH, PaO_2, heart rate and respiration were not affected by clobenpropit treatment in PD mice as compared to untreated PD group on the 8th day (Fig. 2).

5.2.2 Behavioral parameters

Clobenpropit treatment in PD mice significantly enhanced the cognitive and sensory motor functions as displayed on their Rota Rod treadmill performances and also reduced the depressive symptoms as evident from the reduced latency of immobility during forced swimming trial (Fig. 3). Thus, in clobenpropit treated PD mice enhanced their staying latency on the Rota Rod treadmill by 82 ± 8 s that was significantly higher than the untreated PD mice (61 ± 4 s, $P < 0.05$). Likewise the latency of immobility during forced swimming in clobenpropit treated PD mice was reduced to 14 ± 6 s ($P < 0.05$) as compared to the untreated PD group (28 ± 6 s). These observations suggest that the clobenpropit may have additional beneficial effects in PD on behavioral functions (Fig. 3).

5.2.3 Blood-brain barrier permeability

Treatment with clobenpropit in PD mice significantly attenuated the extravasation of EBA (0.72 ± 0.07 mg %, $P < 0.05$, untreated 1.28 ± 0.09 mg%) and radioiodine (0.92 ± 0.06 mg%, $P < 0.05$, untreated 2.04 ± 0.10 mg %) as compared to the untreated PD group. This reduction in the BBB permeability to EBA and radioiodine tracers was superior than BF2649 treated PD mice. Albumin positive cells in the neuropil were significantly reduced in clobenpropit treated group that seems greater than the BF2649 treated PD mice (Fig. 6).

5.2.4 Cerebral blood flow

Clobenpropit treatment was able to significantly increase the CBF in PD mice as compared to the untreated PD group (Fig. 6). Thus, in clobenpropit treated PD mire showed and increased CBF (1.02 ± 0.08 mL/g/min, $P < 0.05$, untreated 0.76 ± 0.08 mL/g/min) as compared to the untreated PD group. This suggests that clobenpropit has superior anti-ischemic effects in PD that BF2649 (Fig. 6).

5.2.5 Brain edema and volume swelling

Treatment with clobenpropit was very effective in reducing brain water content in PD mice as compared to the untreated PD group (Fig. 6). In clobenpropit treated PD mice the brain water content is significantly attenuated to $75.89 \pm 0.08\%$, $P < 0.05$, untreated $77.34 \pm 0.13\%$) as compared to the untreated PD group. Thus, the volume swelling in clobenpropit treated PD mice exhibited only 2% swelling as compared to the untreated D group whether the volume swelling was 8%. This suggests that clobenpropit has superior anti edema effects in PD mice as coma oared to BF2649 treatment (Figs. 6 and 7).

5.2.6 Brain pathology

Clobenpropit treatment in PD mice induces significant neuroprotection in neuronal, glial and axonal damages as compared to the untreated PD group (Fig. 7). Thus, only 76 ± 18 neurons exhibited distorted or damaged cells in clobenpropit treated PD mice as compared to the untreated MPTP treated mice (188 ± 14, $P < 0.05$). Similarly this treatment attenuated the number of GFAP positive cells in PD mice (73 ± 16 cells, $P < 0.05$, untreated 198 ± 21 cells) as compared to the untreated PD group (Fig. 7). Axonal damages and myelin vesiculation were also reduced by clobenpropit in PD mice (58 ± 17 axons, $P < 0.05$, untreated 111 ± 15 axons) as compared to the untreated PD mice (Fig. 7).

These neuropathological observations indicate that clobenpropit induced superior neuroprotection in PD mice as compared to the BF2649 treatment.

5.2.7 Biochemical and immunohistochemical markers

In clobenpropit treated PD mice shoed significant increase in tyrosine hydroxylase immunoreactive cells in the SNpC (5890 ± 102 cells, $P < 0.05$, untreated 4157 ± 93 cells) and in STr (5346 ± 109 cells, $P < 0.05$, untreated 4236 ± 79 cells) as compared to the untreated PD group (Fig. 4). These immunohistochemical data on tyrosine hydroxylase was further confirmed by measurement of the tyrosine hydroxylase enzyme activity in clobenpropit treated PD mice. Thus, clobenpropit treated PD mice exhibited higher content of tyrosine hydroxylase enzyme content in SNpC (0.56 ± 0.12 ng/μg protein, $P<005$, untreated 0.24 ± 0.08 ng/μg protein) and in STr (0.63 ± 0.13 ng/μg protein, $P<0.05$, untreated 0.18 ± 0.06 ng/μg protein) as compared to the untreated PD mice (Fig. 4).

Clobenpropit treatment in PD mice also enhanced the dopamine levels in the SNpC (6.73 ± 0.10 ng/mg, $P<0.05$, untreated 2.32 ± 0.12 ng/mg) and in STr (7.56 ± 0.14 ng/mg, $P<0.05$, untreated 1.86 ± 0.08 ng/mg) as compared to the untreated OD group (Fig. 4).

The DOPAC level was also enhanced significantly in clobenpropit treated OD mice in the SNpC (3.67 ± 0.14 ng/mg, $P<0.05$, untreated 0.68 ± 0.07 ng/mg) and in STr (3.21 ± 0.10 ng/mg, $P<005$, untreated 0.56 ± 0.04 ng/mg) as compared to the untreated PD group (Fig. 4).

The HVA level also increased significantly in clobenpropit treated PD mice in SNpC (1.45 ± 0.08 ng/mg, $P<0.05$, untreated 0.21 ± 0.06) and in STr (1.04 ± 0.07 ng/mg, $P<0.05$, untreated 0.24 ± 0.08 ng/mg) as compared to the untreated PD group (Fig. 4).

These observations indicate that clobenpropit treatment induces superior effects on biochemical and immunohistochemical parameters in PD mice as compared to the BF2649 treatment.

5.2.8 PD hallmarks

The biochemical hallmarks of PD incudes significant rise in ASNC and p-tau in the brain as well as in the CSF (Niu et al., 2020; Sharma et al., 2020). We measured ASNC and p-tau in SNpC and STr of PD mice following clobenpropit treatment (Fig. 5).

Our results showed that in clobenpropit treated PD mice showed significant reduction in the ASNC in SNpC (8.08 ± 0.18 ng/μg, $P<0.05$, untreated 12.34 ± 0.12 ng/μg) as compared to the untreated PD mice (Fig. 5).

Likewise p-tau was also significantly decreased in clobenpropit treated PD mice in the SNpC (35.76 ± 0.25 ng/μg, $P<0.05$, untreated 42.56 ± 0.13 ng/μg) and in STr (37.08 ± 0.17 ng/μg, $P<0.05$, untreated 45.43 ± 0.21 ng/μg) (Fig. 5).

These observations suggest that clobenpropit has the superior effects in reducing ASNC and p-tau in PD brain as compared to the BF2649 treated PD mice.

5.3 Nanowired delivery of histaminergic agents

The histamine H3 inverse agonist BF2649 and Histamine H3 antagonist with partial H4 agonist clobenpropit induced beneficial effects in PD. In this regards, clobenpropit exerted superior effects on neuroprotection in PD. It would be interesting to see whether TiO_2 nanodelivery of BF2649 or clobenpropit may further enhance the neuroprotective effects in MPTP induced PD in mice.

The salient features of nanowired delivery of these histamine agents in PD mouse are described below.

5.3.1 TiO_2 nanowired delivery of BF2649 in PD

Treatment with nanowired BF2649 (NWBF2649) in PD induced superior beneficial effects on physiological, behavioral and biochemical parameters resulting in reducing BBB breakdown and brain edema formation enhancing CBF and reducing brain pathology in PD mice as compared to the conventional BF2649 treatment (Figs. 1, 2, 6 and 7).

Thus, NWBF2649 treatment resulted in significant reduction in the MABP (93 ± 8 Torr) in PD mice and improved blood gases and arterial pH effectively without affecting the other parameters on heart rate, respiration rectal temperature, pain perception or body weight changes (Fig. 2).

The behavioral changes on Rota Rod in NWBF2649 treated mice improved significantly (85 ± 6 s) and immobility latency was markedly reduced (12 ± 4 s) in PD group that was slightly superior to the conventional BF2649 treatment in PD (Fig. 3).

The BBB permeability to EBA (0.68 ± 0.08 mg%) and radioiodine (0.76 ± 0.09%) were considerable reduced in NWBF2649 treated in PD mice as compared to the conventional BF2649 treatment. Albumin positive cells in the neuropil were also significantly reduced as compared to the conventional treatment with BF2649 (Fig. 6). This indicates that nanowired drug delivery induces superior neuroprotective effects in PD.

The CBF values were further enhanced (1.21 ± 0.05 mL/g/min in NWBF2649 treated PD mice as compared to the conventional treatment with BF2649 (0.89 ± 0.07) (Fig. 6).

The brain water content was significantly reduced in NWBF2649 treated PD mice to 76.04 ± 0.08 denoting a volume swelling of 3% as compared to the conventional BF2639 treatment that showed 4% increase in volume swelling in PD mice (Figs. 6 and 7).

The neuropathological changes in NWBF2649 in PD also showed significant reductions in brain pathology in PD as compared to the untreated PD group (Figs. 8–11).

The biochemical changes in tyrosine hydroxylase enzyme, dopamine, DOPAC and HVA in SNpC and STr were considerable reduced in NWBF2649 treated group in PD as compared to the untreated PD mice (Fig. 5).

The number of tyrosine hydroxylase positive cells (Figs. 4 and 5) was significantly enhanced in the SNpC (6503 ± 123 cells) and in STr (5945 ± 108) in NWBF2649 treated PD mice as compared to the conventional BF2649 treatment in PD (Fig. 4). The levels of ASNC in SNpC (7.12 ± 0.08 ng/μg) and in STr (9.34 ± 0.13 ng/μg) were significantly reduced in PD after NWBF2649 treatment as compared to the convectional BF2649 treated mice in PD (Fig. 5). Likewise the p-tau values are much lower in SNpC (34.51 ± 0.17 ng/μg) and in STr (36.18 ± 0.21 ng/μg) in NWBF2649 treated PD mice than the conventional treatment with BF2649 in PD group (Fig. 5).

These observations suggest that nanowired drug delivery induces superior neuroprotection in PD.

5.3.2 TiO_2 nanowired delivery of clobenpropit in PD

Treatment with TiO_2 nanowired clobenpropit (NWCLBPT) induced superior neuroprotection in PD as compared to the conventional delivery of clobenpropit.

Treatment with NWCLBPT significantly reduced MABP (97 ± 4) and restored PaO_2 and $PaCO_2$ values near normal levels together with the arterial pH as compared to the conventional clobenpropit treatment in PD (Fig. 2). The heart rate, respiration, pain perception or body weight were not significantly altered by NWCLBPT as compared to the conventional clobenpropit treatment in PD group (Fig. 2).

NWCLBPT significantly improved behavioral function on Rota Rod treadmill performance and immobility response in forced swimming (Fig. 3). Thus, in NWCLBPT the latency of staying at Rota Rod treadmill was enhanced to 89 ± 8 s as compared to conventional clobenpropit (82 ± 8 s). On the other hand the latency of immobility response during forced swimming was reduced by NWCLBPT to 10 ± 6 s as compared to the conventional clobenpropit treatment (14 ± 6 s) in PD mice (Fig. 3).

Treatment with NWCLBPT in PD resulted in greater reduction in extravasation of EBA (0.59 ± 0.07 mg%) and radioiodine (0.65 ± 0.08%) in PD mice as compared to the conventional BF2649 treated PD group. Albumin positive cells were also significantly reduced in PD as compared to the conventional clobenpropit administration (Fig. 6).

The CBF values were further elevated to 1.38 ± 0.08 mL/g/min in NWCLBPT treated PD mice as compared to the conventional clobenpropit treatment in PD group (Fig. 6).

The values of brain edema was also further reduced in NWCLBPT treated PD mice to 1% visual swelling (brain water content 75.98 ± 0.10) as compared to conventional clobenpropit showing 3% increase in volume swelling after treatment in PD group (Figs. 6 and 7).

The brain pathology also exhibited significant reduction in neuronal (79 ± 13 cells), glial (81 ± 8 cells) and axonal (74 ± 15 axons) pathology in NWCLBPT treated PD mice than the untreated group (Fig. 7).

The biochemical and other histochemical changes were markedly improved in NWCLBPT treated PD mice as compared to the conventional clobenpropit treatment in PD group (Figs. 4 and 5). Thus, the tyrosine hydroxylase positive cells were significantly increase in NWCLBPT treated mice in SNpC (7032 ± 75 cells) and in STr (6052 ± 101 cells) that is far greater than the conventional clobenpropit treated PD group (Fig. 4). The level of ASNC was further reduced in NWCLBPT treated PD mice in SNpC (6.98 ± 0.08 ng/μg) and in STr (8.23 ± 0.12 ng/μg) as compared to the conventional clobenpropit treatment in PD (Fig. 5). Likewise p-tau values ware also considerably reduced by NWCLBPT treatment in mice of SNpC (32.48 ± 0.13 ng/μg) and in STr (34.52 ± 0.16 ng/μg) as compared to the conventional treatment with clobenpropit in PD group (Fig. 5).

The immunohistochemical data on tyrosine hydroxylase was further supported by measurement of the tyrosine hydroxylase enzyme in PD mice treated with NWCLBPT. Our data show that the tyrosine hydroxylase enzyme content was markedly elevated in SNpC (0.67 ± 0.09 ng/μg protein) and in STr (0.70 ± 0.08 ng/μg protein) in PD mice treated with NWCLBPT as compared to the conventional clobenpropit treated PD mice (Figs. 4 and 5).

Dopamine levels in the SNpC (9.23 ± 0.12 ng/mg) and in STr (8.21 ± 0.18 ng/ng) were much greater in PD mice treated with NWCLBPT as compared to the conventional clobenpropit (Fig. 4). The level of DOPAC also showed greater accumulation in SNpC (4.02 ± 0.14 ng/mg) and in STr (3.86 ± 008 ng/mg in NWCLBPT treated PD mice as compared to conventional therapy with clobenpropit (Fig. 4). The level of HVA also showed greater increase in NWCLBPT treated PD mice in SNpC (1.67 ± 0.13 ng/mg) and in STr (1.23 ± 0.09 ng/mg) as compared to the conventional clobenpropit treatment in PD group (Fig. 4).

These observations further support the idea that nanowired delivery of drugs induce superior neuroprotection in PD than the conventional delivery of the identical compound.

5.3.3 TiO_2 nanowired delivery of clobenpropit with monoclonal mouse histamine antibody

Our observations clearly suggest that histamine is actively involved in the pathophysiology of PD. We also found that NWBF2649 and NWCLBPT induce significant neuroprotection in PD. Thus, it would be interesting to combine NWBF2749 and NWCLBPT with nanowired histamine monoclonal antibody (NWHAmAb) for further superior neuroprotective effects in PD.

Our results show that combination of triple agents (NWBF2649 + NWCLBPT + NWHAmAb) induces profound superior neuroprotection in PD (Figs. 8–11). This triple combination of agents resulted in significant reduction in MABP (92 ± 8 Torr) and restored the arterial pH, $PaCO_2$ and PaO_2 near normal values in PD mice (Fig. 2). The other parameters of body weight, body temperature, pain perception, heart and respiration rates are bot significantly affected in PD in these combined treatments (Fig. 1).

These combined treatment of nanowired triple agents significantly enhanced behavioral functions on Rota Rod staying latency to 98 ± 6 s as compared to individual agent treatment in PD. This combined nanowired delivery treatment in PD also reduced the immobility latency to 8 ± 5 s during forced swimming as compared to nanowired delivery of individual agents or their conventional therapy in PD (Fig. 3).

This combined triple treatment in PD mice induced significantly greater reduction in the BBB breakdown to EBA (0.35 ± 0.08 mg%) and radioiodine (0.45 ± 0.08%) as compared to individual treatment of these agents in PD mice. The albumin positive cells in the neuropil were markedly reduced near normal values by this triple agents treatment in PD (Fig. 6).

The CBF values are also enhanced greatly to 1.69 ± 0.08 mL/g/min by combined triple treatment in PD as compared to other treatments alone (Fig. 6). The brain edema formations were almost completely attenuated by these combined triple treatments in PD mice. This is evident with the reduction in brain water content to 75.21 ± 0.14% that is similar to the saline treated control brain water content amounting to no visual swelling in PD brain after these triple treatment through nanowired delivery (Figs. 6 and 7).

The brain pathology in PD mice treated with these agents through nanowired delivery resulted in superior reduction in brain pathology as compared to any agents given individually in PD (Fig. 7). As a result only 26±12 neurons, 24±8 GFAP positive astrocytes and 18±5 axonal deformation were observed in PD mice with the combined treatment of drugs and antibody (Fig. 7) as compared to the any other individual treatments in PD.

Restoration of tyrosine hydroxylase positive cells were the highest in PD mice in triple agents treatment in SNpC (8432±67 cells) and in STr (6842±98 cells) that is much greater in cells in PD after treatment with any individual agents (Fig. 4). This restoration of tyrosine hydroxylase positive cells in PS was further supported by the measurement of tyrosine hydroxylase enzyme in PD treated with triple combination of these agents using nanowired delivery (Fig. 4). Our results showed greater accumulation of tyrosine hydroxylase enzyme in SNpC (0.78±0.14ng/μg protein) and in STr (0.69±0.12ng/μg protein) in PD following combined treatment of agents as compared to any individual therapy in PD (Figs. 4 and 5).

Dopamine content also elevated at a higher level in SNpC (13.34±0.10ng/mg) and in STr (10.28±0.11ng/mg) in PD after triple agent treatment (Fig. 4) as compared to any other individual therapy in PD. DOPAC showed greater accumulation in SNpC (5.64±008ng/mg) and in STr (4.89±0.13ng/mg) in PD with triple agent therapy as compared to any other individual therapy in PD (Fig. 4). The HVA levels in SNpC (1.86±0.05ng/mg) and in STr (1.42±0.07ng/mg) was also highly elevated in PD with triple agents treatment as compared to any other treatment alone in PD (Fig. 4).

These observations suggest that the combined treatment with H3 inverse agonist and partial H4 receptor agonist with histamine antibody induce superior neuroprotection on the pathophysiology of PD in mice.

6 Discussion

The most important salient finding of the present investigation suggest that histamine is actively involved in the pathophysiology of PD. This fact is evident from the findings that a potent histamine H3 inverse agonist compound BF2649 is able to reduce pathophysiology of MPTP induced PD model in mice. In addition, histamine H3 receptor antagonist with partial H4 agonist properties of clobenpropit exerts superior neuroprotective ability in reducing the pathophysiology of PD. This suggests antagonism of histamine H3 receptor and stimulation of H4 receptors induce neuroprotection in PD, not reported earlier.

That histamine is involved in the pathophysiological mechanisms of PD is further confirmed by the use of combined therapy with the monoclonal histamine antibody together with BF2649 and clobenpropit. When these three agents are administered together superior neuroprotection is observed in MPTP induced PD in mice.

These observations clearly indicate that histamine is an important player in PD pathogenesis and treatment. Our observations also suggest that histamine H3 and H4 receptors play important role in PD pathogenesis.

In PD dopaminergic neurodegeneration occurs in the SNpC (Braak and Del Tredici, 2009). Histaminergic neurons emanating from the tuberomamillary nucleus (TMN) send dense histaminergic innervations in SNpC (Panula et al., 1989; Watanabe et al., 1984). For motor functions a close interrelationship exists between SNpC and TMN (Maisonnette et al., 1998; Kononoff Vanhanen et al., 2016). In PD patients, postmortem studies show dense histaminergic innervations with enlarged axonal varicosities and increased histamine levels (Anichtchik et al., 2000b; Rinne et al., 2002). Preclinical studies show that increased histamine levels in the brain is associated with the degeneration of dopaminergic neurons in the SNpC (Liu et al., 2007; Vizuete et al., 2000). In PD model, administration of histamine H3 receptor agonist decreased locomotor dysfunction probably through inhibition of GABA release in the SNpC (Liu et al., 2008). These observations suggest that alterations in histamine level in the brain are associated with pathogenesis leading to PD (Agundez et al., 2008; Palada et al., 2012; Shan et al., 2012).

The role of histamine on PD is further strengthened by the observation of significant alterations in the histamine receptors H3 and H4 mRNAs and protein levels in PD brains at postmortem studies (Shan et al., 2012). Histamine H3 receptor is found within the neurons and astrocytes both at pre and postsynaptic levels located within the STr, hippocampus dentate gyrus and deep layers of cerebral cortex (Anichtchik et al., 2001; Pillot et al., 2002; Pollard et al., 1993). On the other hand, histamine H4 receptors are expressed mainly in neurons located within the cerebral cortex and thalamus (Connelly et al., 2009).

We have used MPTP model to induce PD like symptoms in rodents as described earlier (Niu et al., 2020; Ozkizilcik et al., 2018a, 2019; Sharma et al., 2020). We used 20 mg/kg, intraperitoneal injection of MPTP twice daily within 5 h intervals for 5 days (Ozkizilcik et al., 2018a, 2019). In our hands, this dose and schedule of injections induces PD like symptoms in mice. This is further confirmed by significant loss of tyrosine hydroxylase immunoreactivity in the SNpC and STr (Ozkizilcik et al., 2018a, 2019). In addition, biochemical measurement of dopamine, DOPAC and HVA also showed profound decrease in the SNpC and STr (Ozkizilcik et al., 2018a, 2019). In addition, we measured tyrosine hydroxylase enzyme in mouse model that is also showed significant decline in the SNpC and STr. This observation suggests that the dose and protocol of MPTP administration in mice induces PD symptoms like clinical situations. Thus, the model could be used for exploration of novel therapeutic advances for the treatment strategies of PD.

Previous report from our laboratory suggest that treatment with cerebrolysin-a balanced composition of several neurotrophic factors and active peptide fragments if delivered though nanowired techniques resulted in significant downregulation of the hallmarks of PD and induced marked neuroprotection (Ozkizilcik et al., 2018a, 2019). Thus, significant reduction in ASNC and p-tau was observed after nanowired cerebrolysin treatment in PC (Ozkizilcik et al., 2018a). This treatment also reduced the upregulation of nitric oxide synthase (NOS) enzyme in the brain that produces nitric oxide (NO), the free radical responsible for cell injury (Sharma et al., 1996, 1998a,b,c, 2019a,b; Wiklund et al., 2007).

Several lines of evidences suggest a close interaction between histamine and neurotrophins (Bennedich Kahn et al., 2008; Gotoh et al., 2013; Kumar et al., 2019). Treatment with histamine H3 receptor antagonist reduces depressive symptoms by enhancing brain derived neurotrophic factors (Kumar et al., 2019). Likewise, topical application of neurotrophins or growth factors reduces upregulation of NOS following traumatic spinal cord injury resulting in neuroprotection (Sharma et al., 1997a,b,c, 1998b,c, 2003, 2005, 2006a,b,c,d, 2010a,b,c,d). In addition, histamine receptors modulation induces neuroprotection by downregulating nitric oxide synthase production in spinal cord injury (Sharma et al., 2017a,b). These observations suggest an interaction between histamine, neurotrophins and nitric oxide in neuroprotection.

In present study we use BF2649 a potent histamine receptor inverse agonist that resulted in profound neuroprotection in PD induced brain pathology. This effect of BF2649 was further potentiated when the drug was delivered through nanowired technology. This indicates that nanowired delivery of drugs induces superior beneficial effects as compared to their conventional delivery (Ozkizilcik et al., 2018a; Sharma and Sharma, 2012a,b, 2013). Thus, it is likely that nanowired delivery of BF2649 significantly reduces histamine H3 receptor leading to downregulation of NOS and enhances neurotrophins in PD resulting in greater neuroprotection.

Likewise nanowired delivery of clobenpropit an antagonist of H3 receptor with partial agonistic effects on H4 receptor may also be able to increase neurotrophins, downregulate NOS and induce pronounced neuroprotection in PD. However, to further confirm this point, upregulation of NOS and measurement of neurotrophins is needed in PD. This is a feature that is currently being investigated in our laboratory.

Histamine is a neurotransmitter (Parsons and Ganellin, 2006). Histamine is a well know mediator of BBB and brain edema (Wahl et al., 1988). Histamine infusion induces BBB disruption and brain pathology (Gross et al., 1981, 1982; Wolff et al., 1975). Infusion of histamine induces brain edema and trauma results in increased histamine content in the brain (Joó et al., 1994; Mohanty et al., 1989; Németh et al., 1997; Sharma et al., 2006a,b,c,d). Histamine induces BBB breakdown and brain edema formation that is mediated though histamine H2 receptors in traumatic brain or spinal cord injury and hyperthermic brain damage (Mohanty et al., 1989; Patnaik et al., 2000; Sharma and Cervós-Navarro, 1991; Sharma et al., 1992a,b, 2006a,b,c,d).

In present investigation drugs modulating histamine H3 and H4 receptors reduced BBB breakdown and brain edema formation in PD. This indicates that apart from histamine H2 receptor's antagonism, H3 and H4 receptors modulator drugs are involved in the pathomechanisms of BBB dysfunction and brain edema formation. In spinal cord injury, nanowired delivery of histamine H2 receptors antagonist cimetidine and ranitidine significantly reduces cord pathology, blood-spinal cord barrier breakdown and edema formation (Sharma et al., 2017a). Thus, it would be interesting to investigate whether nanowired delivery of histamine H2 receptor antagonist cimetidine or ranitidine could also affect the BBB breakdown and edema formation in PD (Brimblecombe et al., 2010; Cavanagh et al., 1983; Daly et al., 1981; Malagelada et al., 2004). Alternatively, H2 receptor blockers together with H3 and H4 receptors modulators in combination using nanowired delivery could induce superior neuroprotection on the BBB and brain edema in PD. To test this hypothesis in PD further studies are needed.

Our previous reports on Alzheimer's disease (AD) induced brain pathology are significantly reduced with nanowired delivery of BF2649, clobenpropit together with nanowired monoclonal histamine antibody therapy (Patnaik et al., 2018; Sharma et al., 2017b). Using these individual and/or combined treatment of histamine modulating agents in AD reducing BBB breakdown, edema formation and brain pathology is in line with the idea that the pathogenesis of AD and PD are very similar in nature. Deposition of p-tau, ASNC and amyloid beta peptide in both AD and PD are in line with this idea (Ho et al., 2014; Irwin et al., 2013; Lim et al., 2019; Masliah et al., 2001; Whiten et al., 2020).

Addition of histamine monoclonal antibodies together with histamine H3 and H4 receptor modulating agents antagonized the effects of histamine on brain pathology in PD. This is evident with the results obtained using antibodies to serotonin, histamine, dynorphin, nitric oxide synthase; amyloid beta peptide used as therapy neutralizes the effects of their antigen significantly and reduced brain pathology (Sharma and Sharma, 2008, 2012a,b; Sharma et al., 1996, 1997a,b,c, 2007a,b, 2010a,b,c,d, 2017a, 2018).

Taken together, our investigations in PD using histamine modulating agents and antibodies suggests that these drugs and antibodies could further be developed as suitable therapeutic agents to minimize brain pathology in clinical cases of PD. We found that nanowired delivery of histaminergic agents induce superior neuroprotection. This could be due to the fact that nanowired drugs could penetrate into the brain much faster and deliver their drugs for long time without being catabolized (Sharma et al., 2016a,b). Thus, nanowired delivery of drug may act like a mini infusion pump delivering drugs inside the brain for long time to induce superior neuroprotective effects (Sharma and Sharma, 2012a,b, 2013). These observations clearly support a prominent role of histamine in PD and indicating future therapeutic potentials of histaminergic agents for the treatment of PD.

6.1 Conclusion and future perspective

In summary, our observations indicate a prominent role of BF2649 and clobenpropit either alone or in combination as suitable therapeutic agents for the treatment strategies of PD in clinic. However, further research is needed to find out if histamine H2, H3, and H4 receptor modulating agents in combination may have further superior effects on neuroprotection in PD. Monoclonal antibodies of histamine reduces the pathophysiology of PD in combination with BF2649 and clobenpropit. This indicates elevation of histamine levels in PD brain. To further understand the role of histamine in PD measurement of histamine in SNpC and STr is needed. However, this is still a novel area of investigation and requires further experiments in future.

Acknowledgments

This investigation is supported by grants from the Air Force Office of Scientific Research (EOARD, London, UK), and Air Force Material Command, USAF, under grant number FA8655-05-1-3065; Grants from the Alzheimer's Association (IIRG-09-132087), the National Institutes of Health (R01 AG028679) and the Dr. Robert M. Kohrman Memorial Fund (RJC); Swedish Medical Research Council (Nr 2710-HSS), Göran Gustafsson Foundation, Stockholm, Sweden (HSS), Astra Zeneca, Mölndal, Sweden (HSS/AS), the Ministry of Science & Technology, People Republic of China; The University Grants Commission, New Delhi, India (HSS/AS), Ministry of Science & Technology, Govt. of India (HSS/AS), Indian Medical Research Council, New Delhi, India (HSS/AS) and India-EU Co-operation Program (RP/AS/HSS) and IT-901/16 (JVL), Government of Basque Country and PPG 17/51 (JVL), JVL thanks to the support of the University of the Basque Country (UPV/EHU) PPG 17/51 and 14/08, the Basque Government (IT-901/16 and CS-2203) Basque Country, Spain; and Foundation for Nanoneuroscience and Nanoneuroprotection (FSNN), Romania. Technical and human support provided by Dr. Ricardo Andrade from SGIker (UPV/EHU) is gratefully acknowledged. Dr. Seaab Sahib is supported by Research Fellowship at the University of Arkansas Fayetteville AR by Department of Community Health; Middle Technical University; Wassit; Iraq, and The Higher Committee for Education Development in Iraq; Baghdad; Iraq. We thank Suraj Sharma, Blekinge Inst. Technology, Karlskrona, Sweden and Dr. Saja Alshafeay, University of Arkansas Fayetteville, Fayetteville AR, USA for computer and graphic support. The U.S. Government is authorized to reproduce and distribute reprints for Government purpose notwithstanding any copyright notation thereon. The views and conclusions contained herein are those of the authors and should not be interpreted as necessarily representing the official policies or endorsements, either expressed or implied, of the Air Force Office of Scientific Research or the U.S. Government.

Conflict of interest

There is no conflict of interest between any entity and/or organization mentioned here.

References

Čarman-Kržan, M., Lipnik-Štangelj, M., 2000. Molecular properties of central and peripheral histamine H1 and H2 receptors. Pflugers Arch. 439 (Suppl. 1), r131–r132. https://doi.org/10.1007/s004240000117.

Absinta, M., Lassmann, H., Trapp, B.D., 2020. Mechanisms underlying progression in multiple sclerosis. Curr. Opin. Neurol. 33 (3), 277–285. https://doi.org/10.1097/WCO.0000000000000818. 32324705. PMC7337978.

Agundez, J.A., Luengo, A., Herraez, O., Martinez, C., Alonso-Navarro, H., Jimenez-Jimenez, F.J., Garcia-Martin, E., 2008. Nonsynonymous polymorphisms of histamine-metabolising enzymes in patients with Parkinson's disease. Neuromolecular Med. 10, 10–16.

Ahlskog, J.E., Tyce, G.M., Kelly, P.J., van Heerden, J.A., Stoddard, S.L., Carmichael, S.W., 1989. Cerebrospinal fluid indices of blood-brain barrier permeability following adrenal-brain transplantation in patients with Parkinson's disease. Exp. Neurol. 105 (2), 152–161. https://doi.org/10.1016/0014-4886(89)90114-3.

Al-Bachari, S., Naish, J.H., Parker, G.J.M., Emsley, H.C.A., Parkes, L.M., 2020. Blood-brain barrier leakage is increased in Parkinson's disease. Front. Physiol. 11, 593026. https://doi.org/10.3389/fphys.2020.593026. eCollection 2020.

Alonso, A., Jick, S.S., Hernán, M.A., 2006. Allergy, histamine 1 receptor blockers, and the risk of multiple sclerosis. Neurology 66 (4), 572–575. https://doi.org/10.1212/01.wnl.0000198507.13597.45.

Alvarez, E.O., 2009. The role of histamine on cognition. Behav. Brain Res. 199 (2), 183–189. https://doi.org/10.1016/j.bbr.2008.12.010. Epub 2008 Dec 13.

An, J.H., Oh, B.K., Choi, J.W., 2013. Detection of tyrosine hydroxylase in dopaminergic neuron cell using gold nanoparticles-based barcode DNA. J. Biomed. Nanotechnol. 9 (4), 639–643. https://doi.org/10.1166/jbn.2013.1525.

Anichtchik, O.V., Huotari, M., Peitsaro, N., Haycock, J.W., Männistö, P.T., Panula, P., 2000a. Modulation of histamine H3 receptors in the brain of 6-hydroxydopamine-lesioned rats. Eur. J. Neurosci. 12 (11), 3823–3832. https://doi.org/10.1046/j.1460-9568.2000.00267.x.

Anichtchik, O.V., Rinne, J.O., Kalimo, H., Panula, P., 2000b. An altered histaminergic innervation of the substantia nigra in Parkinson's disease. Exp. Neurol. 163 (1), 20–30. https://doi.org/10.1006/exnr.2000.7362.

Anichtchik, O.V., Peitsaro, N., Rinne, J.O., Kalimo, H., Panula, P., 2001. Distribution and modulation of histamine H(3) receptors in basal ganglia and frontal cortex of healthy controls and patients with Parkinson's disease. Neurobiol. Dis. 8 (4), 707–716. https://doi.org/10.1006/nbdi.2001.0413.

Annamalai, B., Ragu Varman, D., Horton, R.E., Daws, L.C., Jayanthi, L.D., Ramamoorthy, S., 2020. Histamine receptors regulate the activity, surface expression, and phosphorylation of serotonin transporters. ACS Chem. Nerosci. 11 (3), 466–476. https://doi.org/10.1021/acschemneuro.9b00664. Epub 2020 Jan 22.

Aquino-Miranda, G., Molina-Hernández, A., Arias-Montaño, J.A., 2012. Regulation by histamine H3 receptors of neurotransmitter release in the basal ganglia: implications for Parkinson's disease pathophysiology. Gac. Med. Mex. 148 (5), 467–475.

Armstrong, M.J., Okun, M.S., 2020. Diagnosis and treatment of Parkinson disease: a review. JAMA 323 (6), 548–560. https://doi.org/10.1001/jama.2019.22360.

Arrang, J.M., 2007. Histamine and schizophrenia. Int. Rev. Neurobiol. 78, 247–287. https://doi.org/10.1016/S0074-7742(06)78009-6.

Ascherio, A., Schwarzschild, M.A., 2016. The epidemiology of Parkinson's disease: risk factors and prevention. Lancet Neurol. 15 (12), 1257–1272. https://doi.org/10.1016/S1474-4422(16)30230-7. Epub 2016 Oct 11.

Auvinen, S., Panula, P., 1988. Development of histamine-immunoreactive neurons in the rat brain. J. Comp. Neurol. 276 (2), 289–303. https://doi.org/10.1002/cne.902760211.

Axelsen, T.M., Woldbye, D.P.D., 2018. Gene therapy for Parkinson's disease, an update. J. Parkinsons Dis. 8 (2), 195–215. https://doi.org/10.3233/JPD-181331.

Balestrino, R., Schapira, A.H.V., 2020. Parkinson disease. Eur. J. Neurol. 27 (1), 27–42. https://doi.org/10.1111/ene.14108. Epub 2019 Nov 27.

Barata-Antunes, S., Cristóvão, A.C., Pires, J., Rocha, S.M., Bernardino, L., 2017. Dual role of histamine on microglia-induced neurodegeneration. Biochim. Biophys. Acta, Mol. Basis Dis. 1863 (3), 764–769. https://doi.org/10.1016/j.bbadis.2016.12.016.

Bassetti, C.L., Baumann, C.R., Dauvilliers, Y., Croyal, M., Robert, P., Schwartz, J.C., 2010. Cerebrospinal fluid histamine levels are decreased in patients with narcolepsy and excessive daytime sleepiness of other origin. J. Sleep Res. 19 (4), 620–623. https://doi.org/10.1111/j.1365-2869.2010.00819.x. Epub 2010 Sep 15.

Bellou, V., Belbasis, L., Tzoulaki, I., Evangelou, E., Ioannidis, J.P., 2016. Environmental risk factors and Parkinson's disease: an umbrella review of meta-analyses. Parkinsonism Relat. Disord. 23, 1–9. https://doi.org/10.1016/j.parkreldis.2015.12.008. Epub 2015 Dec.

Bellucci, A., Mercuri, N.B., Venneri, A., Faustini, G., Longhena, F., Pizzi, M., Missale, C., Spano, P., 2016. Review: Parkinson's disease: from synaptic loss to connectome dysfunction. Neuropathol. Appl. Neurobiol. 42 (1), 77–94. https://doi.org/10.1111/nan.12297.

Bennedich Kahn, L., Gustafsson, L.E., Olgart, H.C., 2008. Brain-derived neurotrophic factor enhances histamine-induced airway responses and changes levels of exhaled nitric oxide in Guinea pigs in vivo. Eur. J. Pharmacol. 595 (1–3), 78–83. https://doi.org/10.1016/j.ejphar.2008.07.041. Epub 2008 Jul 30.

Besteher, B., Brambilla, P., Nenadić, I., 2020. Twin studies of brain structure and cognition in schizophrenia. Neurosci. Biobehav. Rev. 109, 103–113. https://doi.org/10.1016/j.neubiorev.2019.12.021. Epub 2019 Dec 13.

Biagioni, F., Vivacqua, G., Lazzeri, G., Ferese, R., Iannacone, S., Onori, P., Morini, S., D'Este, L., Fornai, F., 2021. Chronic MPTP in mice damage-specific neuronal phenotypes within dorsal laminae of the spinal cord. Neurotox. Res. 39 (2), 156–169. https://doi.org/10.1007/s12640-020-00313-x. Epub 2020 Nov 18.

Borellini, L., Ardolino, G., Carrabba, G., Locatelli, M., Rampini, P., Sbaraini, S., Scola, E., Avignone, S., Triulzi, F., Barbieri, S., Cogiamanian, F., 2019. Peri-lead edema after deep brain stimulation surgery for Parkinson's disease: a prospective magnetic resonance imaging study. Eur. J. Neurol. 26 (3), 533–539. https://doi.org/10.1111/ene.13852. Epub 2018 Nov 18.

Braak, H., Braak, E., 2000. Pathoanatomy of Parkinson's disease. J. Neurol. 247 (Suppl. 2), II3–10. https://doi.org/10.1007/PL00007758.

Braak, H., Del Tredici, K., 2009. Neuroanatomy and pathology of sporadic Parkinson's disease. Adv. Anat. Embryol. Cell Biol. 201, 1–119.

Braak, H., Rüb, U., Braak, E., 2000. Neuroanatomy of Parkinson disease. Changes in the neuronal cytoskeleton of a few disease-susceptible types of neurons lead to progressive destruction of circumscribed areas in the limbic and motor systems. Nervenarzt 71 (6), 459–469. https://doi.org/10.1007/s001150050607.

Braak, H., Del Tredici, K., Rüb, U., de Vos, R.A., Jansen Steur, E.N., Braak, E., 2003. Staging of brain pathology related to sporadic Parkinson's disease. Neurobiol. Aging 24 (2), 197–211. https://doi.org/10.1016/s0197-4580(02)00065-9.

Brabant, C., Alleva, L., Quertemont, E., Tirelli, E., 2010. Involvement of the brain histaminergic system in addiction and addiction-related behaviors: a comprehensive review with emphasis on the potential therapeutic use of histaminergic compounds in drug dependence. Prog. Neurobiol. 92 (3), 421–441. https://doi.org/10.1016/j.pneurobio.2010.07.002. Epub 2010 Jul 16.

Brandes, L.J., LaBella, F.S., Glavin, G.B., Paraskevas, F., Saxena, S.P., McNicol, A., Gerrard, J.M., 1990. Histamine as an intracellular messenger. Biochem. Pharmacol. 40 (8), 1677–1681. https://doi.org/10.1016/0006-2952(90)90341-h.

Brimblecombe, R.W., Duncan, W.A., Durant, G.J., Ganellin, C.R., Parsons, M.E., Black, J.W., 2010. The pharmacology of cimetidine, a new histamine H2-receptor antagonist. 1975. Br. J. Pharmacol. 160 (Suppl. 1), S52–S53. https://doi.org/10.1111/j.1476-5381.2010.00854.x.

Brioni, J.D., Esbenshade, T.A., Garrison, T.R., Bitner, S.R., Cowart, M.D., 2011. Discovery of histamine H3 antagonists for the treatment of cognitive disorders and Alzheimer's disease. J. Pharmacol. Exp. Ther. 336 (1), 38–46. https://doi.org/10.1124/jpet.110.166876. Epub 2010 Sep 23.

Browman, K.E., Komater, V.A., Curzon, P., Rueter, L.E., Hancock, A.A., Decker, M.W., Fox, G.B., 2004. Enhancement of prepulse inhibition of startle in mice by the H3 receptor antagonists thioperamide and ciproxifan. Behav. Brain Res. 153 (1), 69–76. https://doi.org/10.1016/j.bbr.2003.11.001.

Buckland, K.F., Williams, T.J., Conroy, D.M., 2003. Histamine induces cytoskeletal changes in human eosinophils via the H(4) receptor. Br. J. Pharmacol. 140 (6), 1117–1127. https://doi.org/10.1038/sj.bjp.0705530. Epub 2003 Oct 6.

Cacabelos, R., Yamatodani, A., Niigawa, H., Hariguchi, S., Tada, K., Nishimura, T., Wada, H., Brandeis, L., Pearson, J., 1989. Brain histamine in Alzheimer's disease. Methods Find. Exp. Clin. Pharmacol. 11 (5), 353–360.

Caminiti, S.P., Presotto, L., Baroncini, D., Garibotto, V., Moresco, R.M., Gianolli, L., Volonté, M.A., Antonini, A., Perani, D., 2017. Axonal damage and loss of connectivity in nigrostriatal and mesolimbic dopamine pathways in early Parkinson's disease. Neuroimage Clin. 14, 734–740. https://doi.org/10.1016/j.nicl.2017.03.011. eCollection 2017.

von Campenhausen, S., Bornschein, B., Wick, R., Bötzel, K., Sampaio, C., Poewe, W., Oertel, W., Siebert, U., Berger, K., Dodel, R., 2005. Prevalence and incidence of Parkinson's disease in Europe. Eur. Neuropsychopharmacol. 15 (4), 473–490. https://doi.org/10.1016/j.euroneuro.2005.04.007.

Canevelli, M., Bruno, G., Valletta, M., Fabbrini, A., Vanacore, N., Berardelli, A., Fabbrini, G., 2019. Parkinson's disease among migrants in Europe: estimating the magnitude of an emerging phenomenon. J. Neurol. 266 (5), 1120–1126. https://doi.org/10.1007/s00415-019-09241-z. Epub 2019 Feb 14.

Cavanagh, R.L., Usakewicz, J.J., Buyniski, J.P., 1983. A comparison of some of the pharmacological properties of etintidine, a new histamine H2-receptor antagonist, with those of cimetidine, ranitidine and tiotidine. J. Pharmacol. Exp. Ther. 224 (1), 171–179.

Cervós-Navarro, J., Sharma, H.S., Westman, J., Bongcam-Rudloff, E., 1998. Glial reactions in the central nervous system following heat stress. Prog. Brain Res. 115, 241–274. https://doi.org/10.1016/s0079-6123(08)62039-7.

Chatterjee, K., Roy, A., Banerjee, R., Choudhury, S., Mondal, B., Halder, S., Basu, P., Shubham, S., Dey, S., Kumar, H., 2020. Inflammasome and alpha-synuclein in Parkinson's disease: a cross-sectional study. J. Neuroimmunol. 338, 577089. https://doi.org/10.1016/j.jneuroim.2019.577089. Epub 2019 Oct 25.

Chazot, P.L., 2013. Histamine pharmacology: four years on. Br. J. Pharmacol. 170 (1), 1–3. https://doi.org/10.1111/bph.12319.

Chen, Y.C., Baronio, D., Semenova, S., Abdurakhmanova, S., Panula, P., 2020. Cerebral dopamine neurotrophic factor regulates multiple neuronal subtypes and behavior. J. Neurosci. 40 (32), 6146–6164. https://doi.org/10.1523/JNEUROSCI.2636-19.2020. Epub 2020 Jul 6.

Chinta, S.J., Andersen, J.K., 2005. Dopaminergic neurons. Int. J. Biochem. Cell Biol. 37 (5), 942–946. https://doi.org/10.1016/j.biocel.2004.09.009. Epub 2004 Dec 2.

Collier, T.J., Kanaan, N.M., Kordower, J.H., 2017. Aging and Parkinson's disease: different sides of the same coin? Mov. Disord. 32 (7), 983–990. https://doi.org/10.1002/mds.27037. Epub 2017 May 18.

Connelly, W.M., Shenton, F.C., Lethbridge, N., Leurs, R., Waldvogel, H.J., Faull, R.L., Lees, G., Chazot, P.L., 2009. The histamine H4 receptor is functionally expressed on neurons in the mammalian CNS. Br. J. Pharmacol. 157 (1), 55–63. https://doi.org/10.1111/j.1476-5381.2009.00227.x.

Conrad, M., Söldner, C.A., Miao, Y., Sticht, H., 2020. Agonist binding and G protein coupling in histamine H(2) receptor: a molecular dynamics study. Int. J. Mol. Sci. 21 (18), 6693. https://doi.org/10.3390/ijms21186693.

Corrêa, M.F., dos Santos Fernandes, J.P., 2015. Histamine H4 receptor ligands: future applications and state of art. Chem. Biol. Drug Des. 85 (4), 461–480. https://doi.org/10.1111/cbdd.12431. Epub 2014 Oct 27.

Corrêa, M.F., Balico-Silva, A.L., Kiss, D.J., Fernandes, G.A.B., Maraschin, J.C., Parreiras-E-Silva, L.T., Varela, M.T., Simões, S.C., Bouvier, M., Keserű, G.M., Costa-Neto, C.M., Fernandes, J.P.S., 2021. Novel potent (dihydro)benzofuranyl piperazines as human histamine receptor ligands—functional characterization and modeling studies on H(3) and H(4) receptors. Bioorg. Med. Chem. 30, 115924. https://doi.org/10.1016/j.bmc.2020.115924. Epub 2020 Dec 8.

Cuenca, L., Gil-Martinez, A.L., Cano-Fernandez, L., Sanchez-Rodrigo, C., Estrada, C., Fernandez-Villalba, E., Herrero, M.T., 2019. Parkinson's disease: a short story of 200 years. Histol. Histopathol. 34 (6), 573–591. https://doi.org/10.14670/HH-18-073. Epub 2018 Dec 12.

da Silveira, C.K., Furini, C.R., Benetti, F., Monteiro Sda, C., Izquierdo, I., 2013. The role of histamine receptors in the consolidation of object recognition memory. Neurobiol. Learn. Mem. 103, 64–71. https://doi.org/10.1016/j.nlm.2013.04.001. Epub 2013 Apr 9.

Dai, H., Kaneko, K., Kato, H., Fujii, S., Jing, Y., Xu, A., Sakurai, E., Kato, M., Okamura, N., Kuramasu, A., Yanai, K., 2007. Selective cognitive dysfunction in mice lacking histamine H1 and H2 receptors. Neurosci. Res. 57 (2), 306–313. https://doi.org/10.1016/j.neures.2006.10.020. Epub 2006 Dec 4.

Dale, H.H., Laidlaw, P.P., 1910. The physiological action of β-imidazolylethylamine. J. Physiol. 41, 318–341.

Daly, M.J., Humphray, J.M., Stables, R., 1981. Some in vitro and in vivo actions of the new histamine H2-receptor antagonist, ranitidine. Br. J. Pharmacol. 72 (1), 49–54. https://doi.org/10.1111/j.1476-5381.1981.tb09103.x.

Daubner, S.C., Le, T., Wang, S., 2011. Tyrosine hydroxylase and regulation of dopamine synthesis. Arch. Biochem. Biophys. 508 (1), 1–12. https://doi.org/10.1016/j.abb.2010.12.017. Epub 2010 Dec 19.

Dauvilliers, Y., Bassetti, C., Lammers, G.J., Arnulf, I., Mayer, G., Rodenbeck, A., Lehert, P., Ding, C.L., Lecomte, J.M., Schwartz, J.C., HARMONY I Study Group, 2013. Pitolisant versus placebo or modafinil in patients with narcolepsy: a double-blind, randomised trial. Lancet Neurol. 12 (11), 1068–1075. https://doi.org/10.1016/S1474-4422(13)70225-4. Epub 2013 Oct 7.

De Deurwaerdère, P., Di Giovanni, G., 2020. Serotonin in health and disease. Int. J. Mol. Sci. 21 (10), 3500. https://doi.org/10.3390/ijms21103500.

del Rio, R., Noubade, R., Saligrama, N., Wall, E.H., Krementsov, D.N., Poynter, M.E., Zachary, J.F., Thurmond, R.L., Teuscher, C., 2012. Histamine H4 receptor optimizes

T regulatory cell frequency and facilitates anti-inflammatory responses within the central nervous system. J. Immunol. 188 (2), 541–547. https://doi.org/10.4049/jimmunol.1101498. Epub 2011 Dec 5.

Diez-Garcia, A., Garzon, M., 2017. Regulation of the phases of the sleep-wakefulness cycle with histamine. Rev. Neurol. 64 (6), 267–277 (Spanish).

Drew, D.S., Muhammed, K., Baig, F., Kelly, M., Saleh, Y., Sarangmat, N., Okai, D., Hu, M., Manohar, S., Husain, M., 2020. Dopamine and reward hypersensitivity in Parkinson's disease with impulse control disorder. Brain 143 (8), 2502–2518. https://doi.org/10.1093/brain/awaa198.

Dudek, M., Kuder, K., Kołaczkowski, M., Olczyk, A., Żmudzka, E., Rak, A., Bednarski, M., Pytka, K., Sapa, J., Kieć-Kononowicz, K., 2016. H3 histamine receptor antagonist pitolisant reverses some subchronic disturbances induced by olanzapine in mice. Metab. Brain Dis. 31 (5), 1023–1029. https://doi.org/10.1007/s11011-016-9840-z. Epub 2016 May 24.

Elabi, O., Gaceb, A., Carlsson, R., Padel, T., Soylu-Kucharz, R., Cortijo, I., Li, W., Li, J.Y., Paul, G., 2021. Human alpha-synuclein overexpression in a mouse model of Parkinson's disease leads to vascular pathology, blood brain barrier leakage and pericyte activation. Sci. Rep. 11 (1), 1120. https://doi.org/10.1038/s41598-020-80889-8.

Ellenbroek, B.A., Ghiabi, B., 2015. Do histamine receptor 3 antagonists have a place in the therapy for schizophrenia? Curr. Pharm. Des. 21 (26), 3760–3770. https://doi.org/10.2174/1381612821666150605105325.

Elliott, K.A., Jasper, H., 1949. Measurement of experimentally induced brain swelling and shrinkage. Am. J. Physiol. 157 (1), 122–129. https://doi.org/10.1152/ajplegacy.1949.157.1.122.

Esbenshade, T.A., Browman, K.E., Bitner, R.S., Strakhova, M., Cowart, M.D., Brioni, J.D., 2008. The histamine H3 receptor: an attractive target for the treatment of cognitive disorders. Br. J. Pharmacol. 154 (6), 1166–1181. https://doi.org/10.1038/bjp.2008.147. Epub 2008 May 12.

Fang, Q., Xicoy, H., Shen, J., Luchetti, S., Dai, D., Zhou, P., Qi, X.R., Martens, G.J.M., Huitinga, I., Swaab, D.F., Liu, C., Shan, L., 2021. Histamine-4 receptor antagonist ameliorates Parkinson-like pathology in the striatum. Brain Behav. Immun. 92, 127–138. https://doi.org/10.1016/j.bbi.2020.11.036. Epub 2020 Nov 26.

Farzin, D., Attarzadeh, M., 2000. Influence of different histamine receptor agonists and antagonists on apomorphine-induced licking behavior in rat. Eur. J. Pharmacol. 404 (1–2), 169–174. https://doi.org/10.1016/s0014-2999(00)00608-7.

Fauss, D., Motter, R., Dofiles, L., Rodrigues, M.A., You, M., Diep, L., Yang, Y., Seto, P., Tanaka, K., Baker, J., Bergeron, M., 2013. Development of an enzyme-linked immunosorbent assay (ELISA) to measure the level of tyrosine hydroxylase protein in brain tissue from Parkinson's disease models. J. Neurosci. Methods 215 (2), 245–257. https://doi.org/10.1016/j.jneumeth.2013.03.012. Epub 2013 Mar 26.

Feliszek, M., Speckmann, V., Schacht, D., von Lehe, M., Stark, H., Schlicker, E., 2015. A search for functional histamine H4 receptors in the human, Guinea pig and mouse brain. Naunyn Schmiedebergs Arch. Pharmacol. 388 (1), 11–17. https://doi.org/10.1007/s00210-014-1053-6.

Filippi, M., Preziosa, P., Langdon, D., Lassmann, H., Paul, F., Rovira, À., Schoonheim, M.M., Solari, A., Stankoff, B., Rocca, M.A., 2020. Identifying progression in multiple sclerosis: new perspectives. Ann. Neurol. 88 (3), 438–452. https://doi.org/10.1002/ana.25808. Epub 2020 Jul 6.

Foffani, G., Obeso, J.A., 2018. A cortical pathogenic theory of Parkinson's disease. Neuron 99 (6), 1116–1128. https://doi.org/10.1016/j.neuron.2018.07.028.

Fossati, A., Barone, D., Benvenuti, C., 2001. Binding affinity profile of betahistine and its metabolites for central histamine receptors of rodents. Pharmacol. Res. 43 (4), 389–392. https://doi.org/10.1006/phrs.2000.0795.

Franco, P., Dauvilliers, Y., Inocente, C.O., Guyon, A., Villanueva, C., Raverot, V., Plancoulaine, S., Lin, J.S., 2019. Impaired histaminergic neurotransmission in children with narcolepsy type 1. CNS Neurosci. Ther. 25 (3), 386–395. https://doi.org/10.1111/cns.13057. Epub 2018 Sep 17.

Garbarg, M., Barbin, G., Bischoff, S., Pollard, H., Schwartz, J.C., 1974. Evidence for a specific decarboxylase involved in histamine synthesis in an ascending pathway in rat brain. Agents Actions 4 (3), 181. https://doi.org/10.1007/BF01970260.

Garbarg, M., Javoy-Agid, F., Schwartz, J.C., Agid, Y., 1983. Brain histidine decarboxylase activity in Parkinson's disease. Lancet 1 (8314–5), 74–75. https://doi.org/10.1016/s0140-6736(83)91613-6.

Giannoni, P., Passani, M.B., Nosi, D., Chazot, P.L., Shenton, F.C., Medhurst, A.D., Munari, L., Blandina, P., 2009. Heterogeneity of histaminergic neurons in the tuberomammillary nucleus of the rat. Eur. J. Neurosci. 29 (12), 2363–2374. https://doi.org/10.1111/j.1460-9568.2009.06765.x. Epub 2009 May 22.

Gomez-Ramirez, J., Johnston, T.H., Visanji, N.P., Fox, S.H., Brotchie, J.M., 2006. Histamine H3 receptor agonists reduce L-dopa-induced chorea, but not dystonia, in the MPTP-lesioned nonhuman primate model of Parkinson's disease. Mov. Disord. 21 (6), 839–846. https://doi.org/10.1002/mds.20828.

Goodchild, R.E., Court, J.A., Hobson, I., Piggott, M.A., Perry, R.H., Ince, P., Jaros, E., Perry, E.K., 1999. Distribution of histamine H3-receptor binding in the normal human basal ganglia: comparison with Huntington's and Parkinson's disease cases. Eur. J. Neurosci. 11 (2), 449–456. https://doi.org/10.1046/j.1460-9568.1999.00453.x.

Gordh, T., Chu, H., Sharma, H.S., 2006. Spinal nerve lesion alters blood-spinal cord barrier function and activates astrocytes in the rat. Pain 124 (1–2), 211–221. https://doi.org/10.1016/j.pain.2006.05.020. Epub 2006 Jun 27.

Goris, A., Williams-Gray, C.H., Clark, G.R., Foltynie, T., Lewis, S.J., Brown, J., Ban, M., Spillantini, M.G., Compston, A., Burn, D.J., Chinnery, P.F., Barker, R.A., Sawcer, S.J., 2007. Tau and alpha-synuclein in susceptibility to, and dementia in, Parkinson's disease. Ann. Neurol. 62 (2), 145–153. https://doi.org/10.1002/ana.21192.

Gorton, L.M., Vuckovic, M.G., Vertelkina, N., Petzinger, G.M., Jakowec, M.W., Wood, R.I., 2010. Exercise effects on motor and affective behavior and catecholamine neurochemistry in the MPTP-lesioned mouse. Behav. Brain Res. 213 (2), 253–262. https://doi.org/10.1016/j.bbr.2010.05.009. Epub 2010 May 21.

Gotoh, K., Masaki, T., Chiba, S., Ando, H., Fujiwara, K., Shimasaki, T., Mitsutomi, K., Katsuragi, I., Kakuma, T., Sakata, T., Yoshimatsu, H., 2013. Brain-derived neurotrophic factor, corticotropin-releasing factor, and hypothalamic neuronal histamine interact to regulate feeding behavior. J. Neurochem. 125 (4), 588–598. https://doi.org/10.1111/jnc.12213. Epub 2013 Mar 19.

Gray, M.T., Woulfe, J.M., 2015. Striatal blood-brain barrier permeability in Parkinson's disease. J. Cereb. Blood Flow Metab. 35 (5), 747–750. https://doi.org/10.1038/jcbfm.2015.32. Epub 2015 Mar 11.

Gross, P.M., Teasdale, G.M., Angerson, W.J., Harper, A.M., 1981. H2-receptors mediate increases in permeability of the blood-brain barrier during arterial histamine infusion. Brain Res. 210 (1–2), 396–400. https://doi.org/10.1016/0006-8993(81)90916-1.

Gross, P.M., Teasdale, G.M., Graham, D.I., Angerson, W.J., Harper, A.M., 1982. Intra-arterial histamine increases blood-brain transport in rats. Am. J. Physiol. 243 (2), H307–H317. https://doi.org/10.1152/ajpheart.1982.243.2.H307.

Guo, J.Y., Ragland, J.D., Carter, C.S., 2019. Memory and cognition in schizophrenia. Mol. Psychiatry 24 (5), 633–642. https://doi.org/10.1038/s41380-018-0231-1. Epub 2018 Sep 21.

Guo, M., Wang, J., Zhao, Y., Feng, Y., Han, S., Dong, Q., Cui, M., Tieu, K., 2020. Microglial exosomes facilitate alpha-synuclein transmission in Parkinson's disease. Brain 143 (5), 1476–1497. https://doi.org/10.1093/brain/awaa090.

Haas, H.L., Sergeeva, O.A., Selbach, O., 2008. Histamine in the nervous system. Physiol. Rev. 88 (3), 1183–1241. https://doi.org/10.1152/physrev.00043.2007.

Hadi, F., Akrami, H., Totonchi, M., Barzegar, A., Nabavi, S.M., Shahpasand, K., 2020. Alpha-synuclein abnormalities trigger focal tau pathology, spreading to various brain areas in Parkinson disease. J. Neurochem. https://doi.org/10.1111/jnc.15257. Online ahead of print.

Halpert, A.G., Olmstead, M.C., Beninger, R.J., 2002. Mechanisms and abuse liability of the anti-histamine dimenhydrinate. Neurosci. Biobehav. Rev. 26 (1), 61–67. https://doi.org/10.1016/s0149-7634(01)00038-0.

Hatcher, J.M., Pennell, K.D., Miller, G.W., 2008. Parkinson's disease and pesticides: a toxicological perspective. Trends Pharmacol. Sci. 29 (6), 322–329. https://doi.org/10.1016/j.tips.2008.03.007. Epub 2008 Apr 29.

Hauser, S.L., Cree, B.A.C., 2020. Treatment of multiple sclerosis: a review. Am. J. Med. 133 (12), 1380–1390.e2. https://doi.org/10.1016/j.amjmed.2020.05.049. Epub 2020 Jul 17.

Hishinuma, S., Nozawa, H., Akatsu, C., Shoji, M., 2016. C-terminal of human histamine H(1) receptors regulates their agonist-induced clathrin-mediated internalization and G-protein signaling. J. Neurochem. 139 (4), 552–565. https://doi.org/10.1111/jnc.13834. Epub 2016 Sep 15.

Ho, C.Y., Troncoso, J.C., Knox, D., Stark, W., Eberhart, C.G., 2014. Beta-amyloid, phospho-tau and alpha-synuclein deposits similar to those in the brain are not identified in the eyes of Alzheimer's and Parkinson's disease patients. Brain Pathol. 24 (1), 25–32. https://doi.org/10.1111/bpa.12070. Epub 2013 Jun 28.

Holeček, M., 2020. Histidine in health and disease: metabolism, physiological importance, and use as a supplement. Nutrients 12 (3), 848. https://doi.org/10.3390/nu12030848.

Hu, S., Hu, M., Liu, J., Zhang, B., Zhang, Z., Zhou, F.H., Wang, L., Dong, J., 2020. Phosphorylation of tau and alpha-synuclein induced neurodegeneration in MPTP mouse model of Parkinson's disease. Neuropsychiatr. Dis. Treat. 16, 651–663. https://doi.org/10.2147/NDT.S235562. eCollection 2020.

Iarkov, A., Barreto, G.E., Grizzell, J.A., Echeverria, V., 2020. Strategies for the treatment of Parkinson's disease: beyond dopamine. Front. Aging Neurosci. 12, 4. https://doi.org/10.3389/fnagi.2020.00004. eCollection 2020.

Irwin, D.J., Lee, V.M., Trojanowski, J.Q., 2013. Parkinson's disease dementia: convergence of alpha-synuclein, tau and amyloid-beta pathologies. Nat. Rev. Neurosci. 14 (9), 626–636. https://doi.org/10.1038/nrn3549. Epub 2013 Jul 31.

Ito, C., 2004. The role of the central histaminergic system on schizophrenia. Drug News Perspect. 17 (6), 383–387. https://doi.org/10.1358/dnp.2004.17.6.829029.

Iwabuchi, K., Ito, C., Tashiro, M., Kato, M., Kano, M., Itoh, M., Iwata, R., Matsuoka, H., Sato, M., Yanai, K., 2005. Histamine H1 receptors in schizophrenic patients measured by positron emission tomography. Eur. Neuropsychopharmacol. 15 (2), 185–191. https://doi.org/10.1016/j.euroneuro.2004.10.001. 15695063.

Jaarsma, D., Veenma-van der Duin, L., Korf, J., 1994. N-acetylaspartate and N-acetylaspartylglutamate levels in Alzheimer's disease post-mortem brain tissue. J. Neurol. Sci. 127 (2), 230–233. https://doi.org/10.1016/0022-510x(94)90077-9.

Jackson-Lewis, V., Przedborski, S., 2007. Protocol for the MPTP mouse model of Parkinson's disease. Nat. Protoc. 2 (1), 141–151. https://doi.org/10.1038/nprot.2006.342.

Jadidi-Niaragh, F., Mirshafiey, A., 2010. Histamine and histamine receptors in pathogenesis and treatment of multiple sclerosis. Neuropharmacology 59 (3), 180–189. https://doi.org/10.1016/j.neuropharm.2010.05.005. Epub 2010 May 21.

Jankovic, J., 2008. Parkinson's disease: clinical features and diagnosis. J. Neurol. Neurosurg. Psychiatry 79 (4), 368–376. https://doi.org/10.1136/jnnp.2007.131045.

Jankovic, J., Tan, E.K., 2020. Parkinson's disease: etiopathogenesis and treatment. J. Neurol. Neurosurg. Psychiatry 91 (8), 795–808. https://doi.org/10.1136/jnnp-2019-322338. Epub 2020 Jun 23.

Jiang, P.E., Lang, Q.H., Yu, Q.Y., Tang, X.Y., Liu, Q.Q., Li, X.Y., Feng, X.Z., 2019. Behavioral assessments of spontaneous locomotion in a murine MPTP-induced Parkinson's disease model. J. Vis. Exp. 143. https://doi.org/10.3791/58653.

Joó, F., Kovács, J., Szerdahelyi, P., Temesvári, P., Tósaki, A., 1994. The role of histamine in brain oedema formation. Acta Neurochir. Suppl. (Wien) 60, 76–78. https://doi.org/10.1007/978-3-7091-9334-1_19.

Kalia, L.V., Lang, A.E., 2015. Parkinson's disease. Lancet 386 (9996), 896–912. https://doi.org/10.1016/S0140-6736(14)61393-3. Epub 2015 Apr 19.

Kallweit, U., Aritake, K., Bassetti, C.L., Blumenthal, S., Hayaishi, O., Linnebank, M., Baumann, C.R., Urade, Y., 2013. Elevated CSF histamine levels in multiple sclerosis patients. Fluids Barriers CNS 10, 19. https://doi.org/10.1186/2045-8118-10-19. eCollection 2013.

Kanbayashi, T., Kodama, T., Kondo, H., Satoh, S., Inoue, Y., Chiba, S., Shimizu, T., Nishino, S., 2009. CSF histamine contents in narcolepsy, idiopathic hypersomnia and obstructive sleep apnea syndrome. Sleep 32 (2), 181–187. https://doi.org/10.1093/sleep/32.2.181.

Kempuraj, D., Thangavel, R., Natteru, P.A., Selvakumar, G.P., Saeed, D., Zahoor, H., Zaheer, S., Iyer, S.S., Zaheer, A., 2016. Neuroinflammation induces neurodegeneration. J. Neurol. Neurosurg. Spine 1 (1), 1003. Epub 2016 Nov 18.

Khedkar, T., Koushik, S., Gadhikar, Y., 2012. Expression pattern of histaminergic neurons in the human fetal hypothalamus at second and third trimester. Ann. Neurosci. 19 (3), 116–120. https://doi.org/10.5214/ans.0972.7531.190306.

Kim, C.Y., Alcalay, R.N., 2017. Genetic forms of Parkinson's disease. Semin. Neurol. 37 (2), 135–146. https://doi.org/10.1055/s-0037-1601567. Epub 2017 May 16.

Kim, T., Vemuganti, R., 2017. Mechanisms of Parkinson's disease-related proteins in mediating secondary brain damage after cerebral ischemia. J. Cereb. Blood Flow Metab. 37 (6), 1910–1926. https://doi.org/10.1177/0271678X17694186. Epub 2017 Jan 1.

Kitanaka, J., Kitanaka, N., Hall, F.S., Uhl, G.R., Tatsuta, T., Morita, Y., Tanaka, K., Nishiyama, N., Takemura, M., 2011. Histamine H3 receptor agonists decrease hypothalamic histamine levels and increase stereotypical biting in mice challenged with methamphetamine. Neurochem. Res. 36 (10), 1824–1833. https://doi.org/10.1007/s11064-011-0500-8. Epub 2011 May 15.

Kitanaka, J., Kitanaka, N., Hall, F.S., Uhl, G.R., Takemura, M., 2016. Brain histamine N-methyltransferase as a possible target of treatment for methamphetamine overdose. Drug Target Insights 10, 1–7. https://doi.org/10.4137/DTI.S38342. eCollection 2016.

Kononoff Vanhanen, J., Nuutinen, S., Tuominen, M., Panula, P., 2016. Histamine H3 receptor regulates sensorimotor gating and dopaminergic signaling in the striatum. J. Pharmacol. Exp. Ther. 357 (2), 264–272. https://doi.org/10.1124/jpet.115.230771. Epub 2016 Mar 4.

Koski, S.K., Leino, S., Panula, P., Rannanpää, S., Salminen, O., 2020. Genetic lack of histamine upregulates dopamine neurotransmission and alters rotational behavior but not levodopa-induced dyskinesia in a mouse model of Parkinson's disease. Neurosci. Lett. 729, 134932. https://doi.org/10.1016/j.neulet.2020.134932. Epub 2020 Mar 26.

Kubo, M., Kishi, T., Matsunaga, S., Iwata, N., 2015. Histamine H3 receptor antagonists for Alzheimer's disease: a systematic review and meta-analysis of randomized placebo-controlled trials. J. Alzheimers Dis. 48 (3), 667–671. https://doi.org/10.3233/JAD-150393.

Kumar, A., Dogra, S., Sona, C., Umrao, D., Rashid, M., Singh, S.K., Wahajuddin, M., Yadav, P.N., 2019. Chronic histamine 3 receptor antagonism alleviates depression like conditions in mice via modulation of brain-derived neurotrophic factor and hypothalamus-pituitary adrenal axis. Psychoneuroendocrinology 101, 128–137. https://doi.org/10.1016/j.psyneuen.2018.11.007. Epub 2018 Nov 11.

Kwiatkowski, H., 1941. Observations on the relation of histamine to reactive hyperaemia. J. Physiol. 100 (2), 147–158. https://doi.org/10.1113/jphysiol.1941.sp003931.

Kyle, K., Bronstein, J.M., 2020. Treatment of psychosis in Parkinson's disease and dementia with Lewy bodies: a review. Parkinsonism Relat. Disord. 75, 55–62. https://doi.org/10.1016/j.parkreldis.2020.05.026. Epub 2020.

Lafuente, J.V., Sharma, A., Muresanu, D.F., Ozkizilcik, A., Tian, Z.R., Patnaik, R., Sharma, H.S., 2018. Repeated forced swim exacerbates methamphetamine-induced neurotoxicity: neuroprotective effects of nanowired delivery of 5-HT3-receptor antagonist ondansetron. Mol. Neurobiol. 55 (1), 322–334. https://doi.org/10.1007/s12035-017-0744-7.

Lang, A.E., Espay, A.J., 2018. Disease modification in Parkinson's disease: current approaches, challenges, and future considerations. Mov. Disord. 33 (5), 660–677. https://doi.org/10.1002/mds.27360. Epub 2018 Apr 11.

Lesemann, A., Reinel, C., Hühnchen, P., Pilhatsch, M., Hellweg, R., Klaissle, P., Winter, C., Steiner, B., 2012. MPTP-induced hippocampal effects on serotonin, dopamine, neurotrophins, adult neurogenesis and depression-like behavior are partially influenced by fluoxetine in adult mice. Brain Res. 1457, 51–69. https://doi.org/10.1016/j.brainres.2012.03.046. Epub 2012 Mar 27.

Ligneau, X., Perrin, D., Landais, L., Camelin, J.C., Calmels, T.P., Berrebi-Bertrand, I., Lecomte, J.M., Parmentier, R., Anaclet, C., Lin, J.S., Bertaina-Anglade, V., la Rochelle, C.D., d'Aniello, F., Rouleau, A., Gbahou, F., Arrang, J.M., Ganellin, C.R., Stark, H., Schunack, W., Schwartz, J.C., 2007. BF2.649 (1-{3-(3-(4 Chlorophenyl) propoxy)propyl} piperidine, hydrochloride), a nonimidazole inverse agonist/antagonist at the human histamine H3 receptor: preclinical pharmacology. J. Pharmacol. Exp. Ther. 320 (1), 365–375. https://doi.org/10.1124/jpet.106.111039. Epub 2006 Sep 27.

Lim, E.W., Aarsland, D., Ffytche, D., Taddei, R.N., van Wamelen, D.J., Wan, Y.M., Tan, E.K., Ray Chaudhuri, K., Kings Parcog groupMDS Nonmotor Study Group, 2019. Amyloid-beta and Parkinson's disease. J. Neurol. 266 (11), 2605–2619. https://doi.org/10.1007/s00415-018-9100-8. Epub 2018 Oct 30.

Liu, C., Ma, X., Jiang, X., Wilson, S.J., Hofstra, C.L., Blevitt, J., Pyati, J., Li, X., Chai, W., Carruthers, N., Lovenberg, T.W., 2001. Cloning and pharmacological characterization of a fourth histamine receptor (H(4)) expressed in bone marrow. Mol. Pharmacol. 59 (3), 420–426. https://doi.org/10.1124/mol.59.3.420.

Liu, C.Q., Chen, Z., Liu, F.X., Hu, D.N., Luo, J.H., 2007. Involvement of brain endogenous histamine in the degeneration of dopaminergic neurons in 6-hydroxydopamine-lesioned rats. Neuropharmacology 53 (7), 832–841. https://doi.org/10.1016/j.neuropharm.2007.08.014. Epub 2007 Aug 25 17919665.

Liu, C.Q., Hu, D.N., Liu, F.X., Chen, Z., Luo, J.H., 2008. Apomorphine-induced turning behavior in 6-hydroxydopamine lesioned rats is increased by histidine and decreased by histidine decarboxylase, histamine H1 and H2 receptor antagonists, and an H3 receptor agonist. Pharmacol. Biochem. Behav. 90 (3), 325–330. https://doi.org/10.1016/j.pbb.2008.03.010. Epub 2008 Mar 25.

Liu, W.W., Wei, S.Z., Huang, G.D., Liu, L.B., Gu, C., Shen, Y., Wang, X.H., Xia, S.T., Xie, A. M., Hu, L.F., Wang, F., Liu, C.F., 2020. BMAL1 regulation of microglia-mediated neuroinflammation in MPTP-induced Parkinson's disease mouse model. FASEB J. 34 (5), 6570–6581. https://doi.org/10.1096/fj.201901565RR. Epub 2020 Apr 4.

Logothetis, L., Mylonas, I.A., Baloyannis, S., Pashalidou, M., Orologas, A., Zafeiropoulos, A., Kosta, V., Theoharides, T.C., 2005. A pilot, open label, clinical trial using hydroxyzine in multiple sclerosis. Int. J. Immunopathol. Pharmacol. 18 (4), 771–778. https://doi.org/10.1177/039463200501800421.

Lopiano, L., Modugno, N., Marano, P., Sensi, M., Meco, G., Solla, P., Gusmaroli, G., Tamma, F., Mancini, F., Quatrale, R., Zangaglia, R., Bentivoglio, A., Eleopra, R., Gualberti, G., Melzi, G., Antonini, A., 2019. Motor and non-motor outcomes in patients with advanced Parkinson's disease treated with levodopa/carbidopa intestinal gel: final results of the GREENFIELD observational study. J. Neurol. 266 (9), 2164–2176. https://doi.org/10.1007/s00415-019-09337-6. Epub 2019 May 27.

Lu, C., Diehl, S.A., Noubade, R., Ledoux, J., Nelson, M.T., Spach, K., Zachary, J.F., Blankenhorn, E.P., Teuscher, C., 2010. Endothelial histamine H1 receptor signaling reduces blood-brain barrier permeability and susceptibility to autoimmune encephalomyelitis. Proc. Natl. Acad. Sci. U. S. A. 107 (44), 18967–18972. https://doi.org/10.1073/pnas.1008816107. Epub 2010 Oct 18.

Mahmood, D., Khanam, R., Pillai, K.K., Akhtar, M., 2012. Protective effects of histamine H3-receptor ligands in schizophrenic behaviors in experimental models. Pharmacol. Rep. 64 (1), 191–204. https://doi.org/10.1016/s1734-1140(12)70746-6.

Maisonnette, S., Huston, J.P., Brandao, M., Schwarting, R.K., 1998. Behavioral asymmetries and neurochemical changes after unilateral lesions of tuberomammillary nucleus or substantia nigra. Exp. Brain Res. 120 (3), 273–282. https://doi.org/10.1007/s002210050401. 9628414.

Malagelada, C., Xifró, X., Badiola, N., Sabrià, J., Rodríguez-Alvarez, J., 2004. Histamine H2-receptor antagonist ranitidine protects against neural death induced by oxygen-glucose deprivation. Stroke 35 (10), 2396–2401. https://doi.org/10.1161/01.STR.0000141160.66818.24. Epub 2004 Aug 19.

Marino, B.L.B., de Souza, L.R., Sousa, K.P.A., Ferreira, J.V., Padilha, E.C., da Silva, C.H.T.P., Taft, C.A., Hage-Melim, L.I.S., 2020. Parkinson's disease: a review from pathophysiology to treatment. Mini Rev. Med. Chem. 20 (9), 754–767. https://doi.org/10.2174/1389557519666191104110908.

Martinez-Mir, M.I., Pollard, H., Moreau, J., Arrang, J.M., Ruat, M., Traiffort, E., Schwartz, J.C., Palacios, J.M., 1990. Three histamine receptors (H1, H2 and H3) visualized in the brain of human and non-human primates. Brain Res. 526 (2), 322–327. https://doi.org/10.1016/0006-8993(90)91240-h.

Masliah, E., Rockenstein, E., Veinbergs, I., Sagara, Y., Mallory, M., Hashimoto, M., Mucke, L., 2001. beta-Amyloid peptides enhance alpha-synuclein accumulation and neuronal deficits in a transgenic mouse model linking Alzheimer's disease and Parkinson's disease. Proc. Natl. Acad. Sci. U. S. A. 98 (21), 12245–12250. https://doi.org/10.1073/pnas.211412398. Epub 2001 Sep 25.

Mayer, P.C.M., de Carvalho Neto, M.B., Katz, J.L., 2018. Punishment and the potential for negative reinforcement with histamine injection. J. Exp. Anal. Behav. 109 (2), 365–379. https://doi.org/10.1002/jeab.319. Epub 2018 Feb 27.

McClain, J.L., Mazzotta, E.A., Maradiaga, N., Duque-Wilckens, N., Grants, I., Robison, A.J., Christofi, F.L., Moeser, A.J., Gulbransen, B.D., 2020. Histamine-dependent interactions between mast cells, glia, and neurons are altered following early-life adversity in mice and humans. Am. J. Physiol. Gastrointest. Liver Physiol. 319 (6), G655–G668. https://doi.org/10.1152/ajpgi.00041.2020. Epub 2020 Sep 30.

Medhurst, A.D., Roberts, J.C., Lee, J., Chen, C.P., Brown, S.H., Roman, S., Lai, M.K., 2009. Characterization of histamine H3 receptors in Alzheimer's disease brain and amyloid over-expressing TASTPM mice. Br. J. Pharmacol. 157 (1), 130–138. https://doi.org/10.1111/j.1476-5381.2008.00075.x. Epub 2009 Feb 16.

Minzenberg, M.J., Carter, C.S., 2008. Modafinil: a review of neurochemical actions and effects on cognition. Neuropsychopharmacology 33 (7), 1477–1502. https://doi.org/10.1038/sj.npp.1301534. Epub 2007 Aug 22.

Mobarakeh, J.I., Takahashi, K., Yanai, K., 2009. Enhanced morphine-induced antinociception in histamine H3 receptor gene knockout mice. Neuropharmacology 57 (4), 409–414. https://doi.org/10.1016/j.neuropharm.2009.06.036. Epub 2009 Jul 7.

Mohanty, S., Dey, P.K., Sharma, H.S., Singh, S., Chansouria, J.P., Olsson, Y., 1989. Role of histamine in traumatic brain edema. An experimental study in the rat. J. Neurol. Sci. 90 (1), 87–97. https://doi.org/10.1016/0022-510x(89)90048-8.

Molinari, S.P., Kaminski, R., Di Rocco, A., Yahr, M.D., 1995. The use of famotidine in the treatment of Parkinson's disease: a pilot study. J. Neural. Transm. Park. Dis. Dement. Sect. 9 (2–3), 243–247. https://doi.org/10.1007/BF02259665.

Morales, I., Sanchez, A., Rodriguez-Sabate, C., Rodriguez, M., 2016. The astrocytic response to the dopaminergic denervation of the striatum. J. Neurochem. 139 (1), 81–95. https://doi.org/10.1111/jnc.13684. Epub 2016 Jun 18.

Moreno, E., Moreno-Delgado, D., Navarro, G., Hoffmann, H.M., Fuentes, S., Rosell-Vilar, S., Gasperini, P., Rodríguez-Ruiz, M., Medrano, M., Mallol, J., Cortés, A., Casadó, V., Lluís, C., Ferré, S., Ortiz, J., Canela, E., McCormick, P.J., 2014. Cocaine disrupts histamine H3 receptor modulation of dopamine D1 receptor signaling: sigma1-D1-H3 receptor complexes as key targets for reducing cocaine's effects. J. Neurosci. 34 (10), 3545–3558. https://doi.org/10.1523/JNEUROSCI.4147-13.2014.

Moreno-Delgado, D., Puigdellívol, M., Moreno, E., Rodríguez-Ruiz, M., Botta, J., Gasperini, P., Chiarlone, A., Howell, L.A., Scarselli, M., Casadó, V., Cortés, A., Ferré, S., Guzmán, M., Lluís, C., Alberch, J., Canela, E.I., Ginés, S., McCormick, P.J., 2020. Modulation of dopamine D(1) receptors via histamine H(3) receptors is a novel therapeutic target for Huntington's disease. Elife 9, e51093. https://doi.org/10.7554/eLife.51093.

Moriguchi, T., Takai, J., 2020. Histamine and histidine decarboxylase: immunomodulatory functions and regulatory mechanisms. Genes Cells 25 (7), 443–449. https://doi.org/10.1111/gtc.12774. Epub 2020 May 12.

Moriwaki, C., Chiba, S., Wei, H., Aosa, T., Kitamura, H., Ina, K., Shibata, H., Fujikura, Y., 2015. Distribution of histaminergic neuronal cluster in the rat and mouse hypothalamus. J. Chem. Neuroanat. 68, 1–13. https://doi.org/10.1016/j.jchemneu.2015.07.001. Epub 2015 Jul 9.

Munzar, P., Tanda, G., Justinova, Z., Goldberg, S.R., 2004. Histamine h3 receptor antagonists potentiate methamphetamine self-administration and methamphetamine-induced accumbal dopamine release. Neuropsychopharmacology 29 (4), 705–717. https://doi.org/10.1038/sj.npp.1300380.

Muresanu, D.F., Sharma, A., Sahib, S., Tian, Z.R., Feng, L., Castellani, R.J., Nozari, A., Lafuente, J.V., Buzoianu, A.D., Sjöquist, P.O., Patnaik, R., Wiklund, L., Sharma, H.S., 2020. Diabetes exacerbates brain pathology following a focal blast brain injury: new role of a multimodal drug cerebrolysin and nanomedicine. Prog. Brain Res. 258, 285–367. https://doi.org/10.1016/bs.pbr.2020.09.004. Epub 2020 Nov 9.

Naddafi, F., Mirshafiey, A., 2013. The neglected role of histamine in Alzheimer's disease. Am. J. Alzheimers Dis. Other Demen. 28 (4), 327–336. https://doi.org/10.1177/1533317513488925. Epub 2013 May 15.

Nakai, T., Kitamura, N., Hashimoto, T., Kajimoto, Y., Nishino, N., Mita, T., Tanaka, C., 1991. Decreased histamine H1 receptors in the frontal cortex of brains from patients with chronic schizophrenia. Biol. Psychiatry 30 (4), 349–356. https://doi.org/10.1016/0006-3223(91)90290-3.

National Research Council, 2011. Guide for the Care and Use of Laboratory Animals, eighth ed. The National Academic Press, Washington DC. www.nap.edu.

Nazzaro, J.M., Pahwa, R., Lyons, K.E., 2017. Symptomatic, non-infectious, non-hemorrhagic edema after subthalamic nucleus deep brain stimulation surgery for Parkinson's disease. J. Neurol. Sci. 383, 42–46. https://doi.org/10.1016/j.jns.2017.10.003. Epub 2017 Oct 12.

Németh, L., Szabó, C.A., Deli, M.A., Kovács, J., Krizbai, I.A., Abrahám, C.S., Joó, F., 1997. Intracarotid histamine administration results in dose-dependent vasogenic brain oedema formation in new-born pigs. Inflamm. Res. 46 (Suppl. 1), S45–S46.

Nieto-Alamilla, G., Márquez-Gómez, R., García-Gálvez, A.M., Morales-Figueroa, G.E., Arias-Montaño, J.A., 2016. The histamine H3 receptor: structure, pharmacology, and function. Mol. Pharmacol. 90 (5), 649–673. https://doi.org/10.1124/mol.116.104752. Epub 2016 Aug 25.

Nishino, S., Sakurai, E., Nevsimalova, S., Yoshida, Y., Watanabe, T., Yanai, K., Mignot, E., 2009. Decreased CSF histamine in narcolepsy with and without low CSF hypocretin-1 in comparison to healthy controls. Sleep 32 (2), 175–180. https://doi.org/10.1093/sleep/32.2.175.

Niu, F., Sharma, A., Wang, Z., Feng, L., Muresanu, D.F., Sahib, S., Tian, Z.R., Lafuente, J.V., Buzoianu, A.D., Castellani, R.J., Nozari, A., Patnaik, R., Wiklund, L., Sharma, H.S., 2020. Co-administration of TiO(2)-nanowired dl-3-n-butylphthalide (dl-NBP) and mesenchymal stem cells enhanced neuroprotection in Parkinson's disease exacerbated by concussive head injury. Prog. Brain Res. 258, 101–155. https://doi.org/10.1016/bs.pbr.2020.09.011. Epub 2020 Nov 9.

Nowak, P., Noras, L., Jochem, J., Szkilnik, R., Brus, H., Körossy, E., Drab, J., Kostrzewa, R.M., Brus, R., 2009. Histaminergic activity in a rodent model of Parkinson's disease. Neurotox. Res. 15 (3), 246–251. https://doi.org/10.1007/s12640-009-9025-1. Epub 2009 Feb 28.

Nuutinen, S., Panula, P., 2010. Histamine in neurotransmission and brain diseases. Adv. Exp. Med. Biol. 709, 95–107. https://doi.org/10.1007/978-1-4419-8056-4_10.

Oeckl, P., Halbgebauer, S., Anderl-Straub, S., Steinacker, P., Huss, A.M., Neugebauer, H., von Arnim, C.A.F., Diehl-Schmid, J., Grimmer, T., Kornhuber, J., Lewczuk, P., Danek, A., Consortium for Frontotemporal Lobar Degeneration German, Ludolph, A.C., Otto, M.,

2019. Glial fibrillary acidic protein in serum is increased in Alzheimer's disease and correlates with cognitive impairment. J. Alzheimers Dis. 67 (2), 481–488. https://doi.org/10.3233/JAD-180325.

Okuda, T., Zhang, D., Shao, H., Okamura, N., Takino, N., Iwamura, T., Sakurai, E., Yoshikawa, T., Yanai, K., 2009. Methamphetamine- and 3,4-methylenedioxymethamphetamine-induced behavioral changes in histamine H3-receptor knockout mice. J. Pharmacol. Sci. 111 (2), 167–174. https://doi.org/10.1254/jphs.09024fp. Epub 2009 Sep 26.

Orange, P.R., Heath, P.R., Wright, S.R., Ramchand, C.N., Kolkeiwicz, L., Pearson, R.C., 1996. Individuals with schizophrenia have an increased incidence of the H2R649G allele for the histamine H2 receptor gene. Mol. Psychiatry 1 (6), 466–469.

Owen, S.M., Sturman, G., Freeman, P., 1994. Modulation of morphine-induced antinociception in mice by histamine H3-receptor ligands. Agents Actions 41, C62–C63. https://doi.org/10.1007/BF02007768.

Ozkizilcik, A., Sharma, A., Muresanu, D.F., Lafuente, J.V., Tian, Z.R., Patnaik, R., Mössler, H., Sharma, H.S., 2018a. Timed release of cerebrolysin using drug-loaded titanate nanospheres reduces brain pathology and improves behavioral functions in Parkinson's disease. Mol. Neurobiol. 55 (1), 359–369. https://doi.org/10.1007/s12035-017-0747-4.

Ozkizilcik, A., Williams, R., Tian, Z.R., Muresanu, D.F., Sharma, A., Sharma, H.S., 2018b. Synthesis of biocompatible titanate nanofibers for effective delivery of neuroprotective agents. Methods Mol. Biol. 1727, 433–442. https://doi.org/10.1007/978-1-4939-7571-6_35.

Ozkizilcik, A., Sharma, A., Lafuente, J.V., Muresanu, D.F., Castellani, R.J., Nozari, A., Tian, Z.R., Mössler, H., Sharma, H.S., 2019. Nanodelivery of cerebrolysin reduces pathophysiology of Parkinson's disease. Prog. Brain Res. 245, 201–246. https://doi.org/10.1016/bs.pbr.2019.03.014. Epub 2019 Apr 2.

Palada, V., Terzić, J., Mazzulli, J., Bwala, G., Hagenah, J., Peterlin, B., Hung, A.Y., Klein, C., Krainc, D., 2012. Histamine N-methyltransferase Thr105Ile polymorphism is associated with Parkinson's disease. Neurobiol. Aging 33 (4). https://doi.org/10.1016/j.neurobiolaging.2011.06.015, 836.e1–3. Epub 2011 Jul 27 21794955.

Pallardo-Fernández, I., Muñoz-Rodríguez, J.R., González-Martín, C., Alguacil, L.F., 2020. Histamine H(3) receptor gene variants associated with drug abuse in patients with cocaine use disorder. J. Psychopharmacol. 34 (11), 1326–1330. https://doi.org/10.1177/0269881120961253. Epub 2020 Oct 16.

Panula, P., 2020. Histamine, histamine H(3) receptor, and alcohol use disorder. Br. J. Pharmacol. 177 (3), 634–641. https://doi.org/10.1111/bph.14634. Epub 2019 Apr 10.

Panula, P., Nuutinen, S., 2011. Histamine and H3 receptor in alcohol-related behaviors. J. Pharmacol. Exp. Ther. 336 (1), 9–16. https://doi.org/10.1124/jpet.110.170928. Epub 2010 Sep 23.

Panula, P., Nuutinen, S., 2013. The histaminergic network in the brain: basic organization and role in disease. Nat. Rev. Neurosci. 14 (7), 472–487. https://doi.org/10.1038/nrn3526.

Panula, P., Yang, H.Y., Costa, E., 1984. Histamine-containing neurons in the rat hypothalamus. Proc. Natl. Acad. Sci. U. S. A. 81 (8), 2572–2576. https://doi.org/10.1073/pnas.81.8.2572.

Panula, P., Pirvola, U., Auvinen, S., Airaksinen, M.S., 1989. Histamine-immunoreactive nerve fibers in the rat brain. Neuroscience 28 (3), 585–610. https://doi.org/10.1016/0306-4522(89)90007-9.

Panula, P., Airaksinen, M.S., Pirvola, U., Kotilainen, E., 1990. A histamine-containing neuronal system in human brain. Neuroscience 34 (1), 127–132. https://doi.org/10.1016/0306-4522(90)90307-p.

Panula, P., Chazot, P.L., Cowart, M., Gutzmer, R., Leurs, R., Liu, W.L., Stark, H., Thurmond, R.L., Haas, H.L., 2015. International union of basic and clinical pharmacology. XCVIII. Histamine receptors. Pharmacol. Rev. 67 (3), 601–655. https://doi.org/10.1124/pr.114.010249.

Park, J., Lim, C.S., Seo, H., Park, C.A., Zhuo, M., Kaang, B.K., Lee, K., 2015. Pain perception in acute model mice of Parkinson's disease induced by 1-methyl-4-phenyl-1,2,3,6-tetrahydropyridine (MPTP). Mol. Pain 11, 28. https://doi.org/10.1186/s12990-015-0026-1.

Parsons, M.E., Ganellin, C.R., 2006. Histamine and its receptors. Br. J. Pharmacol. 147 (Suppl. 1), S127–S135. https://doi.org/10.1038/sj.bjp.0706440.

Pasinetti, G.M., Osterburg, H.H., Kelly, A.B., Kohama, S., Morgan, D.G., Reinhard Jr., J.F., Stellwagen, R.H., Finch, C.E., 1992. Slow changes of tyrosine hydroxylase gene expression in dopaminergic brain neurons after neurotoxin lesioning: a model for neuron aging. Brain Res. Mol. Brain Res. 13 (1–2), 63–73. https://doi.org/10.1016/0169-328x(92)90045-d.

Passani, M.B., Benetti, F., Blandina, P., Furini, C.R.G., de Carvalho, M.J., Izquierdo, I., 2017. Histamine regulates memory consolidation. Neurobiol. Learn. Mem. 145, 1–6. https://doi.org/10.1016/j.nlm.2017.08.007. Epub 2017 Aug 23.

Patnaik, R., Mohanty, S., Sharma, H.S., 2000. Blockade of histamine H2 receptors attenuate blood-brain barrier permeability, cerebral blood flow disturbances, edema formation and cell reactions following hyperthermic brain injury in the rat. Acta Neurochir. Suppl. 76, 535–539. https://doi.org/10.1007/978-3-7091-6346-7_112.

Patnaik, R., Sharma, A., Skaper, S.D., Muresanu, D.F., Lafuente, J.V., Castellani, R.J., Nozari, A., Sharma, H.S., 2018. Histamine H3 inverse agonist BF 2649 or antagonist with partial H4 agonist activity clobenpropit reduces amyloid beta peptide-induced brain pathology in Alzheimer's disease. Mol. Neurobiol. 55 (1), 312–321. https://doi.org/10.1007/s12035-017-0743-8.

Peball, M., Krismer, F., Knaus, H.G., Djamshidian, A., Werkmann, M., Carbone, F., Ellmerer, P., Heim, B., Marini, K., Valent, D., Goebel, G., Ulmer, H., Stockner, H., Wenning, G.K., Stolz, R., Krejcy, K., Poewe, W., Seppi, K., Collaborators of the Parkinson's Disease Working Group Innsbruck, 2020. Non-motor symptoms in Parkinson's disease are reduced by nabilone. Ann. Neurol. 88 (4), 712–722. https://doi.org/10.1002/ana.25864. Epub 2020 Aug 31.

Pennington, C., Duncan, G., Ritchie, C., 2020. Altered awareness of motor symptoms in Parkinson's disease and dementia with Lewy bodies: a systematic review. Int. J. Geriatr. Psychiatry 35 (9), 972–981. https://doi.org/10.1002/gps.5362. Epub 2020 Jul 8.

Perdomo-Lampignano, J.A., Pana, T.A., Sleeman, I., Clark, A.B., Sawanyawisuth, K., Tiamkao, S., Myint, P.K., 2020. Parkinson's disease and patient related outcomes in stroke: a matched cohort study. J. Stroke Cerebrovasc. Dis. 29 (7), 104826. https://doi.org/10.1016/j.jstrokecerebrovasdis.2020.104826. Epub 2020 May 10.

Pillot, C., Heron, A., Cochois, V., Tardivel-Lacombe, J., Ligneau, X., Schwartz, J.C., Arrang, J.M., 2002. A detailed mapping of the histamine H(3) receptor and its gene transcripts in rat brain. Neuroscience 114, 173–193.

Pollard, H., Moreau, J., Arrang, J.M., Schwartz, J.C., 1993. A detailed autoradiographic mapping of histamine H3 receptors in rat brain areas. Neuroscience 52, 169–189.

Popiolek-Barczyk, K., Łażewska, D., Latacz, G., Olejarz, A., Makuch, W., Stark, H., Kieć-Kononowicz, K., Mika, J., 2018. Antinociceptive effects of novel histamine H(3) and H(4) receptor antagonists and their influence on morphine analgesia of neuropathic pain in the mouse. Br. J. Pharmacol. 175 (14), 2897–2910. https://doi.org/10.1111/bph.14185. Epub 2018 May 31.

Prell, G.D., Green, J.P., 1986. Histamine as a neuroregulator. Annu. Rev. Neurosci. 9, 209–254. https://doi.org/10.1146/annurev.ne.09.030186.001233.

Provensi, G., Costa, A., Izquierdo, I., Blandina, P., Passani, M.B., 2020. Brain histamine modulates recognition memory: possible implications in major cognitive disorders. Br. J. Pharmacol. 177 (3), 539–556. https://doi.org/10.1111/bph.14478. Epub 2018 Sep 22.

Qian, H., Kang, X., Hu, J., Zhang, D., Liang, Z., Meng, F., Zhang, X., Xue, Y., Maimon, R., Dowdy, S.F., Devaraj, N.K., Zhou, Z., Mobley, W.C., Cleveland, D.W., Fu, X.D., 2020. Reversing a model of Parkinson's disease with in situ converted nigral neurons. Nature 582 (7813), 550–556. https://doi.org/10.1038/s41586-020-2388-4. Epub 2020 Jun 24.

Rafiee Zadeh, A., Falahatian, M., Alsahebfosoul, F., 2018. Serum levels of histamine and diamine oxidase in multiple sclerosis. Am. J. Clin. Exp. Immunol. 7 (6), 100–105. eCollection 2018.

Rapanelli, M., Pittenger, C., 2016. Histamine and histamine receptors in Tourette syndrome and other neuropsychiatric conditions. Neuropharmacology 106, 85–90. https://doi.org/10.1016/j.neuropharm.2015.08.019. Epub 2015 Aug 14.

Reinhard Jr., J.F., O'Callaghan, J.P., 1991. Measurement of tyrosine hydroxylase apoenzyme protein by enzyme-linked immunosorbent assay (ELISA): effects of 1-methyl-4-phenyl-1,2,3,6-tetrahydropyridine (MPTP) on striatal tyrosine hydroxylase activity and content. Anal. Biochem. 196 (2), 296–301. https://doi.org/10.1016/0003-2697(91)90469-a.

Rietdijk, C.D., Perez-Pardo, P., Garssen, J., van Wezel, R.J., Kraneveld, A.D., 2017. Exploring Braak's hypothesis of Parkinson's disease. Front. Neurol. 8, 37. https://doi.org/10.3389/fneur.2017.00037. eCollection 2017.

Rinne, J.O., Anichtchik, O.V., Eriksson, K.S., Kaslin, J., Tuomisto, L., Kalimo, H., Röyttä, M., Panula, P., 2002. Increased brain histamine levels in Parkinson's disease but not in multiple system atrophy. J. Neurochem. 81 (5), 954–960. https://doi.org/10.1046/j.1471-4159.2002.00871.x.

Rodríguez-Ruiz, M., Moreno, E., Moreno-Delgado, D., Navarro, G., Mallol, J., Cortés, A., Lluís, C., Canela, E.I., Casadó, V., McCormick, P.J., Franco, R., 2017. Heteroreceptor complexes formed by dopamine D(1), histamine H(3), and N-methyl-D-aspartate glutamate receptors as targets to prevent neuronal death in Alzheimer's disease. Mol. Neurobiol. 54 (6), 4537–4550. https://doi.org/10.1007/s12035-016-9995-y. Epub 2016 Jul 1.

Roeder, T., 2003. Metabotropic histamine receptors—nothing for invertebrates? Eur. J. Pharmacol. 466 (1–2), 85–90. https://doi.org/10.1016/s0014-2999(03)01553-x.

Rosland, J.H., Hunskaar, S., Broch, O.J., Hole, K., 1992. Acute and long term effects of 1-methyl-4-phenyl-1,2,3,6-tetrahydropyridine (MPTP) in tests of nociception in mice. Pharmacol. Toxicol. 70 (1), 31–37. https://doi.org/10.1111/j.1600-0773.1992.tb00421.x.

Ruppert, M.C., Greuel, A., Tahmasian, M., Schwartz, F., Stürmer, S., Maier, F., Hammes, J., Tittgemeyer, M., Timmermann, L., van Eimeren, T., Drzezga, A., Eggers, C., 2020. Network degeneration in Parkinson's disease: multimodal imaging of nigro-striato-cortical dysfunction. Brain 143 (3), 944–959. https://doi.org/10.1093/brain/awaa019.

Sadek, B., Saad, A., Sadeq, A., Jalal, F., Stark, H., 2016. Histamine H3 receptor as a potential target for cognitive symptoms in neuropsychiatric diseases. Behav. Brain Res. 312, 415–430. https://doi.org/10.1016/j.bbr.2016.06.051. Epub 2016 Jun 27.

Saligrama, N., Noubade, R., Case, L.K., del Rio, R., Teuscher, C., 2012. Combinatorial roles for histamine H1-H2 and H3-H4 receptors in autoimmune inflammatory disease of the central nervous system. Eur. J. Immunol. 42 (6), 1536–1546. https://doi.org/10.1002/eji.201141859.

Samii, A., Nutt, J.G., Ransom, B.R., 2004. Parkinson's disease. Lancet 363 (9423), 1783–1793. https://doi.org/10.1016/S0140-6736(04)16305-8.

Santangelo, A., Passani, M.B., Casarrubea, M., 2017. Brain histamine and behavioral neuroscience. Oncotarget 8 (10), 16107–16108. https://doi.org/10.18632/oncotarget.15365.

Santos-García, D., de Deus Fonticoba, T., Suárez Castro, E., Aneiros Díaz, A., McAfee, D., Catalán, M.J., Alonso-Frech, F., Villanueva, C., Jesús, S., Mir, P., Aguilar, M., Pastor, P., García Caldentey, J., Esltelrich Peyret, E., Planellas, L.L., Martí, M.J., Caballol, N., Hernández Vara, J., Martí Andrés, G., Cabo, I., Ávila Rivera, M.A., López Manzanares, L., Redondo, N., Martinez-Martin, P., COPPADIS Study Group, McAfee, D., 2020. Non-motor symptom burden is strongly correlated to motor complications in patients with Parkinson's disease. Eur. J. Neurol. 27 (7), 1210–1223. https://doi.org/10.1111/ene.14221. Epub 2020 Apr 24.

Scammell, T.E., Jackson, A.C., Franks, N.P., Wisden, W., Dauvilliers, Y., 2019. Histamine: neural circuits and new medications. Sleep 42 (1), zsy183. https://doi.org/10.1093/sleep/zsy183.

Schneider, E.H., 2019. Microglial histamine H(4)R in the pathophysiology of Parkinson's disease-a new actor on the stage? Naunyn Schmiedebergs Arch. Pharmacol. 392 (6), 641–645. https://doi.org/10.1007/s00210-019-01635-0. Epub 2019 Apr 24.

Schneider, E.H., Neumann, D., Seifert, R., 2014. Modulation of behavior by the histaminergic system: lessons from HDC-, H3R- and H4R-deficient mice. Neurosci. Biobehav. Rev. 47, 101–121. https://doi.org/10.1016/j.neubiorev.2014.07.020. Epub 2014 Aug 4.

Schneider, E.H., Neumann, D., Seifert, R., 2015. Histamine H4-receptor expression in the brain? Naunyn Schmiedebergs Arch. Pharmacol. 388 (1), 5–9. https://doi.org/10.1007/s00210-014-1067-0.

Schwartz, J.C., Pollard, H., Quach, T.T., 1980. Histamine as a neurotransmitter in mammalian brain: neurochemical evidence. J. Neurochem. 35 (1), 26–33. https://doi.org/10.1111/j.1471-4159.1980.tb12485.x.

Schweitzer, J.S., Song, B., Herrington, T.M., Park, T.Y., Lee, N., Ko, S., Jeon, J., Cha, Y., Kim, K., Li, Q., Henchcliffe, C., Kaplitt, M., Neff, C., Rapalino, O., Seo, H., Lee, I.H., Kim, J., Kim, T., Petsko, G.A., Ritz, J., Cohen, B.M., Kong, S.W., Leblanc, P., Carter, B.S., Kim, K.S., 2020. Personalized iPSC-derived dopamine progenitor cells for Parkinson's disease. N. Engl. J. Med. 382 (20), 1926–1932. https://doi.org/10.1056/NEJMoa1915872.

Shahnawaz, M., Mukherjee, A., Pritzkow, S., Mendez, N., Rabadia, P., Liu, X., Hu, B., Schmeichel, A., Singer, W., Wu, G., Tsai, A.L., Shirani, H., Nilsson, K.P.R., Low, P.A., Soto, C., 2020. Discriminating alpha-synuclein strains in Parkinson's disease and multiple system atrophy. Nature 578 (7794), 273–277. https://doi.org/10.1038/s41586-020-1984-7. Epub 2020 Feb 5.

Shan, L., Bossers, K., Luchetti, S., Balesar, R., Lethbridge, N., Chazot, P.L., Bao, A.M., Swaab, D.F., 2012. Alterations in the histaminergic system in the substantia nigra and striatum of Parkinson's patients: a postmortem study. Neurobiol. Aging 33 (7). https://doi.org/10.1016/j.neurobiolaging.2011.10.016, 1488.e1–13. Epub 2011 Nov 26.

Shan, L., Swaab, D.F., Bao, A.M., 2013. Neuronal histaminergic system in aging and age-related neurodegenerative disorders. Exp. Gerontol. 48 (7), 603–607. https://doi.org/10.1016/j.exger.2012.08.002. Epub 2012 Aug 11.

Shan, L., Bao, A.M., Swaab, D.F., 2015. The human histaminergic system in neuropsychiatric disorders. Trends Neurosci. 38 (3), 167–177. https://doi.org/10.1016/j.tins.2014.12.008. Epub 2015 Jan 7.

Shan, L., Bao, A.M., Swaab, D.F., 2017. Changes in histidine decarboxylase, histamine N-methyltransferase and histamine receptors in neuropsychiatric disorders. Handb. Exp. Pharmacol. 241, 259–276. https://doi.org/10.1007/164_2016_125.

Sharma, H.S., 1987. Effect of captopril (a converting enzyme inhibitor) on blood-brain barrier permeability and cerebral blood flow in normotensive rats. Neuropharmacology 26 (1), 85–92. https://doi.org/10.1016/0028-3908(87)90049-9.

Sharma, H.S., 2006. Hyperthermia influences excitatory and inhibitory amino acid neurotransmitters in the central nervous system. An experimental study in the rat using behavioural, biochemical, pharmacological, and morphological approaches. J. Neural. Transm. (Vienna) 113 (4), 497–519. https://doi.org/10.1007/s00702-005-0406-1.

Sharma, H.S., 2010. A combination of tumor necrosis factor-alpha and neuronal nitric oxide synthase antibodies applied topically over the traumatized spinal cord enhances neuroprotection and functional recovery in the rat. Ann. N. Y. Acad. Sci. 1199, 175–185. https://doi.org/10.1111/j.1749-6632.2009.05327.x.

Sharma, H.S., 2011a. Blood-CNS barrier, neurodegeneration and neuroprotection: recent therapeutic advancements and nano-drug delivery. J. Neural Transm. (Vienna) 118 (1), 3–6. https://doi.org/10.1007/s00702-010-0542-0.

Sharma, H.S., 2011b. Early microvascular reactions and blood-spinal cord barrier disruption are instrumental in pathophysiology of spinal cord injury and repair: novel therapeutic strategies including nanowired drug delivery to enhance neuroprotection. J. Neural Transm. (Vienna) 118 (1), 155–176. https://doi.org/10.1007/s00702-010-0514-4. Epub 2010 Dec.

Sharma, H.S., Ali, S.F., 2006. Alterations in blood-brain barrier function by morphine and methamphetamine. Ann. N. Y. Acad. Sci. 1074, 198–224. https://doi.org/10.1196/annals.1369.020.

Sharma, H.S., Ali, S.F., 2008. Acute administration of 3,4-methylenedioxymethamphetamine induces profound hyperthermia, blood-brain barrier disruption, brain edema formation, and cell injury. Ann. N. Y. Acad. Sci. 1139, 242–258. https://doi.org/10.1196/annals.1432.052.

Sharma, H.S., Cervós-Navarro, J., 1990. Brain oedema and cellular changes induced by acute heat stress in young rats. Acta Neurochir. Suppl. (Wien) 51, 383–386. https://doi.org/10.1007/978-3-7091-9115-6_129.

Sharma, H.S., Cervós-Navarro, J., 1991. Role of histamine in pathophysiology of heat stress in rats. Agents Actions Suppl. 33, 97–102. https://doi.org/10.1007/978-3-0348-7309-3_8.

Sharma, H.S., Dey, P.K., 1986a. Influence of long-term immobilization stress on regional blood-brain barrier permeability, cerebral blood flow and 5-HT level in conscious normotensive young rats. J. Neurol. Sci. 72 (1), 61–76. https://doi.org/10.1016/0022-510x(86)90036-5.

Sharma, H.S., Dey, P.K., 1986b. Probable involvement of 5-hydroxytryptamine in increased permeability of blood-brain barrier under heat stress in young rats. Neuropharmacology 25 (2), 161–167. https://doi.org/10.1016/0028-3908(86)90037-7.

Sharma, H.S., Dey, P.K., 1987. Influence of long-term acute heat exposure on regional blood-brain barrier permeability, cerebral blood flow and 5-HT level in conscious normotensive young rats. Brain Res. 424 (1), 153–162. https://doi.org/10.1016/0006-8993(87)91205-4.

Sharma, H.S., Dey, P.K., 1988. EEG changes following increased blood-brain barrier permeability under long-term immobilization stress in young rats. Neurosci. Res. 5 (3), 224–239. https://doi.org/10.1016/0168-0102(88)90051-x.

Sharma, H.S., Olsson, Y., 1990. Edema formation and cellular alterations following spinal cord injury in the rat and their modification with p-chlorophenylalanine. Acta Neuropathol. 79 (6), 604–610. https://doi.org/10.1007/BF00294237.

Sharma, H.S., Sharma, A., 2008. Antibodies as promising novel neuroprotective agents in the central nervous system injuries. Cent. Nerv. Syst. Agents Med. Chem. 8 (3), 143–169. https://doi.org/10.2174/187152408785699640.

Sharma, A., Sharma, H.S., 2012a. Monoclonal antibodies as novel neurotherapeutic agents in CNS injury and repair. Int. Rev. Neurobiol. 102, 23–45. https://doi.org/10.1016/B978-0-12-386986-9.00002-8.

Sharma, H.S., Sharma, A., 2012b. Nanowired drug delivery for neuroprotection in central nervous system injuries: modulation by environmental temperature, intoxication of nanoparticles, and comorbidity factors. Wiley Interdiscip. Rev. Nanomed. Nanobiotechnol. 4 (2), 184–203. https://doi.org/10.1002/wnan.172. Epub 2011 Dec 8.

Sharma, H.S., Sharma, A., 2013. New perspectives of nanoneuroprotection, nanoneuropharmacology and nanoneurotoxicity: modulatory role of amino acid neurotransmitters, stress, trauma, and co-morbidity factors in nanomedicine. Amino Acids 45 (5), 1055–1071. https://doi.org/10.1007/s00726-013-1584-z.

Sharma, H.S., Sjöquist, P.O., 2002. A new antioxidant compound H-290/51 modulates glutamate and GABA immunoreactivity in the rat spinal cord following trauma. Amino Acids 23 (1–3), 261–272. https://doi.org/10.1007/s00726-001-0137-z.

Sharma, H.S., Olsson, Y., Dey, P.K., 1990. Changes in blood-brain barrier and cerebral blood flow following elevation of circulating serotonin level in anesthetized rats. Brain Res. 517 (1–2), 215–223. https://doi.org/10.1016/0006-8993(90)91029-g.

Sharma, H.S., Cervós-Navarro, J., Dey, P.K., 1991a. Increased blood-brain barrier permeability following acute short-term swimming exercise in conscious normotensive young rats. Neurosci. Res. 10 (3), 211–221. https://doi.org/10.1016/0168-0102(91)90058-7.

Sharma, H.S., Cervós-Navarro, J., Dey, P.K., 1991b. Acute heat exposure causes cellular alteration in cerebral cortex of young rats. Neuroreport 2 (3), 155–158. https://doi.org/10.1097/00001756-199103000-00012.

Sharma, H.S., Cervós-Navarro, J., Dey, P.K., 1991c. Rearing at high ambient temperature during later phase of the brain development enhances functional plasticity of the CNS and induces tolerance to heat stress. An experimental study in the conscious normotensive young rats. Brain Dysfunct 4, 104–124.

Sharma, H.S., Nyberg, F., Cervos-Navarro, J., Dey, P.K., 1992a. Histamine modulates heat stress-induced changes in blood-brain barrier permeability, cerebral blood flow, brain oedema and serotonin levels: an experimental study in conscious young rats. Neuroscience 50 (2), 445–454. https://doi.org/10.1016/0306-4522(92)90436-6.

Sharma, H.S., Zimmer, C., Westman, J., Cervós-Navarro, J., 1992b. Acute systemic heat stress increases glial fibrillary acidic protein immunoreactivity in brain: experimental observations in conscious normotensive young rats. Neuroscience 48 (4), 889–901. https://doi.org/10.1016/0306-4522(92)90277-9.

Sharma, H.S., Olsson, Y., Cervós-Navarro, J., 1993a. Early perifocal cell changes and edema in traumatic injury of the spinal cord are reduced by indomethacin, an inhibitor of prostaglandin synthesis. Experimental study in the rat. Acta Neuropathol. 85 (2), 145–153. https://doi.org/10.1007/BF00227761.

Sharma, H.S., Olsson, Y., Cervós-Navarro, J., 1993b. p-Chlorophenylalanine, a serotonin synthesis inhibitor, reduces the response of glial fibrillary acidic protein induced by trauma to the spinal cord. An immunohistochemical investigation in the rat. Acta Neuropathol. 86 (5), 422–427. https://doi.org/10.1007/BF00228575.

Sharma, H.S., Westman, J., Nyberg, F., Cervos-Navarro, J., Dey, P.K., 1994. Role of serotonin and prostaglandins in brain edema induced by heat stress. An experimental study in the young rat. Acta Neurochir. Suppl. (Wien) 60, 65–70. https://doi.org/10.1007/978-3-7091-9334-1_17.

Sharma, H.S., Olsson, Y., Persson, S., Nyberg, F., 1995a. Trauma-induced opening of the blood-spinal cord barrier is reduced by indomethacin, an inhibitor of prostaglandin biosynthesis. Experimental observations in the rat using [131I]-sodium, Evans blue and lanthanum as tracers. Restor. Neurol. Neurosci. 7 (4), 207–215. https://doi.org/10.3233/RNN-1994-7403.

Sharma, H.S., Westman, J., Navarro, J.C., Dey, P.K., Nyberg, F., 1995b. Probable involvement of serotonin in the increased permeability of the blood-brain barrier by forced swimming. An experimental study using Evans blue and 131I-sodium tracers in the rat. Behav. Brain Res. 72 (1–2), 189–196. https://doi.org/10.1016/0166-4328(96)00170-2.

Sharma, H.S., Westman, J., Olsson, Y., Alm, P., 1996. Involvement of nitric oxide in acute spinal cord injury: an immunocytochemical study using light and electron microscopy in the rat. Neurosci. Res. 24 (4), 373–384. https://doi.org/10.1016/0168-0102(95)01015-7.

Sharma, H.S., Nyberg, F., Gordh, T., Alm, P., Westman, J., 1997a. Topical application of insulin like growth factor-1 reduces edema and upregulation of neuronal nitric oxide synthase following trauma to the rat spinal cord. Acta Neurochir. Suppl. 70, 130–133. https://doi.org/10.1007/978-3-7091-6837-0_40.

Sharma, H.S., Westman, J., Cervós-Navarro, J., Nyberg, F., 1997b. Role of neurochemicals in brain edema and cell changes following hyperthermic brain injury in the rat. Acta Neurochir. Suppl. 70, 269–274. https://doi.org/10.1007/978-3-7091-6837-0_84.

Sharma, H.S., Westman, J., Nyberg, F., 1997c. Topical application of 5-HT antibodies reduces edema and cell changes following trauma of the rat spinal cord. Acta Neurochir. Suppl. 70, 155–158. https://doi.org/10.1007/978-3-7091-6837-0_47.

Sharma, H.S., Alm, P., Westman, J., 1998a. Nitric oxide and carbon monoxide in the brain pathology of heat stress. Prog. Brain Res. 115, 297–333. https://doi.org/10.1016/s0079-6123(08)62041-5.

Sharma, H.S., Nyberg, F., Westman, J., Alm, P., Gordh, T., Lindholm, D., 1998b. Brain derived neurotrophic factor and insulin like growth factor-1 attenuate upregulation of nitric oxide synthase and cell injury following trauma to the spinal cord. An immunohistochemical study in the rat. Amino Acids 14 (1–3), 121–129. https://doi.org/10.1007/BF01345252.

Sharma, H.S., Westman, J., Nyberg, F., 1998c. Pathophysiology of brain edema and cell changes following hyperthermic brain injury. Prog. Brain Res. 115, 351–412. https://doi.org/10.1016/s0079-6123(08)62043-9.

Sharma, H.S., Winkler, T., Stålberg, E., Gordh, T., Alm, P., Westman, J., 2003. Topical application of TNF-alpha antiserum attenuates spinal cord trauma induced edema formation, microvascular permeability disturbances and cell injury in the rat. Acta Neurochir. Suppl. 86, 407–413. https://doi.org/10.1007/978-3-7091-0651-8_85.

Sharma, H.S., Badgaiyan, R.D., Alm, P., Mohanty, S., Wiklund, L., 2005. Neuroprotective effects of nitric oxide synthase inhibitors in spinal cord injury-induced pathophysiology and motor functions: an experimental study in the rat. Ann. N. Y. Acad. Sci. 1053, 422–434. https://doi.org/10.1111/j.1749-6632.2005.tb00051.x.

Sharma, H.S., Duncan, J.A., Johanson, C.E., 2006a. Whole-body hyperthermia in the rat disrupts the blood-cerebrospinal fluid barrier and induces brain edema. Acta Neurochir. Suppl. 96, 426–431. https://doi.org/10.1007/3-211-30714-1_88.

Sharma, H.S., Lundstedt, T., Boman, A., Lek, P., Seifert, E., Wiklund, L., Ali, S.F., 2006b. A potent serotonin-modulating compound AP-267 attenuates morphine withdrawal-induced blood-brain barrier dysfunction in rats. Ann. N. Y. Acad. Sci. 1074, 482–496. https://doi.org/10.1196/annals.1369.049.

Sharma, H.S., Nyberg, F., Gordh, T., Alm, P., 2006c. Topical application of dynorphin A (1-17) antibodies attenuates neuronal nitric oxide synthase up-regulation, edema formation, and cell injury following focal trauma to the rat spinal cord. Acta Neurochir. Suppl. 96, 309–315. https://doi.org/10.1007/3-211-30714-1_66.

Sharma, H.S., Vannemreddy, P., Patnaik, R., Patnaik, S., Mohanty, S., 2006d. Histamine receptors influence blood-spinal cord barrier permeability, edema formation, and spinal cord blood flow following trauma to the rat spinal cord. Acta Neurochir. Suppl. 96, 316–321. https://doi.org/10.1007/3-211-30714-1_67.

Sharma, H.S., Ali, S.F., Dong, W., Tian, Z.R., Patnaik, R., Patnaik, S., Sharma, A., Boman, A., Lek, P., Seifert, E., Lundstedt, T., 2007a. Drug delivery to the spinal cord tagged with nanowire enhances neuroprotective efficacy and functional recovery following trauma to the rat spinal cord. Ann. N. Y. Acad. Sci. 1122, 197–218. https://doi.org/10.1196/annals.1403.014.

Sharma, H.S., Patnaik, R., Patnaik, S., Mohanty, S., Sharma, A., Vannemreddy, P., 2007b. Antibodies to serotonin attenuate closed head injury induced blood brain barrier disruption and brain pathology. Ann. N. Y. Acad. Sci. 1122, 295–312. https://doi.org/10.1196/annals.1403.022.

Sharma, H.S., Ali, S., Tian, Z.R., Patnaik, R., Patnaik, S., Lek, P., Sharma, A., Lundstedt, T., 2009a. Nano-drug delivery and neuroprotection in spinal cord injury. J. Nanosci. Nanotechnol. 9 (8), 5014–5037. https://doi.org/10.1166/jnn.2009.gr04.

Sharma, H.S., Muresanu, D., Sharma, A., Patnaik, R., 2009b. Cocaine-induced breakdown of the blood-brain barrier and neurotoxicity. Int. Rev. Neurobiol. 88, 297–334. https://doi.org/10.1016/S0074-7742(09)88011-2.

Sharma, H.S., Ali, S.F., Tian, Z.R., Patnaik, R., Patnaik, S., Sharma, A., Boman, A., Lek, P., Seifert, E., Lundstedt, T., 2010a. Nanowired-drug delivery enhances neuroprotective efficacy of compounds and reduces spinal cord edema formation and improves functional outcome following spinal cord injury in the rat. Acta Neurochir. Suppl. 106, 343–350. https://doi.org/10.1007/978-3-211-98811-4_63.

Sharma, H.S., Patnaik, R., Patnaik, S., Sharma, A., Mohanty, S., Vannemreddy, P., 2010b. Antibodies to dynorphin a (1-17) attenuate closed head injury induced blood-brain barrier disruption, brain edema formation and brain pathology in the rat. Acta Neurochir. Suppl. 106, 301–306. https://doi.org/10.1007/978-3-211-98811-4_56.

Sharma, H.S., Sjöquist, P.O., Ali, S.F., 2010c. Alterations in blood-brain barrier function and brain pathology by morphine in the rat. Neuroprotective effects of antioxidant H-290/51. Acta Neurochir. Suppl. 106, 61–66. https://doi.org/10.1007/978-3-211-98811-4_10.

Sharma, H.S., Zimmermann-Meinzingen, S., Johanson, C.E., 2010d. Cerebrolysin reduces blood-cerebrospinal fluid barrier permeability change, brain pathology, and functional deficits following traumatic brain injury in the rat. Ann. N. Y. Acad. Sci. 1199, 125–137. https://doi.org/10.1111/j.1749-6632.2009.05329.x.

Sharma, H.S., Miclescu, A., Wiklund, L., 2011a. Cardiac arrest-induced regional blood-brain barrier breakdown, edema formation and brain pathology: a light and electron microscopic study on a new model for neurodegeneration and neuroprotection in porcine brain. J. Neural Transm. (Vienna) 118 (1), 87–114. https://doi.org/10.1007/s00702-010-0486-4. Epub 2010 Oct 21.

Sharma, H.S., Muresanu, D.F., Patnaik, R., Stan, A.D., Vacaras, V., Perju-Dumbrav, L., Alexandru, B., Buzoianu, A., Opincariu, I., Menon, P.K., Sharma, A., 2011b. Superior neuroprotective effects of cerebrolysin in heat stroke following chronic intoxication of Cu or Ag engineered nanoparticles. A comparative study with other neuroprotective agents using biochemical and morphological approaches in the rat. J. Nanosci. Nanotechnol. 11 (9), 7549–7569. https://doi.org/10.1166/jnn.2011.5114.

Sharma, H.S., Muresanu, D.F., Patnaik, R., Sharma, A., 2013. Exacerbation of brain pathology after partial restraint in hypertensive rats following SiO_2 nanoparticles exposure at high ambient temperature. Mol. Neurobiol. 48 (2), 368–379. https://doi.org/10.1007/s12035-013-8502-y. Epub 2013 Jul 6.

Sharma, H.S., Menon, P., Lafuente, J.V., Muresanu, D.F., Tian, Z.R., Patnaik, R., Sharma, A., 2014. Development of in vivo drug-induced neurotoxicity models. Expert Opin. Drug Metab. Toxicol. 10 (12), 1637–1661. https://doi.org/10.1517/17425255.2014.970168. Epub 2014 Oct 13.

Sharma, H.S., Kiyatkin, E.A., Patnaik, R., Lafuente, J.V., Muresanu, D.F., Sjöquist, P.O., Sharma, A., 2015. Exacerbation of methamphetamine neurotoxicity in cold and hot environments: neuroprotective effects of an antioxidant compound H-290/51. Mol. Neurobiol. 52 (2), 1023–1033. https://doi.org/10.1007/s12035-015-9252-9. Epub 2015 Jun 26.

Sharma, A., Menon, P., Muresanu, D.F., Ozkizilcik, A., Tian, Z.R., Lafuente, J.V., Sharma, H.S., 2016a. Nanowired drug delivery across the blood-brain barrier in central nervous system injury and repair. CNS Neurol. Disord. Drug Targets 15 (9), 1092–1117. https://doi.org/10.2174/1871527315666160819123059.

Sharma, H.S., Muresanu, D.F., Lafuente, J.V., Nozari, A., Patnaik, R., Skaper, S.D., Sharma, A., 2016b. Pathophysiology of blood-brain barrier in brain injury in cold and hot environments: novel drug targets for neuroprotection. CNS Neurol. Disord. Drug Targets 15 (9), 1045–1071. https://doi.org/10.2174/1871527315666160902145145.

Sharma, H.S., Skaper, S.D., Sharma, A., 2016c. Commentary: histaminergic drugs could be novel targets for neuroprotection in CNS disorders. CNS Neurol. Disord. Drug Targets 15 (6), 642–643. https://doi.org/10.2174/1871527315999160606154134.

Sharma, A., Menon, P.K., Patnaik, R., Muresanu, D.F., Lafuente, J.V., Tian, Z.R., Ozkizilcik, A., Castellani, R.J., Mössler, H., Sharma, H.S., 2017a. Novel treatment strategies using TiO(2)-nanowired delivery of histaminergic drugs and antibodies to tau with cerebrolysin for superior neuroprotection in the pathophysiology of Alzheimer's disease. Int. Rev. Neurobiol. 137, 123–165. https://doi.org/10.1016/bs.irn.2017.09.002. Epub 2017 Nov 6.

Sharma, H.S., Patnaik, R., Muresanu, D.F., Lafuente, J.V., Ozkizilcik, A., Tian, Z.R., Nozari, A., Sharma, A., 2017b. Histaminergic receptors modulate spinal cord injury-induced neuronal nitric oxide synthase upregulation and cord pathology: new roles of nanowired drug delivery for neuroprotection. Int. Rev. Neurobiol. 137, 65–98. https://doi.org/10.1016/bs.irn.2017.09.001. Epub 2017 Nov 3.

Sharma, A., Muresanu, D.F., Lafuente, J.V., Sjöquist, P.O., Patnaik, R., Ryan Tian, Z., Ozkizilcik, A., Sharma, H.S., 2018. Cold environment exacerbates brain pathology and oxidative stress following traumatic brain injuries: potential therapeutic effects of nanowired antioxidant compound H-290/51. Mol. Neurobiol. 55 (1), 276–285. https://doi.org/10.1007/s12035-017-0740-y.

Sharma, H.S., Feng, L., Muresanu, D.F., Castellani, R.J., Sharma, A., 2019a. Neuroprotective effects of a potent bradykinin B2 receptor antagonist HOE-140 on microvascular permeability, blood flow disturbances, edema formation, cell injury and nitric oxide synthase upregulation following trauma to the spinal cord. Int. Rev. Neurobiol. 146, 103–152. https://doi.org/10.1016/bs.irn.2019.06.008. Epub 2019 Jul 9.

Sharma, H.S., Muresanu, D.F., Castellani, R.J., Nozari, A., Lafuente, J.V., Tian, Z.R., Ozkizilcik, A., Manzhulo, I., Mössler, H., Sharma, A., 2019b. Nanowired delivery of cerebrolysin with neprilysin and p-Tau antibodies induces superior neuroprotection in Alzheimer's disease. Prog. Brain Res. 245, 145–200. https://doi.org/10.1016/bs.pbr.2019.03.009. Epub 2019 Apr 2.

Sharma, A., Muresanu, D.F., Castellani, R.J., Nozari, A., Lafuente, J.V., Sahib, S., Tian, Z.R., Buzoianu, A.D., Patnaik, R., Wiklund, L., Sharma, H.S., 2020. Mild traumatic brain injury exacerbates Parkinson's disease induced hemeoxygenase-2 expression and brain pathology: neuroprotective effects of co-administration of TiO(2) nanowired mesenchymal stem cells and cerebrolysin. Prog. Brain Res. 258, 157–231. https://doi.org/10.1016/bs.pbr.2020.09.010. Epub 2020 Nov 9.

Smits, R.P., Steinbusch, H.W., Mulder, A.H., 1990. The localization of histidine decarboxylase-immunoreactive cell bodies in the Guinea-pig brain. J. Chem. Neuroanat. 3 (2), 85–100.

Soares-da-Silva, P., Garrett, M.C., 1990. A kinetic study of the rate of formation of dopamine, 3,4-dihydroxyphenylacetic acid (DOPAC) and homovanillic acid (HVA) in the brain of the rat: implications for the origin of DOPAC. Neuropharmacology 29 (10), 869–874. https://doi.org/10.1016/0028-3908(90)90135-e.

Song, I.U., Lee, J.E., Kwon, D.Y., Park, J.H., Ma, H.I., 2017. Parkinson's disease might increase the risk of cerebral ischemic lesions. Int. J. Med. Sci. 14 (4), 319–322. https://doi.org/10.7150/ijms.18025. eCollection 2017.

Sonninen, T.M., Hämäläinen, R.H., Koskuvi, M., Oksanen, M., Shakirzyanova, A., Wojciechowski, S., Puttonen, K., Naumenko, N., Goldsteins, G., Laham-Karam, N., Lehtonen, M., Tavi, P., Koistinaho, J., Lehtonen, Š., 2020. Metabolic alterations in Parkinson's disease astrocytes. Sci. Rep. 10 (1), 14474. https://doi.org/10.1038/s41598-020-71329-8.

Steinbusch, H.W., 1991. Distribution of histaminergic neurons and fibers in rat brain. Comparison with noradrenergic and serotonergic innervation of the vestibular system. Acta Otolaryngol. Suppl. 479, 12–23. https://doi.org/10.3109/00016489109121144.

Steinbusch, H.W., Sauren, Y., Groenewegen, H., Watanabe, T., Mulder, A.H., 1986. Histaminergic projections from the premammillary and posterior hypothalamic region to the caudate-putamen complex in the rat. Brain Res. 368 (2), 389–393. https://doi.org/10.1016/0006-8993(86)90588-3.

Strasser, A., Wittmann, H.J., Buschauer, A., Schneider, E.H., Seifert, R., 2013. Species-dependent activities of G-protein-coupled receptor ligands: lessons from histamine receptor orthologs. Trends Pharmacol. Sci. 34 (1), 13–32. https://doi.org/10.1016/j.tips.2012.10.004. Epub 2012 Dec 8.

Szakacs, Z., Dauvilliers, Y., Mikhaylov, V., Poverennova, I., Krylov, S., Jankovic, S., Sonka, K., Lehert, P., Lecomte, I., Lecomte, J.M., Schwartz, J.C., HARMONY-CTP Study Group, 2017. Safety and efficacy of pitolisant on cataplexy in patients with narcolepsy: a randomised, double-blind, placebo-controlled trial. Lancet Neurol. 16 (3), 200–207. https://doi.org/10.1016/S1474-4422(16)30333-7. Epub 2017 Jan 25.

Tagliaferro, P., Burke, R.E., 2016. Retrograde axonal degeneration in Parkinson disease. J. Parkinsons Dis. 6 (1), 1–15. https://doi.org/10.3233/JPD-150769.

Teismann, P., Schulz, J.B., 2004. Cellular pathology of Parkinson's disease: astrocytes, microglia and inflammation. Cell Tissue Res. 318 (1), 149–161. https://doi.org/10.1007/s00441-004-0944-0. Epub 2004 Aug 24.

Teuscher, C., Subramanian, M., Noubade, R., Gao, J.F., Offner, H., Zachary, J.F., Blankenhorn, E.P., 2007. Central histamine H3 receptor signaling negatively regulates susceptibility to autoimmune inflammatory disease of the CNS. Proc. Natl. Acad. Sci. U. S. A. 104 (24), 10146–10151. https://doi.org/10.1073/pnas.0702291104. Epub 2007 Jun 4.

Thakkar, M.M., 2011. Histamine in the regulation of wakefulness. Sleep Med. Rev. 15 (1), 65–74. https://doi.org/10.1016/j.smrv.2010.06.004. Epub 2010 Sep 20.

Thorpy, M.J., 2020. Recently approved and upcoming treatments for narcolepsy. CNS Drugs 34 (1), 9–27. https://doi.org/10.1007/s40263-019-00689-1.

Thurmond, R.L., 2015. The histamine H4 receptor: from orphan to the clinic. Front. Pharmacol. 6, 65. https://doi.org/10.3389/fphar.2015.00065. eCollection 2015.

Tian, Z.R., Sharma, A., Nozari, A., Subramaniam, R., Lundstedt, T., Sharma, H.S., 2012. Nanowired drug delivery to enhance neuroprotection in spinal cord injury. CNS Neurol. Disord. Drug Targets 11 (1), 86–95. https://doi.org/10.2174/187152712799960727.

Tiligada, E., Ennis, M., 2020. Histamine pharmacology: from Sir Henry Dale to the 21st century. Br. J. Pharmacol. 177 (3), 469–489. https://doi.org/10.1111/bph.14524. Epub 2018 Dec 2.

Tiligada, E., Kyriakidis, K., Chazot, P.L., Passani, M.B., 2011. Histamine pharmacology and new CNS drug targets. CNS Neurosci. Ther. 17 (6), 620–628. https://doi.org/10.1111/j.1755-5949.2010.00212.x. Epub 2010 Nov 22.

Tolosa, E., Wenning, G., Poewe, W., 2006. The diagnosis of Parkinson's disease. Lancet Neurol. 5 (1), 75–86. https://doi.org/10.1016/S1474-4422(05)70285-4.

Trezza, A., Landi, A., Pilleri, M., Antonini, A., Giussani, C., Sganzerla, E.P., 2018. Peri-electrode edema after bilateral subthalamic deep brain stimulation for Parkinson's disease. J. Neurosurg. Sci. 62 (1), 103–105. https://doi.org/10.23736/S0390-5616.16.03421-4.

Tuomisto, L., Kilpeläinen, H., Riekkinen, P., 1983. Histamine and histamine-N-methyltransferase in the CSF of patients with multiple sclerosis. Agents Actions 13 (2–3), 255–257. https://doi.org/10.1007/BF01967346.

Tysnes, O.B., Storstein, A., 2017. Epidemiology of Parkinson's disease. J. Neural Transm. (Vienna) 124 (8), 901–905. https://doi.org/10.1007/s00702-017-1686-y. Epub 2017 Feb 1.

Uguen, M., Perrin, D., Belliard, S., Ligneau, X., Beardsley, P.M., Lecomte, J.M., Schwartz, J.-C., 2013. Preclinical evaluation of the abuse potential of Pitolisant, a histamine H3 receptor inverse agonist/antagonist compared with Modafinil. Br. J. Pharmacol. 169 (3), 632–644. https://doi.org/10.1111/bph.12149.

Umehara, H., Mizuguchi, H., Fukui, H., 2012. Identification of a histaminergic circuit in the caudal hypothalamus: an evidence for functional heterogeneity of histaminergic neurons. Neurochem. Int. 61 (6), 942–947. https://doi.org/10.1016/j.neuint.2012.05.022. Epub 2012 Jun 5.

Vanhanen, J., Nuutinen, S., Lintunen, M., Mäki, T., Rämö, J., Karlstedt, K., Panula, P., 2013. Histamine is required for H3 receptor-mediated alcohol reward inhibition, but not for alcohol consumption or stimulation. Br. J. Pharmacol. 170 (1), 177–187. https://doi.org/10.1111/bph.12170.

Vizuete, M.L., Merino, M., Venero, J.L., Santiago, M., Cano, J., Machado, A., 2000. Histamine infusion induces a selective dopaminergic neuronal death along with an inflammatory reaction in rat substantia nigra. J. Neurochem. 75, 540–552.

Wahl, M., Unterberg, A., Baethmann, A., Schilling, L., 1988. Mediators of blood-brain barrier dysfunction and formation of vasogenic brain edema. J. Cereb. Blood Flow Metab. 8 (5), 621–634. https://doi.org/10.1038/jcbfm.1988.109.

Wanot, B., Jasikowska, K., Niewiadomska, E., Biskupek-Wanot, A., 2018. Cardiovascular effects of H3 histamine receptor inverse agonist/H4 histamine receptor agonist, clobenpropit, in hemorrhage-shocked rats. PLoS One 13 (8). https://doi.org/10.1371/journal.pone.0201519, e0201519. eCollection 2018.

Watanabe, T., Ohtsu, H., 2002. L-histidine decarboxylase as a probe in studies on histamine. Chem. Rec. 2 (6), 369–376. https://doi.org/10.1002/tcr.10036.

Watanabe, T., Taguchi, Y., Hayashi, H., Tanaka, J., Shiosaka, S., Tohyama, M., Kubota, H., Terano, Y., Wada, H., 1983. Evidence for the presence of a histaminergic neuron system in the rat brain: an immunohistochemical analysis. Neurosci. Lett. 39 (3), 249–254. https://doi.org/10.1016/0304-3940(83)90308-7.

Watanabe, T., Taguchi, Y., Shiosaka, S., Tanaka, J., Kubota, H., Terano, Y., Tohyama, M., Wada, H., 1984. Distribution of the histaminergic neuron system in the central nervous system of rats; a fluorescent immunohistochemical analysis with histidine decarboxylase as a marker. Brain Res. 295, 13–25.

Weihe, E., Depboylu, C., Schütz, B., Schäfer, M.K., Eiden, L.E., 2006. Three types of tyrosine hydroxylase-positive CNS neurons distinguished by dopa decarboxylase and VMAT2 co-expression. Cell. Mol. Neurobiol. 26 (4–6), 659–678. https://doi.org/10.1007/s10571-006-9053-9. Epub 2006 May 31.

Weinshilboum, R.M., Otterness, D.M., Szumlanski, C.L., 1999. Methylation pharmacogenetics: catechol O-methyltransferase, thiopurine methyltransferase, and histamine N-methyltransferase. Annu. Rev. Pharmacol. Toxicol. 39, 19–52. https://doi.org/10.1146/annurev.pharmtox.39.1.19.

Weisler, R.H., Pandina, G.J., Daly, E.J., Cooper, K., Gassmann-Mayer, C., 31001074-ATT2001 Study Investigators, 2012. Randomized clinical study of a histamine H3 receptor antagonist for the treatment of adults with attention-deficit hyperactivity disorder. CNS Drugs 26 (5), 421–434. https://doi.org/10.2165/11631990-000000000-00000.

White, T., 1959. Formation and catabolism of histamine in brain tissue in vitro. J. Physiol. 149 (1), 34–42. https://doi.org/10.1113/jphysiol.1959.sp006323.

White, T., 1961. Inhibition of the methylation of histamine in cat brain. J. Physiol. 159 (2), 191–197. https://doi.org/10.1113/jphysiol.1961.sp006801.

Whiten, D.R., Brownjohn, P.W., Moore, S., De, S., Strano, A., Zuo, Y., Haneklaus, M., Klenerman, D., Livesey, F.J., 2020. Tumour necrosis factor induces increased production of extracellular amyloid-beta- and alpha-synuclein-containing aggregates by human Alzheimer's disease neurons. Brain Commun. 2 (2). https://doi.org/10.1093/braincomms/fcaa146, fcaa146. eCollection 2020.

Wiklund, L., Basu, S., Miclescu, A., Wiklund, P., Ronquist, G., Sharma, H.S., 2007. Neuro- and cardioprotective effects of blockade of nitric oxide action by administration of methylene blue. Ann. N. Y. Acad. Sci. 1122, 231–244. https://doi.org/10.1196/annals.1403.016.

Witkin, J.M., Nelson, D.L., 2004. Selective histamine H3 receptor antagonists for treatment of cognitive deficiencies and other disorders of the central nervous system. Pharmacol. Ther. 103 (1), 1–20. https://doi.org/10.1016/j.pharmthera.2004.05.001.

Wolff, J.R., Schieweck, C., Emmenegger, H., Meier-Ruge, W., 1975. Cerebrovascular ultrastructural alterations after intra-arterial infusions of ouabain, scilla-glycosides, heparin and histamine. Acta Neuropathol. 31 (1), 45–58. https://doi.org/10.1007/BF00696886.

Wouterlood, F.G., Sauren, Y.M., Steinbusch, H.W., 1986. Histaminergic neurons in the rat brain: correlative immunocytochemistry, Golgi impregnation, and electron microscopy. J. Comp. Neurol. 252 (2), 227–244. https://doi.org/10.1002/cne.902520207.

Wouterlood, F.G., Steinbusch, H.W., Luiten, P.G., Bol, J.G., 1987. Projection from the prefrontal cortex to histaminergic cell groups in the posterior hypothalamic region of the rat. Anterograde tracing with Phaseolus vulgaris leucoagglutinin combined with immunocytochemistry of histidine decarboxylase. Brain Res. 406 (1–2), 330–336. https://doi.org/10.1016/0006-8993(87)90802-x.

Wouterlood, F.G., Gaykema, R.P., Steinbusch, H.W., Watanabe, T., Wada, H., 1988. The connections between the septum-diagonal band complex and histaminergic neurons in the posterior hypothalamus of the rat. Anterograde tracing with Phaseolus vulgaris-leucoagglutinin combined with immunocytochemistry of histidine decarboxylase. Neuroscience 26 (3), 827–845. https://doi.org/10.1016/0306-4522(88)90103-0.

Yamada, Y., Yoshikawa, T., Naganuma, F., Kikkawa, T., Osumi, N., Yanai, K., 2020. Chronic brain histamine depletion in adult mice induced depression-like behaviours and impaired sleep-wake cycle. Neuropharmacology 175, 108179. https://doi.org/10.1016/j.neuropharm.2020.108179. Epub 2020 Jun 6.

Yang, P., Perlmutter, J.S., Benzinger, T.L.S., Morris, J.C., Xu, J., 2020. Dopamine D3 receptor: a neglected participant in Parkinson disease pathogenesis and treatment? Ageing Res. Rev. 57, 100994. https://doi.org/10.1016/j.arr.2019.100994. Epub 2019 Nov 22.

Yokoyama, H., Onodera, K., Maeyama, K., Sakurai, E., Iinuma, K., Leurs, R., Timmerman, H., Watanabe, T., 1994. Clobenpropit (VUF-9153), a new histamine H3 receptor antagonist, inhibits electrically induced convulsions in mice. Eur. J. Pharmacol. 260 (1), 23–28. https://doi.org/10.1016/0014-2999(94)90005-1.

Yoshikawa, T., Nakamura, T., Yanai, K., 2019. Histamine N-methyltransferase in the brain. Int. J. Mol. Sci. 20 (3), 737. https://doi.org/10.3390/ijms20030737.

Zhou, F.W., Xu, J.J., Zhao, Y., LeDoux, M.S., Zhou, F.M., 2006. Opposite functions of histamine H1 and H2 receptors and H3 receptor in substantia nigra pars reticulata. J. Neurophysiol. 96 (3), 1581–1591. https://doi.org/10.1152/jn.00148.2006. Epub 2006 May 31.

Zhou, P., Homberg, J.R., Fang, Q., Wang, J., Li, W., Meng, X., Shen, J., Luan, Y., Liao, P., Swaab, D.F., Shan, L., Liu, C., 2019. Histamine-4 receptor antagonist JNJ7777120 inhibits pro-inflammatory microglia and prevents the progression of Parkinson-like pathology and behaviour in a rat model. Brain Behav. Immun. 76, 61–73. https://doi.org/10.1016/j.bbi.2018.11.006. Epub 2018 Nov 5.

CHAPTER

2

Ultra early molecular biologic diagnosis of malignant and neurodegenerative diseases by the immunospecific profiles of the proteins markers of the surface of the mobilized autologous hematopoietic stem cells

Andrey S. Bryukhovetskiy[a,*], Lyudmila Y. Grivtsova[b], and Hari Shanker Sharma[c,*]

[a]*ZAO NeuroVita Clinic of the Interventional Restorative Neurology and Therapy, Moscow, Russia*

[b]*A. Tsyb Medical Radiological Center of the FGBU National Medical Research Radiological Center of the Ministry of Health of the Russian Federation, Obninsk, Russia*

[c]*International Experimental Central Nervous System Injury & Repair (IECNSIR), Department of Surgical Sciences, Anesthesiology & Intensive Care Medicine, Uppsala University Hospital, Uppsala University, Uppsala, Sweden*

**Corresponding authors: Tel.: +46-70-20-11-801; Fax: +46-18-24-38-99 (Hari Shanker Sharma); Tel.: +7-495-324-9339; Fax: +7-495-980-1373 (Andrey S. Bryukhovetskiy), e-mail address: sharma@surgsci.uu.se; neurovitaclinic@gmail.com*

Abstract

The comparative study of the protein markers of the membrane surface of the autologous hematopoietic stem cells (CD34+ CD45+) was performed to detect the patterns of the proteomic profiles, if any. The study included five groups. The subpopulations of the hematopoietic stem cells were studied in 569 samples of the mobilized peripheral blood mononuclear cells from adult cancer cases (group 1), 557 samples from children cancer cases (group 2), 66 samples from amyotrophic lateral sclerosis cases (group 3), 5 samples from other neurodegenerative diseases cases (supplementary group 4) and 61 samples isolated from healthy donors (control group 5). The protein markers of the membrane surface of the autologous hematopoietic stem

Progress in Brain Research, Volume 266, ISSN 0079-6123, https://doi.org/10.1016/bs.pbr.2021.06.002
Copyright © 2021 Elsevier B.V. All rights reserved.

cells were mapped and profiled with flow cytometry. The specific patterns of the proteomic profiles characteristic for cancer and neurodegenerative disease and different from those of healthy donors were identified. The authors suppose that as far as the HSC is the parent cell of the immune system cells, the modification of its proteomic profile speaks of specific immune insufficiency that not only accompanies the disease but presents the key cause for its onset and precedes its clinical manifestation. Thus, the evaluation of the protein markers of the membrane surface of hematopoietic stem cells can be used for ultra early diagnosis of these diseases.

Keywords

Early diagnosis of cancer, Amyotrophic lateral sclerosis, Protein markers of membrane surface, Hematopoietic stem cells, Proteome, Proteome mapping and profiling, Alzheimer's disease, Parkinson's disease

1 Introduction

Despite significant progress of the global medicine concerning many of the civilization diseases (CDs), there exist a group of incurable diseases that are very hard to diagnose at an early stage. Here belong malignant diseases (MDs) and neurodegenerative diseases (NDDs). Some types of MDs, such as lung cancer, liver cancer, pancreatic cancer as well as other MDs, like hemoblastosis, glioblastoma multiforme, melanoma and other appeared to be incurable and fatal (Miller et al., 2019). There are no methods of early molecular diagnostics of most of the NDDs, and the methods to avoid rapid lethal outcome are absent. It can be best illustrated with the example of amyotrophic lateral sclerosis (ALS) which today is often called the motor neuron disease (MND). The life span of the patients with established diagnosis of incurable MDs varies from 1 to 3 years; about 10–15% of the patients live up to 5 years and rare live more (Chiò et al., 2009). The life span of the patients with the diagnosed ALS varies from 2 to 3 years and none of contemporary therapies is effective and is able to prevent lethal outcome. These diagnoses mean death sentence with no clemency as so far the researchers have not established the real causes and pathogenetic mechanisms of these disease and thus no effective therapy has been found.

If we imagine a line where all known mortal diseases are distributed, the MDs would be located on one side, while the NDDs on the opposite. These two polar points are two opposites of the human mortal diseases. ALS, Alzheimer's disease (AD) and Parkinson's disease (PD) are accompanied by massive neural degeneration, hypotrophy and atrophy the muscles and development of the "minus tissue" syndrome. On the contrary, the malignant diseases are accompanied by proliferation, tissue swelling, growing, migration of the cells and by the "plus-tissue" syndrome. So we have supposed that comparative analysis of these two opposite phenomena might give the key to the early diagnostics of them and other low curable NDDs.

Early diagnostics of any incurable disease gives additional chances to prolong life time and even permit long-term survival due to the promptly started therapy.

Contemporary oncology can boast early molecular-biological diagnostics of cancer based on the detection of cancer-specific molecules also known as tumor markers in such biological fluids of the patients as blood, urine, cerebrospinal fluid (CSF) and other. Tumor markers permit detection of malignancies at stage 1 or 2 and after adequate therapy provide for 5-year survival. It has been shown that the survival rates in cancer cases increase two or three times thanks to early diagnostics of the cancer with tumor markers (Nagpal et al., 2016). However, contemporary criteria of early diagnosis of cancer permits diagnosing the carcinogenic process when it has already acquired tissue specificity and definite localization in the organ, and in some cases this might be late as the cancer gives multiple metastases while having minimal primary focus by high grade malignancy. Detection of the tumor markers implies presence of tissue specific molecules of the tumor in the body, induces large scale screening and monitoring of the disease during the therapy. High levels of any tumor markers require additional tests and without them the diagnosis is unauthorized (Nagpal et al., 2016).

Such NDD, as ALS, is diagnosed only by the clinical picture and by electroneuromyographic test when it is already impossible to save the patient and only from 10 to 15% of motor neurons of the brain and spinal cord remain intact (Zavalishina, 2009). There is no early diagnosis and no effective therapy of ALS. The diagnosis of ALS by the neurofilament molecules is only possible at the stage of massive degeneration of motor neurons (Zavalishina, 2009). It is possible to verify AD through tau proteins or PD through Levi bodies only when the cortical neurons or striatal system have been severely destroyed by that time (Yakhno, 2005). There have been detected some early tumor for cancer diagnosis but there are no early markers for ALS, PD or AD and accordingly, the prognosis in ALS diagnosis sounds even more tragic than in cancer. Certain types of cancer are quite treatable, while the AD, PD and ALS are not. Death in cancer usually is less tortuous as it is accompanied by loss of consciousness and coma induced by cancer intoxication, while the ALS patient dies from paralysis of breath with continuing heartbeats and in full awareness of what is going on.

The inefficiency of the diagnosis and treatment of these diseases seems to be explained by the absence of understanding of the fundamental molecular-biological mechanisms of the onset and progress of these diseases. Simultaneously, we know a lot about the general molecular processes in the specialized cells in these diseases. The NDDs and cancer are united by the accumulation of the pathological proteins in highly differentiated cells that are involved into pathological process. These are cancer-specific proteins (CSP) in cancer, and tau-proteins, Lewy bodies, SOD1-proteins and other proteins in NDDs. It is typical both for cancer and for NDDs. When we consider ALS to be the diseases of motor neuron, we deprive the patients of the chance for cure as the existing paradigm permits diagnosis of the disease only at its final stage. It would be sensible to detect the system-forming pattern that triggers these diseases and then we are able to diagnose them and to cure.

The role of the immune system in the development of cancer, other malignant disorders and NDDs is obvious, and the pathology of HSCs is a system-forming link in common of these polar CDs. It is known that a selective functional insufficiency of the immune system (IIS) is common to these lethal diseases and it is of key importance for pathogenesis and pathomorphology of these diseases. So, the immune system of the patients remains tolerant to the tumor growth, huge primary tumor nodes and metastases in many advanced malignancies. The tumor cells easily avoid the elimination by natural killers of the innate immunity and cytotoxic T-cells of the acquired immunity, and these mechanisms are described in detail (Sharma et al., 2017). Meanwhile, only anti-tumor function of the immunity disorders in cancer, while other functions, such as anti-microbial, anti-viral, anti-fungal remain intact. The indicants of immune status of cancer patients correspond with the normal levels or close to them. The situation with the immune system in the NDDs is equal in magnitude but opposite in sign to the situation with the immune system in cancer. Although, the NDDs have been long considered autoimmune disorders, no quantitative changes in the cellular and humoral immunity have been detected in ALS and AD. Moreover, recently, the autoimmune genesis of the ALS is considered unreasoned because the effect from the standard therapy for the autoimmune diseases is not observed in the ALS: (1) no remission after glucocorticoids; (2) no effects from the transplantation of autologous bone marrow as in multiple sclerosis or other autoimmune diseases of connective tissue; (3) no results from using the interleukin blockers and other medications aiming to reduce the systemic inflammation and tissue destruction (Yakhno, 2005; Yeghorkina and Gaponov, 2007; Zavalishina, 2009). These arguments have been considered decisive to exclude the autoimmune genesis in most of the NDDs.

Uniformity of the type and nonspecific polarity of the immunological reactions of the organism let us suppose that the pathogenetic mechanisms of their development are similar with minimal variations. Comparison of cancer and NDDs clearly shows that these diseases have confirmed multigenetic deficiency and statistically significant epigenetic, transcriptomic and proteomic defects that permitted assumption that their molecular-biological defect is based on the multi-vector damage of the proteome of the HSCs, as the originators of all cells of hematopoiesis and immunity. The HSCs are a low count but heterogeneous population that integrates several types of subpopulations of the cells that differ by the levels of differentiation and ability to proliferate. The non-differentiated and almost non-dividing stem cells (SCs) as well as committed (with limited differentiation direction) precursor cells are found there. Normally, the concentration of SCs in peripheral blood is very low, less than 0.01% and it prevents from their examination even with the most sensitive instruments (Grivtsova and Tupitsyn, 2016). However, the HSCs have the longest cell cycle of 360 days among all cells of a human body and there represent the main regulatory and controlling system in the available hierarchy of all cell systems of the organism. They first react to the mutations in the genes and onset of aseptic inflammation in the pathological organs and tissues, migrate there to the gradient of inflammation concentration from the bone marrow, adhere to the pathological cells and induce either differentiation or apoptosis through bystander effect. The HSCs perform vertical and horizontal information exchange of the cytoplasmic proteins with the pathological cells (Milkina et al., 2016). Depending on the degree of injury of the target cell, the epigenomic-proteomic damage of HSCs is launched and the cancer stem cells (CSCs), or the cells that have been immunized with pathospecific proteins, start developing from them (Bryukhovetskiy, 2019). Hence, the goal of this work is to study the proteomic markers of the HSCs surface in such diseases as malignant disorders and ALS (as the NDD model) to detect the immunospecific characteristics of molecular landscape of the HSCs surface as the fundamental molecular-biological markers for early diagnostics of these incurable diseases.

2 Materials and methods

Group 1 included 167 adult cancer cases. The cases of oncohematological diseases prevailed in the group and totaled 158, i.e., 94.7%: Hodgkin lymphoma—58 cases (34.7%), non-Hodgkin lymphoma (NHL) (diffuse large B-cell lymphoma, precursor B-cell lymphoblastic lymphoma and 2 cases of Burkett lymphoma)—57 cases (34.1); 40 cases of multiple myeloma (23.9%) and 3 cases of acute myeloid leukemia (AML). Non-hematopoietic tumors proportion was small, it included only 9 cases (5.3%): 3 cases of breast cancer, 2 cases of small cell lung cancer, 1 case of testicular germ cell tumor, 1 case of testicular non-germ cell tumor and 2 cases of Ewing's sarcoma. Average age was 33 years (median of 32 years, from 16 to 64). The weight varied from 43–113kg. The group included 113 males and 54 females (32.2%).

Group 2 included 263 children cancer cases. The group included 48 cases of relapsed or chemotherapy resistant neuroblastoma, 41 case of other soft-tissue sarcomas (14—rhabdomyosarcoma, 7—synovial sarcoma, 20—retinoblastoma of high risk), 24 cases of primitive neuroectodermal tumor, 63 cases of high risk Ewing sarcoma, 22 cases of high risk acute myeloid leukemia (AML), 16 cases of Hodgkin lymphoma, 6 cases of non-Hodgkin lymphoma, 11 cases of the tumors of central nervous system (glioma, atypical teratoid rhabdoid tumor, ependymoma), 28 cases of medulloblastoma, 1 case of metastatic osteosarcoma and 3 cases of germ cell tumors. Average age is 7.88 ± 0.28 years (7 years median, from 0.6 to 19 years), average weight 29.3 ± 0.98 (median 23.0, from 7.0 to 87.0) kg. The group included 123 females and 140 males.

Group 3 included 60 ALS cases, three of which had a family type ALS and 57 had a sporadic type. Average age was 56 years (median 54 years, from 42 to 68 years). The number of males was 45, the females—15. Average weight of ALS patients was 74kg in males and 56 in females.

Group 4 was supplementary and included five cases of other NDDs: three cases of AD and two cases of PD. The group included two females with AD and one male with AD. Average age was 72.6. The AD was diagnosed in the psychiatric hospital. Two other females had akinetic-rigid form of PD with systemic cortical atrophy; aged 66 and 69.

Group 5 included 50 healthy donors ALS-free: average age of healthy donors was 37.6 (from 26 to 44); there were 31 males (62%), and 19 females (38%). Average weight of male donors was 86kg, average weight of female donors was 61.6.

The subpopulations of the HSCs were studied in 569 samples of the mobilized peripheral blood mononuclear cells (PBMCs) from adult cancer cases, 557 samples of PBMCs of children cancer cases, 61 samples of healthy donors, 66 samples of ALS cases and 5 samples of other NDD cases.

The granulocyte colony stimulating factor (G-CSF) (Neupogen, Amgen Europe B.V., the Netherlands) was administered for 4 days, the PBMCs were harvested by the standard procedure (Baxter, Cobe Spectra, USA). Two to three milliliters of the PBMCs were taken for the examination, or, alternatively, the same volume of the cryopreserved PBMCs stored in the bank of Blokhin Cancer Research Center were thawed and examined. Granulocyte-macrophage colony-stimulating factor (GM-CSF) was given in four cases of children cancer.

The preceding treatment varied from 2 to 22 courses of chemotherapy in adult cancer cases. The chemotherapeutic regimens varied depending on the diagnosis. DEXA-BEAM and BEACOPP schemes were used in the cases of Hodgkin lymphoma and some cases of non-Hodgkin lymphoma. R-CHOP scheme was used in some cases of B-cell lymphoma; VAD regimen was given for multiple myeloma and CAF for breast cancer.

In children cancer cases, 95% (420 of 442) of leukocytapheresis was conducted after mobilization regimen that included chemotherapy with consequent administration.

By the beginning of leukapheresis, positive response (full or partial remission) was achieved in all cancer cases and no tumor metastases had been detected in the bone marrow samples.

The granulocyte colony stimulating factor (G-CSF) (Neupogen, Amgen Europe B.V., the Netherlands) was administered for 4 days, the PBMCs were harvested by the standard procedure (Baxter, Cobe Spectra, USA). Two to three milliliters of the PBMCs were taken for the examination, or, the same volume of the cryopreserved PBMCs stored in the bank of Blokhin Cancer Research Center was thawed and examined. Alternatively, the donors underwent sternal puncture under local anesthesia with 0.5–1% of novocaine or lidocaine and 2–3mL of bone marrow was isolated.

All donors tolerated the G-CSF stimulation and sternal punctures satisfactorily.

Membrane phenotype of HSCs was evaluated by the multiparameter flow cytometry on the basis of detection of CD38, CD71, CD90, CD56, CD19, CD61, CD117, CD10 and CD2 antigens (Table 1) within CD34+ pool and the expression of the main markers of the cell surface in various HSCs populations was profiled.

Further, the scope of the studied protein markers of the membrane surface (PMMS) of HSCs was expanded. In case of clinical and neurophysiologic signs of ALS, the PMMS of HSCs were profiled and the specific molecular-biologic

Table 1 The antigens and monoclonal antibodies used to detect the expression of the HSC surface markers.

Antigen (differentiation cluster)	Clone/class	Manufacturer
Gp105-120 (CD34)	HPCA-2/8G12/IgG_1 k	Becton Dickinson, USA
Leukocyte common antigen (CD45)	HI30/IgG_1 k	Becton Dickinson, USA
ICAM-3 (CD50)	TU41/IgG^2b k	Immunotech, Coulter, France
HLA-DR	L243IgG_{2a}, k	Immunotech, Coulter, France Becton Dickinson, USA
CD38	HLT2, HB7/IgG_1 k	Becton Dickinson, USA
Thy-1 (CD90)	5E10/IgG_1 k	Immunotech, Coulter, France
c-kit (CD117)	YB5.B8/IgG_1, k	Becton Dickinson, USA
N-CAM (CD56)	B159/IgG_1 k	Becton Dickinson, USA
Transferrin receptor CD71	M-A712/IgG_{2a}, k	Immunotech, Coulter, France Becton Dickinson, USA
CD13	L138/IgG_1, k	Becton Dickinson, USA
CD33	P67.6 IgG_1, k	Becton Dickinson, USA Immunotech, Coulter, France
Integrinβ_3 chain (CD61)	RUU-PL7F12/IgG	Becton Dickinson, USA
CD19	4G7	Immunotech, Coulter, France
CD7	M-T701/IgG_1 k	Immunotech, Coulter, France
CD57	HNK-1/IgM	Becton Dickinson, USA
LFA-2	RPA-2.10/IgG_1	Becton Dickinson, USA
CD10	HI10a/IgG_{2a}	Becton Dickinson, USA
Glycophorin A (CD236a)	GA-R2 (HIR2)/IgG_{2b}, k	Immunotech, Coulter, France

modifications of the profile of membrane proteins and defined the detected changes of the PMMS of HSCs as the neurospecific profile of the molecular landscape of the HSCs surface in ALS.

Further, the expression of 16 standard protein membrane markers of HSCs (CD34+CD45+HLA DR+) were evaluated with FACScan flow cytometer using standard antibodies (Table 1). The expression of the main markers of the HSC surface (CD38, CD71, CD90, CD56, CD19, CD61, CD 117, CD10 and CD2) was profiled. When the acquired profile of the HSC membrane proteins coincided by 6–9 parameters with the cancer-specific profile of expression of the HSC membrane markers, the conclusion of the ongoing malignant process and detected chronic insufficiency of the anti-tumor function of the immune system was made. In the case of clinical and neurophysiological signs of ALS and NDDs, the specific molecular-biological changes of the membrane proteins profile were interpreted as the neuro-specific profile of the molecular landscape of HSC surface.

The data were processed with version 17 of the SPSS software for Windows. We used the frequency function to describe the consecutive measurements of variables, mean observations, median, standard error of the means, the variance of variable values; the function of bivariate correlation to determine the coefficient of correlation of two independent variables, correlation is accurate at $P \leq 0.05$; the function of the contrast of means of independent and dependent variables, confidence interval 95% and over; one-hot encoding according to the limit of the selected variable.

3 Results

The membrane expression of antigens associated with the stem hematopoietic cells at different stages of differentiation have been evaluated during the study of 1258 samples of the mononuclear fraction of blood of the stimulated peripheral blood or bone marrow samples.

Table 2 shows the main changes of HSC markers expression in 1187 samples of cancer patients and healthy donors. Same data are presented in the diagram at Fig. 1 that demonstrates that the intercellular information protein exchange between the CCs and HSCs significantly alters the proteomic composition of the expressed membranes of HSCs in cancer cases. The comparative analysis of the HSC surface markers in healthy donors and cancer cases (children and adults) shows downregulation such cell surface markers as CD38, CD71, CD90, CD56, CD19, CD61 twofold and more in adult cancer cases. Also, these cases demonstrate high concentration of CD117 and newly emerged cell surface markers CD10 and CD2 in HSCs that normally have not ever been found in healthy donors.

Table 3. shows the main fundamental changes in the expression of HSC surface markers in the 1187 cell samples of cancer cases. 54 samples from healthy donors and 66 samples from ALS cases.

In the children's cancer cases we observed the general tendency for the same protein cell surface markers (CD38, CD71, CD90, CD117, CD56, CD19, CD61) levels to decrease along with the record of the previously unknown new cell surface markers, such as CD2, CD10. The CD45 marker is significantly upregulated.

As it can be seen at the diagrams (Figs. 2 and 3) all ALS cases exhibit significant downregulation and almost complete absence of CD38 marker; the CD1217 and CD71 are downregulated, while the HSCs of the cancer cases demonstrate significant downregulation of the CD33, CD71 and CD90 (Thy-1). ALS is a proteomic disease of the HSCs when its progeny, the pathological immune cells, launches local inflammation in the lateral columns of the spinal cord inducing the degenerative process. In the case of cancer, the cells of the immune system do not control growth and

Table 2 The CD34+ subpopulation of PBMCs (percentage of antigen positive cells within HSCs pool) in cancer cases, ALS and NDD cases healthy donors.

Antigen	Group	Average ± average error	Median	Deviation	N	*P*
CD45	Children	79.2 ± 1.2	88.5	1.1–100.0	392	0.049
	Adults	74.1 ± 1.3	83.0	1.2–100.0	390	
	Healthy donors	70.4 ± 3.4	77.9	1.5–100.0	54	
HLA-DR	Children	93.8 ± 0.51	95.5	75.7–100.0	128	0.087
	Adults	93.7 ± 0.6	95.4	80.7–100.0	77	
	Healthy donors	92.9 ± 1.1	94.0	74.4–99.9	29	
CD38	Children	58.1 ± 2.63	54.2	0.5–100.0	113	0.001
	Adults	62.9 ± 3.2	66.5	5.1–99.3	71	
	Healthy donors	65.0 ± 4.6	71.5	22.2–97.6	32	
CD33	Children	64.3 ± 3.05	73.2	0.9–99.8	101	0.0001
	Adults	43.7 ± 4.4	34.4	0.4–99.6	63	
	Healthy donors	67.6 ± 4.7	75.5	7.1–99.0	24	
CD13	Children	85.7 ± 1.97	92.95	1.0–100.0	104	0.009
	Adults	80.0 ± 2.9	91.7	1.6–99.4	69	
	Healthy donors	90.3 ± 1.9	92.4	50.0–99.5	28	
CD71	Children	49.1 ± 2.11	45.7	8.8–99.2	121	0.0001
	Adults	35.7 ± 1.9	36.2	4.0–74.7	78	
	Healthy donors	78.7 ± 4.9	88.9	9.8–98.4	25	
CD117	Children	69.0 ± 3.35	75.4	13.1–97.5	48	0.003
	Adults	83.1 ± 3.1	89.8	37.7–99.1	25	
	Healthy donors	79.6 ± 3.9	82.4	48.4–97.5	16	
CD90 (Thy-1)	Children	26.4 ± 3.26	10.4	0.1–99.8	91	0.02
	Adults	21.8 ± 2.8	16.5	0.0–85.6	54	
	Healthy donors	41.9 ± 6.8	28.4	0.4–92.4	26	
CD50	Children	98.1 ± 0.34	99.0	84.9–100.0	63	0.2
	Adults	98.9 ± 0.3	99.4	89.2–100.0	36	
	Healthy donors	98.1 ± 0.6	98.9	92.4–100.0	18	
CD56	Children	3.24 ± 1.09	1.2	0.0–60.0	60	0.03
	Adults	0.8 ± 0.3	0.0	0.0–10.2	37	
	Healthy donors	28.7 ± 12.8	5.4	0–92.6	9	
CD19	Children	3.01 ± 0.56	1.4	0.0–36.5	89	0.01
	Adults	2.1 ± 0.4	0.0	0.0–16.0	60	
	Healthy donors	15.4 ± 4.4	6.3	0–86.1	26	
CD61	Children	2.7 ± 0.3	1.8	0.0–10.7	56	0.006
	Adults	3.9 ± 0.7	2.4	0.0–15.8	29	
	Healthy donors	10.6 ± 2.3	6.4	2.7–28.0	13	
CD7	Children	3.5 ± 0.4	2.3	0.0–31.0	109	0.06
	Adults	4.1 ± 0.8	2.5	0.0–41.4	57	
	Healthy donors	11.2 ± 3.8	4.5	0–98.1	28	
CD10	Children	2.11 ± 0.46	1.1	0.0–13.1	34	
	Adults	0.5 ± 0.1	0.4	0.1–1.4	9	
CD2	Children	3.24 ± 0.93	1.7	0.4–32.4	34	0.01
	Adults	1.4 ± 0.4	1.1	0.4–3.6	7	

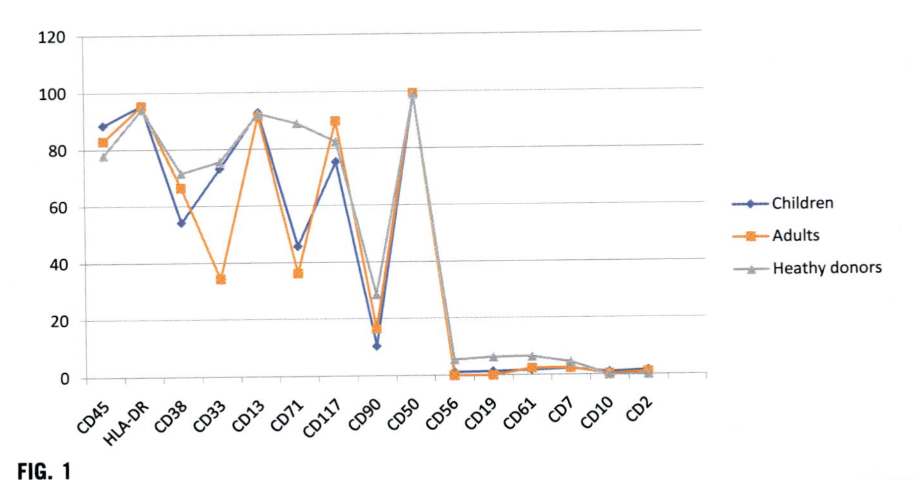

FIG. 1

The proteomic profiles of the cells surface markers of hematopoietic stem cells in cancer cases (orange line—adults, blue line—children) and healthy donors (gray line).

proliferation of the tumor cells and this results in the uncontrollable tumor growth. In both cases, the specific changes in the profiles of expression of the HSCs membrane markers are obvious.

We profiled the markers of the proteins of the HSCs membrane surface for family ALS and detected the same pattern as in sporadic ALS. All cases of the family ALS showed immunospecific features of the antigens of HSCs surface and the difference between the ALS patient and their healthy relatives are minimal. The immunophenotypic profiles of the HSCs of a healthy donor, who is a sibling of the ALS patient and the ALS patients demonstrate slight dissimilation in the antigens. The healthy sibling shows rather high levels of CD117+ CD33+ CD28+CD300+CD11b +CD123+ markers. The difference in the levels of CD38+ precursors was detected, but both the healthy donor and the ALS patients had weak expression of the antigen. The HSC population that prevails in healthy donor implies monocytic differentiation of the precursor cells. In other words, the diagram of the immunophenotypic profile can point to the diagnosis and predict the expected manifestation of the disease long before the clinical manifestation of the family ALS. The analysis of the immunophenotypic profile of the HSCs of sporadic ALS cases detected definitive immunospecific changes of the proteomic profile of the antigens of HSCs surface. The sporadic ALS cases demonstrated higher level of CD117+ CD33+ CD28+CD300+CD11b +CD123+ precursor cells which is different from healthy donors. The diagram Fig. 4 shows the features of the membrane markers profile in the patients with AD and PD as compared to the profiles of other studied cases.

Table 3 Subpopulations of CD34+ cells of leukapheresis product (percentage of AG^+ cells in the pool of HSCs) in the patients with ALS, malignant diseases and healthy donors.

Antigen	Group	Average ± average error	Median	Deviation	N	*P*
CD45	ALS	89.2 ± 1.4	87.5	80.3–100.0	60	0.05
	Malignant disorders	74.1 ± 1.3	83.0	1.2–100.0	390	
	Healthy donors	70.4 ± 3.4	77.9	1.5–100.0	54	
HLA-DR	ALS	95.1 ± 0.31	95.0	85.7–98.0	60	0.001
	Malignant disorders	93.7 ± 0.6	95.4	80.7–100.0	390	
	Healthy donors	92.9 ± 1.1	94.0	74.4–99.9	29	
CD38	ALS	20.1 ± 2.63	20.0	14.5–26.0	60	0.001
	Malignant disorders	62.9 ± 3.2	66.5	5.1–99.3	71	
	Healthy donors	65.0 ± 4.6	71.5	22.2–97.6	32	
CD33	ALS	24.3 ± 3.05	25.0	13.6–44.8	15	0.0001
	Malignant disorders	43.7 ± 4.4	25.4	0.4–99.6	63	
	Healthy donors	67.6 ± 4.7	75.5	7.1–99.0	24	
CD13	ALS	78.7 ± 1.97	78.1	1.0–100.0	60	0.009
	Malignant disorders	80.0 ± 2.9	91.7	1.6–99.4	69	
	Healthy donors	90.3 ± 1.9	92.4	50.0–99.5	28	
CD71	ALS	16.1 ± 1.12	16.0	10.6–44.3	58	0.0001
	Malignant disorders	35.7 ± 1.9	36.2	4.0–74.7	78	
	Healthy donors	78.7 ± 4.9	88.9	9.8–98.4	25	
CD117	ALS	69.0 ± 3.35	24.1	13.1–97.5	60	0.003
	Malignant disorders	83.1 ± 3.1	89.8	37.7–99.1	25	
	Healthy donors	79.6 ± 3.9	82.4	48.4–97.5	16	
CD90 (Thy-1)	ALS	26.4 ± 3.26	24.0	0.1–49.5	60	0.02
	Malignant disorders	21.8 ± 2.8	16.5	0.0–85.6	54	
	Healthy donors	41.9 ± 6.8	28.4	0.4–92.4	26	
CD50	ALS	98.1 ± 0.34	99.0	84.9–100.0	55	0.001
	Malignant disorders	98.9 ± 0.3	99.4	89.2–100.0	36	
	Healthy donors	98.1 ± 0.6	98.9	92.4–100.0	18	
CD56	ALS	0.6 ± 1.09	0.6	0.0–5.0	55	0.01
	Malignant disorders	0.8 ± 0.3	0.0	0.0–10.2	37	
	Healthy donors	28.7 ± 12.8	5.4	0–92.6	9	
CD19	ALS	3.01 ± 0.56	2.5	0.4–6.2	60	0.01
	Malignant disorders	2.1 ± 0.4	0.0	0.0–16.0	60	
	Healthy donors	15.4 ± 4.4	6.3	0–86.1	26	
CD61	ALS	2.47 ± 0.3	24.5	0.0–40.3	60	0.005
	Malignant disorders	3.9 ± 0.7	2.4	0.0–15.8	29	
	Healthy donors	10.6 ± 2.3	6.4	2.7–28.0	13	
CD7	ALS	1.5 ± 0.4	2.3	0.0–4.0	60	0.05
	Malignant disorders	4.1 ± 0.8	2.5	0.0–41.4	57	
	Healthy donors	11.2 ± 3.8	4.5	0–98.1	28	

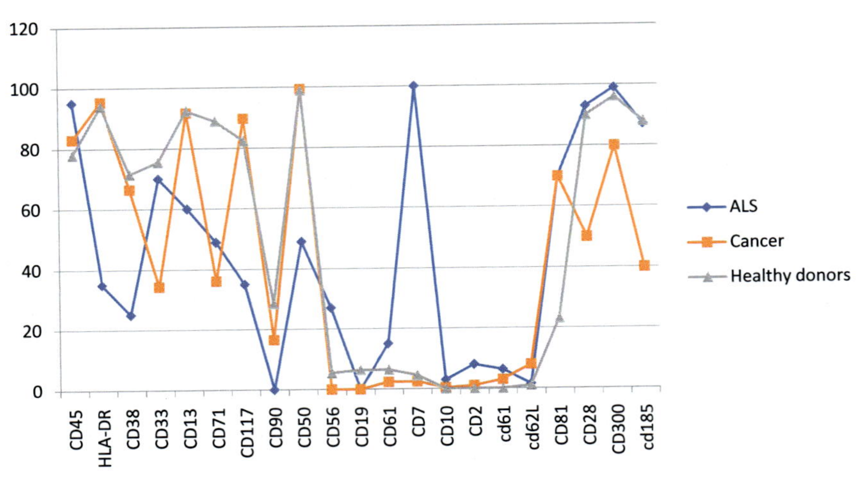

FIG. 2

The proteomic profiles of the cells surface markers of hematopoietic stem cells in 50 healthy donors (gray line), in 150 cancer cases (orange line) and 15 ALS cases (blue line).

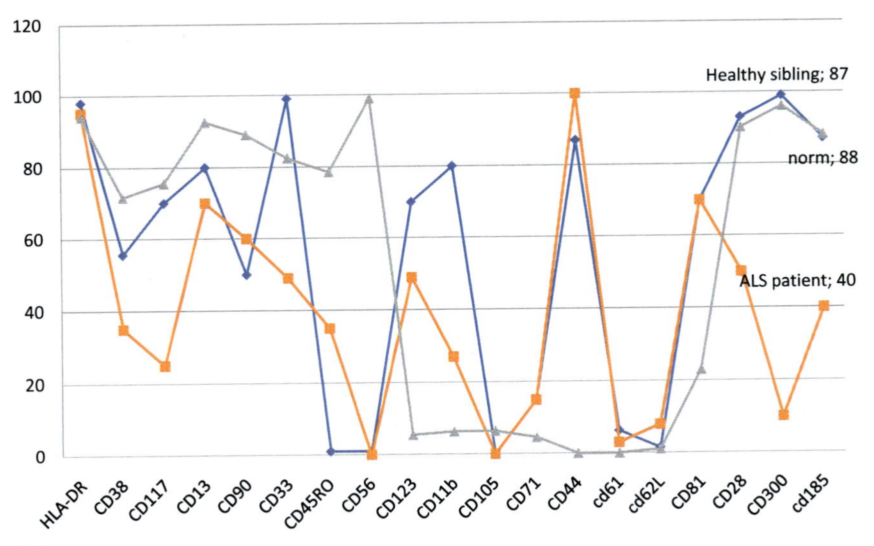

FIG. 3

The immunophenotypic profiles of the cells surface markers of hematopoietic stem cells in family ALS (blue line—healthy donor, sibling of ALS patient; red line—ALS patient, gray line—norm for healthy donors). The deviations of protein profile of the HSC surface markers in family ALS are obvious.

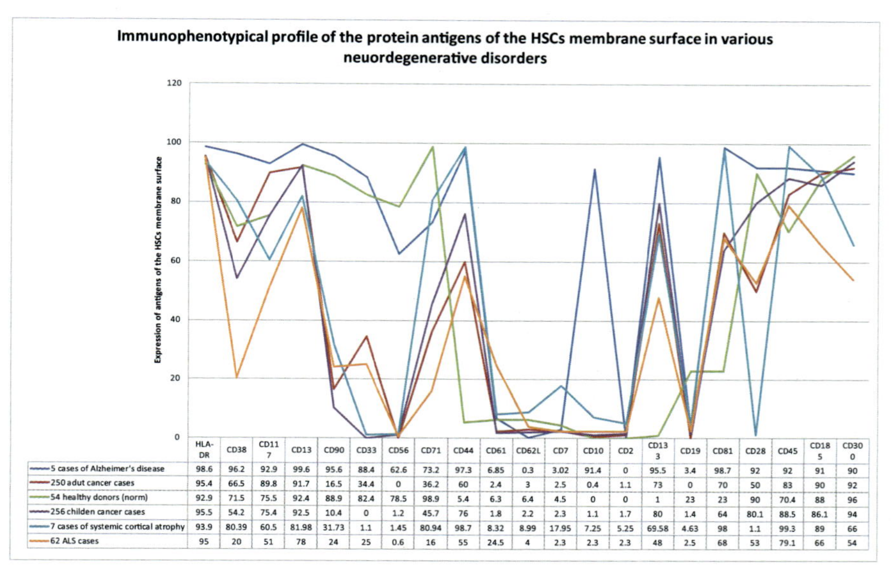

	HLA-DR	CD38	CD117	CD13	CD90	CD33	CD56	CD71	CD44	CD61	CD62L	CD7	CD10	CD2	CD133	CD19	CD81	CD28	CD45	CD185	CD300
5 cases of Alzheimer's disease	98.6	96.2	92.9	99.6	95.6	88.4	62.6	73.2	97.3	6.85	0.3	3.02	91.4	0	95.5	3.4	98.7	92	92	91	90
250 adut cancer cases	95.4	66.5	89.8	91.7	16.5	34.4	0	36.2	60	2.4	3	2.5	0.4	1.1	73	0	70	50	83	90	92
54 healthy donors (norm)	92.9	71.5	75.5	92.4	88.9	82.4	78.5	98.9	5.4	6.3	6.4	4.5	0	0	1	23	23	90	70.4	88	96
256 childen cancer cases	95.5	54.2	75.4	92.5	10.4	0	1.2	45.7	76	1.8	2.2	2.3	1.1	1.7	80	1.4	64	80.1	88.5	86.1	94
7 cases of systemic cortical atrophy	93.9	80.39	60.5	81.98	31.73	1.1	1.45	80.94	98.7	8.32	8.99	17.95	7.25	5.25	69.58	4.63	98	1.1	99.3	89	66
62 ALS cases	95	20	51	78	24	25	0.6	16	55	24.5	4	2.3	2.3	2.3	48	2.5	68	53	79.1	66	54

FIG. 4

The averaged immunophenotypical profiles of the antigen expression of the HSCs surface: 5 cases of Alzheimer's disease, 7 cases of the systemic cortical atrophy in Parkinson's disease, 65 cases of ALS, 250 cases of cancer in adults, 265 cases of cancer in children and 54 healthy donors of bone marrow.

4 Discussion

The research was done in small samples; nevertheless, their size let us avoid the bias of the small numbers. The number of the ALS was not large but the results prompt significant conclusions. First, we would like to discuss the results of the cancer cases testing. Previously, our experiment showed that the cancer stem cells (CSCs) exhibit changes at the level of the proteins of the cell membranes, cytoplasmic and nuclear proteins (Bryukhovetskiy, 2019). However, the proteomic mapping of the HSCs and profiling of their proteins by mass-spectrometry to detect all ranges of cancer-specific changes is too expensive and takes a lot of time, thus being used only for the research purposes. However, it can be hardly used in clinical practice and become the tool for regular diagnosis. Practice demands a simple and safe system of the markers for ultra early diagnostics of cancer. Hence, the main goal of the method that we proposed is to develop a simple, safe and clear method of cancer diagnosis at an extremely early stage. The proposed method of ultra early diagnostics of malignant diseases and NDDs is protected by the Patent of the Russian Federation №2706 714 C1.

The key evidence that underlies the proposed ultra early diagnostics is cancer specific modifications of the expression profile of the protein markers of HSCs surface that have been detected in the cases of cancer as the earliest objective signs of cancer onset and anti-tumor insufficiency of the immune system and that can become the main early markers of cancer. The evidence is proven by our own significant data about the cancer-specific modification of the proteomic profile of the HSCs membrane surface in cancer cases that permit comparative analysis of the data of the expression of membrane proteins of HSCs in cancer cases and healthy donors. This research permitted detection of the specific set of fundamental differences in the expression of HSCs surface markers in children and adult cancer cases and healthy donors who had neither clinical nor laboratory signs of cancer.

The acquired data on the immune-specific proteomic changes in HSCs membrane protein profiles (MPP) of a cancer case speak of cancer-specific local change of the proteomic profile (oncoproteome) of MPP HSCs and selective modification of the part of the peptide chart of HSCs in cancer cases. Possibly, the selective weakness of the function of the immune system, such as to eliminate the cancer cells (CCs) leaving other functions of the immune system, such as anti-fungal, anti-infectious, anti-viral, intact. The immune system of the humans where all cells are the progeny of HSCs with altered cancer-specific MPP ceases to react to the large numbers of the CCs in the body. This forms anti-tumor insufficiency of the immune system of the cancer patient. It is underlain by qualitative molecular biological restructuring of the proteomic structure of the immunocompetent cells (ICCs) and not the qualitative deficit of the cells of the immune system.

The immune system is known to mobilize the ICCs of the acquired immunity (dendritic cells and cytotoxic cells) and the cells of the innate immunity (macrophages, NK-cells, NKT-cells, γδ T lymphocyte, B1 cells) and HSCs to combat the CCs. Normally, healthy ICCs detect CCs, express killer cytokines, kill CCs and continue circulation, patrolling and controlling the organism. In case the ICCs equipped with cytokines and microRNA are not able to eliminate large numbers of the CCs, they receive their dose of cytoplasmic CSPs in the horizontal and vertical protein exchange (Bryukhovetskiy, 2019). This information exchange results in activation or/and development of the focus of aseptic inflammation of the CCs in the tissue, while the ICCs become tolerant to the CSPs of this tumor. The HSCs and tissue specific stem cells (tsSCs) arrive to the tissue of the organ that was the source of the CCs following the concentration gradient of systemic inflammation The HSCs and/or tsSCs initiate apoptosis in the CCs and return to their niches in the tissues and bone marrow. Having arrived at the inflammation focus, the HSCs and tsSCs of the damaged organ exchange their cytoplasmic proteins with the CCs and the CSPs invade into their cytoplasm (Bryukhovetskiy, 2019). If the number of the HSCs and tsSCs is sufficient, no less than three HSCs per one CSC, they suppress the function of the CSCs. In case their number is insufficient for regulatory or effector apoptotic effect on the CCs and CSCs, the HSCs and tsSCs receive sort of "immunization" with the CSPs, return to their niches and begin to produce their progeny, the ICCs, that are tolerant to the tumor. This seems to be the main molecular-biological mechanism of the development of HSCs with cancer-specific proteomic profile. One of the variants is that the first meeting of the HSCs and CCs and their intercellular interaction results in the development of the tissue-specific tumor SC

(CSC) with the phenotype changed from normal tumor which is typical for the tumors of solid organs and brain. The second variant is that the HSCs proteome that has been replaced with oncoproteome can lead to the development of the tumor phenotype of HSCs and become the reason for hemoblastosis. The third and the most frequent variant of the modification of the HSCs and tsSCs proteome under the effect of the CSPs of the CCs is the development of the CSC from the tsSC (neural SC, mesenchymal stromal SC, liver SC and other) and partial change of the proteomic profile of the protein markers of HSC surface with the onset of the immunotolerance of all HSCs progeny (and, accordingly, of all cells of the immune system) to the CCs in the body. This fully complies with our research of the relations of HSCs with CCs of lung cancer, breast cancer and glioblastoma (Bryukhovetskiy, 2019).

Recently, the work of the team of Michael H. Sieweke (de Laval et al., 2020) showed that the proteomic changes in the HSCs are conditioned mostly by the epigenetic accumulations of the C/EBP b enhancer proteins around the genome DNA of these cells. The study establishes the central role of the blood HSCs in the immune memory and the role of these proteins in the epigenetic structural heredity. The modeling in the artificial bacteria showed that the HSCs with modified proteome could stimulate a quicker and more effective response if they had been previously affected by an infectious antigen. The researchers found that the memory of all infectious agents is stored as the kind of a bookmark of enhancer proteins opposite the genome DNA of the HSCs. These marks help generate an immediate immune response and generate the sufficient number of the specialized immune cells (macrophages, dendritic cells, cytotoxic lymphocytes). The first effect of the antigen leaves proteomic marks on the DNA of the stem cells, right around the genes that are important for the immune response. These marks on the DNA guarantee that the genes will be easily detected, accessed, and activated for quick reaction in case of repeated infection with this agent. The epigenetic memory has been shown to be inscribed near DNA and the enhancer protein appeared to be the key figure for emergency immune reactions. These proteins change all HSC proteome and, accordingly, all their regulatory and supervisory function. Previously, the dogma of immunology claimed the HSCs to be solely non-specific cells for blood cells proliferation that are blind for such external signals as infections. It has been considered that only specialized daughter immune cells are able to perceive these signals and activate the immune response. The conventional development of vaccines from viral and bacterial infections is built on the immunization of the peripheral immune cells and formation of the acquired immunity to the pathological antigen. A lot of immunologists stick to this point of view on the role of the HSCs in the organism till today. But the work of the laboratory of Michael H. Sieweke (de Laval et al., 2020) demonstrates that this dogmatic notion is erroneous and the HSCs are able to perceive the external factors to specifically reproduce the subtypes of the immune cells "on demand" to combat the infectious agents. That is, the HSCs possess evolutionary and acquirable immune memory that provide for improved reactions to the recurrent infectious agents, but every new agent modifies its proteomic structure that can be seen in the molecular landscape of the HSCs membranes.

When it became obvious that the markers of the cell surface of the HSCs of cancer cases reflected the state of the immune system of the patient and signified cancer we presumed that in ALS these MPP variations could be radically different from the similar indicants of HSCs in cancer and HSCs in health. In the light of the detected specificity of the information exchange between the CCs and HSCs that includes both horizontal and vertical exchange with the cytoplasmic proteins, we could not exclude similar information exchange between the neural cells with SOD-1 proteins neural cells and healthy HSCs or HSC with SOD-1 proteins and a healthy neuron, as well as between the lymphocytes of peripheral blood containing SOD-1 proteins in neurons. We came to the conclusion that mechanism of intercellular information exchanges that we described for mutant CCs and HSCs were universal and could fully reflect the process of the information interaction in ALS, AD, PD where the proteome modified $HSCs_{SOD-1}$ met the neural cells in the site of blood brain barrier injury and all processes of intercellular exchange were identical to the processes in malignant diseases. We presume that we detected a unique systemic information-commutation molecular-biological mechanism of the HSCs interaction with the pathological cells as exemplified by the tumors. Possibly, we can detect the analogous profile specific for other diseases, such as cardiovascular or/endocrine disorders. In case this mechanism is an important tool of the CCs and HSCs interaction, it will also work with SOD-1 containing lymphocytes and HSCs that contain these proteomic changes. But as we have seen, any proteomic changes induce specific changes in the HSCs genome and development of the immunospecific transcriptomic and proteomic modifications in the mapped protein profile of the cell, and it will necessarily manifest in the profiling of the PMCS expression in ALS. Our hypothesis is confirmed and we detected a very specific immunophenotypical profile of molecular proteins of the HSCs membrane surface in the ALS cases that we named neurospecific and the IIS in ALS is noted as neurodegenerative IIS (Bryukhovetskiy et al., 2020). Fig. 2 shows the differences of the HSCs membrane surface in health, in cancer and in ALS. Fig. 5 demonstrates the specificity of the HSC protein markers profiling on AD and PD.

This evidence seems to be very promising for clinical practice. First, the detected changes concern a rather large number of cancer patients with different nosologies as these proteomic changes of the cell surface markers are integral markers of the accomplished tumor-specific proteomic modification of the HSC membrane surface. That is, the detection of these specific changes of proteomic landscape of HSCs *registers the onset of the chronic insufficiency of the function of anti-tumor immune protection*. Accordingly, if such changes on the HSC's membrane surface have been detected, the cancer is already initiated in the organism; it is not detected yet, but can manifest any time. Hence, we receive another set of alternative information, no less important than the tumor markers that helps diagnose cancer at the earliest possible stage when the tumor markers are not yet identifiable or the patient is not fully examined or cancer has not yet acquired the tissue specificity. Hence, such patient requires intensive dynamic monitoring. Second, the detected general patterns of the transformation of protein markers reflect the consistency of the HSCs damage in a

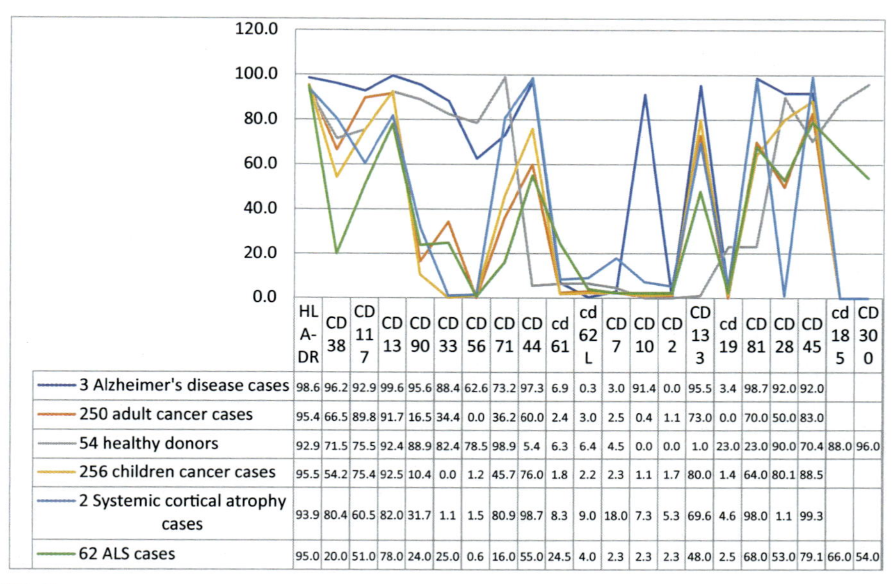

	HLA-DR	CD38	CD117	CD13	CD90	CD33	CD56	CD71	CD44	cd61	cd62L	CD7	CD10	CD2	CD133	cd19	CD81	CD28	CD45	cd185	CD300
3 Alzheimer's disease cases	98.6	96.2	92.9	99.6	95.6	88.4	62.6	73.2	97.3	6.9	0.3	3.0	91.4	0.0	95.5	3.4	98.7	92.0	92.0		
250 adult cancer cases	95.4	66.5	89.8	91.7	16.5	34.4	0.0	36.2	60.0	2.4	3.0	2.5	0.4	1.1	73.0	0.0	70.0	50.0	83.0		
54 healthy donors	92.9	71.5	75.5	92.4	88.9	82.4	78.5	98.9	5.4	6.3	6.4	4.5	0.0	0.0	1.0	23.0	23.0	90.0	70.4	88.0	96.0
256 children cancer cases	95.5	54.2	75.4	92.5	10.4	0.0	1.2	45.7	76.0	1.8	2.2	2.3	1.1	1.7	80.0	1.4	64.0	80.1	88.5		
2 Systemic cortical atrophy cases	93.9	80.4	60.5	82.0	31.7	1.1	1.5	80.9	98.7	8.3	9.0	18.0	7.3	5.3	69.6	4.6	98.0	1.1	99.3		
62 ALS cases	95.0	20.0	51.0	78.0	24.0	25.0	0.6	16.0	55.0	24.5	4.0	2.3	2.3	2.3	48.0	2.5	68.0	53.0	79.1	66.0	54.0

FIG. 5

The averaged immunophenotypic profiles of expression of HSC membrane antigens of 3 HSCs samples from Alzheimer's disease cases, 2 samples from system cortical atrophy cases in Parkinson's disease, average values of 66 samples of ALS cases, 250 samples from adult cancer cases 256 samples from children cancer cases and 54 samples from healthy donors.

cancer patient that do not depend on the immune-tissular and histochemical structure of the tumor. This specific set of the cell surface markers of the human HSCs reflects the inconsistency of the anti-tumor function of the immune system independent of the localization of the tumor. Third, these proteomic changes of expression of the cell membrane landscape of the HSCs of cancer patients are the manifestation of the changed proteomic structure of the HSC and, accordingly, the reason for the CC's and CSC's ability to escape the supervision of the cells of the innate immunity. The immune cells that form the primary response (NK cells, NKT cells, γδ T cells, B cells and other) and those that generate the secondary immune response (cytotoxic lymphocytes, dendritic cells, macrophages) are the progeny of the HSCs with tumor modified proteome in cancer cases. These pathological HSCs and their progeny (all cells of the immune system) have already received the "graft of the immune tolerance" to the specific tumor and will not eliminate it even if intensively activated.

The analysis of the expression of the markers of HSC membrane surface can also present a fundamental criterion for the scientific validation of the immune therapy in cancer cases with the confirmed diagnosis of malignant disease. If profile of the HSCs surface markers is identical to the protein expression profile of a healthy HSC, the immune therapy is not necessary. In this case, the resources and opportunities of the body are sufficient and do not require reinforcement. If the tests of MPP of HSCs detect typical cancer specific proteomic profile, which means that immunocompetent cells are unable to provide anti-cancer effect, the replacement immune therapy must become an indispensable component of the anti-tumor therapy.

It becomes obvious that detection of the morpho-functional biological differences of the landscape of PMMS of HSCs of a cancer patient are conditioned by the epigenomic and post-genomic changes in the structure of the DNA and RNA network, as well as the proteome of these cells, the originators of all cells of the immune system. It becomes clear that these cells, the progeny of cancer-modified HSCs are unable to perform their effector antitumor functions in the immune system of a host and demand specific rigidly targeted correction at the level of epigenome and post-genomic levels (transcriptome, proteome and metabolome).

It is of key importance to understand to what extent the input of every subpopulation of HSC in the cancer-specific modification is definitive as well as it is necessary to evaluate the input of every HSC population to the immunotherapy of cancer. Predominance of some of the populations can be of crucial importance for early diagnostics of different types of cancer and this issue requires additional research.

Our research of PMMS of HSCs show that the range of HSC surface markers in a cancer case help predict systemic anti-tumor insufficiency: weak immunogenicity of the CCs for effector cell systems of the innate immunity of a cancer patient, insufficient regulatory function of the patients HSCs toward the CSCs and CCs as well as predict weak primary and secondary immune response to the tumor antigens. That is, we are able to detect a credible evidence of the modified repertoire of the PMMS profile of HSCs that speaks of high probability of cancer onset before any clinical or paraclinical signs of the malignant disorder. It still has to be understood how we can detect the organ or tissue that will become the site of the immune system disruption in the specific case. However, the identification of the cancer-specific MPP of the HSCs implies high probability of the tumor onset even if the tumor cannot be diagnosed clinically or detected by the contemporary imaging methods, such as PET, PET/CT radioisotopic scanning and other.

Undoubtedly, the presented data on possible early diagnostics of ALS, PD and AD on the basis of the PMMS require further profound research and fundamental analysis. We have presented very few cases as we are a small hospital. However, these results demonstrate the potential of this method in the diagnostics of these diseases as well as family ALS and help predict the opportunity of its development in the family members carrying SOD-1 protein in their HSCs. Accumulation of the SOD-1 conditioned proteins in the HSCs in family ALS cases intensifies proteomic differences of the PMMS of HSCs and the cytofluorometry of the markers of these cells permit diagnostics of the disease in the bud before the degeneration of motor neurons manifests with clinical malfunction so that it can be verified as the motor neuron damage (Bryukhovetskiy et al., 2020). We presume that accurate and profound study of this phenomenon and its verification has the potential to become the cornerstone of early ALS diagnosis and to permit a new viewpoint on the disease along with the opportunity to monitor the degeneration in the CNS from the very early stages and hope for arrest of the disease development at early stage.

The established evidence of proteomic modifications in the HSCs of the family ALS cases permits saying that the epigenomic-proteomic changes in the HSCs that are diagnosed in the healthy relatives of the ALS patients are primary; and the neurodegenerative changes in the motor neurons are observed later. The epigenomic-proteomic changes in the HSCs are also primary in the sporadic ALS and emerge due to the effect of the etiological factors of the disease on the bone marrow and HSCs circulating in the blood of the patient.

If we accept the hypothesis that the ALS is the epigenomic-proteomic disease of the HSCs and bot the motor neuron disease, then the inefficiency of the therapy of ALS as of the autoimmune disease and NDD becomes understandable. The HSCs in such NDDs as multiple sclerosis, encephalomyelitis, systemic lupus erythematosus, rheumatoid arthritis, are healthy and its proteome is intact, while the autoimmune reactivity of the lymphocytes to the neurospecific proteins of the neural tissue forms at the periphery (in neural tissue) in adult and differentiated lymphocytes. So, their suppression with glucocorticoids and immune suppressants as well as blocking their secretomes represented by the toxic cytokines with monoclonal antibodies reduces their levels in blood, the level of excitotoxicity and promotes their replacement with healthy cell systems that are the progeny of a healthy HSC. This is the reason why the transplantation of the autologous bone marrow is effective and leads to full remission. However, ALS and some other similar diseases are underlain by the changed HSCs with the pathological proteome that keeps producing autoreactive and aggressive to the neural tissue ICCs. Hence, the transplantation of the autologous bone marrow is ineffective, just as the immune suppression. ALS is the epigenetically inherited or determined acquired proteome-based disease of the autologous HSC and this explains for the ineffectiveness of the standard autoimmune therapy of ALS.

5 Conclusion

The performed research detected the patterns of the PMCSs profiles specific for the malignant diseases, ALS and other NDDs. The evidence requires further research as it might provide the solution both for the therapy of these diseases and for their diagnosis. Special attention should be given to the promise of the opportunity to ultra early diagnosis of cancer, ALS and some other diseases. The method of early diagnosis of malignancies, ALS, AD and PD help structure new approaches to the epigenomic and post-genomic technologies in the therapy of the above mentioned disorders and permit detection of the molecular-biological target in the therapy. Potentially, the progress of cancer and ALS can be arrested by the standard autologous transplantation in case the PBMCs have been harvested before the disease manifestation. In ALS the transplantation of allogeneic bone marrow might be able to arrest the progress of the disease and help avoid the lethal outcome. The consequent restoration of the damaged motor neurons can be achieved with contemporary biomedical cell products for neuroregeneration. Using cell products without arrest of the disease progress is ineffective and methodologically erroneous.

It seems that the future of the therapy of cancer, other malignancies and ALS belongs to the harvest of healthy HSCs and their transplantation after the onset of the disease. In future, these technologies might become the biotechnological platform to cure this disease, and, possible, such other disabling neurodegenerative diseases as AD and PD.

Conflict of interests

No conflict of interest.

References

Bryukhovetskiy, A.S., 2019. Clinical Oncoproteomics: Proteome-Based Personalized Anti-Cancer Therapy. Nova Publishers, New York.

Bryukhovetskiy, A.S., Grivtsova, L.Y., Sharma, H.S., 2020. Is the ALS a motor neuron disease or a hematopoietic stem cell disease? Prog. Brain Res. 258, 381–396. https://doi.org/10.1016/bs.pbr.2020.09.005. Epub 2020 Oct 24 33223039.

Chiò, A., Logroscino, G., Hardiman, O., et al., 2009. Prognostic factors in ALS: a critical review. Amyotroph. Lateral Scler. 10 (5–6), 310–323. https://doi.org/10.3109/17482960802566824.

de Laval, B., Maurizio, J., Kandalla, P.K., Brisou, G., Simonnet, L., Huber, C., Gimenez, G., Matcovitch-Natan, O., Reinhardt, S., David, E., Mildner, A., Leutz, A., Nadel, B., Bordi, C., Amit, I., Sarrazin, S., Sieweke, M.H., 2020. C/EBPβ-dependent epigenetic memory induces trained immunity in hematopoietic stem cells. Cell Stem Cell 26 (5), 657–674.e8. https://doi.org/10.1016/j.stem.2020.01.017. Epub 2020 Mar 12. Erratum in: Cell Stem Cell. 2020 May 7; 26(5):793 32169166.

Grivtsova, L.Y., Tupitsyn, N.N., 2016. Mobilized hemopoietic stem cells: autologous and allogeneic transplantation in oncology practice. Haematopoiesis Immunol. 1, 6–73. http://www.imhaemo.ru/media/documents/2016-09-12/HI_2016_1_1.pdf. (Accessed January 20, 2021).

Milkina, E.V., Miscshenko, P.V., Zaytsev, S.V., Bryukhovetskiy, I.S., Bryukhovetskiy, A.S., Duizen, I.V., Khotimchenko, Y.S., 2016. Osobennosti vzaimodeystviya mezhdu gemopoeticheskimi stvolovymi i opukholevymi kletkami razlichnikh liniy in vitro [Features of interaction between hematopoietic stem and tumor cells of different lines in vitro]. Gheny & Kletky 11 (3), 63–71 (Russian).

Miller, K.D., Nogueira, L., Mariotto, A.B., Rowland, J.H., Yabroff, K.R., Alfano, C.M., Jemal, A., Kramer, J.L., Siegel, R.L., 2019. Cancer treatment and survivorship statistics, 2019. CA Cancer J. Clin. 69, 363–385. https://doi.org/10.3322/caac.21565.

Nagpal, M., Singh, S., Singh, P., Chauhan, P., Zaidi, M.A., 2016. Tumor markers: a diagnostic tool. Natl. J. Maxillofac. Surg. 7 (1), 17–20. https://doi.org/10.4103/0975-5950.196135.

Sharma, P., Kumar, P., Sharma, R., 2017. Natural killer cells—their role in tumour immunosurveillance. J. Clin. Diagn. Res. 11 (8), BE01–BE05. https://doi.org/10.7860/JCDR/2017/26748.10469.

Yakhno, N.N. (Ed.), 2005. Bolezni nervnoy sistemi [The Diseases of Nervous System]. vol. 1. Meditsina, Moscow, pp. 649–658 (Russian).

Yeghorkina, O.V., Gaponov, I.K., 2007. Klinicheskiy podhod k lecheniyu neirodeghenerativnykh zabolevaniy s dementsiyey. [Clinical approach to the therapy of the neurodegenerative disease with dementia]. Mezhdunar. Nevrol. Zhurn. 1, 111–117 (Russian).
Zavalishina, I.A. (Ed.), 2009. Bokovoy amiotroficheskiy skleros. [Amyotrophic Lateral Sclerosis]. Gheotarmedia, Moscow (Russian).

CHAPTER

3

Neuroprotective effects of insulin like growth factor-1 on engineered metal nanoparticles Ag, Cu and Al induced blood-brain barrier breakdown, edema formation, oxidative stress, upregulation of neuronal nitric oxide synthase and brain pathology

Hari Shanker Sharma[a,*], José Vicente Lafuente[b], Dafin F. Muresanu[c,d], Seaab Sahib[e], Z. Ryan Tian[e], Preeti K. Menon[k], Rudy J. Castellani[f], Ala Nozari[g], Anca D. Buzoianu[h], Per-Ove Sjöquist[i], Ranjana Patnaik[j], Lars Wiklund[a], and Aruna Sharma[a,*]

[a]*International Experimental Central Nervous System Injury & Repair (IECNSIR), Department of Surgical Sciences, Anesthesiology & Intensive Care Medicine, Uppsala University Hospital, Uppsala University, Uppsala, Sweden*

[b]*LaNCE, Department of Neuroscience, University of the Basque Country (UPV/EHU), Leioa, Bizkaia, Spain*

[c]*Department of Clinical Neurosciences, University of Medicine & Pharmacy, Cluj-Napoca, Romania*

[d]*"RoNeuro" Institute for Neurological Research and Diagnostic, Cluj-Napoca, Romania*

[e]*Department of Chemistry & Biochemistry, University of Arkansas, Fayetteville, AR, United States*

[f]*Department of Pathology, University of Maryland, Baltimore, MD, United States*

[g]*Anesthesiology & Intensive Care, Massachusetts General Hospital, Boston, MA, United States*

[h]*Department of Clinical Pharmacology and Toxicology, "Iuliu Hatieganu" University of Medicine and Pharmacy, Cluj-Napoca, Romania*

Progress in Brain Research, Volume 266, ISSN 0079-6123, https://doi.org/10.1016/bs.pbr.2021.06.005
Copyright © 2021 Elsevier B.V. All rights reserved.

[i]Division of Cardiology, Department of Medicine, Karolinska Institutet, Karolinska University Hospital, Stockholm, Sweden
[j]Department of Biomaterials, School of Biomedical Engineering, Indian Institute of Technology, Banaras Hindu University, Varanasi, India
[k]Department of Biochemistry and Biophysics, Stockholm University, Stockholm, Sweden
*Corresponding authors: Tel. & Fax: +46-18-24-38-99; Email: sharma@surgsci.uu.se; harishanker_sharma55@icloud.com; hssharma@aol.com; aruna.sharma@surgsci.uu.se

Abstract

Military personnel are vulnerable to environmental or industrial exposure of engineered nanoparticles (NPs) from metals. Long-term exposure of NPs from various sources affect sensory-motor or cognitive brain functions. Thus, a possibility exists that chronic exposure of NPs affect blood-brain barrier (BBB) breakdown and brain pathology by inducing oxidative stress and/or nitric oxide production. This hypothesis was examined in the rat intoxicated with Ag, Cu or Al (50–60 nm) nanoparticles (50 mg/kg, i.p. once daily) for 7 days. In these NPs treated rats the BBB permeability, brain edema, neuronal nitric oxide synthase (nNOS) immunoreactivity and brain oxidants levels, e.g., myeloperoxidase (MP), malondialdehyde (MD) and glutathione (GT) was examined on the 8th day. Cu and Ag but not Al nanoparticles increased the MP and MD levels by twofold in the brain although, GT showed 50% decline. At this time increase in brain water content and BBB breakdown to protein tracers were seen in areas exhibiting nNOS positive neurons and cell injuries. Pretreatment with insulin like growth factor-1 (IGF-1) in high doses (1 μg/kg, i.v. but not 0.5 μg/kg daily for 7 days) together with NPs significantly reduced the oxidative stress, nNOS upregulation, BBB breakdown, edema formation and cell injuries. These novel observations demonstrate that (i) NPs depending on their metal constituent (Cu, Ag but not Al) induce oxidative stress and nNOS expression leading to BBB disruption, brain edema and cell damage, and (ii) IGF-1 depending on doses exerts powerful neuroprotection against nanoneurotoxicity, not reported earlier.

Keywords

Nanoparticles, Oxidative stress, Nitric oxide, Cell injury, Insulin like growth factor-1 (IGF-1), Brain edema, Cu, Al, Ag

1 Introduction

Long-term exposure of nanoparticles from environment, contaminated food or water, industrial by products, chemicals or explosive materials affects human health adversely and could cause brain dysfunction (Chasapis et al., 2020; Ferdous and Nemmar, 2020; Herrero et al., 2020; Lafuente et al., 2012; Li and Cummins,

2020; Lojk et al., 2020; Lu et al., 2015; Köerich et al., 2020; Parsai and Kumar, 2021; Sharma and Sharma, 2007, 2012a,b; Sousa and Teixeira, 2020; Sharma et al., 2009b, 2010a,b,c; Tortella et al., 2020; Wu and Kong, 2020). The possible link between nanoparticles exposure and brain dysfunction are still not well known. Thus, further studies are needed to find out the possible mechanisms of potential adverse effects of nanoparticles on the central nervous system (CNS) using pharmacological and biochemical approaches.

Previous works from our laboratory show that chronic exposure of engineered nanoparticles from metals induces profound nanoneurotoxicity by inducing breakdown of the blood-brain barrier (BBB) function (Menon et al., 2017; Sharma, 2009a,b; Sharma and Sharma, 2007; Sharma et al., 2009a,b,c,d,e, 2010a,b, 2011a,b,c, 2013a,b, 2014, 2015a,b,c,Sharma et al., 2015c. However, the detailed mechanisms of nanoparticles induced neurotoxicity are still not well known. Available evidences suggest that oxidative stress induced by nanoparticles could play major roles in cell or organ toxicity (Di Gioacchino et al., 2011; Kamikubo et al., 2019; Murugadoss et al., 2017; Nemmar et al., 2019; Paunovic et al., 2020; Sarkar et al., 2014; Yoshitomi and Nagasaki, 2011; Zhou et al., 2019; Zuberek and Grzelak, 2018). However, the role of oxidative stress caused by nanoparticles on brain damage is not investigated in greater details. Oxidative stress either following nanoparticle intoxication, ischemia, trauma or hyperthermia will generate lipid peroxidation and formation or free radicals together with increased nitric oxide (NO) production (Araujo and Welch, 2006; Arimon et al., 2015; Del Rio et al., 2005; Di Gioacchino et al., 2011; Huun et al., 2018; Jones, 2008; Lushchak, 2014; Poprac et al., 2017; Ramana et al., 2017; Rush and Ford, 2007; Sahni et al., 2018; Sharma and Alm, 2002, 2004; Sharma and Sjöquist, 2002; Tejero et al., 2019; Tomita et al., 2011; Tsikas, 2017; Yoshitomi and Nagasaki, 2011; Yui et al., 2016). These oxygen and NO radicals may directly damage the cell membranes of neurons, glial and the endothelial cells resulting in the brain injury (Cordeiro, 2018; Dröge, 2002; Sharma et al., 1996, 1997, 1998a,b,c,d, 2006a,b,c,d; Torreggiani et al., 2019; Valko et al., 2016). Obviously, damage of endothelial cell membranes result in BBB breakdown and allow penetration of serum proteins and other elements into the brain fluid microenvironment (Kadry et al., 2020; Kaplan et al., 2020; Sharma and Kiyatkin, 2009; Sharma et al., 2011b). Exposure of these injury factors will lead to edema formation and cell or tissue damages in the CNS (Semenas et al., 2011; Sharma, 1999, 2004, 2009b, 2010b; Sharma and Alm, 2002, 2004; Sharma and Sharma, 2007; Sharma and Westman, 2004; Sharma et al., 2000a).

Keeping these views in consideration in present investigation we measured different oxidative stress parameters and nitric oxide (NO) production in the brain following chronic intoxication of engineered nanoparticles from three different metals, e.g., Ag, Cu or Al of similar sizes (50–60nm) (Sharma and Sharma, 2007, 2012a,b; Sharma et al., 2010a,b). Furthermore, the influence of exogenous supplement of insulin like growth factor-1 (IGF-1), a known neuroprotective agent (Higashi et al., 2010; Muresanu et al., 2015; Sharma, 2003, 2005, 2007, 2010; Sharma and Johanson, 2007; Sharma et al., 1997, 1998a,b, 2000a,b; Serhan et al., 2020;

Wang et al., 2019) was examined on these oxidative stress parameters and NO production in the brain following nanoparticles intoxication in our rat model.

2 Materials and methods

2.1 Animals

Experiments were carried out on male Sprague Dawley rats (200–350 g) housed at controlled ambient temperature (21 ± 1 °C) with 12 h light and dark schedule. Rat food pellets and tap water were supplied ad libitum. All experiments were conducted according to the National Institute of Health Guidelines for Care and Handling of Animals and approved by the Local Institutional Ethics Committee (National Research Council (US), 2011).

2.2 Administration of nanoparticles

Nanoparticles from Ag, Cu and Al (50–60 nm, Sigma Chemical Co., St. Louis, USA) was administered (50 mg/kg, i.p.) once daily for 7 days in a suspension of Tween 80 as described earlier. On the 8th day different parameters on brain function were examined (Sharma and Sharma, 2007, 2012a,b, 2013; Sharma et al., 2009a,b,c,d,e).

2.3 Insulin like growth factor-1 (IGF-1) administration

IGF-1 (Sigma Chemical Co., St. Louis, USA) was administered 0.5 μg or 1 μg/kg, i.v. once daily for 7 days through an indwelling polythene cannula into the right jugular vein aseptically implanted 6 days before the experiment (Sharma et al., 1997, 1998a,b). These injections were commenced 2 h after nanoparticles administration intraperitoneally every day.

2.4 Measurement of oxidative stress parameters in the brain

Myeloperoxidase (MP), malondialdehyde (MD) and glutathione (GH) were measured in brain tissue homogenate according to standard biochemical and enzymatic procedures using spectrophotometer as described elsewhere (Sharma et al., 2018; Wiklund et al., 2018). For this process brain tissue were homogenized in ice-cold 50 mM potassium phosphate buffer (KpB, pH 6.0) and processed for enzymatic activity for MP measured at 460 nm in a spectrophotometer. The MD and GH levels were assayed for products of lipid peroxidation (see 1, Sharma et al., 2018).

2.5 Blood-brain barrier and neuronal injury

The blood-brain barrier (BBB) integrity was assessed by leakage of [131]-Iodine and Evans blue as describe earlier (Sharma, 1987; Sharma and Dey, 1986a, 1986b, 1987; Sharma et al., 1990a,b). In addition, leakage of endogenous serum albumin in the brain was also examined in the brain using immunohistochemical methods as described in details earlier. Monoclonal antibody to albumin (1:500) was used. Neural injury was evaluated by histological methods using Nissl or Haematoxylin and Eosin (H&E) staining (Sharma and Sharma, 2007; Sharma et al., 2011a,b).

2.6 Brain edema formation

Brain water content was used to measure edema formation. For this purpose, brains were removed after decapitation and small tissue pieces from the desired area were removed, weighed and placed in an incubator at 90 °C to obtain dry weight of the samples. After 72 h, when the dry weight became constant was recorded and percentage brain water was calculated from the differences between wet and dry weight as described earlier (Kiyatkin and Sharma, 2011; Sharma and Cervós-Navarro, 1990; Sharma and Kiyatkin, 2009; Sharma and Olsson, 1990; Sharma and Sharma, 2007; Sharma and Westman, 1997).

2.7 Morphological analysis

For morphological evaluation and immunohistochemical study, after the experiment, animals were anesthetized deeply with Equithesin and perfused with 4% neutral paraformaldehyde through cardiac puncture after incising right auricle preceded with a brief saline rinse (Sharma and Sharma, 2007; Sharma et al., 1991a,b). After perfusion, the brains were dissected out and placed in the same fixative at 4 °C for 1 week. After that coronal section from the brain was cut passing through thee hippocampal area and embedded in paraffin. About 3-μm thick section were cut from each block and processed for different histological and immunochemical assays as described below (Sharma and Kiyatkin, 2009; Sharma and Sharma, 2007; Sharma et al., 1998c,d).

2.8 Nitric oxide activity

The nitric oxide upregulation was assessed using immunohistochemistry of neuronal nitric oxide synthase (nNOS) antibody (1:500) on paraffin sections as described earlier (Muresanu et al., 2019; Sharma and Alm, 2002; Sharma et al., 1996, 1997, 1998a,b,c,d, 2005, 2019a,b).

2.9 Statistical analyses of data

Quantitative data obtained were analyzed using ANOVA followed by Dunnett's test for multiple group comparison from one control group suing commercial software StatView 5 on a Macintosh Computer in Classic environment. Semiquantitative data obtained from morphological analyses were analyzed for statistical significance using non-parametric Chi-Square test. A *p*-value less than 0.05 is considered significant.

3 Results

3.1 Nanoparticles intoxication induces oxidative stress

Chronic treatment with nanoparticles significantly enhanced the MP (Fig. 1) and MD levels (Fig. 2) in the cortex by more than twofold from the control group, whereas GH level showed about 50% decline from saline treated rats on the 8th day (Fig. 3). In general Ag and Cu treatment was most effective in inducing oxidative stress as

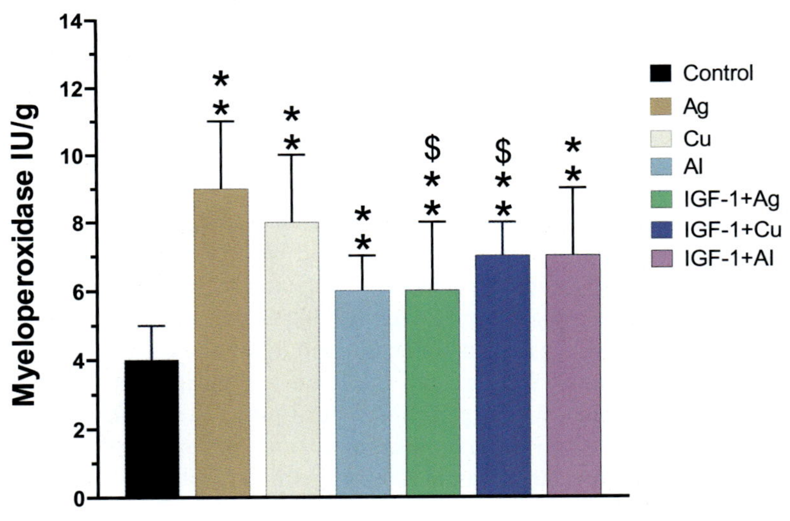

FIG. 1

Measurement of myeloperoxidase (MP) activity in control and in Ag, Cu, or Al nanoparticles (NPs) in the rat and its modification with insulin like growth factor-1 (IGF-1) in the brain. Engineered metal NPs significantly increased MP activity in the cerebral cortex and IGF-1 treatment induced significant attenuation in MP activity that was the most marked in IGF-1 + Ag NPs. The ability of metal NPs to induce oxidative stress, as seen using increase in MP activity was the highest by Ag NPs followed by Cu NPs and Al NPs. For details see text.Values are Mean ± SD of 5–6 rats at each data point. ** $P<0.01$ from control. $ $P<0.05$ from respective Metal NPs treatment. ANOVA followed by Dunnett's test for multiple group comparison using one control.

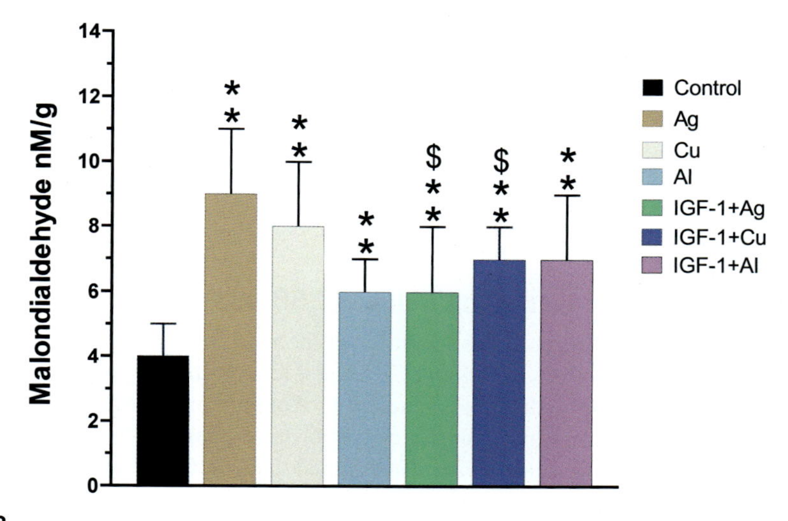

FIG. 2

Measurement of malondialdehyde (MD) activity in control and its increase by Ag, Cu, or Al nanoparticles (NPs) in the rat and its modification with insulin like growth factor-1 (IGF-1) in the brain. Engineered metal NPs significantly increased MD activity in the cerebral cortex and IGF-1 treatment reduced significantly this increase in MD that was the most marked in IGF-1 + Ag NPs. The ability of metal NPs to induce oxidative stress, as seen using increase in MD activity was the highest by Ag NPs followed by Cu NPs and Al NPs. For details see text. Values are Mean $\pm$ SD of 5–6 rats at each data point. ** $P<0.01$ from control. $ $P<0.05$ from respective Metal NPs treatment. ANOVA followed by Dunnett's test for multiple group comparison using one control.

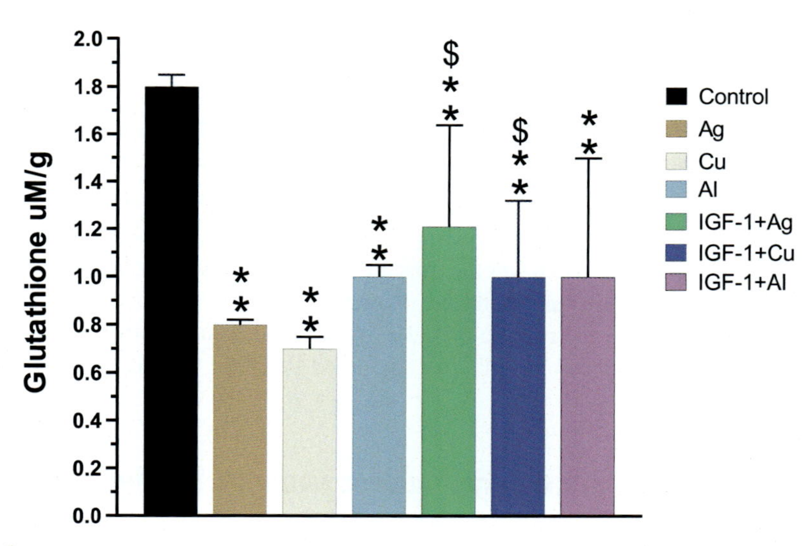

FIG. 3

Measurement of glutathione (GH) activity in control and its reduction by Ag, Cu, or Al nanoparticles (NPs) in the rat and its modification with insulin like growth factor-1 (IGF-1) in the brain. Engineered metal NPs significantly decreased GH activity in the cerebral cortex and IGF-1 treatment induced significant increase in GH that was the most marked in IGF-1 + Ag NPs. The ability of metal NPs to induce oxidative stress, as seen using decrease in GH activity was the highest by Ag NPs followed by Cu NPs and Al NPs. For details see text. Values are Mean $\pm$ SD of 5–6 rats at each data point. ** $P<0.01$ from control. $ $P<0.05$ from respective Metal NPs treatment. ANOVA followed by Dunnett's test for multiple group comparison using one control.

compared to Al treatment that exhibited relatively slight changes than that of the saline control group (Figs. 1–3).

3.2 Nanoparticles intoxication induces BBB disruption

Morphological studies demonstrated a significant increase in albumin extravasation in the neuropil and in neurons of Ag and Cu treated rats on the 8th day (Fig. 4). Al treatment should only sporadic mild increase in albumin immunoreactivity. The areas showing albumin immunostaining also showed damage to neuropil where several neurons are damaged and general sponginess and edema was more frequent (Fig. 5).

Measurement of the BBB breakdown induced by NPs showed significant increases of Evans blue albumin (EBA) and radioiodine extravasation (Table 1). The magnitude and intensity of BBB disruption was the most marked in AG NPs intoxication followed by Cu and Al NPs (Table 1).

3.3 Nanoparticles intoxication induces brain edema

Measurement of water content further confirmed that areas showing neuronal damage and albumin leakage are indeed induce edematous swelling in the brain (Table 1). Thus a significant increase in brain water was seen in Ag and Cu treated rats in the cortex. Al treatment exhibited only a mild increase in brain water from the saline treated control group (Table 1).

3.4 Nanoparticles intoxication induces neuronal nitric oxide synthase upregulation

Ag and Cu nanoparticles induced moderate to massive upregulation of nNOS immunoreactivity in the brain areas showing neuronal damage and albumin leakage (Figs. 4–6). The nNOS activity was the most prominent in those neurons that showed perineuronal edema, dark and distorted neuronal structure and chromatolysis. Al treatment was also able to enhance nNOS activity but the magnitude and intensity of nNOS upregulation was only mild as compared to saline treated group (Figs. 6 and 7).

3.5 Nanoparticles intoxication induces neuronal injuries

Marked neuronal damages were seen in Ag and Cu treated rats on the day 8th day (Figs. 5 and 8). Thus, swollen or shrunken neurons with perineuronal edema are quite frequent in Ag and Cu treated rats. Al treatment induced similar but very mild damage to the nerve cells (Fig. 5). In several neurons the karyoplasm was condensed and the nucleolus was eccentric. Interestingly few neurons have normal appearance also within the same area where many nerve cells are showing neurodegenerative changes.

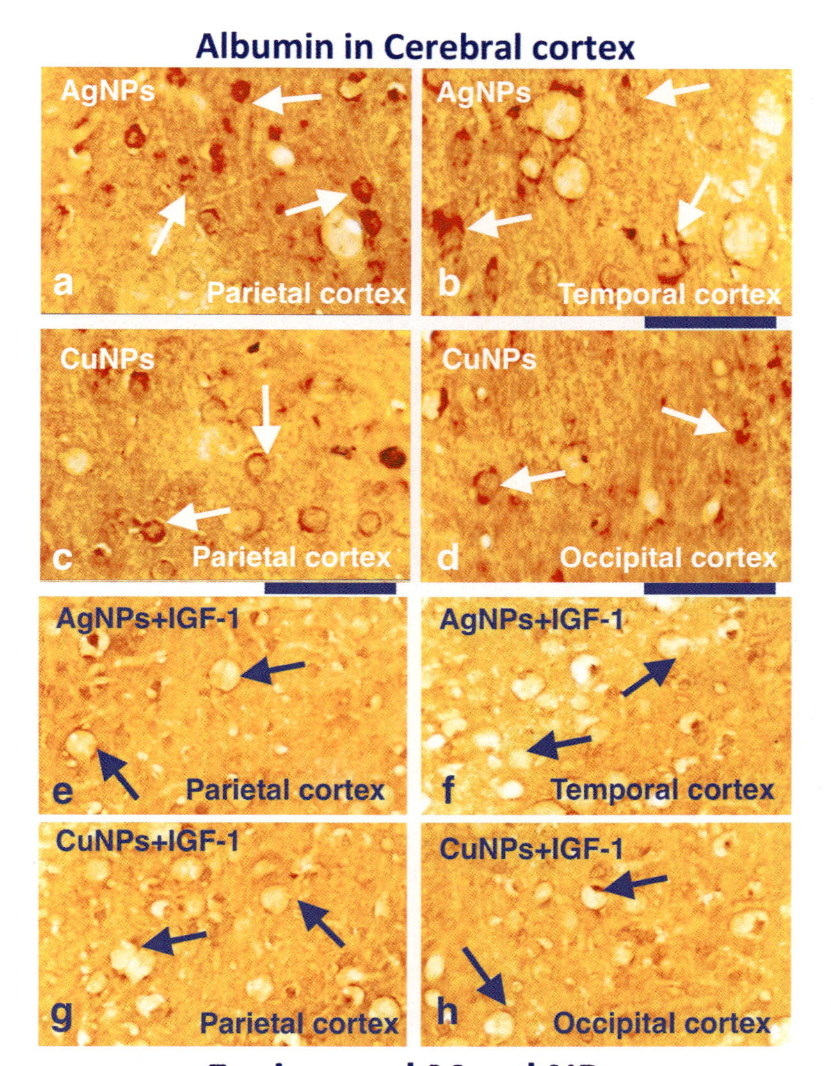

FIG. 4

Immunohistochemistry of albumin leakage in the parietal (A and C), temporal (B) and occipital cortex (D) wish engineered metal nanoparticles (NPs) intoxication of AgNPs (A and B) and CuNPs (C and D) in the rat and its modification with insulin like factor-1 (IGF-1) on AgNPs (E and F) and CuNPs (G and H) in parietal cortex (E and G), temporal cortex (F) and occipital cortex (H). Several albumin positive cells are seen in Metal NPs intoxicated group (A–D, arrows) and IGF-1 treatment markedly reduced the albumin leakage in the cerebral cortex following their intoxication (E–H, arrows). Bar = 40 μm.

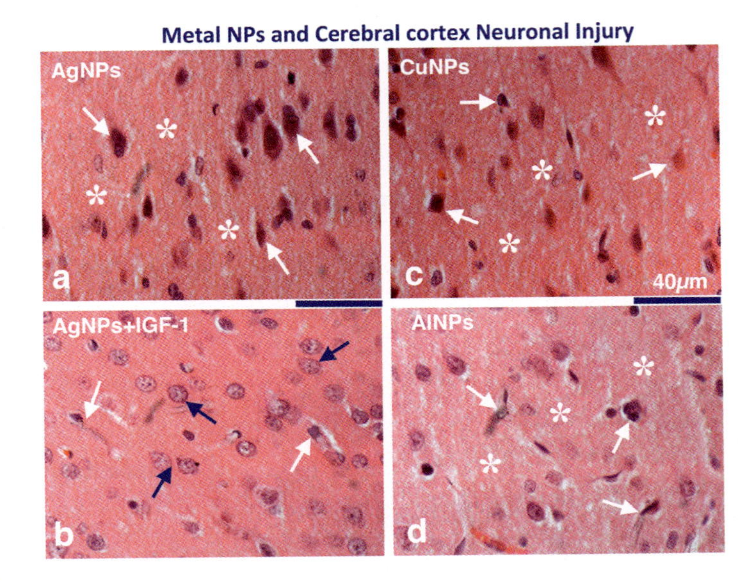

FIG. 5

Light micrograph of Haematoxylin & Eosin (H&E) on Paraffin sections (3-μm thick) from the cerebral cortex showing neuronal damages by AgNPs (A), CuNPs (C) and AlNPs (D) and neuroprotection by IGF-1 treatment in AgNPs (B). In Metal NPs intoxication several damaged and distorted cerebral cortical neurons are seen (arrows, A, C and D) together with perineuronal edema, sponginess (*) and expansion of the neuropil. In the parietal cerebral cortex, the magnitude and severity of neuronal damages were the most marked in AgNPs group (A) followed by CuNPs (C) and AlNPs (D). Treatment with IGF-1 in AgNPs markedly reduced these neuronal damages as seen from several healthy looking nerve cells (blue arrows) whereas, some neurons still show mild damage (white arrows), perineuronal edema and some sponginess (B). Bar = 40 μm.

Table 1 Blood-brain barrier breakdown, edema formation and volume swelling in brain following Ag, Cu and Al nanoparticles and their modification with insulin like growth factor-1 treatment.

	BBB breakdown		Brain edema	Volume Swelling
Type of Experiment	**EBA mg %**	**[131]-I %**	**Brain water %**	**% *f***
Control	0.22 ± 0.06	0.30 ± 0.08	74.56 ± 0.15	Nil
Ag NPs	0.70 ± 0.05^{b}	0.82 ± 0.06^{b}	75.68 ± 0.23^{b}	4%
Cu NPs	0.62 ± 0.08^{b}	0.74 ± 0.09^{b}	75.12 ± 0.28^{b}	2%
Al NPs	0.51 ± 0.05^{b}	0.62 ± 0.07^{b}	74.98 ± 0.15^{b}	1.50%
IGF-1 + Ag NPS	$0.54 \pm 0.08^{a,c}$	$0.60 \pm 0.06^{a,c}$	$75.09 \pm 0.24^{a,c}$	2%
IGF-1 + Cu NPs	$0.52 \pm 0.04^{a,c}$	$0.58 \pm 0.06^{a,c}$	$74.89 \pm 0.21^{a,c}$	1%
IGF-1 + Al NPs	$0.43 \pm 0.04^{a,c}$	$0.51 \pm 0.05^{a,c}$	$74.69 \pm 0.10^{a,c}$	0.50%

Values are Mean ± SD of 5–6 rats at each data point.
[a]P < 0.05
[b]P < 0.01 from control
[c]P < 0.05 from respective NPs treatment. ANOVA followed by Dunnett's test for multiple group comparison from one control. Volume swelling (% f) was calculated from changes in brain water content from control group according to formula of Elliott & Jasper (1949). About 1% change in brain water content is equal to about % increase in volume swelling. The values are approximated to nearest value for simplicity. For details, see text.

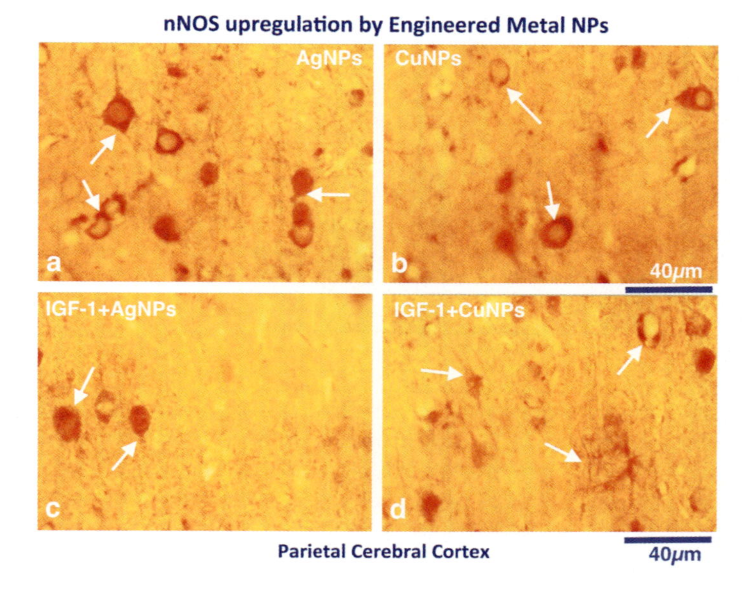

FIG. 6

Neuronal nitric oxide synthase (nNOS) immunohistochemistry in the Parietal cerebral cortex of Ag (A) and Cu (B) nanoparticles (NPs) intoxication and its modification with IGF-1 treatment following Ag (C) and Cu (D) NPs. AgNPs induced the most massive upregulation of nNOS positive cells in the parietal cortex (arrows) followed by CuNPs (B, arrows). Both cytoplasm and karyoplasm of nerve cells show nNOS immunostaining (A and B). Treatment with IGF-1 in AgNPs (C) and CuNPs (D) markedly attenuated upregulation of nNOS positive cells in the parietal cortex (arrows). This effect was the most pronounced in AgNPs treatment (C) as compared to the CuNPs (D). Bar = 40 μm.

3.6 Effect of insulin like growth factor-1 (IGF-1) on nanoneurotoxicity

Repeated intravenous administration of IGF-1 daily for 7 days together with metal nanoparticles intoxication almost neutralized their harmful effects in the brain. Thus, IGF-1 was able to thwart increase in MP and MD levels in the brain and also attenuated the GH decrease on the 8th day (Figs. 1–3). This effect of IGF-1 was dose related, as 1 μg dose was most effective in reducing these changes whereas 0.5 μg/kg was not that effective. The beneficial effects of IGF-1 were most marked in Ag and Cu treated group as compared to Al nanoparticles treatment (Figs. 1–3).

Reduction is oxidative stress parameters by IGF-1 in high doses was also associated with a marked decrease in nNOS activity and neuronal distortion (Fig. 6). Thus in IGF-1 treated animals did not exhibit much nNOS upregulation or neuronal injuries in nanoparticles treated rats (Figs. 5–8). In these animals increase in BBB to EBA and radioiodine (Table 1) or bran water content and volume swelling

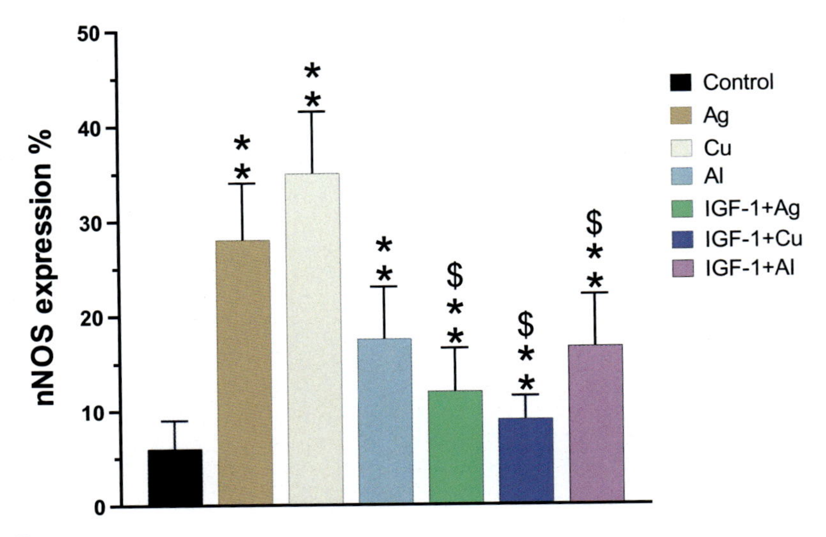

FIG. 7

Semiquantitative data on neuronal nitric oxide synthase (nNOS) positive cells in the cerebral cortex in control and following Ag, Cu, or Al nanoparticles (NPs) in toxication in the rat and its modification with insulin like growth factor-1 (IGF-1) in the cerebral cortex. Engineered metal NPs significantly upregulated nNOS positive cell activity in the cerebral cortex and IGF-1 treatment induced significantly the increase in nNOS positive neurons that was the most marked in IGF-1 + Ag NPs. The ability of metal NPs to induce oxidative stress, as seen using increase in nNOS positive activity was the highest by Ag NPs followed by Cu NPs and Al NPs. For details see text.Values are Mean ± SD of 5–6 rats at each data point. ** $P<0.01$ from control. $ $P<0.05$ from respective Metal NPs treatment. Non-parametric Chi-Square test.

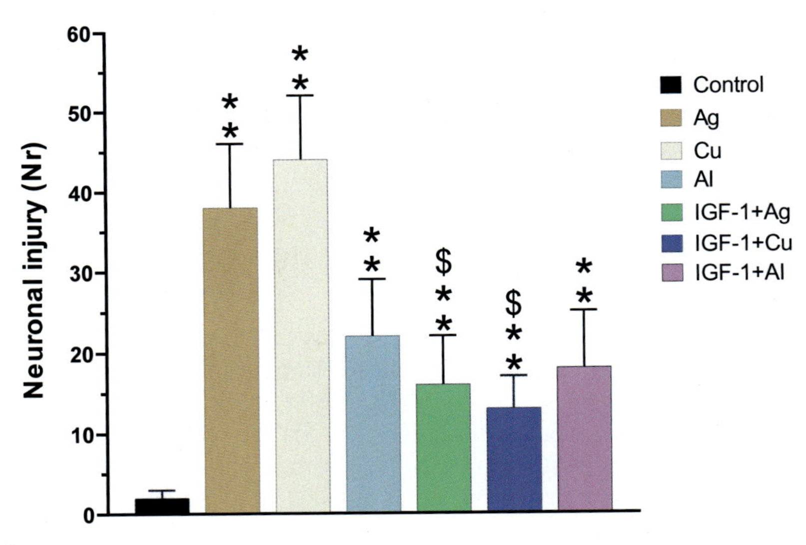

FIG. 8

Semiquantitative data on neuronal damages in the cerebral cortex in control and following Ag, Cu, or Al nanoparticles (NPs) intoxication in the rat and its modification with insulin like growth factor-1 (IGF-1). Engineered metal NPs significantly increased neuronal distortion in the cerebral cortex and IGF-1 treatment reduced significantly ths damage of neurons that was the most marked in IGF-1 + Ag NPs. The ability of metal NPs to induce neuronal damages was the highest by Ag NPs followed by Cu NPs and Al NPs. For details see text.Values are Mean ± SD of 5–6 rats at each data point. ** $P<0.01$ from control. $ $P<0.05$ from respective Metal NPs treatment. Non-parametric Chi-Square test.

(Table 1) was significantly reduced in Ag, Cu or Al NPs. Interestingly, the albumin leakage were also thwarted markedly as compared to untreated group given identical nanoparticles treatment (Fig. 4).

4 Discussion

The present results for the first time demonstrate that chronic administration of engineered nanoparticles from metals induces profound oxidative stress in the brain. However, the magnitude and intensity of oxidative stress caused by nanoparticles depend on the inherent properties of the metal nanoparticles rather than their sizes (Kovacic and Somanathan, 2010, 2013; Migliore et al., 2015; Zhou et al., 2019). In this study we examined only one dose of nanoparticles thus it is still unclear whether induction of oxidative stress by different metal nanoparticles could be dose dependent, a feature that is currently being investigated in our laboratory.

The reasons behind such a selective increase in oxidative stress by Ag and Cu nanoparticles as compared to Al is not known from this study. However, it appears that depending on the type of metal nanoparticles and their biophysical behavior, e.g., ionic charges, zeta potentials in biological fluids could play some roles in inducing different magnitude of neurotoxicity (Lojk et al., 2020; Sharma and Sharma, 2007; Sharma et al., 2009a,b,c, 2010b,c, 2011a,b,c). Based on our results it appears that Al nanoparticles are less toxic in similar concentrations in vivo situations than Ag and Cu (Sharma and Sharma, 2012a,b; Sharma et al., 2009a,b). This observation is in line with our previous findings, which showed that intracerebroventricular, intravenous or intraperitoneal administration of Ag, and Cu nanoparticles were more effective in BBB breakdown to proteins than Al nanoparticles (Sharma et al., 2009a,b,c,d,e, 2010a,c). It is quite likely that a less intense production of oxidative stress by Al nanoparticles as compared to Ag and Cu nanoparticles could be responsible for less neurotoxicity or BBB breakdown in vivo situations. Obviously, a less intensive oxidative stress is associated with less production of free radicals are subsequent BBB disruption and/or brain damage (Sharma et al., 2000a, 2009b).

Another novel finding of this study suggests that co-administration of IGF-1 with nanoparticles resulted in profound neuroprotection and attenuation in oxidative stress parameters in the brain. However, this neuroprotective effect of IGF-1 is dose dependent. This suggests that sufficient amount of exogenous supplement of neurotrophic factors, e.g., IGF-1 could neutralize or attenuates the harmful effects of nanoparticles on brain function (Menon et al., 2017; Sharma et al., 2011b,c, 2014). It appears that IGF-1 potentiates neurotrophic activity and reduce neuroinflammation are crucial for neuroprotection (Bianchi et al., 2017; Castilla-Cortázar et al., 2019; Menon et al., 2012; Sádaba et al., 2016; Serhan et al., 2020; Sohrabji, 2015; Spielman et al., 2014; Wang et al., 2019). However, the possible mechanisms of IGF-1 induced neuroprotection are still not well understood. It appears that IGF-1 induced reduction in cellular or molecular stress caused by nanoparticles could play important roles (Kiyatkin and Sharma, 2011; Sharma et al., 1997, 2000b, 2011a). This is further evident from the fact that the IGF-1 treatment attenuated oxidative stress parameters

(Ayadi et al., 2016; Guan et al., 2018; Higashi et al., 2010; Su and Huang, 2018). This suggests that IGF-1 is able to alter the cellular and/or molecular response of stress caused by nanoparticles and thereby reducing the adverse effects in the brain (Allahdadi et al., 2019; Ogundele et al., 2017; Pharaoh et al., 2020; Riis et al., 2020). To verify this hypothesis, effect of IGF-1 on expression of heat shock protein (HSP) 72 kDa, a stress protein response is needed that is currently being investigated in our laboratory.

Ischemia, trauma, stress when imposed on the central nervous system (CNS) results in cascade of events depending on the magnitude, intensity and duration of the primary noxious insult (Niu et al., 2019, 2020; Sharp and Sagar, 1994; Sharma et al., 2006a,b,c, 2011c, 2019a,b, 2020a,b; Wiklund et al., 2007, 2018). These noxious insults cause cellular and molecular stress in the tissues or organs initiating neurochemical release, depletion of endogenous growth factors to maintain homeostasis (Sharma and Alm, 2002; Sharma and Sjöquist, 2002; Sharma and Winkler, 2002; Sharma et al., 1998a,b,c,d, 2011a). However, continuous presence of noxious stimulus or high insanity of primary stimulus could deplete the endogenous sources of growth factors and other endogenous neuroprotective elements. This would result in cell toxicity. Thus, exogenous supplement of these factors may rescue the nerve cells and maintain homeostasis of the CNS leading to attenuation of neurotoxicity or repair of brain damages. This idea is clearly supported by the IGF-1 treatment in our study. Our results clearly show that high doses of IGF-1 supplement lead to neuroprotection following nanoparticles induce noxious cellular and molecular stress in the brain. These intense noxious stressors could disrupt the BBB function probably by inducing oxidative stress and free radical formation (Di Gioacchino et al., 2011; Lafuente et al., 2018; Kiyatkin and Sharma, 2011; Semenas et al., 2011; Sharma et al., 2013a,b; Tomita et al., 2011; Yoshitomi and Nagasaki, 2011). Stress caused by emotional, physical or environmental heat or cold factors are well known to disrupt BBB function and induce brain damage (Pandey et al., 2012; Sharp and Sagar, 1994; Sharma et al., 1995, 2006a,b, 2009a,c, 2011c). Our results further show that nanoparticles stress could also be sufficient to induce breakdown of the BBB and induce brain damage.

In conclusion, our observations show that nanoparticles deepening of the type are able to induce BBB breakdown probably by induction of oxidative stress. Breakdown of the BBB appears to be instrumental in inducing brain damage. Exogenous supplement of IGF-1 in these situations attenuated BBB disruption, oxidative stress and brain damage. It will be interesting to see whether IGF-1 alone or in combination with other factors in different doses if given several days after nanoparticle intoxication could also be able to attenuate brain damage and/or oxidative stress production. This is a subject that is currently being investigated in our laboratory.

Acknowledgments

This investigation is supported by grants from the Air Force Office of Scientific Research (EOARD, London, UK), and Air Force Material Command, USAF, under grant number FA8655-05-1-3065; Grants from the Alzheimer's Association (IIRG-09-132087), the

National Institutes of Health (R01 AG028679) and the Dr. Robert M. Kohrman Memorial Fund (RJC); Swedish Medical Research Council (Nr 2710-HSS), Göran Gustafsson Foundation, Stockholm, Sweden (HSS), Astra Zeneca, Mölndal, Sweden (HSS/AS), Alexander von Humboldt Foundation, Bonn, Germany (HSS), The University Grants Commission, New Delhi, India (HSS/AS), Ministry of Science & Technology, Govt. of India (HSS/AS), Indian Medical Research Council, New Delhi, India (HSS/AS) and India-EU Co-operation Program (RP/AS/HSS) and IT-901/16 (JVL), Government of Basque Country and PPG 17/51 (JVL), JVL thanks to the support of the University of the Basque Country (UPV/EHU) PPG 17/51 and 14/08, the Basque Government (IT-901/16 and CS-2203) Basque Country, Spain; and Foundation for Nanoneuroscience and Nanoneuroprotection (FSNN), Romania. Technical assistance of Kärstin Flink, Ingmarie Olsson, Uppsala University and Franzisca Drum, Katja Deparade, Free University Berlin, Germany are highly appreciated. Technical and human support provided by Dr. Ricardo Andrade from SGIker (UPV/EHU) is gratefully acknowledged. Dr. Seaab Sahib is supported by Research Fellowship at University of Arkansas Fayetteville AR, USA by Department of Community Health; Middle Technical University; Wassit; Iraq, and The Higher Committee for Education Development in Iraq; Baghdad; Iraq. We thank Suraj Sharma, Blekinge Inst. Technology, Karlskrona, Sweden and Dr. Saja Alshafeay, University of Arkansas, Fayetteville, AR, USA affiliated with University of Baghdad, Baghdad IQ, for computer and graphic support. The U.S. Government is authorized to reproduce and distribute reprints for Government purpose notwithstanding any copyright notation thereon. The views and conclusions contained herein are those of the authors and should not be interpreted as necessarily representing the official policies or endorsements, either expressed or implied, of the Air Force Office of Scientific Research or the U.S. Government.

Conflict of interest

The authors have no conflict of interests with any funding agency or entity reported here.

References

Allahdadi, K.J., de Santana, T.A., Santos, G.C., Azevedo, C.M., Mota, R.A., Nonaka, C.K., Silva, D.N., Valim, C.X.R., Figueira, C.P., Dos Santos, W.L.C., do Espirito Santo, R.F., Evangelista, A.F., Villarreal, C.F., Dos Santos, R.R., de Souza, B.S.F., Soares, M.B.P., 2019. IGF-1 overexpression improves mesenchymal stem cell survival and promotes neurological recovery after spinal cord injury. Stem Cell Res. Ther. 10 (1), 146. https://doi.org/10.1186/s13287-019-1223-z.

Araujo, M., Welch, W.J., 2006. Oxidative stress and nitric oxide in kidney function. Curr. Opin. Nephrol. Hypertens. 15 (1), 72–77. https://doi.org/10.1097/01.mnh.0000191912.65281.e9.

Arimon, M., Takeda, S., Post, K.L., Svirsky, S., Hyman, B.T., Berezovska, O., 2015. Oxidative stress and lipid peroxidation are upstream of amyloid pathology. Neurobiol. Dis. 84, 109–119. https://doi.org/10.1016/j.nbd.2015.06.013. Epub 2015 Jun 21.

Ayadi, A.E., Zigmond, M.J., Smith, A.D., 2016. IGF-1 protects dopamine neurons against oxidative stress: association with changes in phosphokinases. Exp. Brain Res. 234 (7), 1863–1873. https://doi.org/10.1007/s00221-016-4572-1. Epub 2016 Feb 19.

Bianchi, V.E., Locatelli, V., Rizzi, L., 2017. Neurotrophic and neuroregenerative effects of GH/IGF1. Int. J. Mol. Sci. 18 (11), 2441. https://doi.org/10.3390/ijms18112441.

Castilla-Cortázar, I., Iturrieta, I., García-Magariño, M., Puche, J.E., Martín-Estal, I., Aguirre, G.A., Femat-Roldan, G., Cantu-Martinez, L., Muñoz, Ú., 2019. Neurotrophic factors and their receptors are altered by the mere partial IGF-1 deficiency. Neuroscience 404, 445–458. https://doi.org/10.1016/j.neuroscience.2019.01.041. Epub 2019 Jan.

Chasapis, C.T., Ntoupa, P.A., Spiliopoulou, C.A., Stefanidou, M.E., 2020. Recent aspects of the effects of zinc on human health. Arch. Toxicol. 94 (5), 1443–1460. https://doi.org/10.1007/s00204-020-02702-9. Epub 2020 May 12.

Cordeiro, R.M., 2018. Reactive oxygen and nitrogen species at phospholipid bilayers: peroxynitrous acid and its homolysis products. J. Phys. Chem. B 122 (34), 8211–8219. https://doi.org/10.1021/acs.jpcb.8b07158. Epub 2018 Aug 14.

Del Rio, D., Stewart, A.J., Pellegrini, N., 2005. A review of recent studies on malondialdehyde as toxic molecule and biological marker of oxidative stress. Nutr. Metab. Cardiovasc. Dis. 15 (4), 316–328. https://doi.org/10.1016/j.numecd.2005.05.003.

Di Gioacchino, M., Petrarca, C., Lazzarin, F., Di Giampaolo, L., Sabbioni, E., Boscolo, P., Mariani-Costantini, R., Bernardini, G., 2011. Immunotoxicity of nanoparticles. Int. J. Immunopathol. Pharmacol. 24 (Suppl. 1), 65S–71S. Review.

Dröge, W., 2002. Free radicals in the physiological control of cell function. Physiol. Rev. 82 (1), 47–95. https://doi.org/10.1152/physrev.00018.2001.

Ferdous, Z., Nemmar, A., 2020. Health impact of silver nanoparticles: a review of the biodistribution and toxicity following various routes of exposure. Int. J. Mol. Sci. 21 (7), 2375. https://doi.org/10.3390/ijms21072375.

Guan, C.P., Li, Q.T., Jiang, H., Geng, Q.W., Xu, W., Li, L.Y., Xu, A.E., 2018. IGF-1 resist oxidative damage to HaCaT and depigmentation in mice treated with H(2)O(2). Biochem. Biophys. Res. Commun. 503 (4), 2485–2492. https://doi.org/10.1016/j.bbrc.2018.07.004. Epub 2018 Jul 4.

Herrero, M., Rovira, J., Esplugas, R., Nadal, M., Domingo, J.L., 2020. Human exposure to trace elements, aromatic amines and formaldehyde in swimsuits: assessment of the health risks. Environ. Res. 181, 108951. https://doi.org/10.1016/j.envres.2019.108951. Epub 2019 Nov 22.

Higashi, Y., Sukhanov, S., Anwar, A., Shai, S.Y., Delafontaine, P., 2010. IGF-1, oxidative stress and atheroprotection. Trends Endocrinol. Metab. 21 (4), 245–254. https://doi.org/10.1016/j.tem.2009.12.005. Epub 2010 Jan 12.

Huun, M.U., Garberg, H.T., Buonocore, G., Longini, M., Belvisi, E., Bazzini, F., Proietti, F., Saugstad, O.D., Solberg, R., 2018. Regional differences of hypothermia on oxidative stress following hypoxia-ischemia: a study of DHA and hypothermia on brain lipid peroxidation in newborn piglets. J. Perinat. Med. 47 (1), 82–89. https://doi.org/10.1515/jpm-2017-0355.

Jones, D.P., 2008. Radical-free biology of oxidative stress. Am. J. Physiol. Cell Physiol. 295 (4), C849–C868. https://doi.org/10.1152/ajpcell.00283.2008. Epub 2008 Aug 6.

Kadry, H., Noorani, B., Cucullo, L., 2020. A blood-brain barrier overview on structure, function, impairment, and biomarkers of integrity. Fluids Barriers CNS 17 (1), 69. https://doi.org/10.1186/s12987-020-00230-3.

Kamikubo, Y., Yamana, T., Hashimoto, Y., Sakurai, T., 2019. Induction of oxidative stress and cell death in neural cells by silica nanoparticles. ACS Chem. Nerosci. 10 (1), 304–312. https://doi.org/10.1021/acschemneuro.8b00248. Epub 2018 Oct 5.

Kaplan, L., Chow, B.W., Gu, C., 2020. Neuronal regulation of the blood-brain barrier and neurovascular coupling. Nat. Rev. Neurosci. 21 (8), 416–432. https://doi.org/10.1038/s41583-020-0322-2. Epub 2020 Jul 7.

Kiyatkin, E.A., Sharma, H.S., 2011. Expression of heat shock protein (HSP 72 kDa) during acute methamphetamine intoxication depends on brain hyperthermia: neurotoxicity or neuroprotection? J. Neural Transm. 118 (1), 47–60. Epub 2010 Oct 8.

Köerich, J.S., Nogueira, D.J., Vaz, V.P., Simioni, C., Silva, M.L.N.D., Ouriques, L.C., Vicentini, D.S., Matias, W.G., 2020. Toxicity of binary mixtures of Al(2)O(3) and ZnO nanoparticles toward fibroblast and bronchial epithelium cells. J. Toxicol. Environ. Health A 83 (9), 363–377. https://doi.org/10.1080/15287394.2020.1761496. Epub 2020 May 15.

Kovacic, P., Somanathan, R., 2010. Biomechanisms of nanoparticles (toxicants, antioxidants and therapeutics): electron transfer and reactive oxygen species. J. Nanosci. Nanotechnol. 10 (12), 7919–7930. https://doi.org/10.1166/jnn.2010.3028.

Kovacic, P., Somanathan, R., 2013. Nanoparticles: toxicity, radicals, electron transfer, and antioxidants. Methods Mol. Biol. 1028, 15–35. https://doi.org/10.1007/978-1-62703-475-3_2.

Lafuente, J.V., Sharma, A., Patnaik, R., Muresanu, D.F., Sharma, H.S., 2012. Diabetes exacerbates nanoparticles induced brain pathology. CNS Neurol. Disord. Drug Targets 11 (1), 26–39. https://doi.org/10.2174/187152712799960808.

Lafuente, J.V., Sharma, A., Muresanu, D.F., Ozkizilcik, A., Tian, Z.R., Patnaik, R., Sharma, H.S., 2018. Repeated forced swim exacerbates methamphetamine-induced neurotoxicity: neuroprotective effects of nanowired delivery of 5-HT3-receptor antagonist ondansetron. Mol. Neurobiol. 55 (1), 322–334. https://doi.org/10.1007/s12035-017-0744-7.

Li, Y., Cummins, E., 2020. Hazard characterization of silver nanoparticles for human exposure routes. J. Environ. Sci. Health A Tox. Hazard. Subst. Environ. Eng. 55 (6), 704–725. https://doi.org/10.1080/10934529.2020.1735852. Epub 2020 Mar 13.

Lojk, J., Repas, J., Veranič, P., Bregar, V.B., Pavlin, M., 2020. Toxicity mechanisms of selected engineered nanoparticles on human neural cells in vitro. Toxicology 432, 152364. https://doi.org/10.1016/j.tox.2020.152364. Epub 2020 Jan 9.

Lu, S., Zhang, W., Zhang, R., Liu, P., Wang, Q., Shang, Y., Wu, M., Donaldson, K., Wang, Q., 2015. Comparison of cellular toxicity caused by ambient ultrafine particles and engineered metal oxide nanoparticles. Part. Fibre Toxicol. 12, 5. https://doi.org/10.1186/s12989-015-0082-8.

Lushchak, V.I., 2014. Free radicals, reactive oxygen species, oxidative stress and its classification. Chem. Biol. Interact. 224, 164–175. https://doi.org/10.1016/j.cbi.2014.10.016. Epub 2014 Oct 28.

Menon, P.K., Muresanu, D.F., Sharma, A., Mössler, H., Sharma, H.S., 2012. Cerebrolysin, a mixture of neurotrophic factors induces marked neuroprotection in spinal cord injury following intoxication of engineered nanoparticles from metals. CNS Neurol. Disord. Drug Targets 11 (1), 40–49. https://doi.org/10.2174/187152712799960781.

Menon, P.K., Sharma, A., Lafuente, J.V., Muresanu, D.F., Aguilar, Z.P., Wang, Y.A., Patnaik, R., Mössler, H., Sharma, H.S., 2017. Intravenous administration of functionalized magnetic iron oxide nanoparticles does not induce CNS injury in the rat: influence of spinal cord trauma and cerebrolysin treatment. Int. Rev. Neurobiol. 137, 47–63. https://doi.org/10.1016/bs.irn.2017.08.005. Epub 2017 Nov 3.

Migliore, L., Uboldi, C., Di Bucchianico, S., Coppedè, F., 2015. Nanomaterials and neurodegeneration. Environ. Mol. Mutagen. 56 (2), 149–170. https://doi.org/10.1002/em.21931. Epub 2015 Jan 28.

Muresanu, D.F., Sharma, A., Lafuente, J.V., Patnaik, R., Tian, Z.R., Nyberg, F., Sharma, H.S., 2015. Nanowired delivery of growth hormone attenuates pathophysiology of spinal cord injury and enhances insulin-like growth factor-1 concentration in the plasma and the spinal cord. Mol. Neurobiol. 52 (2), 837–845. https://doi.org/10.1007/s12035-015-9298-8. Epub 2015 Jul 1.

Muresanu, D.F., Sharma, A., Patnaik, R., Menon, P.K., Mössler, H., Sharma, H.S., 2019. Exacerbation of blood-brain barrier breakdown, edema formation, nitric oxide synthase upregulation and brain pathology after heat stroke in diabetic and hypertensive rats. Potential neuroprotection with cerebrolysin treatment. Int. Rev. Neurobiol. 146, 83–102. https://doi.org/10.1016/bs.irn.2019.06.007. Epub 2019 Jul 18.

Murugadoss, S., Lison, D., Godderis, L., Van Den Brule, S., Mast, J., Brassinne, F., Sebaihi, N., Hoet, P.H., 2017. Toxicology of silica nanoparticles: an update. Arch. Toxicol. 91 (9), 2967–3010. https://doi.org/10.1007/s00204-017-1993-y. Epub 2017 Jun 1.

National Research Council (US), 2011. Committee for the Update of the Guide for the Care and Use of Laboratory Animals. Guide for the Care and Use of Laboratory Animals, eigth ed. National Academies Press (US), Washington (DC).

Nemmar, A., Al-Salam, S., Beegam, S., Yuvaraju, P., Ali, B.H., 2019. Aortic oxidative stress, inflammation and DNA damage following pulmonary exposure to cerium oxide nanoparticles in a rat model of vascular injury. Biomolecules 9 (8), 376. https://doi.org/10.3390/biom9080376.

Niu, F., Sharma, A., Feng, L., Ozkizilcik, A., Muresanu, D.F., Lafuente, J.V., Tian, Z.R., Nozari, A., Sharma, H.S., 2019. Nanowired delivery of DL-3-n-butylphthalide induces superior neuroprotection in concussive head injury. Prog. Brain Res. 245, 89–118. https://doi.org/10.1016/bs.pbr.2019.03.008. Epub 2019 Apr 2.

Niu, F., Sharma, A., Wang, Z., Feng, L., Muresanu, D.F., Sahib, S., Tian, Z.R., Lafuente, J.V., Buzoianu, A.D., Castellani, R.J., Nozari, A., Patnaik, R., Wiklund, L., Sharma, H.S., 2020. Co-administration of TiO(2)-nanowired dl-3-n-butylphthalide (dl-NBP) and mesenchymal stem cells enhanced neuroprotection in Parkinson's disease exacerbated by concussive head injury. Prog. Brain Res. 258, 101–155. https://doi.org/10.1016/bs.pbr.2020.09.011. Epub 2020 Nov 9.

Ogundele, O.M., Ebenezer, P.J., Lee, C.C., 2017. Francis stress-altered synaptic plasticity and DAMP signaling in the hippocampus-PFC axis; elucidating the significance of IGF-1/IGF-1R/CaMKIIalpha expression in neural changes associated with a prolonged exposure therapy. J. Neurosci. 353, 147–165. https://doi.org/10.1016/j.neuroscience.2017.04.008. Epub 2017 Apr.

Pandey, A.K., Patnaik, R., Muresanu, D.F., Sharma, A., Sharma, H.S., 2012. Quercetin in hypoxia-induced oxidative stress: novel target for neuroprotection. Int. Rev. Neurobiol. 102, 107–146. https://doi.org/10.1016/B978-0-12-386986-9.00005-3.

Parsai, T., Kumar, A., 2021. Weight-of-evidence process for assessing human health risk of mixture of metal oxide nanoparticles and corresponding ions in aquatic matrices. Chemosphere 263, 128289. https://doi.org/10.1016/j.chemosphere.2020.128289. Epub 2020 Sep 10.

Paunovic, J., Vucevic, D., Radosavljevic, T., Mandić-Rajčević, S., Pantic, I., 2020. Iron-based nanoparticles and their potential toxicity: focus on oxidative stress and apoptosis. Chem. Biol. Interact. 316, 108935. https://doi.org/10.1016/j.cbi.2019.108935. Epub 2019 Dec 21.

Pharaoh, G., Owen, D., Yeganeh, A., Premkumar, P., Farley, J., Bhaskaran, S., Ashpole, N., Kinter, M., Van Remmen, H., Logan, S., 2020. Disparate central and peripheral effects of circulating IGF-1 deficiency on tissue mitochondrial function. Mol. Neurobiol. 57 (3), 1317–1331. https://doi.org/10.1007/s12035-019-01821-4. Epub 2019 Nov 15.

Poprac, P., Jomova, K., Simunkova, M., Kollar, V., Rhodes, C.J., Valko, M., 2017. Targeting free radicals in oxidative stress-related human diseases. Trends Pharmacol. Sci. 38 (7), 592–607. https://doi.org/10.1016/j.tips.2017.04.005. Epub 2017 May 24.

Ramana, K.V., Srivastava, S., Singhal, S.S., 2017. Lipid peroxidation products in human health and disease 2016. Oxid. Med. Cell. Longev. 2017, 2163285. https://doi.org/10.1155/2017/2163285. Epub 2017 Feb 28.

Riis, S., Murray, J.B., O'Connor, R., 2020. IGF-1 signalling regulates mitochondria dynamics and turnover through a conserved GSK-3beta-Nrf2-BNIP3 pathway. Cell 9 (1), 147. https://doi.org/10.3390/cells9010147.

Rush, J.W., Ford, R.J., 2007. Nitric oxide, oxidative stress and vascular endothelium in health and hypertension. Clin. Hemorheol. Microcirc. 37 (1–2), 185–192.

Sádaba, M.C., Martín-Estal, I., Puche, J.E., Castilla-Cortázar, I., 2016. Insulin-like growth factor 1 (IGF-1) therapy: mitochondrial dysfunction and diseases. Biochim. Biophys. Acta 1862 (7), 1267–1278. https://doi.org/10.1016/j.bbadis.2016.03.010. Epub 2016 Mar 25.

Sahni, S., Hickok, J.R., Thomas, D.D., 2018. Nitric oxide reduces oxidative stress in cancer cells by forming dinitrosyliron complexes. Nitric Oxide 76, 37–44. https://doi.org/10.1016/j.niox.2018.03.003. Epub 2018 Mar 6.

Sarkar, A., Ghosh, M., Sil, P.C., 2014. Nanotoxicity: oxidative stress mediated toxicity of metal and metal oxide nanoparticles. J. Nanosci. Nanotechnol. 14 (1), 730–743. https://doi.org/10.1166/jnn.2014.8752.

Semenas, E., Sharma, H.S., Nozari, A., Basu, S., Wiklund, L., 2011. Neuroprotective effects of 17β-estradiol after hypovolemic cardiac arrest in immature piglets: the role of nitric oxide and peroxidation. Shock 36 (1), 30–37.

Serhan, A., Aerts, J.L., Boddeke, E.W.G.M., Kooijman, R., 2020. Neuroprotection by insulin-like growth factor-1 in rats with ischemic stroke is associated with microglial changes and a reduction in neuroinflammation. Neuroscience 426, 101–114. https://doi.org/10.1016/j.neuroscience.2019.11.035. Epub 2019 Dec 14.

Sharma, H.S., 1987. Effect of captopril (a converting enzyme inhibitor) on blood-brain barrier permeability and cerebral blood flow in normotensive rats. Neuropharmacology 26 (1), 85–92. https://doi.org/10.1016/0028-3908(87)90049-9.

Sharma, H.S., 1999. Pathophysiology of Blood-Brain Barrier, Brain Edema and Cell Injury Following Hyperthermia: New Role of Heat Shock Protein, Nitric Oxide and Carbon Monoxide. An Experimental Study in the Rat Using Light and Electron Microscopy. vol. 830 Acta Universitatis Upsaliensis, pp. 1–94.

Sharma, H.S., 2003. Neurotrophic factors attenuate microvascular permeability disturbances and axonal injury following trauma to the rat spinal cord. Acta Neurochir. Suppl. 86, 383–388. https://doi.org/10.1007/978-3-7091-0651-8_81.

Sharma, H.S., 2004. Blood-brain and spinal cord barriers in stress. In: Sharma, H.S., Westman, J. (Eds.), The Blood-Spinal Cord and Brain Barriers in Health and Disease. Elsevier Academic Press, San Diego, pp. 231–298.

Sharma, H.S., 2005. Neuroprotective effects of neurotrophins and melanocortins in spinal cord injury: an experimental study in the rat using pharmacological and morphological approaches. Ann. N. Y. Acad. Sci. 1053, 407–421. https://doi.org/10.1111/j.1749-6632.2005.tb00050.x.

Sharma, H.S., 2007. A select combination of neurotrophins enhances neuroprotection and functional recovery following spinal cord injury. Ann. N. Y. Acad. Sci. 1122, 95–111. https://doi.org/10.1196/annals.1403.007.

Sharma, H.S., 2009a. Blood–central nervous system barriers: the gateway to neurodegeneration, neuroprotection and neuroregeneration. In: Lajtha, A., Banik, N., Ray, S.K. (Eds.), Handbook of Neurochemistry and Molecular Neurobiology: Brain and Spinal Cord Trauma. Springer Verlag, Berlin, Heidelberg, New York, pp. 363–457.

Sharma, H.S., 2009b. Preface. Nanoneuropharmacology and nanoneurotoxicology. Prog. Brain Res. 180, vii–ix. Epub 2009 Dec 8.

Sharma, H.S., 2010a. Selected combination of neurotrophins potentiate neuroprotection and functional recovery following spinal cord injury in the rat. Acta Neurochir. Suppl. 106, 295–300. https://doi.org/10.1007/978-3-211-98811-4_55.

Sharma, H.S., 2010b. A combination of tumor necrosis factor-alpha and neuronal nitric oxide synthase antibodies applied topically over the traumatized spinal cord enhances neuroprotection and functional recovery in the rat. Ann. N. Y. Acad. Sci. 1199, 175–185.

Sharma, H.S., Alm, P., 2002. Nitric oxide synthase inhibitors influence dynorphin A (1-17) immunoreactivity in the rat brain following hyperthermia. Amino Acids 23 (1–3), 247–259. https://doi.org/10.1007/s00726-001-0136-0.

Sharma, H.S., Alm, P., 2004. Role of nitric oxide on the blood-brain and the spinal cord barriers. In: Sharma, H.S., Westman, J. (Eds.), The Blood-Spinal Cord and Brain Barriers in Health and Disease. Elsevier Academic Press, San Diego, pp. 191–230.

Sharma, H.S., Cervós-Navarro, J., 1990. Brain oedema and cellular changes induced by acute heat stress in young rats. Acta Neurochir. Suppl. (Wien) 51, 383–386. https://doi.org/10.1007/978-3-7091-9115-6_129.

Sharma, H.S., Dey, P.K., 1986a. Influence of long-term immobilization stress on regional blood-brain barrier permeability, cerebral blood flow and 5-HT level in conscious normotensive young rats. J. Neurol. Sci. 72 (1), 61–76. https://doi.org/10.1016/0022-510x(86)90036-5.

Sharma, H.S., Dey, P.K., 1986b. Probable involvement of 5-hydroxytryptamine in increased permeability of blood-brain barrier under heat stress in young rats. Neuropharmacology 25 (2), 161–167. https://doi.org/10.1016/0028-3908(86)90037-7.

Sharma, H.S., Dey, P.K., 1987. Influence of long-term acute heat exposure on regional blood-brain barrier permeability, cerebral blood flow and 5-HT level in conscious normotensive young rats. Brain Res. 424 (1), 153–162. https://doi.org/10.1016/0006-8993(87)91205-4.

Sharma, H.S., Johanson, C.E., 2007. Intracerebroventricularly administered neurotrophins attenuate blood cerebrospinal fluid barrier breakdown and brain pathology following whole-body hyperthermia: an experimental study in the rat using biochemical and morphological approaches. Ann. N. Y. Acad. Sci. 1122, 112–129. https://doi.org/10.1196/annals.1403.008.

Sharma, H.S., Kiyatkin, E.A., 2009. Rapid morphological brain abnormalities during acute methamphetamine intoxication in the rat: an experimental study using light and electron microscopy. J. Chem. Neuroanat. 37 (1), 18–32. Epub 2008 Aug 19.

Sharma, H.S., Olsson, Y., 1990. Edema formation and cellular alterations following spinal cord injury in the rat and their modification with p-chlorophenylalanine. Acta Neuropathol. 79 (6), 604–610. https://doi.org/10.1007/BF00294237.

Sharma, H.S., Sharma, A., 2007. Nanoparticles aggravate heat stress induced cognitive deficits, blood-brain barrier disruption, edema formation and brain pathology. Prog. Brain Res. 162, 245–273. Review.

Sharma, H.S., Sharma, A., 2012a. Nanowired drug delivery for neuroprotection in central nervous system injuries: modulation by environmental temperature, intoxication of nanoparticles, and comorbidity factors. Wiley Interdiscip. Rev. Nanomed. Nanobiotechnol. 4 (2), 184–203. https://doi.org/10.1002/wnan.172. Epub 2011 Dec 8.

Sharma, H.S., Sharma, A., 2012b. Neurotoxicity of engineered nanoparticles from metals. CNS Neurol. Disord. Drug Targets 11 (1), 65–80. https://doi.org/10.2174/187152712799960817.

Sharma, H.S., Sharma, A., 2013. New perspectives of nanoneuroprotection, nanoneuropharmacology and nanoneurotoxicity: modulatory role of amino acid neurotransmitters, stress, trauma, and co-morbidity factors in nanomedicine. Amino Acids 45 (5), 1055–1071. https://doi.org/10.1007/s00726-013-1584-z.

Sharma, H.S., Sjöquist, P.O., 2002. A new antioxidant compound H-290/51 modulates glutamate and GABA immunoreactivity in the rat spinal cord following trauma. Amino Acids 23 (1–3), 261–272. https://doi.org/10.1007/s00726-001-0137-z.

Sharma, H.S., Westman, J., 1997. Prostaglandins modulate constitutive isoform of heat shock protein (72 kD) response following trauma to the rat spinal cord. Acta Neurochir. Suppl. 70, 134–137.

Sharma, H.S., Westman, J., 2004. The Blood-Spinal Cord and Brain Barriers in Health and Disease. Academic Press, San Diego, pp. 1–617. Release date: Nov. 9, 2003.

Sharma, H.S., Winkler, T., 2002. Assessment of spinal cord pathology following trauma using spinal cord evoked potentials. A pharmacological and morphological study in the rat. Muscle Nerve Suppl. 11, S83–S91.

Sharma, H.S., Olsson, Y., Dey, P.K., 1990a. Changes in blood-brain barrier and cerebral blood flow following elevation of circulating serotonin level in anesthetized rats. Brain Res. 517 (1–2), 215–223. https://doi.org/10.1016/0006-8993(90)91029-g.

Sharma, H.S., Olsson, Y., Dey, P.K., 1990b. Early accumulation of serotonin in rat spinal cord subjected to traumatic injury. Relation to edema and blood flow changes. Neuroscience 36 (3), 725–730. https://doi.org/10.1016/0306-4522(90)90014-u.

Sharma, H.S., Cervós-Navarro, J., Dey, P.K., 1991a. Acute heat exposure causes cellular alteration in cerebral cortex of young rats. Neuroreport 2 (3), 155–158. https://doi.org/10.1097/00001756-199103000-00012.

Sharma, H.S., Cervós-Navarro, J., Dey, P.K., 1991b. Increased blood-brain barrier permeability following acute short-term swimming exercise in conscious normotensive young rats. Neurosci. Res. 10 (3), 211–221. https://doi.org/10.1016/0168-0102(91)90058-7.

Sharma, H.S., Westman, J., Navarro, J.C., Dey, P.K., Nyberg, F., 1995. Probable involvement of serotonin in the increased permeability of the blood-brain barrier by forced swimming. An experimental study using Evans blue and 131I-sodium tracers in the rat. Behav. Brain Res. 72 (1–2), 189–196. https://doi.org/10.1016/0166-4328(96)00170-2.

Sharma, H.S., Westman, J., Olsson, Y., Alm, P., 1996. Involvement of nitric oxide in acute spinal cord injury: an immunocytochemical study using light and electron microscopy in the rat. Neurosci. Res. 24 (4), 373–384.

Sharma, H.S., Nyberg, F., Gordh, T., Alm, P., Westman, J., 1997. Topical application of insulin like growth factor-1 reduces edema and upregulation of neuronal nitric oxide synthase following trauma to the rat spinal cord. Acta Neurochir. Suppl. 70, 130–133. https://doi.org/10.1007/978-3-7091-6837-0_40.

Sharma, H.S., Nyberg, F., Westman, J., Alm, P., Gordh, T., Lindholm, D., 1998a. Brain derived neurotrophic factor and insulin like growth factor-1 attenuate upregulation of nitric oxide synthase and cell injury following trauma to the spinal cord. An immunohistochemical study in the rat. Amino Acids 14 (1–3), 121–129. https://doi.org/10.1007/BF01345252.

Sharma, H.S., Nyberg, F., Gordh, T., Alm, P., Westman, J., Stålberg, E., Sharma, H.S., Olsson, Y., 1998b. Neurotrophic factors attenuate neuronal nitric oxide synthase upregulation, microvascular permeability disturbances, edema formation and cell injury in the spinal cord following trauma. In: Spinal Cord Monitoring. Basic Principles, Regeneration, Pathophysiology and Clinical Aspects. Springer, Wien New York, pp. 118–148.

Sharma, H.S., Westman, J., Nyberg, F., 1998c. Pathophysiology of brain edema and cell changes following hyperthermic brain injury. Prog. Brain Res. 115, 351–412. https://doi.org/10.1016/s0079-6123(08)62043-9.

Sharma, H.S., Alm, P., Westman, J., 1998d. Nitric oxide and carbon monoxide in the brain pathology of heat stress. Prog. Brain Res. 115, 297–333. https://doi.org/10.1016/s0079-6123(08)62041-5.

Sharma, H.S., Nyberg, F., Gordh, T., Alm, P., Westman, J., 2000a. Neurotrophic factors influence upregulation of constitutive isoform of heme oxygenase and cellular stress response in the spinal cord following trauma. An experimental study using immunohistochemistry in the rat. Amino Acids 19 (1), 351–361.

Sharma, H.S., Drieu, K., Alm, P., Westman, J., 2000b. Role of nitric oxide in blood-brain barrier permeability, brain edema and cell damage following hyperthermic brain injury. An experimental study using EGB-761 and Gingkolide B pretreatment in the rat. Acta Neurochir. Suppl. 76, 81–86.

Sharma, H.S., Badgaiyan, R.D., Alm, P., Mohanty, S., Wiklund, L., 2005. Neuroprotective effects of nitric oxide synthase inhibitors in spinal cord injury-induced pathophysiology and motor functions: an experimental study in the rat. Ann. N. Y. Acad. Sci. 1053, 422–434. https://doi.org/10.1111/j.1749-6632.2005.tb00051.x.

Sharma, H.S., Sjöquist, P.O., Mohanty, S., Wiklund, L., 2006a. Post-injury treatment with a new antioxidant compound H-290/51 attenuates spinal cord trauma-induced c-fos expression, motor dysfunction, edema formation, and cell injury in the rat. Acta Neurochir. Suppl. 96, 322–328. https://doi.org/10.1007/3-211-30714-1_68.

Sharma, H.S., Wiklund, L., Badgaiyan, R.D., Mohanty, S., Alm, P., 2006b. Intracerebral administration of neuronal nitric oxide synthase antiserum attenuates traumatic brain injury-induced blood-brain barrier permeability, brain edema formation, and sensory motor disturbances in the rat. Acta Neurochir. Suppl. 96, 288–294. https://doi.org/10.1007/3-211-30714-1_62.

Sharma, H.S., Nyberg, F., Gordh, T., Alm, P., 2006c. Topical application of dynorphin A (1-17) antibodies attenuates neuronal nitric oxide synthase up-regulation, edema formation, and cell injury following focal trauma to the rat spinal cord. Acta Neurochir. Suppl. 96, 309–315.

Sharma, H.S., Gordh, T., Wiklund, L., Mohanty, S., Sjöquist, P.O., 2006d. Spinal cord injury induced heat shock protein expression is reduced by an antioxidant compound H-290/51. An experimental study using light and electron microscopy in the rat. J. Neural Transm. (Vienna) 113 (4), 521–536. https://doi.org/10.1007/s00702-005-0405-2.

Sharma, H.S., Muresanu, D.F., Sharma, A., Patnaik, R., Lafuente, J.V., 2009a. Chapter 9—Nanoparticles influence pathophysiology of spinal cord injury and repair. Prog. Brain Res. 180, 154–180. https://doi.org/10.1016/S0079-6123(08)80009-X. Epub 2009e Dec 8.

Sharma, H.S., Patnaik, R., Sharma, A., Sjöquist, P.O., Lafuente, J.V., 2009b. Silicon dioxide nanoparticles (SiO2, 40-50 nm) exacerbate pathophysiology of traumatic spinal cord injury and deteriorate functional outcome in the rat. An experimental study using pharmacological and morphological approaches. J. Nanosci. Nanotechnol. 9 (8), 4970–4980. https://doi.org/10.1166/jnn.2009.1717.

Sharma, H.S., Ali, S.F., Hussain, S.M., Schlager, J.J., Sharma, A., 2009c. Influence of engineered nanoparticles from metals on the blood-brain barrier permeability, cerebral blood flow, brain edema and neurotoxicity. An experimental study in the rat and mice using biochemical and morphological approaches. J. Nanosci. Nanotechnol. 9 (8), 5055–5072. https://doi.org/10.1166/jnn.2009.gr09.

Sharma, H.S., Ali, S.F., Tian, Z.R., Hussain, S.M., Schlager, J.J., Sjöquist, P.O., Sharma, A., Muresanu, D.F., 2009d. Chronic treatment with nanoparticles exacerbate hyperthermia induced blood-brain barrier breakdown, cognitive dysfunction and brain pathology in the rat. Neuroprotective effects of nanowired-antioxidant compound H-290/51. J. Nanosci. Nanotechnol. 9 (8), 5073–5090. https://doi.org/10.1166/jnn.2009.gr10.

Sharma, H.S., Ali, S., Tian, Z.R., Patnaik, R., Patnaik, S., Lek, P., Sharma, A., Lundstedt, T., 2009e. Nano-drug delivery and neuroprotection in spinal cord injury. J. Nanosci. Nanotechnol. 9 (8), 5014–5037. https://doi.org/10.1166/jnn.2009.gr04.

Sharma, H.S., Patnaik, R., Sharma, A., 2010a. Diabetes aggravates nanoparticles induced breakdown of the blood-brain barrier permeability, brain edema formation, alterations in cerebral blood flow and neuronal injury. An experimental study using physiological and morphological investigations in the rat. J. Nanosci. Nanotechnol. 10 (12), 7931–7945.

Sharma, H.S., Hussain, S., Schlager, J., Ali, S.F., Sharma, A., 2010b. Influence of nanoparticles on blood-brain barrier permeability and brain edema formation in rats. Acta Neurochir. Suppl. 106, 359–364. https://doi.org/10.1007/978-3-211-98811-4_65.

Sharma, H.S., Sharma, A., Hussain, S., Schlager, J., Sjöquist, P.O., Muresanu, D., 2010c. A new antioxidant compound H-290/51 attenuates nanoparticle induced neurotoxicity and enhances neurorepair in hyperthermia. Acta Neurochir. Suppl. 106, 351–357.

Sharma, H.S., Miclescu, A., Wiklund, L., 2011a. Cardiac arrest-induced regional blood-brain barrier breakdown, edema formation and brain pathology: a light and electron microscopic study on a new model for neurodegeneration and neuroprotection in porcine brain. J. Neural Transm. 118 (1), 87–114. https://doi.org/10.1007/s00702-010-0486-4. Epub 2010 Oct 21.

Sharma, H.S., Muresanu, D.F., Patnaik, R., Stan, A.D., Vacaras, V., Perju-Dumbrava, L., Alexandru, B., Buzoianu, A., Opincariu, I., Menon, P.K., Sharma, A., 2011b. Superior neuroprotective effects of cerebrolysin in heat stroke following chronic intoxication of Cu or Ag engineered nanoparticles. A comparative study with other neuroprotective agents using biochemical and morphological approaches in the rat. J. Nanosci. Nanotechnol. 11 (9), 7549–7569.

Sharma, H.S., Ali, S.F., Patnaik, R., Zimmermann-Meinzingen, S., Sharma, A., Muresanu, D.F., 2011c. Cerebrolysin attenuates heat shock protein (HSP 72 KD) expression in the rat spinal cord following morphine dependence and withdrawal: possible new therapy for pain management. Curr. Neuropharmacol. 9 (1), 223–235.

Sharma, A., Muresanu, D.F., Patnaik, R., Sharma, H.S., 2013a. Size- and age-dependent neurotoxicity of engineered metal nanoparticles in rats. Mol. Neurobiol. 48 (2), 386–396. https://doi.org/10.1007/s12035-013-8500-0. Epub 2013 Jul 3.

Sharma, H.S., Muresanu, D.F., Patnaik, R., Sharma, A., 2013b. Exacerbation of brain pathology after partial restraint in hypertensive rats following SiO2 nanoparticles exposure at high ambient temperature. Mol. Neurobiol. 48 (2), 368–379. https://doi.org/10.1007/s12035-013-8502-y. Epub 2013 Jul 6.

Sharma, H.S., Menon, P.K., Lafuente, J.V., Aguilar, Z.P., Wang, Y.A., Muresanu, D.F., Mössler, H., Patnaik, R., Sharma, A., 2014. The role of functionalized magnetic iron oxide nanoparticles in the central nervous system injury and repair: new potentials for neuroprotection with Cerebrolysin therapy. J. Nanosci. Nanotechnol. 14 (1), 577–595. https://doi.org/10.1166/jnn.2014.9213.

Sharma, H.S., Muresanu, D.F., Lafuente, J.V., Sjöquist, P.O., Patnaik, R., Sharma, A., 2015a. Nanoparticles exacerbate both ubiquitin and heat shock protein expressions in spinal cord injury: neuroprotective effects of the proteasome inhibitor carfilzomib and the antioxidant compound H-290/51. Mol. Neurobiol. 52 (2), 882–898. https://doi.org/10.1007/s12035-015-9297-9. Epub 2015 Jul 1.

Sharma, H.S., Kiyatkin, E.A., Patnaik, R., Lafuente, J.V., Muresanu, D.F., Sjöquist, P.O., Sharma, A., 2015b. Exacerbation of methamphetamine neurotoxicity in cold and hot environments: neuroprotective effects of an antioxidant compound H-290/51. Mol. Neurobiol. 52 (2), 1023–1033. https://doi.org/10.1007/s12035-015-9252-9. Epub 2015 Jun 26.

Sharma, A., Muresanu, D.F., Lafuente, J.V., Patnaik, R., Tian, Z.R., Buzoianu, A.D., Sharma, H.S., 2015c. Sleep deprivation-induced blood-brain barrier breakdown and brain dysfunction are exacerbated by size-related exposure to Ag and Cu nanoparticles. neuroprotective effects of a 5-HT3 receptor antagonist ondansetron. Mol. Neurobiol. 52 (2), 867–881. https://doi.org/10.1007/s12035-015-9236-9.

Sharma, A., Muresanu, D.F., Lafuente, J.V., Sjöquist, P.O., Patnaik, R., Ryan Tian, Z., Ozkizilcik, A., Sharma, H.S., 2018. Cold environment exacerbates brain pathology and oxidative stress following traumatic brain injuries: potential therapeutic effects of nanowired antioxidant compound H-290/51. Mol. Neurobiol. 55 (1), 276–285. https://doi.org/10.1007/s12035-017-0740-y.

Sharma, H.S., Muresanu, D.F., Nozari, A., Castellani, R.J., Dey, P.K., Wiklund, L., Sharma, A., 2019a. Anesthetics influence concussive head injury induced blood-brain barrier breakdown, brain edema formation, cerebral blood flow, serotonin levels, brain pathology and functional outcome. Int. Rev. Neurobiol. 146, 45–81. https://doi.org/10.1016/bs.irn.2019.06.006. Epub 2019 Jul 8.

Sharma, H.S., Feng, L., Muresanu, D.F., Castellani, R.J., Sharma, A., 2019b. Neuroprotective effects of a potent bradykinin B2 receptor antagonist HOE-140 on microvascular permeability, blood flow disturbances, edema formation, cell injury and nitric oxide synthase upregulation following trauma to the spinal cord. Int. Rev. Neurobiol. 146, 103–152. https://doi.org/10.1016/bs.irn.2019.06.008. Epub 2019 Jul 9.

Sharma, A., Muresanu, D.F., Castellani, R.J., Nozari, A., Lafuente, J.V., Sahib, S., Tian, Z.R., Buzoianu, A.D., Patnaik, R., Wiklund, L., Sharma, H.S., 2020a. Mild traumatic brain injury exacerbates Parkinson's disease induced hemeoxygenase-2 expression and brain pathology: neuroprotective effects of co-administration of TiO_2 nanowired mesenchymal stem cells and cerebrolysin. Prog. Brain Res. 258, 157–231. https://doi.org/10.1016/bs.pbr.2020.09.010. Epub 2020 Nov 9.

Sharma, A., Muresanu, D.F., Sahib, S., Tian, Z.R., Castellani, R.J., Nozari, A., Lafuente, J.V., Buzoianu, A.D., Bryukhovetskiy, I., Manzhulo, I., Patnaik, R., Wiklund, L., Sharma, H.S., 2020b. Concussive head injury exacerbates neuropathology of sleep deprivation: superior neuroprotection by co-administration of TiO(2)-nanowired cerebrolysin, alpha-melanocyte-stimulating hormone, and mesenchymal stem cells. Prog. Brain Res. 258, 1–77. https://doi.org/10.1016/bs.pbr.2020.09.003. Epub 2020 Nov 2.

Sharp, F.R., Sagar, S.M., 1994. Alterations in gene expression as an index of neuronal injury: heat shock and the immediate early gene response. Neurotoxicology 15 (1), 51–59. Review.

Sohrabji, F., 2015. Estrogen-IGF-1 interactions in neuroprotection: ischemic stroke as a case study. Front. Neuroendocrinol. 36, 1–14. https://doi.org/10.1016/j.yfrne.2014.05.003. Epub 2014 May 29.

Sousa, V.S., Teixeira, M.R., 2020. Metal-based engineered nanoparticles in the drinking water treatment systems: a critical review. Sci. Total Environ. 707, 136077. https://doi.org/10.1016/j.scitotenv.2019.136077. Epub 2019 Dec 13.

Spielman, L.J., Little, J.P., Klegeris, A., 2014. Inflammation and insulin/IGF-1 resistance as the possible link between obesity and neurodegeneration. J. Neuroimmunol. 273 (1–2), 8–21. https://doi.org/10.1016/j.jneuroim.2014.06.004. Epub 2014 Jun.

Su, J., Huang, M., 2018. Etidronate protects chronic ocular hypertension induced retinal oxidative stress and promotes retinal ganglion cells growth through IGF-1 signaling pathway. Eur. J. Pharmacol. 841, 75–81. https://doi.org/10.1016/j.ejphar.2018.10.002. Epub 2018 Oct 13.

Tejero, J., Shiva, S., Gladwin, M.T., 2019. Sources of vascular nitric oxide and reactive oxygen species and their regulation. Physiol. Rev. 99 (1), 311–379. https://doi.org/10.1152/physrev.00036.2017.

Tomita, Y., Rikimaru-Kaneko, A., Hashiguchi, K., Shirotake, S., 2011. Effect of anionic and cationic n-butylcyanoacrylate nanoparticles on NO and cytokine production in Raw264.7 cells. Immunopharmacol. Immunotoxicol. 33 (4), 730–737. Epub ahead of print.

Torreggiani, A., Tinti, A., Jurasekova, Z., Capdevila, M., Saracino, M., Foggia, M.D., 2019. Structural lesions of proteins connected to lipid membrane damages caused by radical stress: assessment by biomimetic systems and Raman spectroscopy. Biomolecules 9 (12), 794. https://doi.org/10.3390/biom9120794.

Tortella, G.R., Rubilar, O., Durán, N., Diez, M.C., Martínez, M., Parada, J., Seabra, A.B., 2020. Silver nanoparticles: toxicity in model organisms as an overview of its hazard for human health and the environment. J. Hazard. Mater. 390, 121974. https://doi.org/10.1016/j.jhazmat.2019.121974. Epub 2019 Dec 24 32062374. Review.

Tsikas, D., 2017. Assessment of lipid peroxidation by measuring malondialdehyde (MDA) and relatives in biological samples: analytical and biological challenges. Anal. Biochem. 524, 13–30. https://doi.org/10.1016/j.ab.2016.10.021. Epub 2016 Oct 24.

Valko, M., Jomova, K., Rhodes, C.J., Kuča, K., Musílek, K., 2016. Redox- and non-redox-metal-induced formation of free radicals and their role in human disease. Arch. Toxicol. 90 (1), 1–37. https://doi.org/10.1007/s00204-015-1579-5. Epub 2015 Sep 7.

Wang, Y.J., Wong, H.S., Wu, C.C., Chiang, Y.H., Chiu, W.T., Chen, K.Y., Chang, W.C., 2019. The functional roles of IGF-1 variants in the susceptibility and clinical outcomes of mild traumatic brain injury. J. Biomed. Sci. 26 (1), 94. https://doi.org/10.1186/s12929-019-0587-9.

Wiklund, L., Basu, S., Miclescu, A., Wiklund, P., Ronquist, G., Sharma, H.S., 2007. Neuro- and cardioprotective effects of blockade of nitric oxide action by administration of methylene blue. Ann. N. Y. Acad. Sci. 1122, 231–244. https://doi.org/10.1196/annals.1403.016.

Wiklund, L., Patnaik, R., Sharma, A., Miclescu, A., Sharma, H.S., 2018. Cerebral tissue oxidative ischemia-reperfusion injury in connection with experimental cardiac arrest and cardiopulmonary resuscitation: effect of mild hypothermia and methylene blue. Mol. Neurobiol. 55 (1), 115–121. https://doi.org/10.1007/s12035-017-0723-z.

Wu, Y., Kong, L., 2020. Advance on toxicity of metal nickel nanoparticles. Environ. Geochem. Health 42 (7), 2277–2286. https://doi.org/10.1007/s10653-019-00491-4. Epub 2020 Jan 1.

Yoshitomi, T., Nagasaki, Y., 2011. Nitroxyl radical-containing nanoparticles for novel nanomedicine against oxidative stress injury. Nanomedicine (Lond.) 6 (3), 509–518. Review.

Yui, K., Kawasaki, Y., Yamada, H., Ogawa, S., 2016. Oxidative stress and nitric oxide in autism spectrum disorder and other neuropsychiatric disorders. CNS Neurol. Disord. Drug Targets 15 (5), 587–596. https://doi.org/10.2174/1871527315666160413121751.

Zhou, F., Liao, F., Chen, L., Liu, Y., Wang, W., Feng, S., 2019. The size-dependent genotoxicity and oxidative stress of silica nanoparticles on endothelial cells. Environ. Sci. Pollut. Res. Int. 26 (2), 1911–1920. https://doi.org/10.1007/s11356-018-3695-2. Epub 2018 Nov 20.

Zuberek, M., Grzelak, A., 2018. Nanoparticles-caused oxidative imbalance. Adv. Exp. Med. Biol. 1048, 85–98. https://doi.org/10.1007/978-3-319-72041-8_6.

CHAPTER

4

Methamphetamine exacerbates pathophysiology of traumatic brain injury at high altitude. Neuroprotective effects of nanodelivery of a potent antioxidant compound H-290/51

Hari Shanker Sharma[a,*], José Vicente Lafuente[b], Lianyuan Feng[c], Dafin F. Muresanu[d,e], Preeti K. Menon[l], Rudy J. Castellani[f], Ala Nozari[g], Seaab Sahib[h], Z. Ryan Tian[h], Anca D. Buzoianu[i], Per-Ove Sjöquist[j], Ranjana Patnaik[k], Lars Wiklund[a], and Aruna Sharma[a,*]

[a]*International Experimental Central Nervous System Injury & Repair (IECNSIR), Department of Surgical Sciences, Anesthesiology & Intensive Care Medicine, Uppsala University Hospital, Uppsala University, Uppsala, Sweden*

[b]*LaNCE, Department of Neuroscience, University of the Basque Country (UPV/EHU), Leioa, Bizkaia, Spain*

[c]*Department of Neurology, Bethune International Peace Hospital, Shijiazhuang, Hebei Province, China*

[d]*Department of Clinical Neurosciences, University of Medicine & Pharmacy, Cluj-Napoca, Romania*

[e]*"RoNeuro" Institute for Neurological Research and Diagnostic, Cluj-Napoca, Romania*

[f]*Department of Pathology, University of Maryland, Baltimore, MD, United States*

[g]*Anesthesiology & Intensive Care, Massachusetts General Hospital, Boston, MA, United States*

[h]*Department of Chemistry & Biochemistry, University of Arkansas, Fayetteville, AR, United States*

[i]*Department of Clinical Pharmacology and Toxicology, "Iuliu Hatieganu" University of Medicine and Pharmacy, Cluj-Napoca, Romania*

[j]*Division of Cardiology, Department of Medicine, Karolinska Institutet, Karolinska University Hospital, Stockholm, Sweden*

[k]*Department of Biomaterials, School of Biomedical Engineering, Indian Institute of Technology, Banaras Hindu University, Varanasi, India*

Progress in Brain Research, Volume 266, ISSN 0079-6123, https://doi.org/10.1016/bs.pbr.2021.06.008
Copyright © 2021 Elsevier B.V. All rights reserved.

[1]Department of Biochemistry and Biophysics, Stockholm University, Stockholm, Sweden
*Corresponding authors: Tel./Fax: +46-18-24-38-99,
e-mail addresses: harishanker_sharma55@icloud.com; hssharma@aol.com;
sharma@surgsci.uu.se; aruna.sharma@surgsci.uu.se

Abstract

Military personnel are often exposed to high altitude (HA, ca. 4500–5000m) for combat operations associated with neurological dysfunctions. HA is a severe stressful situation and people frequently use methamphetamine (METH) or other psychostimulants to cope stress. Since military personnel are prone to different kinds of traumatic brain injury (TBI), in this review we discuss possible effects of METH on concussive head injury (CHI) at HA based on our own observations. METH exposure at HA exacerbates pathophysiology of CHI as compared to normobaric laboratory environment comparable to sea level. Increased blood-brain barrier (BBB) breakdown, edema formation and reductions in the cerebral blood flow (CBF) following CHI were exacerbated by METH intoxication at HA. Damage to cerebral microvasculature and expression of beta catenin was also exacerbated following CHI in METH treated group at HA. TiO2-nanowired delivery of H-290/51 (150mg/kg, i.p.), a potent chain-breaking antioxidant significantly enhanced CBF and reduced BBB breakdown, edema formation, beta catenin expression and brain pathology in METH exposed rats after CHI at HA. These observations are the first to point out that METH exposure in CHI exacerbated brain pathology at HA and this appears to be related with greater production of oxidative stress induced brain pathology, not reported earlier.

Keywords

High altitude, Brain pathology, Methamphetamine, Traumatic brain injury, Brain edema, Cerebral blood flow, Antioxidant, Nanowired delivery, H-290/51, Military medicine

1 Introduction

Military personnel are often engaged in combat operation or peace-keeping activities at high altitude (HA) ranging from 3000 to more than 5000m from sea levels (Bakker-Dyos et al., 2016; Hochachka et al., 1999; Liu et al., 2018; McLaughlin et al., 2017; Muza, 2007; Rodway and Muza, 2011). At this HA adaptation takes longer periods to acclimatize physiologically (Sikri et al., 2019). However, several neurological functions and biomarkers show massive alterations at (HA) (Lefferts et al., 2019; Mellor et al., 2014; Naeije, 2010; Paul et al., 2018; Sharma et al., 2019a,b,c; Yang et al., 2016). To cope with HA induced alterations in the physiological system people often use several ways to adapt to new situations. Among these substance use is quite common to feel better at HA (Biondich and Joslin, 2015; Chaturvedi and Mahanta, 2004; Donegani et al., 2016; Evans and Witt, 1966; Kerry et al., 2016; Kim et al., 2014; Pavlovic et al., 2019; Pitman and Harris, 2012; Quiñones-Laveriano et al., 2016; Schinder and Ruder, 1989). Several lines of evidences suggests that use of recreational drugs such as methamphetamine (METH), amphetamine, cocaine, ecstasy (3,4-Methylenedioxymethamphetamine, MDMA), cannabis, opioids, alcohol, and other psychoactive drugs are affecting central nervous system (CNS) abnormalities and induce brain dysfunction after their chronic use (Le Berre et al., 2017; Loftis et al., 2019; Mizoguchi and Yamada, 2019; Oña and Bouso, 2021; Palmer, 1983; Razavi et al., 2020; Schrot and Hubbard, 2016).

Out of these drugs of abuse, METH is the most widely used in people at HA (see Kim et al., 2014; Klette et al., 2006). METH alone is well known to induce neurotoxicity and alter psychological and behavioral abnormalities (Halpin et al., 2014; Moszczynska and Callan, 2017; Shaerzadeh et al., 2018). If METH user receives another physical injury (TBI) to the brain for example traumatic head or brain injury, it appears that brain pathology in METH with TBI will be exacerbated (Duong et al., 2018; El Hayek et al., 2020; Gold et al., 2009; Vandenbark et al., 2019; Warren et al., 2006). Thus, HA alone induces severe brain pathology including edema, cellular dysfunction and alterations in the brain fluid environment disrupting the blood-brain barrier (BBB), it is likely that METH use at HA may further worsens brain dysfunction, edema and cell injury (Hoiland et al., 2018; Lafuente et al., 2016; Lin et al., 2012; Mellor and Woods, 2014; Winter et al., 2016). Furthermore, additional TBI to METH users at HA will further exacerbate pathology of brain damage (Ismailov and Lytle, 2016; Ma et al., 2019; Moore et al., 2014; Olson-Madden et al., 2012; Pugh et al., 2019; Wang et al., 2019; Wei et al., 2020).

Previous reports from our laboratory show that METH, MDMA, cocaine and morphine abuse leads to profound disruption of the BBB inducing brain edema and cell injuries (Kiyatkin et al., 2007; Kiyatkin and Sharma, 2009, 2012, 2015, 2016, 2019; Sharma and Ali, 2006, 2008; Sharma and Kiyatkin, 2009; Kiyatkin and Sharma, 2012; Sharma et al., 2004, 2006a,b,c, 2007a,b, 2009a,b,c, 2010, 2011a,b,c, 2014, 2019a,b,c). These neuropathological changes are widespread in the cortex, hippocampus, cerebellum, thalamus, hypothalamus, brain stem and spinal cord. These cell changes occur during both phases of drug dependence and withdrawal (Sharma and Kiyatkin, 2009; Sharma et al., 2007a,b). Pretreatment or repeated treatment with naloxone, H-290/51 or cerebrolysin induced significant neuroprotection following substance abuse dependence and withdrawal (Sharma et al., 2007a,b, 2010, 2011a,b,c, 2015a,b,c). Also a select combination of neurotrophic factors such as brain derived neurotrophic factor (BDNF), glial derived neurotrophic factor (GDNF), nerve growth factor (NGF) and ciliary neurotrophic factor (CNTF) in combination reduces brain pathology of METH and morphine withdrawal and dependence (Sharma H, unpublished observations). Pretreatment with a serotonin (5-hydroxytryptamine, 5-HT) 5-HT3-receptor antagonist ondansetron is able to provide marked neuroprotection in morphine-induced neuropathology

(Lafuente et al., 2018; Sharma et al., 2019a,b,c). Thus, it appears that substance abuse induce brain pathology involves oxidative stress, neurotrophic factors, and serotonin mediated pathways (Bayazit et al., 2020; Müller and Homberg, 2015; Sharma et al., 2006a,b,c, 2007a,b, 2010, 2011a,b,c; Weizman and Weizman, 2000).

Brain pathology in TBI is also using identical pathways of increased oxidative stress and serotonin metabolism as well as a reduction in neurotrophic factors (Alm et al., 1998, 2000; Cornelius et al., 2013; Duan et al., 2016; Menon et al., 2012; Muresanu et al., 2012, 2019; Olsson et al., 1990; Otero-Losada and Capani, 2019; Sharma and Sjöquist, 2002; Sharma et al., 1990a,b, 1993a,b, 2000a,b, 2001, 2003, 2006a,b,c, 2007a,b, 2009a,b,c, 2011a,b, 2012a,b, 2015a,b, 2018a,b,c, 2019a,b; Thörnwall et al., 1997). Thus, it is quite likely that a combination of substance abuse and TBI will exacerbate brain pathology. There are several reports indicating that HA alone induced brain pathology and alters metabolism of oxidative stress, serotonin and neurotrophins metabolism (Askew, 2002; Bai et al., 2015; Das et al., 2018; Dosek et al., 2007; Janocha et al., 2017; Kious et al., 2018; León-López et al., 2018; Quindry et al., 2016). Considering these facts into the consideration it is likely that METH treatment will induce higher brain pathology at HA. With additional TBI in METH at HA will further exaggerate brain dysfunction and pathology.

Military personnel that are engaged in combat operation at HA ate quite prone to TBI resulting in possibly greater brain pathology (Armistead-Jehle et al., 2017; Bhattrai et al., 2019; Moore et al., 2020; Swanson et al., 2017). If such situations are accompanied with METH exposure the brain pathology is likely to be seen even at a higher scale and profound brain dysfunctions. These aspects require further investigation from a strategic point of view. It is imperative that the drug therapy under these situations requires modification in dose or delivery procedures for new strategies due to presence of several co-morbidity factors (Friedrich et al., 2019). It appears that nanomedicine could be one potential option for superior neuroprotection in multiple injury factors such as HA, TBI and substance abuse. The, the need of the hour is to consider nanodelivery of suitable drugs, e.g., antioxidant compounds in multiple injury factors at HA (Sharma, 2007; Sharma and Sharma, 2012, 2013; Sharma et al., 2020a,b). Keeping these views in mind we discuss the effects of METH on HA and TBI in relation to brain pathology in this review. In addition, nanowired delivery of a potent antioxidant compound H-290/51 for superior neuroprotection in combination of METH, HA and TBI is discussed based on our own investigation. The functional significance and treatment potential of nanomedicine in HA with TBI and METH exposure is discussed.

1.1 Brain function at high altitude

Military personnel while combat duty are often subjected to HA that affects their psychological and physiological health (Chen et al., 2020b; Clemente-Suárez et al., 2017; Marriott and Carlson, 1996; Mishra and Ganju, 2010; Nation et al., 2017). When TBI occurs at HA the pathogenesis is far greater than the identical trauma at sea level due to a variety of altered factors. Thus, the pathophysiology of HA with reference to brain function at healthy state and following TBI needs to be addressed in order to expand our knowledge in this emerging new area of neurotrauma.

Potential adverse effects of HA may initiate from 1500 to 3500 m and affect brain functions moderately (Arias-Reyes et al., 2020; Hassan et al., 2019; Khodaee et al., 2016). Further elevation of 3500–5500 m HA results in severe dysfunction in brain function (Han et al., 2020; Hill et al., 2011; San et al., 2013). Extreme effects of HA on brain functions could occur above 5500 m elevation and require special attention (Adeva et al., 2012; Ainslie and Subudhi, 2014; Bailey et al., 2006; Berger et al., 2015; Hoiland et al., 2018; Hornbein, 1992; Jason et al., 1989; Pun et al., 2019; Xie et al., 2020). This is true for both civilians including mountain climbers, aviators and military personnel involved in conflicts or combat operation.

Rapid ascent to HA results in mountain sickness (MS) and may lead to development of HA cerebral edema (HACE) (Dawadi and Adhikari, 2019; Imray et al., 2010; Joyce et al., 2018; Sagoo et al., 2017; Zafren et al., 2017). The HA induced CE is due to swelling of brain and increased intracranial pressure (ICP) (Fagenholz et al., 2007; Lawley et al., 2016; Wilson et al., 2014; Wright et al., 1995; Yanamandra et al., 2014). In addition cold environment and lack of oxygen at HA results in hypoxia leading to serious brain dysfunction in healthy individuals (Bilo et al., 2019; Fornasiero et al., 2020; Leissner and Mahmood, 2009; Li et al., 2019; Whayne, 2014; Zhou et al., 2018). All these factors will affect other physiological variables such as mean arterial blood pressure (MABP), fluid and electrolyte balance and changes in the ICP in healthy persons causing brain and cognitive dysfunction (Aryal et al., 2019; Bilo et al., 2019; Naeije, 2010; Nation et al., 2017; Parati et al., 2015; Pun et al., 2019). Most common effects of HACE and mountain sickness (MS) include headache, fatigue, dizziness, nausea and/or vomiting and sleep disturbances (Hackett and Roach, 2004; Hackett et al., 2019; Hamilton et al., 1986; Villafuerte and Corante, 2016). These factors are further complicated with exhaustion, dehydration, migraine and hypothermia (Hashmi et al., 1998; Richalet et al., 2020). In some cases these symptoms are developed at HA of 2400 m however, more than 54% individuals develop these syndromes at HA of 4000–5000 m (Bellis et al., 2005; Kamat and Banerji, 1972). Most often HACE is also accompanied with high altitude pulmonary edema (HAPE) that is more serious and life threatening (Bärtsch, 1999; Hackett and Rennie, 2002). Symptoms of HAPE include dyspnea, tachypnea, tachycardia, dry cough and fever (Bhagi et al., 2014; Ebert-Santos, 2017; Jensen and Vincent, 2020). HACE symptoms are mainly dizziness, intense fatigue, tingling, ataxia, altered level of consciousness, retinal hemorrhage, papilledema and focal neurological deficits (Hsieh et al., 2020; Speidel et al., 2021; Wu et al., 2006). The basic cause of HACE is largely due to systemic and cerebral dysfunction. These include altered cerebral blood flow (CBF), disruption of the blood-brain barrier (BBB) permeability, hypoxia, vasogenic edema and raised ICP (Jansen et al., 2000; Lassen, 1992; Van Osta et al., 2005). HA induced focal neurological symptoms include transient ischemic attacks (TIA), double vision, scotomas and cerebral venous thrombosis that are independent of MS and HACE (Basnyat et al., 2004; Richalet et al., 2020; Sagoo et al., 2017). Pathological symptoms of HACE known as "end stage" include brain herniation, coma and eventually death (Hackett, 1999).

1.1.1 Mood and cognitive functions at high altitude

Adverse impairment of mood and cognitive functions occur at HA exposure above 3000m. HA exposure to individuals at 3500–4300m for sometime make them quarrelsome, anxious, easily irritable and apathetic due to altered emotional states (de Aquino et al., 2012; Heinrich et al., 2019). Above 4300m individuals become less friendly, confused, dizzy, aggressive, sleepy and unhappy (Chao et al., 2019; Shukitt and Banderet, 1988). These emotional states and mood disturbances are adversely affected by duration and level of HA (West, 2016a,b; Truesdell and Wilson, 2006).

In a study on mountaineering on Stanford, Alaska in US army personnel adverse changes in tension, anger, depression, fatigue and confusion were apparent even at HA of 3630m during a 7-day sojourn (Rosen et al., 2002; Shukitt-Hale et al., 1990, 1996, 1998). These data suggest that HA alters mood states largely due to physical exertion or exercise that enhance greater demand of oxygen in climbers and make them hypoxic under such situations.

Apart from mood changes, cognitive performances deteriorated also at HA of 3000m and onwards that are long lasting. Even after return to lower altitudes these cognitive disturbances are persisting for up to 1 year or longer (Shukitt-Hale and Liebermann, 1996). Impairment of cognitive functions at HA above 3000m is decline in mental skills, memory, vigilance, logical reasoning and reaction time (Shukitt-Hale et al., 1990). Alterations in cognitive dysfunctions are more susceptible to HA exposure than psychomotor performances (Shukitt-Hale et al., 1991, 1996). Thus, impaired functions at HA lead to slowing of performances, decision-making ability and increased error as well as combination of these factors (Shukitt-Hale and Liebermann, 1996). There are reasons to believe that these impairments are related to decrease in oxygen availability and resulting hypoxia (Shukitt-Hale et al., 1996). This is further supported by that fact that when 23 male subjects were exposed to HA in hypobaric chambers equivalent to 4206 and 4725m simulating hypoxic conditions their simple cognitive task of choice reaction time, vigilance and attentiveness were decreased in a graded manner (Shukitt-Hale and Liebermann, 1996; Shukitt-Hale et al., 1994). These observations suggest that cognitive performances are adversely affected with increase in HA exposure.

1.1.2 Behavior and neurochemistry at high altitude

As evident with HA induced cognitive impairment largely due to hypoxia there are reasons to believe that central mechanisms and neurotransmitters are responsible for most of these changes (Shukitt-Hale and Liebermann, 1996). Several lines of evidences indicate that mild hypoxia in the brain reflects altered neurotransmission in the central nervous system (CNS) (Amann and Kayser, 2009; Gibson et al., 1981a,b; Holtzer et al., 2001; Ray et al., 2011). These changes in neurotransmitters function are also related to behavioral alterations seen at HA (Kious et al., 2019; Reno et al., 2018; Virués-Ortega et al., 2006). Since oxygen is needed for the synthesis of neurotransmitters, thus, alterations in their metabolism by HA induced hypoxia affects neurochemistry and behavior of individuals (Heinrich et al., 2019; Hoiland et al., 2019; Julian, 2017; Julian and Moore, 2019).

Mild or moderate hypoxia affects cholinergic, aminergic and amino acid neurotransmitters metabolism affecting learning, memory and psychological processes (Forsling and Aziz, 1983; Gibson et al., 1981a,b; Kogure et al., 1977; Zanella et al., 2014). The cholinergic neurotransmission is the most vulnerable in HA induced hypoxia and is primarily related to cognitive dysfunction (Colangelo et al., 2019; Lim et al., 2015; Martinelli et al., 2017). Other psychological impairment such as depression, suicidal thought, mood alterations, vigilance and attention are related to monoamine neurotransmitters metabolism (Elsworth et al., 2012; Kehagia et al., 2010; Pehrson and Sanchez, 2014). Neuronal dysfunction is thus caused by the impairment of several neurotransmitters in the brain due to lack of oxygen availability at HA (Critchley et al., 2015; Ott et al., 2020).

Several laboratory studies show that cholinergic transmission is worst affected in HA induced learning and memory performances (Ji et al., 2020; Maiti et al., 2008; Roy et al., 2018). For this purpose, trained rats were exposed to hypobaric chambers simulated at 5500, 5950 and 6400m for 2 to 6h and tested in Morris Water Maze for their escape latencies (Shukitt-Hale et al., 1994). The results showed that hypobaric hypoxia deteriorated their performances adversely related to elevation of HA. Thus, with elevation of HA from 5500 to 6400m the escape latencies increased significantly in Morris Water Maze tests. In addition, the increased path length and decreased swim speeds was also noted in rats with increase at elevation of 5950m HA to 6400m (Bahrke and Shukitt-Hale, 1993; Shukitt-Hale et al., 1994). Interestingly, these parameters were not affected at HA elevation of 5500m. These observations suggest that spatial memory impairment is related to HA elevation (Shukitt-Hale and Liebermann, 1996). It is believed that hippocampal cholinergic function together with other neurotransmitters is involved in spatial memory impairments at HA (Maiti et al., 2008; Sharma et al., 2019a,b,c). However, further studies are needed to explore HA related cognitive decline and treatment strategies in future.

1.1.3 Oxidative stress at high altitude

Exposure to HA results in oxygen deficiency and reduction in blood oxygen saturation (Ando et al., 2020; Otani et al., 2020; Simonson et al., 2014; Wang et al., 2018). Oxidative stress and generation of free radicals are associated with low oxygen supply (Giordano, 2005; Trinity et al., 2016). Since HA is associated with oxygen deficiency and hypoxia thus production of oxidative stress and generation of free radicals are key factors in inducing behavioral, cognitive and functional deficits at HA (Bakonyi and Radak, 2004; Dosek et al., 2007; Gaur et al., 2020). Hypoxic cells are highly susceptible to oxidative stress (Herst et al., 2017). If in such situations further oxygen deficiency induces radical chain reaction leading to cell and membrane damages. About 1% reduction in blood oxygen saturation occurs at every 100m above 1500m (Goldberg et al., 2012; Lewis et al., 2017; Shukitt-Hale et al., 1996, 1998).

Data from mountaineers climbed at HA of 5000m and stayed there for 1–3 weeks exhibited significant increase in oxidative stress as measured using amount of thiobarbituric acid reactive substances (TBARS) in the blood plasma (Jefferson et al., 2004; Magalhães et al., 2005). A tremendous increase in TBARS was observed in these individuals indicating that oxidative stress is severely increased at HA (Shukitt-Hale et al., 1996; Wozniak et al., 2001). When these individuals were supplemented with daily vitamin E, a powerful antioxidant this treatment counteracted the increased TBARS activity at HA exposure (Engwa et al., 2020; Pfeiffer et al., 1999; Vij et al., 2005). These observations confirm the role of oxidative stress in HA induced alteration in cognitive, psychological and behavioral activities (Shukitt-Hale and Liebermann, 1996).

2 Methamphetamine at high altitude

There is a long historical association of methamphetamine (METH) use in military and at HA since 1927 (Anglin et al., 2000; Defalque and Wright, 2011; Vearrier et al., 2012). Decrease in barometric pressure with increasing HA results in hypobaric hypoxia (Degache et al., 2020; Mrakic-Sposta et al., 2021; Murray, 2016). The hypobaric hypoxia increases brain dopamine neurotransmission that is responsible for euphoric and rewarding effect of drugs (Cymerman et al., 1972; Ray et al., 2011; Wustmann et al., 1982). This is one of the reasons that residents of HA use very frequently recreational drugs like cocaine and METH. In military during World War II METH was used for hyper-alertness and vigilance (Vearrier et al., 2012).

2.1 History of methamphetamine

Lazăr Edeleanu was the first to synthesize amphetamine from plant-derived ephedrine in 1887 at University of Berlin in Germany (Edeleano, 1887). In 1893 Nagayoshi Nagai synthesized methamphetamine from ephedrine at Tokyo Imperial University, Japan (Hillstrom, 2015; Lock, 1984). In 1919 Akira Ogata synthesized methamphetamine hydrochloride known as Crystal Meth from reduction of ephedrine at Humboldt University of Berlin. (Hartmann, 2003; Tamura, 1989). In 1935 first scientific study on amphetamine and/or methamphetamine was done in Los Angeles, California, USA using 55 subjects who described the effect of drug as sense of well being and exhilaration together with lessened fatigue (Rasmussen, 2006). During World War II, METH was used by Allied and Axis forces to select personnel for performance-enhancing effects (Rasmussen, 2011). In 1960 recreational drug was used in United Kingdom for all nigh dances and in Manchester Twisted Wheel club

(Edmundson et al., 2018; Rasmussen, 2006, 2011, 2015). In 1965 Food and Drug Administration (FDA, USA) regulated METH use strictly as controlled substance because of its highly addictive and dependence properties (Hunt et al., 1996; Rasmussen, 2006, 2015).

2.1.1 High altitude related variations in methamphetamine use

Hypobaric hypoxia is one of the basic reasons for some people consuming METH at HA to feel good from depression, fatigue and other stress-related disorders (Barati et al., 2014; Lee, 2011; McKetin et al., 2006; Sabrini et al., 2019). Living at HA affects brain function of residents by hypobaric hypoxia that reduces the synthesis of neurotransmitters in the brain (Cymerman et al., 1972; Ray et al., 2011; Wustmann et al., 1982). This could be one of the main reasons of METH use in the United States at HA by local individuals depending on their psychoactive, genetic and environmental factors (Lahiri et al., 1976; Li et al., 2020a,b). It appears that genetic factors could account 33% of cases and environmental factors are responsible for 67% cases of METH usage at HA in US mountain regions (Galvin et al., 2020; May et al., 2015). The brain requires 20% of the whole body oxygen consumption for maintaining its normal neurophysiological functions (Hoyer, 1982). The rate-limiting enzymes for synthesis of several neurotransmitters including dopamine, norepinephrine and serotonin are highly oxygen-dependent (Clauss, 2014). Thus, even a moderate oxygen deficiency caused by hypoxia will impair synthesis of the neurotransmitters in brain. As a result neurophysiological and psychological functions are remarkable impaired.

Exposure to HA causes an increase in dopamine and decrease in serotonin levels in the brain with high intensity in the frontal cortex (Zhong et al., 2021). Altered dopamine and serotonin levels in the brain plays major role in METH induced syndromes and lethality (Shaerzadeh et al., 2018). Dopaminergic hyperactivity and serotonergic hypoactivity is related to chronic stress, personality disorders, and behavioral dysfunctions (Solomon et al., 2010). METH is a powerful psychostimulant causing euphoria, increased alertness, energy and heightened emotions (Mizoguchi and Yamada, 2019). Thus METH use at HA in some population is quite frequent. Interestingly, the use of METH seems to be influenced with HA as an important environmental factor (Kerry et al., 2016; Kim et al., 2014).

Epidemiological studies show that increased incidences of depression and related psychiatric problems including suicide are somehow related to residents of HA (Betz et al., 2011; Kerry et al., 2016; Kious et al., 2018; Reno et al., 2018; Sabic et al., 2019). These phenomena are also related to increased dopamine and decrease in serotonin levels in the brain. Severe psychological distress and major depressive illnesses encourage cocaine and METH use (Ishikawa et al., 2016; Kious et al., 2019). Young adults seeking pleasure from HA environment and psychiatric illnesses are prone to METH and cocaine use (Biondich and Joslin, 2015). Clinical data suggests that some adults with genetic predisposition toward high level of dopamine activity in brain are also likely to use these psychopharmacological agents related to novelty-seeking behavior (Kim et al., 2014; Ma et al., 2020; Wang et al., 2013). These effects of METH and cocaine use show a positive correlation with HA (Kim et al., 2014). This is evident with the finding showing different states of US use of METH positively correlates with each state HA. These observations support the idea of variation in METH use is related to changes in HA in US (Kim et al., 2014). This is quite likely that this relationship is also prevailing in other parts of the World. However, further studies are needed to ascertain this fact in future. It will be interesting to find out if alterations in neurotransmitters level in the brain is also related to HA, a feature that require additional investigation.

2.1.2 Methamphetamine neurotoxicity

METH is a powerful neurotoxic agent that induced neuronal, glial and axonal dysfunction within short periods of time (Yasaei and Saadabadi, 2020). METH consumption enhanced dopamine concentration within the synapse and release within the brain causing psychomotor and addiction effects rapidly (Richards and Laurin, 2020). Severe dysfunction of dopaminergic neuronal network and imbalance is one of the important factors in METH neurotoxicity (Paulus and Stewart, 2020). METH releases dopamine from synaptic vesicles into cytosol where its metabolism and oxidation tales place (Pubill et al., 2005). This will result in induction of oxidative stress, generation of superoxide anions, hydrogen peroxide and reactive oxygen species (ROS) (Jang et al., 2017; Kita et al., 2009). Excessive production of ROS damages cell membranes and induces mitochondrial dysfunction leading to neurotoxicity (Sharma et al., 2007a,b).

Experiments carried out in our laboratory show that small amount of METH (0.9 mg/kg, s.c.) when administered into the rat induced profound hyperthermia (<40 °C) within 3–4 h at room temperature. When the same amount of METH was administered at higher ambient temperature greater hyperthermia (<41 °C) occurred within 2–3 h in the animals (Kiyatkin et al., 2007; Sharma and Kiyatkin, 2009). Breakdown of the blood-brain barrier (BBB), brain edema and neuronal, glial and axonal injuries are seen in these METH treated rats within less than 4 h period (Sharma and Ali, 2006; Sharma and Kiyatkin, 2009). Likewise 4 h after METH (40 mg/kg, i.p.) administration into mice at room temperature resulted in core body hyperthermia greater than 41 °C associated with BBB leakage, vasogenic brain edema and brain cell injuries (Sharma and Ali, 2006). Interestingly, these METH induced changes in rodents were aggravated by environmental temperature and/or nanoparticles intoxication. Administration of METH (0.9 mg/kg, s.c.) in chronically exposed rats to engineered nanoparticles from metals (Ag, Cu) or from silica (SiO2) for 1 week acclimatized at either cold (4 °C) or hot (34 °C) environments resulted in exacerbation of neurotoxicity as compared to identical experiments at room temperature (Sharma et al., 2015a,b,c). SiO2 exposed rats showed most pronounced METH neurotoxicity followed by Ag and Cu treated animals (Sharma et al., 2015a,b,c).

Interestingly both cold and hot exposed rats exhibited greater METH neurotoxicity as compared to animals at thermoneutral ambient temperature (Sharma et al., 2015a,b,c). This suggests that nanoparticles and environmental pollutants and temperature all could influence METH induced neurotoxicity. Treatment with a potent antioxidant H-290/51 significantly induced neuroprotection following METH intoxication irrespective of ambient temperature or nanoparticles administration (Sharma et al., 2015a,b,c). This suggests that METH induced oxidative stress is primarily responsible for the neurotoxicity in these animals.

3 Traumatic brain injury at high altitude

Military personnel are often subjected to traumatic brain injury (TBI) on combat operation or peacekeeping mission across the Globe including high altitude mountains (Midla, 2004; Korzeniewski et al., 2013; Rodway and Muza, 2011). TBI results from either penetration of external sharp objects into the brain or head injury by blunt forces (Concussive head injury, CHI) (Capizzi et al., 2020; Holsinger et al., 2002; Knapik et al., 2014; Vella et al., 2017). In addition, explosion or blast forces cause sudden brain acceleration or deceleration, rotational movements, tear, laceration or tissue loss is most prominent forms of TBI or CHI (Fievisohn et al., 2018; Kinch et al., 2019; Ling et al., 2015; Yamamoto et al., 2018). TBI is classified into mild, moderate or severe. Several forms of TBI and CHI are seen in United States armed forces during conflicts in Iraq and Afghanistan where brain injuries occurs in more than 60% cases largely because of blast injuries (Hasoon, 2017; Stewart et al., 2019). More than 28% army personnel suffered some kinds of mild TBI in these armed conflicts (Kennedy et al., 2019; Silverberg et al., 2020). Available data suggest that outcome of TBI and progression or persistence of brain pathology is adversely correlated with the altitude of high 1500–3500 m, very high 3500–5500 m and extremely high altitudes 5500–7500 m (Ma et al., 2020; Smith et al., 2013; Wang et al., 2019).

However, military conflicts leading to TBI also occur like civilians at ground levels and at HA (Ismailov and Lytle, 2016). TBI at HA induced severe adverse effects on the pathophysiology and management of patient care, days of hospital admission, morbidity, and mortality and recovery of victims (Berger et al., 2015; Das et al., 2018). In Afghanistan average elevation is 1200 m and Bagram air base is located at 1491 m altitude where the outcome on brain health and performance after TBI are the cause of concern for United States military personnel (Gohy et al., 2016). The Afghan capital Kabul is located at HA of 1800 m and military operation are often takes place at 2000–3000 m HA. Apart from military personnel ground personnel for health and logistics are also working at 3000–5000 m HA. At this altitude unacclimatized individuals show early signs of hypoxia that could complicate the outcome of TBI with enhanced disability and mortality (Bordes et al., 2017). In case of severe CHI at this HA the disability and mortality rates in patients could be higher than

50–52% of cases (Patel et al., 2019). Thus, further research is needed to explore novel therapeutic strategies using nanomedicine to treat these CH cases that are adversely affected at HA.

3.1 Secondary brain injury processes are affected at high altitude

Adverse effects on TBI at HA are largely determined by the factors associated with the secondary injury processes. Several secondary injury processes including mechanistic, physiologic, metabolic and neuroinflammatory cascades occurring within minutes after TBI and continue for hours to days and weeks of injury (Elder and Cristian, 2009; Kasper, 2015; Spielberg et al., 2015). Metabolic alterations after TBI plays important role in recovery at both ground level and at HA (Clark et al., 2020). Depressed glucose metabolism is seen in patients of mild TBI that are asymptomatic but cellular injury is progressing with time (Carteron et al., 2018; Jalloh et al., 2015; Lozano et al., 2020).

Several lines of evidences suggest that TBI at HA alters secondary injury processes that are responsible for adverse outcome and greater brain pathology in 40% of TBI victims (Hornbein, 2001; Li et al., 2013). These include brain injury induced intracranial hemorrhage, elevated intracranial pressure (ICP), reduction in local cerebral blood flow (CBF), BBB permeability, vasogenic or cytotoxic edema formation, ischemia and hypoxia (Fayed et al., 2006; Li et al., 2020a,b; Maa, 2010; Sareban et al., 2021; Wilson et al., 2014). Release of neurochemicals, cellular stress, oxidative stress, tissue damage and brain herniation are other pathologies that aggravates brain pathology at HA (Connolly et al., 2018; Finnoff, 2008; Riech et al., 2015; Wilson et al., 2009). Brain hypoxia is a major issue at HA that is positively related to the magnitude of elevation (Aksel et al., 2019; Hoiland et al., 2018; Zhang et al., 2020). Thus, most of the secondary injury cascades exacerbate the pathophysiology of TBI at HA.

The most prominent secondary injury cascade of TBI at HA causing adverse effects are brain hypoxia and systemic hypotension that are present in more than 40–50% of cases (Gao et al., 2020). Those patients who develop hypoxia and hypotension after TBI at HA before their hospital admission are at greater risk of disability and death (Kauser et al., 2016). Although brain response to HA after TBI is not well known yet, but available evidences indicate that altitude related hypoxia and aggravation of secondary injury cascades are primarily responsible for increased risk of brain pathology (Wettervik et al., 2020).

Brain is the most oxygen-dependent organ and neurons could response to oxygen deprivation within milliseconds time (Bailey, 2019; Leu et al., 2019; Stepanova et al., 2019). Oxygen concentration in the air is reduced by 50% at HA of 5500m and at 8800m oxygen concentration remains only one third to ground levels (Lichtenauer et al., 2015; West, 2016a,b). TBI inflicted at HA where hypobaric hypoxia is prevalent due to altitude induced several secondary injury factors such as brain edema, ICP, ischemia, systemic hypotension and apoptosis (Banchero et al., 1966; Smith et al., 2013; Wilson et al., 2014). This is the main reason of increased risk of death and disability of TBI patients at HA.

4 Our observation on methamphetamine and traumatic brain injury at high altitude

TBI induces severe brain pathology and oxidative stress (Kaplan et al., 2018). Hypoxic conditions at HA aggravate brain pathology probably via increased oxidative stress (Luan et al., 2019). METH induce brain pathology is also dependent on psychostimulant induced oxidative stress mechanisms (Shin et al., 2018). The secondary injury cascades of free radical formation, and generation of oxidative stress is common in TBI, METH intoxication and HA exposure (Bakonyi and Radak, 2004; Jørgensen et al., 1997; Mori et al., 2007). Thus, it would be interesting to find out whether METH intoxication enhances TBI induced brain pathology at HA. In addition, whether treatment with a potent antioxidant H-290/51 is able to reduce brain pathology with METH intoxication at HA and TBI. In addition, whether nanodelivery of the antioxidant H-290/51 has superior neuroprotective effects in METH induced aggravation of brain pathology in TBI at HA.

We have undertaken a series of investigations to address these issues to expand our knowledge for better therapeutic approaches to military personnel who are prone to TBI at HA with or without METH exposure. The novel data emerged from our investigation is summarized below.

4.1 Methodological consideration

The experiments were carried out to evaluate METH effects of TBI at HA. Our investigations on brain pathology and neuroprotection is descried briefly below.

4.1.1 Animals

Experiments were carried out on in bread male Wistar Rats (Age 15–20 weeks, body weight 350–440 g) housed at controlled ambient temperature (21 ± 1 °C) with 12 h light and 12 h dark schedule. The food and tap water were provided ad libitum.

All experiments are conducted according to Guidelines and Care or Handling of experimental animals according to national Instate of Health (NIH) (National Research Council, 2011) and approved by local Ethics Committee.

4.1.2 Exposure to high altitude

Animals were exposed to environmental controlled simulating hypobaric High Altitude Chamber (Envisys, Miami Beach, FL, USA) equivalent to an altitude of 5000 m for 1 week. The oxygen concentration was automatically reduced to HA of 5000 m to about 11.4% and atmospheric pressure 50 kPa with relative humidity maintained to 55–60%. Wind velocity 7–10 cm/s and chamber temperature 10–15 °C. Rats were acclimatized to the test chamber for 3 days (daily exposure to 4 h per day) before placing them in the HA chamber for 1 week continuously. Rats were placed in a group of 2 rats per cage with food and tap water ad libitum maintaining 12 h light and 12 h dark schedule. The cages were cleaned every 2nd alternate day for short period of 20–30 min in order to avoid variables in the HA chamber as described earlier (Muresanu et al., 2017; Sharma et al., 2018a,b,c).

4.1.3 Control group

The control group of rats was identically kept at Altitude 200 m from sea level in the normobaric (ca. 1003 hPa) laboratory environment. The room temperature was 21 ± 1 °C, wind velocity 15–20 cm/s, relative humidity 45–55%. All experiments carried out at normobaric environment means sea level barometric pressure at the laboratory environment.

4.2 Concussive head injury

Animals were subjected to CHI under Equithesin anesthesia using a weight drop technique over the right parietal skull as described earlier (Dey and Sharma, 1984; Sharma et al., 2007a,b, 2016a,b). In brief, under Equithesin anesthesia (3 mL/kg, i.p.) the animal's head was fixed in the rat stereotaxic instrument (Harvard Apparatus, Holliston, MA, USA) and the parietal skull bone was exposed after an incision over the skin and underlying muscles were scrapped (Dey and Sharma, 1984). A stainless steel tapers rod weighing 114.6 g was then dropped over the parietal bone through an aluminum tube from the 20 cm height. Cotton wool padding was placed under the head of the rat to avoid any head movement (Sharma et al., 2007a,b). This weight and height adjustment delivers an impact force of 0.224N over the skull surface.

Immediately after the impact the rats were taken out from the stereotaxic apparatus their skin were sutured and sprayed with Xylocaine (London, UK) and Boric acid powered (BDH, London, UK) over the wound. When the animals were recovered and could walk normally they are subjected to HA exposure.

4.2.1 Methamphetamine intoxication

Methamphetamine hydrochloride (CAS nr. 51-57-0, Sigma Aldrich, St Lois, MO, USA) was dissolved in sterile water and administered in the rats at the dose of 9 mg/kg, s.c. in control, HA exposed group with or without brain injury. In CHI group METH was administered after 4 h of insult and placed in hypobaric test chamber. The CHI rats were placed in individual cages in HA chamber (Sharma and Kiyatkin, 2009; Sharma et al., 2018a,b,c). Control groups were administered 0.9% physiological saline instead of METH.

4.2.2 Drug treatment

Animals were treated with a powerful chain-breaking antioxidant compound H-290/51 (Astra Zeneca, Mölndal, Sweden) in a dose of 100 or 150 mg/kg, i.p. in control, CHI with or without HA exposure intoxicated with METH.

To further enhance the neuroprotective effects of H-290/51 TiO2-nanowired H-290/51 (NWH-290/51) was used in a dose of 150 mg/kg, i.p. in HA group treated with METH as compared to normal control group in identical manner. The H-290/51 was nanowired as described in details earlier (Sharma et al., 2009a,b,c, 2015a,b, 2018a,b,c).

4.2.3 Physiological evaluation of animals

All animals including normobaric or hypobaric exposure groups were examined for their body weight, rectal temperature, heart rate, respiration and pain sensitivity to assess their normal well being daily morning between 8:00 and 9:00 AM using standard procedures.

4.2.4 Physiological variables

The mean arterial blood pressure (MABP), Arterial pH, and Arterial PaO2 and PaCO2 were analyzed according to standard protocol. For this purpose, a polythene cannula PE 10 was implanted into the left common carotid artery in the rats under Equithesin anesthesia 1 week before the experiments under aseptic conditions. At the time of the experiments, the common carotid cannula retrogradely toward heart was connected to a Strain Gauge Pressure Transducer Statham P23 (New York, NY, USA) that is connected to a chart recorder (Electromed, London, UK). Immediately before connecting the arterial cannula to pressure transducer, about 1 mL of whole blood was withdrawn from the cannula for later determination of arterial pH and blood gases using Radiometer Apparatus (Copenhagen, Denmark) (Sharma, 1987).

The heart rate and respiration cycle was also measured using proper electrodes and recorded on the same chart recorder along with the blood pressure (Sharma and Dey, 1986a,b, 1987).

4.2.5 Body weight, rectal temperature and pain perception

In all animals, rectal temperature was recorded between 8:00 and 9:00 AM using a12-channel digital telethermometer (Harvard Apparatus, Holliston, MA, USA) using rat rectal temperature thermistor probes (Yellow Springfield, NJ, USA). These thermistor probes were inserted into the rectum connected to the telethermometer and placed in position with the tail with adhesive tape to stabilize the movement. Rats were handled for this purpose about 1 week before the experiment daily so that their rectal temperature is stable at the time of the measurement without stress.

The body weights of the animals were recorded once daily between 8:00 and 9:00 AM in the morning using the sensitive animal weighing scale (Mettler Toledo SB16001DR, Bradford, MA, USA).

For pain perception, both thermal and noxious nociception was measured using mechanical and radiant heat stimulus according to standard protocol as described earlier (Sharma et al., 1998a).

4.3 Pathophysiology of brain injury

The following parameters are measured to assess pathophysiological changes in animals at normobaric or hypobaric groups with CHI intoxicated with METH or saline (Richalet, 2020).

4.3.1 Blood brain barrier breakdown

Breakdown of the BBB was determined using two exogenous tracers Evans blue albumin (EBA) and radioiodine [131]-Iodine administered intravenously into the right femoral vein through an indwelling cannula (PE10) implanted ascetically 1 week before the experiments. The EBA (3 mL/kg, i.v.) and radioiodine 100 μCi/kg (i.v.) was administered into the systemic circulation and allowed to circulate for 10 min at the end of the experiment. After that the animals were deeply anesthetized with Equithesin (3 mL/kg, i.p.) and the intravascular tracer was washed out using cold physiological saline (0.9% NaCl) solution perfusion through heart (90 Torr) as described earlier (Sharma, 1987). Immediately before intracardiac perfusion about 1 mL of whole blood was collected through left ventricle puncture for later analysis of radioactivity and EBA (Sharma and Dey, 1986a,b).

After blood washout, the brains were dissected out and examined for Evans blue dye leakage on the dorsal and ventral surfaces of the brain. After a mid sagittal section, penetration of the dye was examined into the cerebral ventricle and other areas inside the brain and noted down. After this inspection of EBA leakage, the brains were dissected into the desired regions and weighed immediately and counted radioactivity in a 3-in Gamma Counter (Packard, USA).

Extravasation of radioactivity in the brain was expressed as percentage leakage over the blood radioactivity as described earlier (Sharma, 1987). Alter counting the radioactivity, the brain samples were homogenized in to a mixture of 0.5% sodium sulfate and Pure Grade Acetone to extract Evans blue dye entered into the brain tissues. The samples were centrifuged at $900 \times g$ and the supernatant were measured in a spectrophotometer (Harvard Apparatus, Holliston, MA, USA) at 620 nm. The concentration of EBA entered into the brain was calculated against standard curve of the EBA (Sharma, 1987; Sharma and Dey, 1981, 1984, 1986a,b, 1987).

4.3.2 Cerebral blood flow measurement

Cerebral blood flow was measured using radiolabeled carbonized microspheres (o.d. 15 ± 0.6 μm) using peripheral reference blood flow as described earlier (Sharma and Dey, 1986a, 1987, Sharma et al., 1990a,b). In brief about 90–100 k [125]-I-labeled carbonized microspheres were administered as a bolus into the left common carotid artery cannula (PE 10) implanted 1 week before the experiments retrogradely toward the heart within 45 s. A reference blood was withdrawn from the femoral artery cannula implanted aseptically earlier at the rate of 0.8 mL/min starting from 45 s before microspheres injection at every 30 s until 90 min after completion of the intracarotid administration.

After that the brains were taken out and placed on cold saline wetted filter paper. The large superficial blood vessels and blood clot if any are removed immediately. After that the desired brain areas were dissected out and weighed immediately and the radioactivity counted in a 3-in Gamma counter (Packard, USA). The blood samples taken out periodically and all were counted for radioactivity as described earlier (Sharma, 1987). The CNF was calculated from the regional brain radioactivity and the whole blood radioactivity in the blood samples according to formula described earlier (Sharma et al., 1990a,b; Sharma and Dey, 1986a, 1987). The CBF was expressed as ml/g/min.

4.3.3 Brain edema formation

Brain edema formation was measured using brain water content and associated volume swelling (%*f*) as described earlier (Sharma et al., 1991; Sharma and Cervós-Navarro, 1990; Sharma and Olsson, 1990). In brief, after the end of the experiments, the brain were dissected out and placed on the cold saline wetted filter paper. After discarding large superficial blood vessels and blood clots, if any the desired brain areas were dissected and placed on preweighed filter paper and weighed immediately.

After obtaining wet weight of the brain samples they were placed in an incubator maintained at 90 °C for 72 h to obtain the dry weight of the tissues. The dry wet of the brain samples were taken until they become constant in at least 3 determinations on a sensitive electronic balance (Mettler Toledo, Sensitivity 0.1 mg, Columbus, OH, USA). The brain water percentage was calculated from the differences between wet and dry weights of the samples.

Volume swelling was calculated from the differences between control and experimental brain tissues according to formula of Elliott and Jasper (1949) as described earlier (Sharma and Cervós-Navarro, 1990).

4.3.4 Beta catenin measurement

Beta catenin maintains the regulation of BBB and also regulates local CBF through Wnt gene (Liebner et al., 2008). Accordingly, beta catenin was measured in the cerebral cortex of CHI inflicted animals at HA or room temperature according to standard protocol. The beta catenin was measured using Sandwich ELISA rat beta catenin kit (MBS 843456, My BioSource, San Diego, CA, USA, sensitivity 0.005 ng/g) according to commercial protocol.

4.4 Brain pathology

For morphological examination, standard histopathological techniques were used to study neuronal, glial and myelin changes using light and electron microscopy (Sharma and Cervós-Navarro, 1990; Sharma and Olsson, 1990; Sharma et al., 1992a,b, 1993a,b, 1995, 1996).

4.4.1 Perfusion and fixation

For morphological purposes, at the end of the experiment, animals were deeply anesthetized with Equithesin (3 mL/kg, i.p.) and chest was rapidly opened and heart was exposed. The right auricle was cut and a butterfly needle (21G) was inserted into the left ventricle and connected to a Peristaltic Pump (Harvard Apparatus, Holliston, MA, USA) connected to a bottle of 150 mL of 0.1 M phosphate buffer saline (PBS, pH 7.1) for perfusion at 90 Torr to wash out intravascular blood. Immediately after PBS perfusion, about 250 mL of 4% buffered paraformaldehyde (pH 7.4) containing 0.25% picric acid was perfused at identical pressure (90 Torr) (Sharma, 2011; Sharma et al., 1995).

After in situ perfusion, the animals were wrapped in an aluminum foil and kept in a refrigerator at 4 °C for overnight. On the next day, the brain was removed and several serial coronal sections were cut and processed for histological purposes using automatic Tissue processor (FineTek, Sakura, Torrance, CA, USA) and embedded in paraffin (Sharma, 2011; Sharma et al., 2011a,b,c).

4.4.2 Light microcopy

For light microscopy the paraffin embedded tissue blocks were cut 3-μm thick sections and processed for staining with Nissl, Hematoxylin & Eosin (H&E), Toluidine blue, Luxol Fast Blue for standard histopathological staining protocol (Sharma et al., 2009a,b,c, 2011a,b,c, 2016a,b, 2018a,b,c, 2019a,b,c).

For immunohistochemistry to study astrocytes glial fibrillary acidic protein (GFAP) was used according to commercial protocol (Sharma et al., 1992a,b). In brief, monoclonal GFAP mouse antibody (Anti-GFAP antibody 2A5; ab4648, abcam, Cambridge, MA, USA) was used in a dilution of 1:500 and immunoreaction product was visualized using Avidin-Biotin Complex (ABC) technique (Burlingame, CA, USA) as describe earlier (Sharma et al., 1992a,b, 1993a,b).

The light microscopy images were visualized in an Car Zeiss Inverted Bright Field Microscope attached with a digital camera and stored in an Apple Macintosh Power Book (System Mc Os El Capitan, 10.11.6) and processed using commercial software Adobe Photoshop Extension vs 12.4 using identical brightness, contrast and color filters (Sharma and Sjöquist, 2002).

4.4.3 Transmission electron microscopy

For ultrastructural investigation, transmission electron microscopy (TEM) was used. For this purpose, small tissue pieces of paraformaldehyde fixed tissues containing picric acid were cut and post fixed in osmium tetraoxide (OsO4) and embedded in plastic (Epon 812).

About 1-μm thick semithin sections were cut and stained with Toluidine blue and examined under a bench microscope for identifying the desired areas for TEM analysis. The desired area after identification the tissue Epon blocks were trimmed and ultrathin sections (50 nm) were cut on an LKB Ultramicrotome (Stockholm, Sweden) using a diamond knife (Diatome, USA). The serial sections were then collected on one hole copper grid and counterstained with uranyl acetate and lead citrate. Some sections were used unstained. Thee TEM grids were analyzed in a Phillips 400 Electron Microscope (The Netherlands) attached with a digital camera (Gatan System, USA). The images were stored in an Apple Macintosh Computer with original magnifications of 4–12 k magnification. The images from control and experimental groups were processed using commercial Adobe Photoshop extension vs, 12.4. using identical contrast and brightness filters (Sharma et al., 1998a,b; Sharma and Sjöquist, 2002).

4.5 Statistical analysis

The quantitative data were analyzed using commercial software Abacus concept StatView 5 on a Macintosh computer in classical environment. ANOVA followed by Dunnett's test for multiple group comparison using one control was used to find statistical significance between the data obtained. The semiquantitative analysis was done using non-parametric Chi Square test. A p-value less than 0.05 was considered significant.

5 Results

5.1 Physiological parameters at normobaric control group

Subjection of animals to normobaric group at laboratory room temperature showed body weight 386 ± 6 g with body temperature 36.52 ± 0.16° (Fig. 1). These animals show their pain threshold at thermal nociception of 5 ± 2 s and at mechanical nociception 3 ± 2 s (Fig. 1). The MABP shows 120 ± 4 Torr and the arterial pH was 7.38 ± 0.02 (Fig. 1). The PaCO2 was 34.45 ± 0.23 Torr and the PaO2 was80.56 ± 0.43 Torr. The heart rate was 320 ± 12 beats per min and respiration was 35 ± 5 cycles per min (Fig. 1).

METH treatment at room temperature showed body weight of 376 ± 8 g and exhibited significant hyperthermia (39.21 ± 0.23 °C, P < 0.05). The METH treated animas showed profound analgesia to thermal nociception (8 ± 2 s, P < 0.05) and mechanical nociception (9 ± 3 s, P < 0.05) (Fig. 1).

The MABP was significantly lowered to 94 ± 6 Torr (P < 0.05) and the arterial pH was slightly decreased to 7.33 ± 0.12. The arterial PaCO2 significantly increased to 34.89 ± 0.10 Torr (P < 0.05) and arterial PaO2 showed slight but significantly enhanced (80.78 ± 0.12 Torr, P < 0.05). The heart rate decreased significantly (280 ± 12 bpm) and respiratory cycles also decreased significantly (28 ± 7 cpm, P < 0.05) (Fig. 2).

5.1.1 Physiological parameters in CHI at normobaric control group

Subjection of CHI at normobaric environment at laboratory showed body weight (360 ± 6 g, P < 0.05) and the body temperature was 37,12 ± 0.12 (P < 0.05). The thermal nociception was 7 ± 3 s and mechanical nociception was 6 ± 4 s (Fig. 1). The MABP significantly decreased to 80 ± 8 Torr (P < 0.05) whereas the arterial pH was 7.36 ± 0.05. The arterial PaCO2 increased significantly to 35.36 ± 0.12 Torr (P < 005) and Pa = 2 also showed slight elevation to 81.34 ± 0.18 Torr (P < 0.05). The heart rate significantly decreased to 288 ± 12 bpm (P < 0.05) and respiration rate also decreased to 25 ± 10 cpm (P < 0.05) (Fig. 2).

CHI in METH treatment the body weight was 370 ± 7 g (P < 0.05) and the body temperature was significantly enhanced to 39.87 ± 0.21 °C (P < 0.05). The thermal nociception was 16 ± 3 s (P < 0.05) and the mechanical nociception to 18 ± 8 s (P < 0.05) (Fig. 1). The MABP was 88 ± 5 Torr (P < 0.05) and the arterial pH was

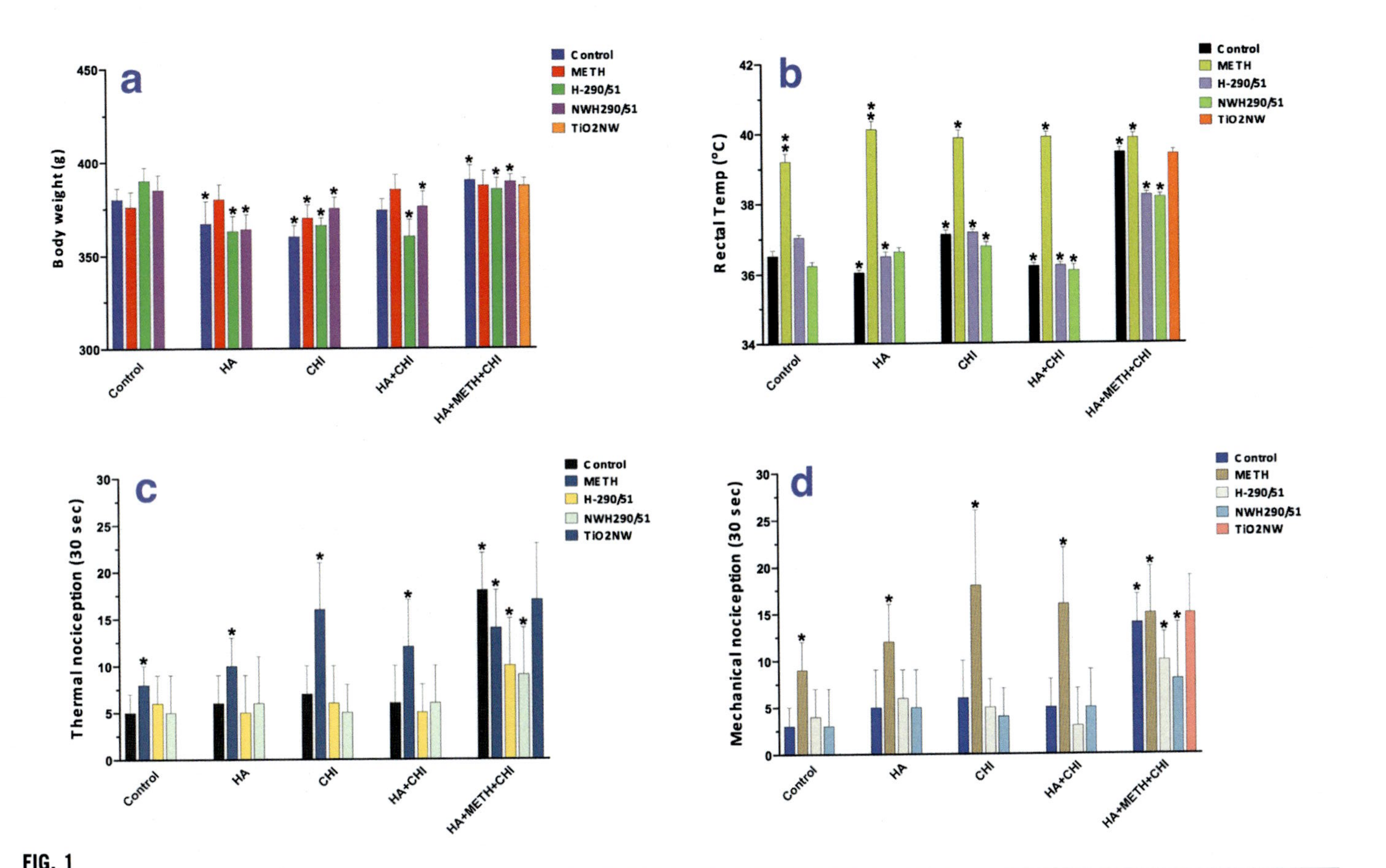

FIG. 1

Changes in body weight (A) rectal temperature (B), thermal nociception (C) and mechanical nociception (D) after High Altitude (HA) exposure in normal and methamphetamine (METH) exposed rats with or without concussive head injury (CHI). All the data were measured after 1 week of exposure to high altitude (HA), concussive head injury (CHI), methamphetamine (METH) with or without H-290/51 or NWH-290/51 and/or vehicle TiO2 nanowire (TiO2NW) treatment. Values are Mean ± SD of 6–8 rats at ach point. * = $P < 0.05$ from control; ANOVA followed by Dunnett's test for multiple group comparison from one control. High Altitude exposure was simulated to 5000 m in Altitude Chamber. Concussive head injury (CHI) was delivered 0.224 N impact over the right parietal skull bone using weight drop technique under Equithesin anesthesia. RH = right hemisphere, LH = left hemisphere; METH was administered 9 ng/kg, s.c. H-290/51 or TiO2 nanowired H-290/51 was administered in a dose of 150 mg/kg, i.p. 4 h after CHI or HA exposure. The duration of HA exposure was 1 week. CHI was also inflicted in gropes after 4 h HA exposure that was continued for 1 week. All parameters were measured at the end of HA exposure with or without CHI or METH. For details see text.

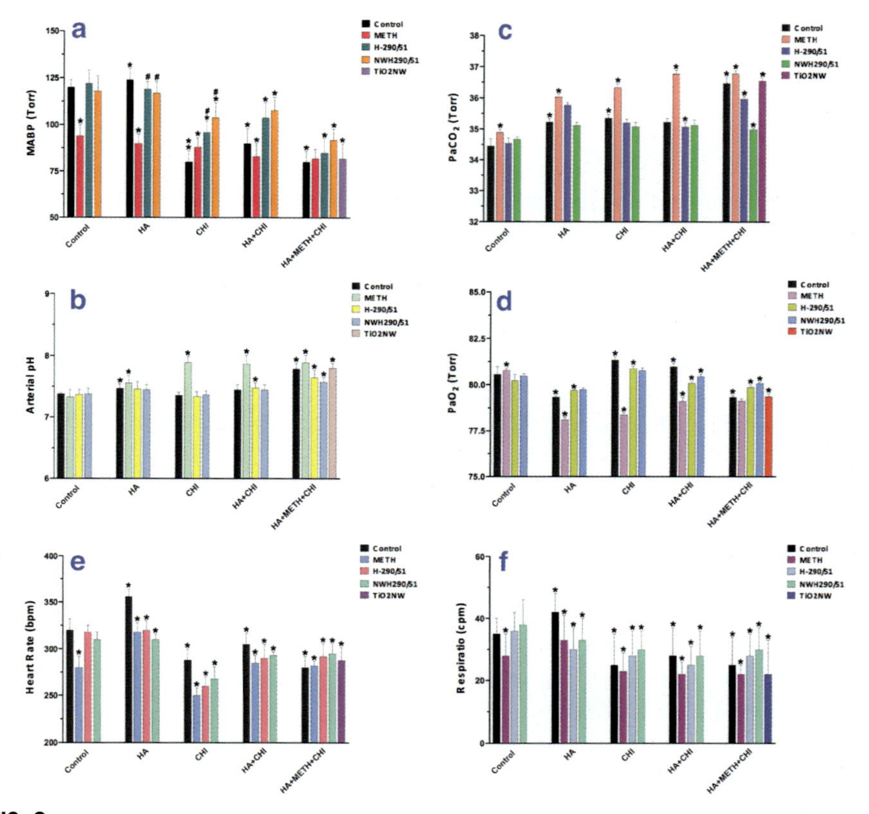

FIG. 2

Changes in mean arterial blood pressure (MABP) (A) arterial pH (B), PaCO2 (C) PaO2 (D) heart rate (E) and respiration rate (F) after High Altitude (HA) exposure in normal and methamphetamine (METH) exposed rats with or without concussive head injury (CHI). All the data were measured after 1 week of exposure to high altitude (HA), concussive head injury (CHI), methamphetamine (METH) with or without H-290/51 or NWH-290/51 and/or vehicle TiO2 nanowire (TiO2NW) treatment. Values are Mean ± SD of 6–8 rats at ach point. * = $P < 0.05$ from control; ANOVA followed by Dunnett's test for multiple group comparison from one control. High Altitude exposure was simulated to 5000 m in Altitude Chamber. Concussive head injury (CHI) was delivered 0.224 N impact over the right parietal skull bone using weight drop technique under Equithesin anesthesia. RH = right hemisphere, LH = left hemisphere; METH was administered 9 ng/kg, s.c. H-290/51 or TiO2 nanowired H-290/51 was administered in a dose of 150 mg/kg, i.p. 4 h after CHI or HA exposure. The duration of HA exposure was 1 week. CHI was also inflicted in gropes after 4 h HA exposure that was continued for 1 week. All parameters were measured at the end of HA exposure with or without CHI or METH. For details see text.

7.56 ± 0.12 ($P < 0.05$) (Fig. 2). The arterial PaCO2 was 36.04 ± 0.09 Torr ($P < 0.05$) and the arterial PaO2 showed 78.10 ± 0.14 Torr ($P < 0.05$). The heart rate was significantly down to 250 ± 8 bpm ($P < 0.05$) and respiration was 23 ± 6 cpm ($P < 0.05$) (Fig. 2).

5.1.2 Physiological parameters at high altitude group

Subjection of animals at high altitude showed a body weight of 367 ± 12 g ($P < 0.05$) and the body temperature of 36.04 ± 0.08 °C ($P < 0.05$) (Fig. 1). The thermal nociception was 6 ± 3 s and mechanical nociception to 5 ± 3 s (Fig. 1). The MABP was 124 ± 6 Torr ($P < 0.05$) and the arterial pH was 7.47 ± 0.07 ($P < 0.05$). The arterial PaCO2 level was 35.23 ± 0.12 Torr ($P < 0.05$) and the PaO2 value significantly decreased to 79.34 ± 0.18 Torr ($P < 0.05$) (Fig. 2). The heart rate significantly decreased to 356 ± 8 bpm ($P < 0.05$) and the respiration rate was 42 ± 6 cpm ($P < 0.05$) (Fig. 2).

In METH treated group the body weight was 380 ± 8 g and body temperature showed significant hyperthermia (40.12 ± 0.23 °C, $P < 0.05$). The thermal nociception was increased significantly to 10 ± 3 s ($P < 0.05$) and the mechanical nociception was 16 ± 6 s ($P < 0.05$) (Fig. 1). The MABP was significant lower (90 ± 5 Torr, $P < 0.05$) and the arterial pH was higher to 7.56 ± 0.12 ($P < 0.05$). The arterial PaCO2 was significantly increased to 36.04 ± 0.09 Torr ($P < 0.05$) and the arterial PaO2 was 78.10 ± 0.14 Torr ($P < 0.05$). The heart rate was 318 ± 10 bpm ($P < 0.05$) and respiration cycle was 33 ± 8 cpm ($P < 0.05$) (Fig. 2).

5.1.3 Physiological parameters in CHI at high altitude group

Subjection of CHI at high altitude showed body weight of 374 ± 6 g and the body temperature of 36.21 ± 0.08 °C ($P < 0.05$). The thermal nociception was 6 ± 4 s and the mechanical nociception values were 5 ± 3 s (Fig. 1). The MABP showed significant reduction (90 ± 8 Torr, $P < 0.05$) and the arterial pH was 7.45 ± 0.08 (Fig. 2). The PaCO2 values were 35.23 ± 0.12 Torr and the arterial PaO2 was 79.98 ± 0.18 Torr ($P < 0.05$). The heart rate was 305 ± 12 ($P < 0.05$) and respiration rate was 28 ± 9 cpm ($P < 0.05$) (Fig. 2).

CHI in METH treated rats the body weight was 385 ± 8 g and the body temperature was 39.89 ± 0.14 ($P < 0.05$). The thermal nociception was increased significantly to 12 ± 5 s ($P < 0.05$) and mechanical nociception values were 16 ± 6 s ($P < 0.05$) (Fig. 1). The MABP was significantly down to 83 ± 7 Torr ($P < 0.05$) and the arterial pH was 7.87 ± 0.14 ($P < 0.05$). The arterial PaCO2 was 36.78 ± 0.12 Torr ($P < 0.05$) and the arterial PaO2 values were 79.12 ± 0.21 ($P < 0.05$). The heart rate was significantly down (285 ± 8 bpm, $P < 0.05$) and the respiration rate was 22 ± 4 cpm ($P < 0.05$) (Fig. 2).

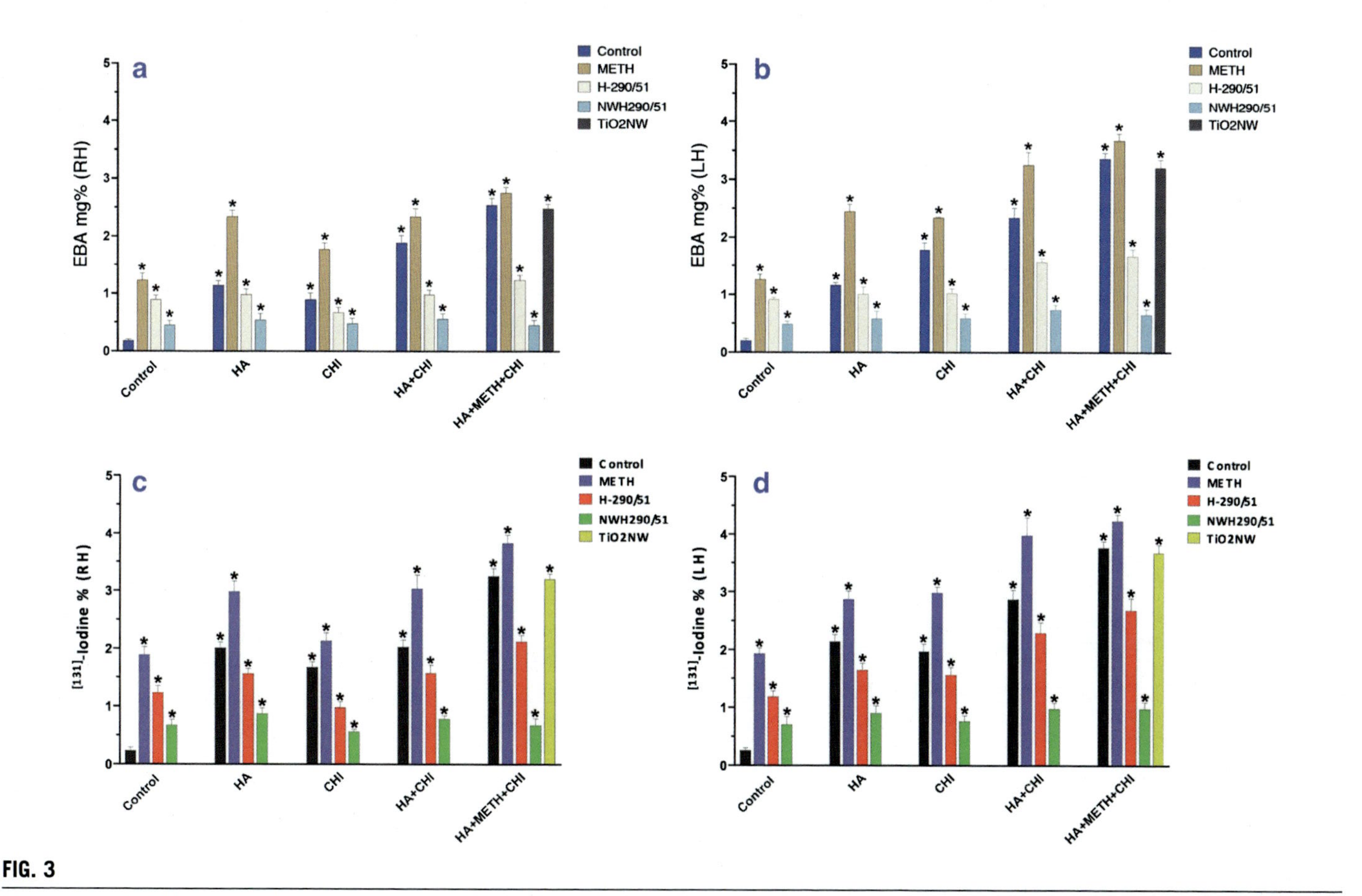

FIG. 3

Measurement of blood-brain barrier breakdown of Evans blue albumin (A and B) and radioiodine (C and D) after High Altitude (HA) exposure in normal and methamphetamine (METH) exposed rats with or without concussive head injury (CHI). All the data were measured after 1 week of exposure to high altitude (HA), concussive head injury (CHI), methamphetamine (METH) with or without H-290/51 or NWH-290/51 and/or vehicle TiO2 nanowire (TiO2NW) treatment. Values are Mean ± SD of 6–8 rats at ach point. * = $P < 0.05$ from control; ANOVA followed by Dunnett's test for multiple group comparison from one control. High Altitude exposure was simulated to 5000 m in Altitude Chamber. Concussive head injury (CHI) was delivered 0.224 N impact over the right parietal skull bone using weight drop technique under Equithesin anesthesia. RH = right hemisphere, LH = left hemisphere; METH was administered 9 ng/kg, s.c. H-290/51 or TiO2 nanowired H-290/51 was administered in a dose of 150 mg/kg, i.p. 4 h after CHI or HA exposure. The duration of HA exposure was 1 week. CHI was also inflicted in gropes after 4 h HA exposure that was continued for 1 week. All parameters were measured at the end of HA exposure with or without CHI or METH. For details see text.

5.2 Blood-brain barrier permeability in normobaric group

In normobaric group, the BBB permeability to EBA and radioiodine was very low. Thus, the EBA permeability was 0.18±0.03mg % in RH and in LH was only 0.20±0.04mg % (Fig. 2). The radioiodine leakage was slightly more in RH 0.23≈ 0.06% and in LH 0.26±0.04% (Fig. 3).

METH treatment at normobaric grope resulted in a significant increase in the BBB breakdown to EBA in RH 1.23±0.12mg % ($P<0.05$) and in LH 1.26±0.09mg % ($P<0.05$). For radioiodine the value showed slightly more increase in RH 1.89±0.14% ($P<0.05$) and in LH it was 1.93±0.10% ($P<0.05$) (see Fig. 3).

5.2.1 Blood-brain barrier permeability in CHI at normobaric control group

Subjection of animals to CHI at normobaric environment the BBB to EBA showed significant increase that was higher in the LH 1.78±0.13mg % ($P<0.05$) as compared to the RH (0.89±0.12mg %, $P<0.05$). In terms of radioiodine leakage the pattern was similar but slightly higher than EBA. Thus, radioiodine leakage in LH was 1.97±0.13% in LH and in RH the value was 1.67±0.10% ($P<0.05$) (Fig. 3).

METH treatment further enhanced the extravasation following CHI in EBA and radioiodine. Thus, the EBA values in LH was 2.34±0.12mg % ($P<0.05$) and in RH it was 1.78±0.11mg % ($P<0.05$) after CHI. The radioiodine leakage was also higher after CHI in METH group LH 2.98±0.10% ($P<0.05$) and in RH 2.13±0.14% ($P<0.05$) at normobaric environment (Fig. 3).

5.2.2 Blood-brain barrier permeability at high altitude group

The BBB to EBA and radioiodine in animals at high altitude was significantly higher in normal animals. Thus, EBA showed extravasation of 1.14±0.08mg % ($P<0.05$) in the RH and in LH the value was 1.16±0.05mg % ($P<0.05$) (Fig. 3). The radioiodine showed greater extravasation than in EBA. Thus, the radioiodine leakage in RH was 2.11±0.10% ($P<0.05$) and in LH was 2–14±0.12% ($P<0.05$) (Fig. 3).

In METH treated normal animals displayed EBA leakage in RH 2–34±0.11mg % EBA and in LH the value of EBA was 2.45±0.12mg % ($P<0.05$) (Fig. 3). The radioiodine extravasation was higher in RH 2.98±0.18% ($P<0.05$) and in LH was 2.87±0.14% ($P<0.05$) (Fig. 3).

5.2.3 Blood-brain barrier permeability in CHI at high altitude group

CHI in normal or METH treated animals was significantly higher at high altitude group as compared to the normal group. Thus, CHI resulted in EBA extravasation in RH to 1.89±0.13mg % ($P<0.05$) and in LH was 2.34±0.16mg % ($P<0.05$) (Fig. 3). The radioiodine extravasation in RH was 2.03±0.12% ($P<0.05$) and in LH was 2.87±0.16% ($P<0.05$) in normal animals after CHI (Fig. 3).

In METH treated group CHI cased higher BBB permeability to EBA and radioiodine. Thus, the EBA extravasation after CHI in METH group was

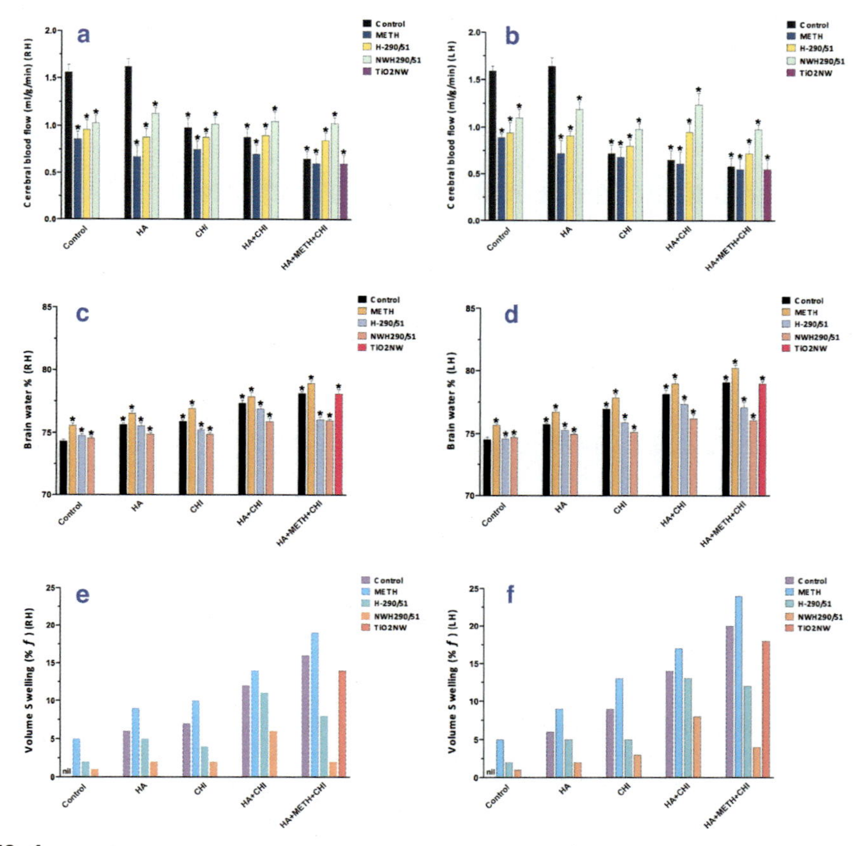

FIG. 4

Measurement of cerebral blood flow (CBF, A and B), brain edema (C and D) and volume swelling (% f, E and F) after High Altitude (HA) exposure in normal and methamphetamine (METH) exposed rats with or without concussive head injury (CHI). All the data were measured after 1 week of exposure to high altitude (HA), concussive head injury (CHI), methamphetamine (METH) with or without H-290/51 or NWH-290/51 and/or vehicle TiO2 nanowire (TiO2NW) treatment. Values are Mean ± SD of 6–8 rats at ach point. * = $P < 0.05$ from control; ANOVA followed by Dunnett's test for multiple group comparison from one control. High Altitude exposure was simulated to 5000 m in Altitude Chamber. Concussive head injury (CHI) was delivered 0.224 N impact over the right parietal skull bone using weight drop technique under Equithesin anesthesia. RH = right hemisphere, LH = left hemisphere; METH was administered 9 ng/kg, s.c. H-290/51 or TiO2 nanowired H-290/51 was administered in a dose of 150 mg/kg, i.p. 4 h after CHI or HA exposure. The duration of HA exposure was 1 week. CHI was also inflicted in gropes after 4 h HA exposure that was continued for 1 week. All parameters were measured at the end of HA exposure with or without CHI or METH. For details see text.

2–34 ± 0.14 mg % ($P < 0.05$) and in LH was 3.25 ± 0.22 mg % ($P < 0.05$). The radio-iodine leakage was 3.03 ± 0.24% ($P < 0.05$) in RH and in LH the value was 3–98 ± 0.31% ($P < 0.05$) (Fig. 3).

5.3 Cerebral blood flow in control at normobaric group

Cerebral blood flow (CBF) was 1.56 ≈ 0.08 mL/g/min in RH and 1.59 ± 0.05 mL/g/min in LH in normal control group at laboratory normobaric environment (Fig. 4). In METH treated normal animals showed a significant decrease in CBF that was in RH 0.86 ± 0.08 mL/g/min ($P < 0.05$) and in LH this was 0.89 ± 0.04 mL/g/min ($P < 0.05$) (Fig. 4).

5.3.1 Cerebral blood flow in CHI at normobaric group

Control group of animals after CHI at normobaric laboratory environment showed significant reduction in CBF in RH (0.98 ± 0.10 mL/g/min, $P < 0.05$) and further reduction was exhibited in the LH (0.72 ± 0.09, $P < 0.05$) (Fig. 4).

In METH treated CHI group at high altitude showed greater decrease in CBF in RH (0.75 ± 0.09 mL/g/min, $P < 0.05$) and in LH the values are even more decreased after CHI (0.68 ± 0.11 mL/g/min, $P < 0.05$) (Fig. 4).

5.3.2 Cerebral blood flow in control at high altitude group

Animals exposed to high altitude showed a value of the CBF in RH 1.02 ± 0.08 mL/g/min and in LH 1.04 ± 0.09 mL/g/min (Fig. 4). In METH treated group the CBF values declined significantly in RH to 0.67 ± 0.12 mL/g/min ($P < 0.05$) and in LH 0.72 ± 0.14 mL/g/min ($P < 0.05$) (Fig. 4).

5.3.3 Cerebral blood flow in CHI at high altitude group

CHI inflicted at high altitude showed significant decline in the CBF to RH 0.88 ± 0.09 mL/g/min ($P < 0.05$) and in LH declined further to 0.65 ± 0.10 mL/g/min ($P < 0.05$) (Fig. 4). In METH treated CHI group the value of CBF further declined in the RH to 0.70 ± 0.10 mL/g/min ($P < 0.05$) and in LH the value decreased further to 0.61 ± 0.13 mL/g/min ($P < 0.05$) (Fig. 4).

5.4 Brain edema and volume swelling in control at normobaric group

Brain water content in normal animals at normobaric laboratory environment was 74.34 ± 0.13% in RH and 74.53 ± 0.18% in LH. In METH treatment the brain water content increased in RH to 75.56 ± 0.16% ($P < 0.05$) and in LH 75.67 ± 0.12% ($P < 0.05$) representing a volume swelling of about 5% in both RH and in LH (Fig. 4).

5.4.1 Brain edema and volume swelling in CHI at normobaric group

After CHI at normobaric laboratory environment, the brain water content increased in RH to 75.89±0.12% ($P<0.05$) and in LH 76.96±0.12% ($P<0.05$) in LH equivalent to a volume swelling of 7% in RH and 9% in LH (Fig. 4). In METH treated CHI group revealed greater edema formation in RH 76.92±0.21% ($P<0.05$) and in LH 77.86±0.24% ($P<0.05$) comprising about 10% volume swelling in RH and about 13% in LH (Fig. 4).

5.4.2 Brain edema and volume swelling in control at high altitude group

In animal's exposure to high altitude increased brain water content by 75.63±0.12% ($P<0.05$) in RH and LH showed brain water 75.74±0.11% ($P<0.05$). This is equivalent to a volume swelling of 6% each in RH and in LH (Fig. 4). In METH treated normal group exposed to high altitude exhibited further increase in brain water in RH to 76.53±0.18% ($P<0.05$) and in LH to 76.73±0.16% ($P<0.05$). This amounts to 9% swelling in the RH and LH (Fig. 4).

5.4.3 Brain edema and volume swelling in CHI at high altitude group

CHI normal group at high altitude exhibited brain water content significantly higher in RH 77.34±0.23% ($P<0.05$) and in LH 78.14±0.31% ($P<0.05$). This comprises about 12% volume swelling in LH and 14% volume swelling in the LH (Fig. 4). In METH treated CHI group at high altitude the brain water content further increased to 77.87±0.31% ($P<0.05$) in RH and 78.98±0.34% ($P<0.05$) in LH (Fig. 4). The volume swelling showed 14% increase in RH and 17% increase in LH (Fig. 4).

6 Biochemical changes

Beta catenin is needed to maintain BBB structure and function (Benz and Liebner, 2020; Tran et al., 2016). Alterations in beta catenin occur during brain pathology and at high altitude (Jia et al., 2019; Wang et al., 2020). This suggests that beta catenin is altered during high altitude induced changes in brain pathology.

6.1 Beta catenin level at normobaric group

Beta catenin was measured in the control group of brain showed 5.67±0.01 mg/g in RH and in LH this was 5.70±0.04 pg/g (Fig. 5). In METH treated rats the values of beta catenin increased to 12.45±0.28 pg/g ($P<0.05$) in RH whereas in LH this was 12.63±0.14 pg/g ($P<0.05$) (Fig. 5).

6.1.1 Beta catenin level at high altitude group

Subjection of rats to high altitude enhanced the beta catenin levels significantly. Thus, at high altitude, the RH shows 8.34±0.12 pg/g ($P<0.05$) and in LH the values of beta catenin was 8.47±0.08 pg/g ($P<0.05$) (Fig. 5). METH treatment has enhanced the beta catenin level further at high altitude in RH to 19.56±0.15 pg/g ($P<0.05$) and in LH beta catenin was 22.34±0.14 ($P<0.05$) (Fig. 5).

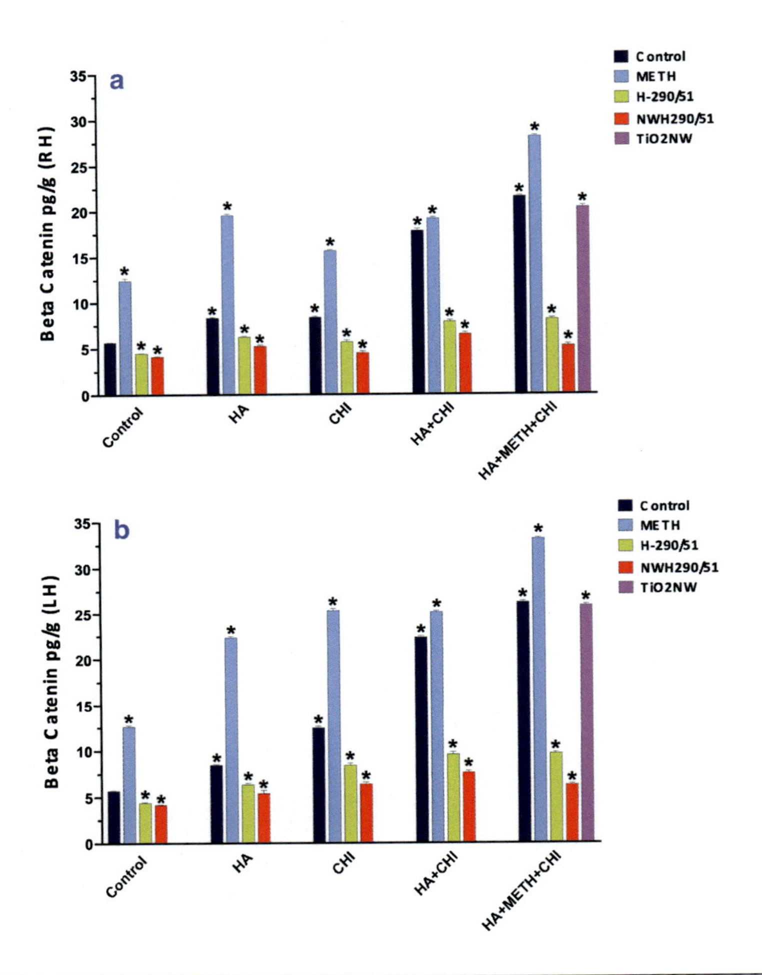

FIG. 5

Biochemical measurement of beta catenin following High Altitude (HA) exposure in normal and methamphetamine (METH) intoxicated rats with or without concussive head injury (CHI). All the data were measured after 1 week of exposure to high altitude (HA), concussive head injury (CHI), methamphetamine (METH) with or without H-290/51 or NWH-290/51 and/or vehicle TiO2 nanowire (TiO2NW) treatment. Values are Mean ± SD of 6–8 rats at ach point. * = $P < 0.05$ from control; ANOVA followed by Dunnett's test for multiple group comparison from one control. High Altitude exposure was simulated to 5000 m in Altitude Chamber. Concussive head injury (CHI) was delivered 0.224 N impact over the right parietal skull bone using weight drop technique under Equithesin anesthesia. RH = right hemisphere, LH = left hemisphere; METH was administered 9 ng/kg, s.c. H-290/51 or TiO2 nanowired H-290/51 was administered in a dose of 150 mg/kg, i.p. 4 h after CHI or HA exposure. The duration of HA exposure was 1 week. CHI was also inflicted in gropes after 4 h HA exposure that was continued for 1 week. All parameters were measured at the end of HA exposure with or without CHI or METH. For details see text.

6.1.2 Beta catenin level in CHI at normobaric group

Rats subjected to CHI at normobaric laboratory environment showed an increase in beta catenin in RH to 8.37 ± 0.12 pg/g ($P < 0.05$) and in LH 12.48 ± 0.16 pg/g ($P < 0.05$). In METH treated CHI showed farther increase in beta catenin in RH to 15.67 ± 0.08 pg/g ($P < 0.05$) and in LH 25.32 ± 0.22 pg/g ($P < 0.05$) (Fig. 5).

6.1.3 Beta catenin level in CHI at high altitude group

At high altitude rats subjected to CHI showed high beta catenin in RH (17.82 ± 0.23 pg/g, $P < 0.05$) and in LH the value was 22.34 ± 0.14 pg/g ($P < 0.05$). In METH treated CHI group at high altitude the RH showed significant increase in beta catenin to 19.18 ± 0.14 pg/g ($P < 0.05$) whereas the LH exhibited additional increase of 25.15 ± 0.12 pg/g ($P < 0.05$) (Fig. 5).

7 Brain pathology

Brain pathology was examined using light and electron microscopy at high altitude exposure of CHI in normal and in METH treated group.

7.1 Neuronal injury at normobaric group

Control group of rats at normobaric laboratory environment did not exhibit neuronal damages (Fig. 6). Only few neurons show distortion in RH (3 ± 4 cells) and in LH (2 ± 3 cells). However, METH treatment resulted in significant increase in neuronal distortion in RH (36 ± 8 cells, $P < 0.05$) and in LH 38 ± 8 cells ($P < 0.05$) (Fig. 6).

However, in METH treated rats showed profound neuronal damages as evident in RH (36 ± 8 cells, $P > 0.05$) and in LH 38 ± 6 cells ($P < 005$) (Fig. 6).

7.1.1 Neuronal injury at high altitude group

Subjection of animals to high altitude environment resulted in profound neuronal injuries in RH (58 ± 6 cells, $P < 0.05$) and in LH 60 ± 6 cells ($P < 0.05$) (Fig. 6). In METH treated group exposure to high altitude resulted in further increase in neuronal injury significantly. Thus, METH treated high altitude exposed group revealed neuronal injuries in RH 87 ± 9 cells ($P < 0.05$) and in LH 92 ± 8 cells ($P < 0.05$) (Fig. 6).

7.1.2 Neuronal injury in CHI at normobaric group

Animals subjected to CHI at normobaric laboratory environment showed significant neuronal injuries in RH (78 ± 11 cells, $P < 0.05$ and in LH 108 ± 13 cells, $P < 0.05$) (Fig. 6). In METH treated group CHI resulted in greater cell injury. Thus, CHI in METH showed 108 ± 16 injured cells in RH ($P < 0.05$) and 142 ± 10 cells in LH ($P < 0.05$) (Fig. 6).

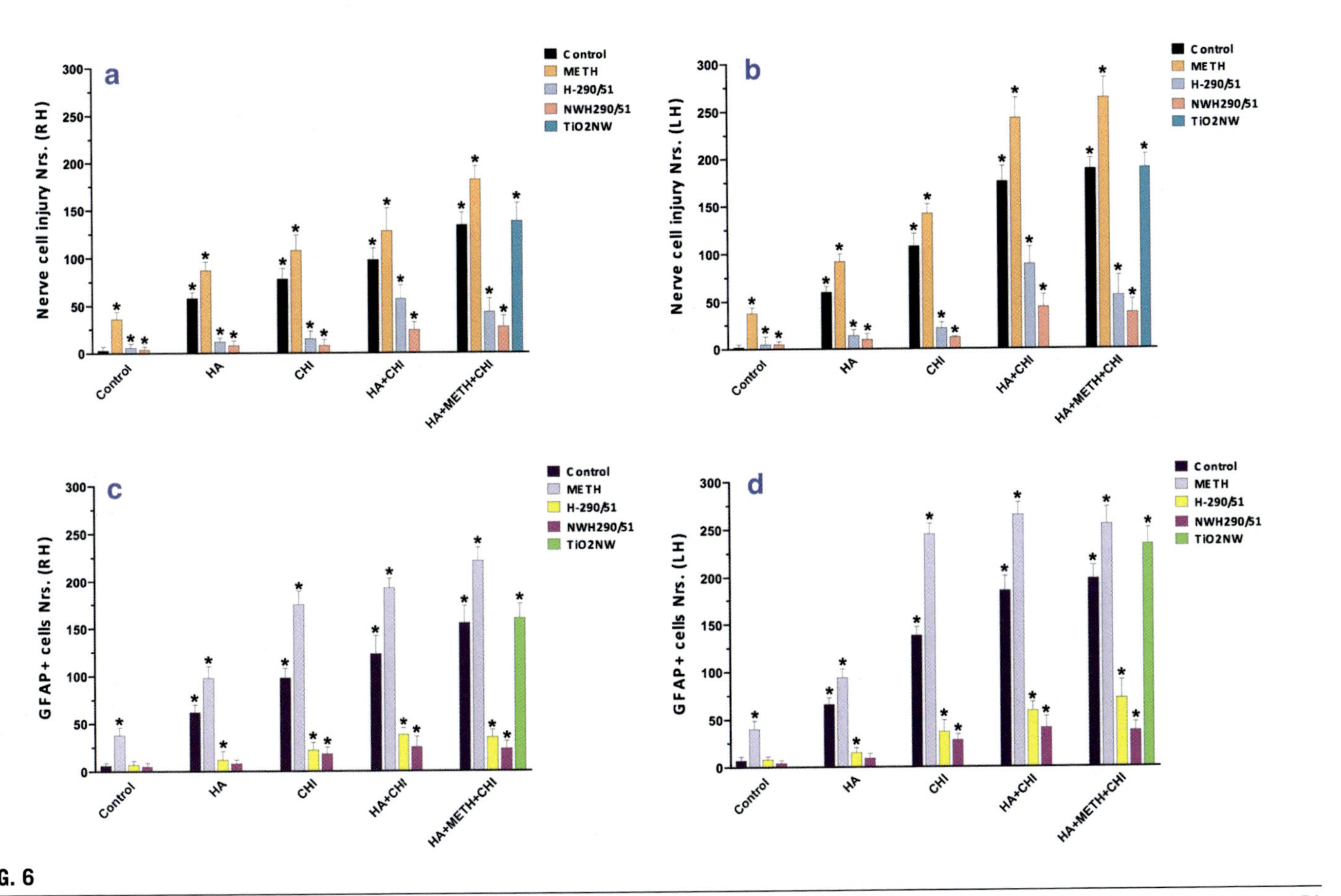

FIG. 6

Semiquantitative analysis of neuronal injury at light microcopy (LM) (A and B) and immunoreactivity of glial fibrillary acidic protein (GFAP) (C and D) following High Altitude (HA) exposure in normal and methamphetamine (METH) intoxicated rats with or without concussive head injury (CHI). All the data were measured after 1 week of exposure to high altitude (HA), concussive head injury (CHI), methamphetamine (METH) with or without H-290/51 or NWH-290/51 and/or vehicle TiO2 nanowire (TiO2NW) treatment. Values are Mean ± SD of 6 to 8 rats at ach point. * = $P < 0.05$ from control; Non parametric Chi Square test. High Altitude exposure was simulated to 5000 m in Altitude Chamber. Concussive head injury (CHI) was delivered 0.224 N impact over the right parietal skull bone using weight drop technique under Equithesin anesthesia. RH = right hemisphere, LH = left hemisphere; METH was administered 9 ng/kg, s.c. H-290/51 or TiO2 nanowired H-290/51 was administered in a dose of 150 mg/kg, i.p. 4 h after CHI or HA exposure. The duration of HA exposure was 1 week. CHI was also inflicted in gropes after 4 h HA exposure that was continued for 1 week. All parameters were measured at the end of HA exposure with or without CHI or METH. For details see text.

7.1.3 Neuronal injury in CHI at high altitude group

CHI at high altitude significantly enhanced neuronal injury. Thus, about 98 ± 12 cells ($P < 0.05$) was seen in RH and 176 ± 16 cells ($P < 0.05$) showed neuronal damage in LH (Fig. 6). In METH treated group CHI resulted in higher cell damage in RH 128 ± 24 cells ($P < 0.05$) and in LH 243 ± 21 cells ($P < 0.05$) were seen distorted or damaged (Fig. 6).

8 Glial fibrillary acidic protein immunoreactivity

Changes in astrocytes reactivity was examined using glial fibrillary acidic protein (GFAP) immunoreactivity in CHI exposed to normobaric or high altitude.

8.1 GFAP reactivity in control at normobaric group

Only a few GFAP positive astrocytes were seen in normal animals at normobaric laboratory environment. Thus, control group exhibited only 6 ± 3 cells in RH and in LH 7 ± 4 GFAP positive astrocytes were seen (Fig. 6). In METH treatment the number of astrocytes positive cells were significantly increased. Thus, about 38 ± 8 cells ($P < 0.05$) were positive in RH whereas the LH showed 40 ± 9 cells ($P < 0.05$) that were GFAP positive in METH treated group (Fig. 6).

8.1.1 GFAP reactivity at high altitude group

When animals were exposed to high altitude the number of GFAP positive cells were significantly enhanced. Thus, 62 ± 8 cells were GFAP positive in RH ($P < 005$) and in LH 66 ± 7 cells ($P < 0.05$) were positive to GFAP immunoreactivity in animals exposed to high altitude (Fig. 6). When METH treated animals were exposed to high altitude the number of GFAP positive cells were further increased significantly. Thus, there were 92 ± 12 GFAP positive cells in RH ($P < 0.05$) and about 94 ± 9 cells ($P < 0.06$) in LH were GFAP positive in METH treated rats exposed to high altitude (Fig. 6).

8.1.2 GFAP reactivity in CHI at normobaric group

CHI in animals at normobaric laboratory environment resulted in increased GFAP positive cells that were significantly higher in LH as compared to the RH side. Thus, there were 138 ± 9 GFAP positive cells ($P < 0.05$) in LH after CHI rats and the RH side that showed 98 ± 10 cells GFAP positive ($P < 0.05$) (Fig. 6). This increase in GFAP positive cells were further enhanced in both RH and LH side in METH treated rats with CHI. Thus, about 175 ± 14 cells were GFAP positive in RH ($P < 0.05$) and about 245 ± 11 GFAP positive cells were seen in LH ($P < 0.05$) after CHI in METH treated rats (Fig. 6).

8.1.3 GFAP reactivity in CHI at high altitude group

Subjection of rats with CHI at high altitude significantly increased GFAP positive astrocytes in identical manner. Thus, the RH side exhibited 123 ± 19 cells GFAP positive ($P < 0.05$) and in LH side 185 ± 16 cells ($P < 0.05$) were GFAP positive after CHI at high altitude (Fig. 6).

In METH treated rats exposed to high altitude CHI further exacerbated the number of GFAP positive cells in RH and LH. Thus about 243 ± 21 cells ($P < 0.05$) were GFAP positive in RH and in LH about 265 ± 13 cells ($P < 0.05$) were GFAP positive in METH treated CHI group (Fig. 6).

Brain pathology at ultrastructural level seen at transmission electron microscopy (TEM) showed profound alterations in myelin damage and vesiculation along with endothelial cells deformity in CHI at normobaric and at high altitude environment as described below.

9 Myelin damage of control at normobaric group

Myelin deformity, damage and vesiculation were minimal in control group at normobaric laboratory environment. Thus, only 4 ± 3 axons exhibited myelin vesiculation or damage in RH and 5 ± 3 axons displayed myelin vesiculation in LH (Fig. 7). In METH treated rats exhibited higher numbers of myelin vesiculation and damage at normobaric environment. Thus, about 18 ± 5 axons ($P < 0.05$) showed damage with vesiculation in RH and 19 ± 6 axons ($P < 0.05$) exhibited myelin damage in LH after METH treatment (Fig. 7).

9.1 Myelin damage in CHI at normobaric group

When CHI was inflicted in normal rats the number of myelin damage was increased. Thus, at normobaric environment CHI resulted in 38 ± 6 axons ($P < 0.05$) damage with myelin vesiculation in RH and about 51 ± 8 axons ($P < 0.05$) are seen with myelin vesiculation in LH (Fig. 7). In METH treated rats subjected to CHI myelin damage was exacerbated at normobaric laboratory environment. Thus, in RH about 54 ± 7 axons ($P < 0.05$) was seen and in LH 86 ± 12 axons ($P < 0.05$) were damaged in METH treatment after CHI (Fig. 7).

9.2 Myelin damage at high altitude group

When normal animals were exposed to high altitude environment myelin damage was seen in RH in 22 ± 6 axons ($P < 0.05$) and in LH 23 ± 8 axons ($P < 0.05$) were distorted or damaged (Fig. 7). At high altitude METH treatment exacerbated this myelin damage. Thus, in RH 38 ± 8 axons ($P < 0.05$) were seen and in LH 40 ± 9 axons ($P < 0.05$) showed myelin vesiculation and deformity after METH treatment (Fig. 7).

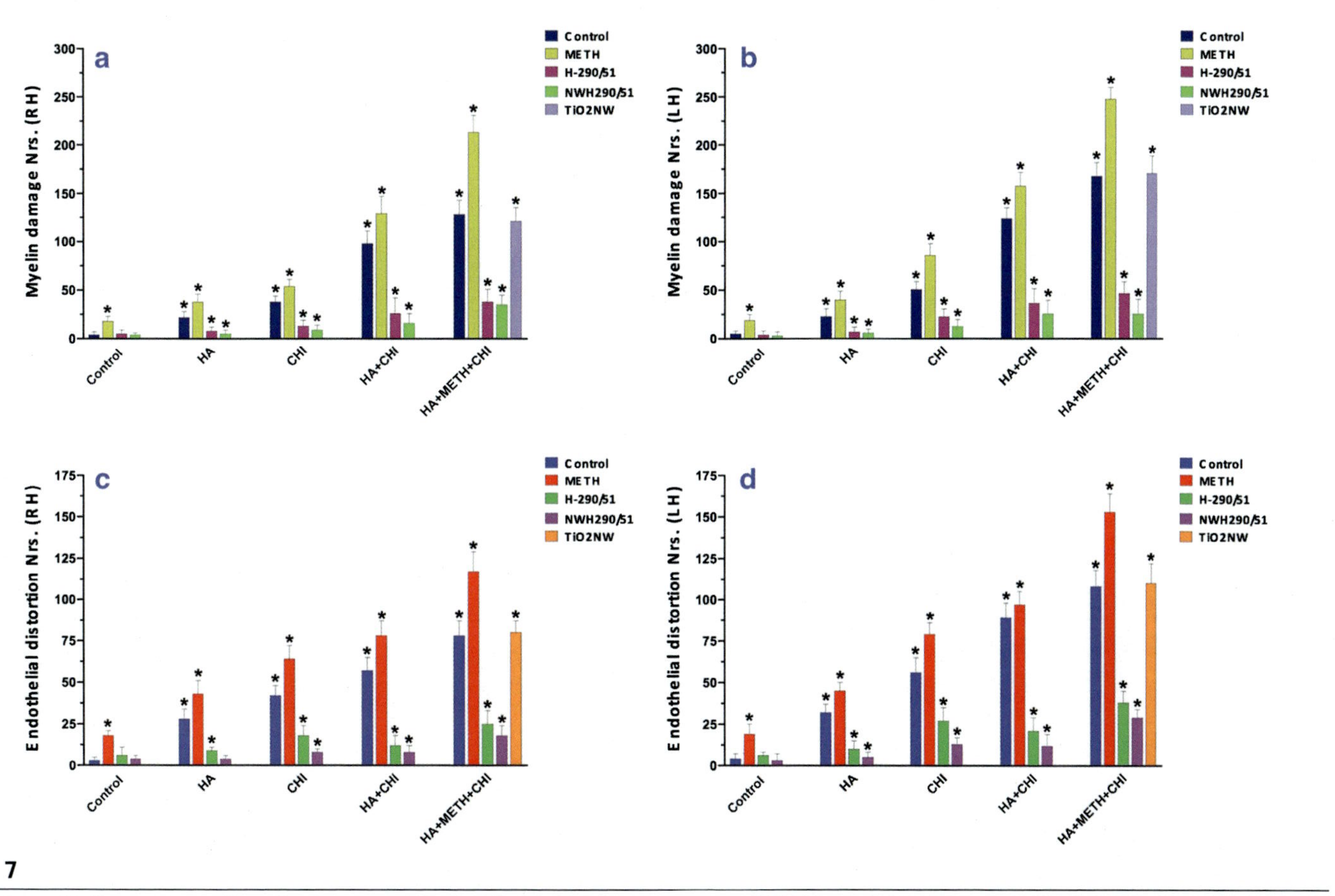

FIG. 7

Semiquantitative analysis of myelin damage at Transmission electron microcopy (TEM) (A and B) and endothelial cell deformity (C and D) following High Altitude (HA) exposure in normal and methamphetamine (METH) intoxicated rats with or without concussive head injury (CHI). All the data were measured after 1 week of exposure to high altitude (HA), concussive head injury (CHI), methamphetamine (METH) with or without H-290/51 or NWH-290/51 and/or vehicle TiO2 nanowire (TiO2NW) treatment. Values are Mean ± SD of 6–8 rats at ach point. * = $P < 0.05$ from control; Non parametric Chi Square test. High Altitude exposure was simulated to 5000 m in Altitude Chamber. Concussive head injury (CHI) was delivered 0.224 N impact over the right parietal skull bone using weight drop technique under Equithesin anesthesia. RH = right hemisphere, LH = left hemisphere; METH was administered 9 ng/kg, s.c. H-290/51 or TiO2 nanowired H-290/51 was administered in a dose of 150 mg/kg, i.p. 4 h after CHI or HA exposure. The duration of HA exposure was 1 week. CHI was also inflicted in gropes after 4 h HA exposure that was continued for 1 week. All parameters were measured at the end of HA exposure with or without CHI or METH. For details see text.

9.3 Myelin damage in CHI at high altitude group

When CHI was done at high altitude the number of damaged axons and vesiculation of myelin has increased further. Thus, in CHI at high altitude 98 ± 13 axons (P < 0.05) were damaged in RH while 124 ± 118 axons (P < 0.05) exhibited myelin vesiculation and deformity in LH (Fig. 7). In METH treated CHI animals at high altitude further exacerbated this myelin damage. Thus, about 129 ± 18 axons (P < 0.05) showed damage or distortion in RH and 158 ± 14 axons (P < 0.05) exhibited myelin damage in LH after CHI in METH treated group at high altitude (Fig. 7).

10 Endothelial distortion at normobaric group

Control animals did not show endothelial cell deformity in brain at normobaric laboratory environment. Thus, only 3 ± 2 microvessels in RH and about 4 ± 3 capillaries in LH show some deformation at TEM (Fig. 7). In METH treated group the number of endothelial deformity has increased significantly. Thus, about 18 ± 3 microvessels (P < 0.05) showed deformity in RH and about 19 ± 6 cerebral vessels (P < 0.05) were deformed in LH (Fig. 7).

10.1 Endothelial distortion in CHI at normobaric group

Subjection to CHI in rats at normobaric laboratory environment the number of vessels showing deformity is enhanced. Thus, about 42 ± 6 microvessels (P < 0.05) in RH and 56 ± 9 vessels (P < 0.05) were damaged in LH after CHI at laboratory normobaric environment (Fig. 7). In METH treated CHI group the number of deformed vessels increases significantly. Accordingly, the RH showed 64 ± 8 microvessels (P < 0.05) deformed and the LH exhibited 79 ± 7 microvessels (P < 0.05) exhibiting partially collapsed and distorted after CHI in METH treated group (Fig. 7).

10.2 Endothelial distortion at high altitude group

Exposure to high altitude in normal animals resulted in microvessels deformity in RH 28 ± 6 vessels (P < 005) and in LH 32 ± 5 vessels (P < 0.05) with distorted appearances (Fig. 7). In METH treated rates exposure at high altitude resulted in greater numbers of microvessel deformity. Thus, in RH 43 ± 8 vessels (P < 005) and in LH 45 ± 5 vessels (P < 0.05) were deformed after CHI in METH treated rats (Fig. 7).

10.3 Endothelial distortion in CHI at high altitude group

CHI at high altitude increased the endothelial cell deformity. Thus, the number of endothelial cell deformity was increased in RH at high altitude after CHI was 57 ± 8 microvessels (P < 005) and in LH this was 89 ± 9 microvessels (P < 0.05) (Fig. 7). In METH treated group after CHI the nr of vessel deformity increased further to 78 ± 9 microvessels in RH (P < 0.05) and in LH this was 97 ± 8 microvessels (P < 0.05) (Fig. 7).

11 Treatment with antioxidant drug H-290/51

Treatment with H-290/51 a potent antioxidant significantly attenuated brain pathology in CHI at normobaric environment and/or following high altitude exposure (Fig. 8). This treatment also thwarted brain pathology exacerbated by METH (Fig. 8).

Thus, in H-290/51 treated groups in CHI showed significant reductions in the BBB permeability, brain edema and enhanced cerebral blood flow following high altitude exposure (Figs. 1 and 2). This effect was also pronounced in METH treated high altitude (Figs. 3 and 4).

A significant reduction in beta catenin was also observed in METH or CHI group exposed at high altitude with H-290/51 (Fig. 5). The cellular changes in neurons, GFAP, myelin vesiculation and endothelial ell deformity was also significantly reduce in H-290/51 treated groups at high altitude with CHI or METH (Fig. 5).

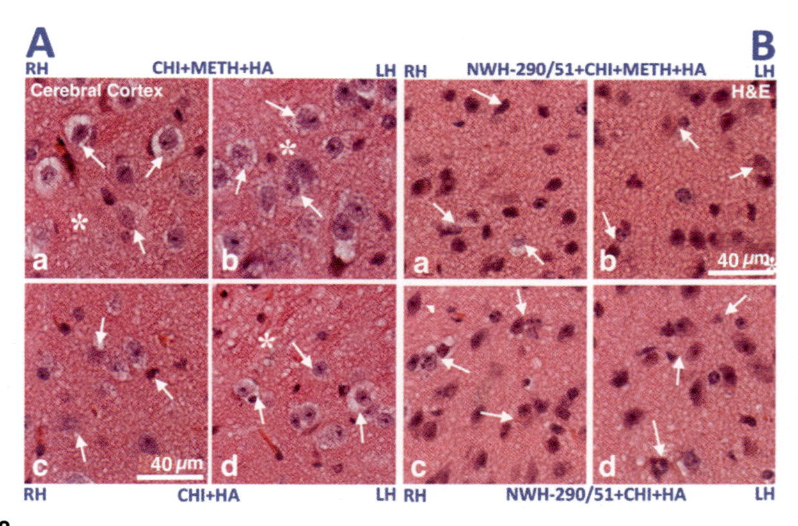

FIG. 8

Light microscopy of Hematoxylin and Eosin (H&E) staining on paraffin sections (3 µm thick) showing neuronal damages in parietal cerebral cortex from concussive head injury (CHI) at high altitude (HA, A. c, d) and CHI at HA with methamphetamine (METH) intoxication (A. A, b) and their modification with nanowired (NW) delivered H-290/51 treatment (B) in the injured right hemisphere (RH, A. a, c; B. a, c) and in the uninjured left hemisphere (LH, A. b, d; B. b, d). Several neurons (A) showed degeneration and perineuronal edema (arrows) is prominent in CHI at HA (A. c, d) that was further exacerbated after METH intoxication (A. a, b). Sponginess and edematous expansion (*) of the neuropil is clearly seen in CHI at HA with or without METH intoxication. Treatment wish NWH-290/51 markedly reduced perineuronal edema and extensive neuronal degeneration (Arrows, B) in CHI at HA (B. c, d) as well as CHI at HA with METH intoxication (B. a, b). Sponginess and edematous expansion of the neuropil appear to be also reduced. Bar = 40 µm.

The other physiological variables like body temperature, nociception, MABP and blood gases were mildly affected by H-290/51 treatment in METH or CHI group (Figs. 1 and 2).

11.1 Treatment with nanowired H-290/51

Treatment with NWH-290/51 significantly attenuated BBB permeability to EBA and radioiodine and brain edema formation or volume swelling at high altitude induced by CHI or METH. This effect is more pronounced by NWH-290/51 as compared to H-290/51 (Figs. 3 and 4).

The reduction in beta catenin was most marked in NWH-290/51 treated CHI or METH group at normobaric of high altitude exposed group (Fig. 5). The neuronal damage, GFAP activation, myelin vesiculation and endothelial deformity were significantly attenuated by NWH-290/51 treatment (Figs. 8–11).

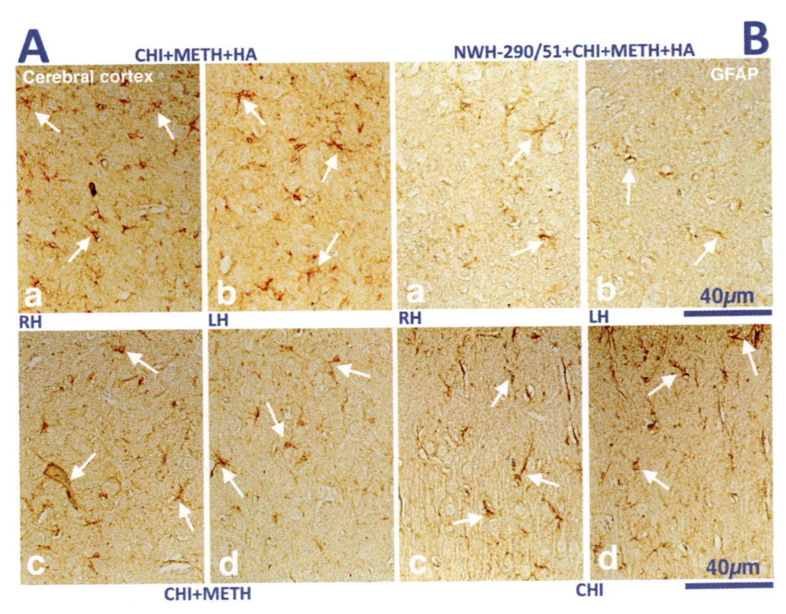

FIG. 9

Immunohistochemistry of glial fibrillary acidic protein (GFAP) showing activation of astrocytes in concussive head injury (CHI, B. c, d), CHI with methamphetamine (METH) intoxication (A. c, d) and CHI with METH at high altitude (HA, A. a, b) and its modification with nanowired (NW) delivery of H-290/51 in CHI with METH at HA (B. a, b) in the parietal cerebral cortex. GFAP positive cells are located across the microvessels and astrocytes (arrows) in the neuropil that is most prominent in the uninjured left hemisphere (LH) than the injured right hemisphere (RH) in almost all groups (arrows). Exposure of CHI with METH at HA exhibited mot marked GFAP positive cells in the neuropil (A. a, b) and treatment with NWH-290/51 markedly reduced the GFAP activation (B. a, b). Paraffin section 3-μm, Bar = 40 μm.

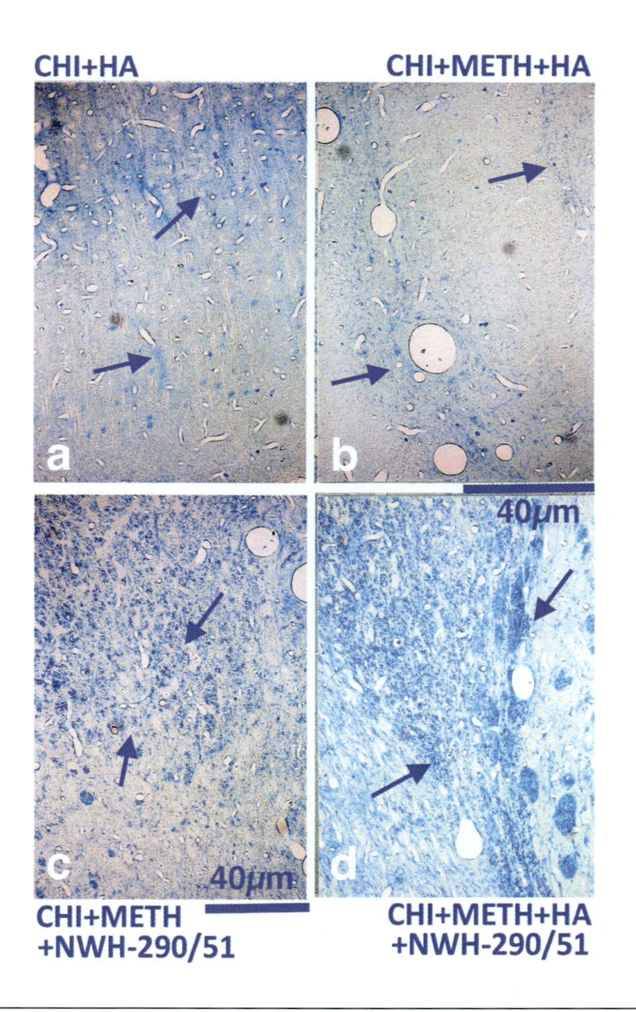

FIG. 10

Luxol Fast Blue (LFB) staining of parietal cerebral cortex showing myelin degeneration in concussive head injury (CHI) at high altitude (HA) (A) and following methamphetamine (METH) intoxication (B). Myelin degeneration is aggravated by METH intoxication in CHI at HA (arrows, B). Treatment with nanowired (NW) delivery of H-290/51 induced neuroprotection by thwarting myelin degeneration in CHI with METH intoxication (C) and even exposure to HA in-group of CHI with METH (D). LFB staining of myelin fibers are strong in NWH-290/51 treated group (C and D). Bar = 40 μm.

The body temperature, nociception and MABP values are significantly improved by NWH-290/51 treated CHI or METH group at normobaric of high altitude group after METH or CHI as compared to the untreated groups (Figs. 1 and 2).

These observations clearly show that nanowired delivery of H-290/51 is far more superior in neuroprotection as compared to the conventional H-290/51 therapy.

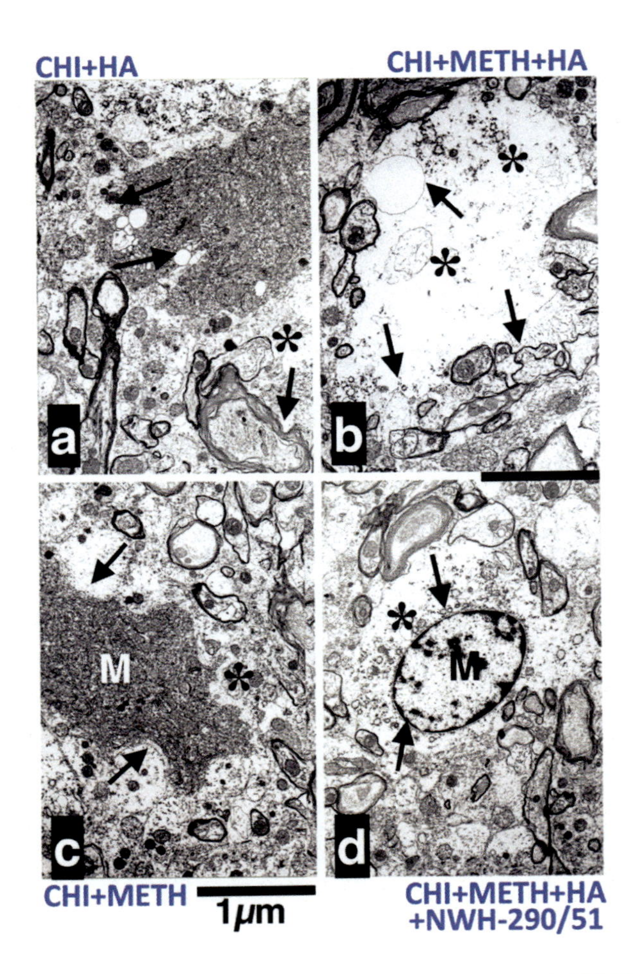

FIG. 11

High power Transmission electron micrograph showing degenerated nerve cell nucleus with condensed karyoplasm, vacuolation (arrows) after concussive head injury (CHI) at high altitude (HA) (A) and exacerbation of nerve cell injury and membrane vacuolation (arrow) after methamphetamine (METH) intoxication in CHI at HA (B). CHI induced exacerbation of nerve cell injury and neuropil after METH intoxication (C) and neuroprotection by nanowired (NW) H-290/51 treatment in CHI + METH + HA. Nerve cell protection with healthy nucleus after METH intoxication (M) is seen (arrow) (D). Vacuolation (*) and vesiculation of myelin and expansion of neuropil is clearly seen (A–C). Bar = 1 μm.

12 Discussion

The salient new findings of the present investigation show that high altitude significantly enhances brain pathology of CHI as compared to the identical injury performed at normobaric conditions at sea level. Furthermore our study is the first to show that METH intoxication at high altitude induced pronounced brain edema and cell injury along with reduced CBF as compared to the identical METH exposure at sea level. In addition, METH intoxicated animals when subjected to CHI at high altitude the brain pathology exacerbates as compared to the identical injury in METH intoxicated group at sea level, not reported earlier.

These observations suggest that high altitude and lack of oxygen are crucial factors in enhancing CHI induced pathology. In addition, when METH is consumed the pathology of brain injury is much more higher than the situation without METH exposure. Since METH is widely consumed at high altitude, it is likely that when these persons get addition brain injury their outcome will be much more serious than the cases without METH users.

The effect of METH on brain trauma suggests that METH exacerbates brain pathology after trauma and this effect is further potentiated at high altitude, not reported earlier. This indicates that military personnel who are engaged in combat operations at high altitude are more vulnerable to trauma induced brain pathology than their counterpart engaged in similar activities at lower altitude or at sea level (Hu et al., 2010; Wang et al., 2019; Zhou et al., 2017). Military personnel stationed at high altitude are often prone head injury (Banderet and Burse, 1992) and thus their pathophysiological symptoms are much worse than those at lower altitude or at sea level. Due to several anxieties, sleeplessness and other psychological problems at high altitude psychostimulants are needed in military personnel to stay active (Buguet et al., 2003; Eliyahu et al., 2007; Zhang et al., 2016). These psychostimulants are quite similar to derivative of amphetamine and has better effects on health parameters at high altitude (Brühl et al., 2019; Roberts et al., 2020). Thus, it is possible that these drug uses could adversely affect traumatic brain injury induced brain pathology.

Apart from methamphetamine various other psychostimulants are used in minor populations in army and vigilance officials to keep them alert (Agaku et al., 2020; Costa et al., 2015; CDC, 1995; Defalque and Wright, 2011; Lacy et al., 2008; Loeffler et al., 2012; Platteborze et al., 2013; Wiegmann et al., 1996). This is still unclear whether like methamphetamine, other psychostimulants also adversely affect brain pathology following traumatic brain injury or concussive head injury. Our laboratory is currently investigating the effects of other psychostimulants on brain injury induce adverse pathology at high altitude and at sea level.

METH is the second most psychostimulant used by more than 16 millions users worldwide after cannabis (United Nations Office on Drugs and Crime, 2007). Immediately after consuming METH a feeling of euphoria, increased productivity, decreased anxiety and increased energy occurs in users for long period of time (Homer et al., 2008). METH has a half-life of action ranging from 10 to 12 h (Volkow et al., 2010). However, METH induces hyperthermia and induces aggression, hypertension and altered behavioral and neurological functions (Moszczynska and Callan, 2017; Potvin et al., 2018; Sharma and Ali, 2006).

METH indices breakdown of the BBB permeability and edema formation (Kiyatkin et al., 2007; Sharma and Kiyatkin, 2009). These METH induced changes are mediated through generation of reactive oxygen species (ROS) and free radical formation (Cadet and Krasnova, 2009; Jeng et al., 2006; Krasnova and Cadet, 2009). METH is consumed at high altitude more frequently die to its euphoric effect (Kim et al., 2014). Thus, METH at high altitude changes the brain function that is crucial for brain injury resulting in adverse brain pathology (Chen et al., 2020a,b; Farnia et al., 2020; Huang et al., 2017; Kogachi et al., 2017).

Exposure to high altitude induced several changes in brain function (Sharma et al., 2019a,b,c). Thus, changes in synaptic plasticity, astrocytic dysfunction and memory alterations occur at high altitude exposure. In addition, neuronal and vascular deficits also occur after high altitude exposure. A reduction in the CBF and increased volume of ventricles is seen using MRI. Expansion of neurovascular network, activation of microglial and demyelination in several brain structures occur (Cramer et al., 2019). High altitude cerebral edema (HACE) is another factor that worsen brain pathology (Hackett, 1999). High altitude exposure also induces oxidative stress that is primarily responsible for brain damage and morphological changes (Lin et al., 2012). Thus, a combination of brain injury, METH exposure and high altitude has cumulative effects on brain pathology.

This is further evident from the findings of drug treatment. When we used H-290/51 treatment it reduces METH induced brain pathology. This supports our findings of oxidative stress induced brain damage in METH treatment (Sharma et al., 2007a,b). This was further potentiated when NWH-290/51 was used. Nanowired drug delivery is shown to induce superior neuroprotection in brain diseases (Sharma et al., 2016a,b). Nanowired drug could easily penetrate into the brain and release its content into the extracellular or intracellular fluid microenvironment for longer periods (Sharma et al., 2016a,b). This would enhance superior effects of the drug in then brain. Our observations clearly show that BWH-290/51 is superior in inducing neuroprotection in brain injury at high altitude and thus reduces brain pathology. This observation suggests that nanodelivery of drugs are needed to reduce brain pathology at high altitude.

Another important information came out from our study that beta catenin that maintains BBB and neuronal integrity (Hübner et al., 2018; Laksitorini et al., 2019; Luo et al., 2020) is elevated in CHI at high altitude. This suggests that beta catenin increases after brain injury in order to maintain the BBB structure and function. However, in spite of increased beta catenin the BBB breakdown is higher in CHI both at the sea level and at high altitude. When H-290/51 or NWH-290/51 is given to CHI or METH intoxicated rats the level of beta catenin was normalized. This indicates that beta catenin playing key roles in protecting BBB breakdown in CHI. Also METH intoxication alone that disturbs BBB function; the beta catenin is elevated to restore the BBB function. When H-290/51 or NWH-290/51 was delivered in METH intoxicated group the beta catenin was normalized again. This suggests that beta catenin is crucial in maintaining BBB function in health and disease. However, further studies are needed to understand the physiological function of beta catenin in CHI induced brain pathology.

Our study further shows that TiO2 nanowires alone could not influence brain function in CHI alone or intoxicated with METH at high altitude. This suggests that nanowired itself doesn't have any neuroprotective function in brain injury. Thus, the drug that is labeled through nanowires is playing important functions in inducing neuroprotection. This study thus further supports the idea of oxidative stress in inducing brain pathology. To further confirm this findings measurement of oxidative stress at high altitude and brain injury is needed. This is a feature currently being examined in our laboratory. Brain injury either caused by blunt trauma or perforating objects results in oxidative stress (Abdul-Muneer et al., 2015; Cornelius et al., 2013; Eghwrudjakpor and Allison, 2010). Oxidative stress in brain injury caused imbalance between antioxidant production and reactive oxygen species (ROS) and nitrogen species (RNS) and leads to BBB impairment through activation of matrix metalloproteinases (MMPs) (Ali et al., 2019; Amin et al., 2016). Oxidative stress activates inflammatory cytokines such as interleukin-1β (IL-1β), tumor necrosis factor-α (TNF-α), and transforming growth factor-beta (TGF-β) (Kim et al., 2016; Tian et al., 2020). Degradation of endothelial vascular endothelial growth factor receptor-2 (VEGFR-2) elevates cellular/serum VEGF leading to neuroinflammation through activation of caspase-1/3 (Bowman et al., 2018; Zhang et al., 2018).

Our previous reports indicate that H-290/51 is able to induce profound neuroprotection following hyperthermic brain injury, nanoparticles intoxication and brain or spinal cord injury (Sharma et al., 2003, 2006a,b,c, 2009a,b, 2015a,b). Also METH treatment and morphine induced BBB and brain pathology was reduced by H-290/51 (Sharma et al., 2007b, 2015a,b). This suggests that lipid peroxidation and oxidative stress is important determinates of BBB breakdown, edema formation and brain injury under a wide variety of noxious insults to the brain. Since H-290/51 is a chain breaking antioxidant (Westerlund et al., 1996) it appears to be very potent compound in preventing oxidative stress in brain injury. Nanowiring of H-290/51 has further enhanced its potential in exerting superior neuroprotection in brain injury (Sharma et al., 2009a,b, 2015a,b).

In summary, our observations for the first time show that METH treatment enhances brain injury induced pathophysiology at high altitude. This means that our military personnel posted at high altitude in combat operations are vulnerable to brain injury and enhanced brain pathology. Use of psychostimulants further complicates the situation of brain injury induced pathological outcome. In these situations treatment with antioxidants like H-290/51 is neuroprotective. Also use of nanowired delivery of H-290/51 that induces superior Neuroprotection in brain injury could be utilized. Our results are thus the first to point out that brain injury at high altitude is further complicated with psychostimulants use and these residents who are using METH at high altitude are the most vulnerable to additional brain injury at high altitude.

13 Future perspectives

Further research is needed to understand the role of psychostimulants on brain injury at high altitude. Whether other substance abuse could also enhance brain pathology following brain injury at high altitude requires further investigation. It is believed that cannabis; alcohol, ecstasy, cocaine, heroin, morphine and related substance abuse could further enhance brain pathology of CHI or traumatic brain injury at high altitude. Experimental studies using these individual drugs or a combination of them could be used to assess the brain pathology following CHI or perforating brain injury. These data will be a useful in further assessing brain pathology at high altitude for bath the civilians and military in brain injury.

Acknowledgment

This investigation is supported by grants from the Air Force Office of Scientific Research (EOARD, London, UK), and Air Force Material Command, USAF, under grant number FA8655-05-1-3065; Grants from the Alzheimer's Association (IIRG-09-132087), the National Institutes of Health (R01 AG028679) and the Dr. Robert M. Kohrman Memorial Fund (RJC); Swedish Medical Research Council (Nr 2710-HSS), the Ministry of Science & Technology, People Republic of China, Göran Gustafsson Foundation, Stockholm, Sweden (HSS), Astra Zeneca, Mölndal, Sweden (HSS/AS), The University Grants Commission, New Delhi, India (HSS/AS), Ministry of Science & Technology, Govt. of India (HSS/AS), Indian Medical Research Council, New Delhi, India (HSS/AS) and India-EU Co-operation Program (RP/AS/HSS) and IT-901/16 (JVL), Government of Basque Country and PPG 17/51 (JVL), JVL thanks to the support of the University of the Basque Country (UPV/EHU) PPG 17/51 and 14/08, the Basque Government (IT-901/16 and CS-2203) Basque Country, Spain; and Foundation for Nanoneuroscience and Nanoneuroprotection (FSNN), Romania. Technical and human support provided by Dr. Ricardo Andrade from SGIker (UPV/EHU) is gratefully acknowledged. Dr. Seaab Sahib is supported by Research Fellowship at the University of Arkansas Fayetteville AR by Department of Community Health; Middle Technical University; Wassit; Iraq, and The Higher Committee for Education Development in Iraq; Baghdad; Iraq. We thank Suraj Sharma, Blekinge Inst. Technology, Karlskrona, Sweden and Dr. Saja Alshafeay, University of Arkansas Fayetteville, Fayetteville AR, USA for computer and graphic support. The U.S. Government is authorized to reproduce and distribute reprints for Government purpose notwithstanding any copyright notation thereon. The views and conclusions contained herein are those of the authors and should not be interpreted as necessarily representing the official policies or endorsements, either expressed or implied, of the Air Force Office of Scientific Research or the U.S. Government.

Conflict of interest

There is no conflict of interest between any entity and/or organization mentioned here.

References

Abdul-Muneer, P.M., Chandra, N., Haorah, J., 2015. Interactions of oxidative stress and neurovascular inflammation in the pathogenesis of traumatic brain injury. Mol. Neurobiol. 51 (3), 966–979. https://doi.org/10.1007/s12035-014-8752-3. (Epub 2014 May 28).

Adeva, M.M., Souto, G., Donapetry, C., Portals, M., Rodriguez, A., Lamas, D., 2012. Brain edema in diseases of different etiology. Neurochem. Int. 61 (2), 166–174. https://doi.org/10.1016/j.neuint.2012.05.007. (Epub 2012 May 9).

Agaku, I., Odani, S., Nelson, J.R., 2020. U.S. military veteran versus nonveteran use of licit and illicit substances. Am. J. Prev. Med. 59 (5), 733–741. https://doi.org/10.1016/j.amepre.2020.04.027. (Epub 2020 Oct 2).

Ainslie, P.N., Subudhi, A.W., 2014. Cerebral blood flow at high altitude. High Alt. Med. Biol. 15 (2), 133–140. https://doi.org/10.1089/ham.2013.1138.

Aksel, G., Çorbacıoğlu, Ş.K., Özen, C., Turk, J., 2019. High-altitude illness: management approach. Emerg. Med. 19 (4), 121–126. https://doi.org/10.1016/j.tjem.2019.09.002. eCollection 2019 Oct.

Ali, M.M., Mahmoud, A.M., Le Master, E., Levitan, I., Phillips, S.A., 2019. Role of matrix metalloproteinases and histone deacetylase in oxidative stress-induced degradation of the endothelial glycocalyx. Am. J. Physiol. Heart Circ. Physiol. 316 (3), H647–H663. https://doi.org/10.1152/ajpheart.00090.2018.

Alm, P., Sharma, H.S., Hedlund, S., Sjöquist, P.O., Westman, J., 1998. Nitric oxide in the pathophysiology of hyperthermic brain injury. Influence of a new anti-oxidant compound H-290/51. A pharmacological study using immunohistochemistry in the rat. Amino Acids 14 (1–3), 95–103. https://doi.org/10.1007/BF01345249.

Alm, P., Sharma, H.S., Sjöquist, P.O., Westman, J.A., 2000. New antioxidant compound H-290/51 attenuates nitric oxide synthase and heme oxygenase expression following hyperthermic brain injury. An experimental study using immunohistochemistry in the rat. Amino Acids 19 (1), 383–394. https://doi.org/10.1007/s007260070069.

Amann, M., Kayser, B., 2009. Nervous system function during exercise in hypoxia. High Alt. Med. Biol. 10 (2), 149–164. https://doi.org/10.1089/ham.2008.1105.

Amin, M., Pushpakumar, S., Muradashvili, N., Kundu, S., Tyagi, S.C., Sen, U., 2016. Regulation and involvement of matrix metalloproteinases in vascular diseases. Front. Biosci. (Landmark Ed) 21, 89–118. https://doi.org/10.2741/4378.

Ando, S., Komiyama, T., Sudo, M., Higaki, Y., Ishida, K., Costello, J.T., Katayama, K., 2020. The interactive effects of acute exercise and hypoxia on cognitive performance: a narrative review. Scand. J. Med. Sci. Sports 30 (3), 384–398. https://doi.org/10.1111/sms.13573. (Epub 2019 Oct 27).

Anglin, M.D., Burke, C., Perrochet, B., Stamper, E., Dawud-Noursi, S., 2000. History of the methamphetamine problem. J. Psychoactive Drugs 32 (2), 137–141. https://doi.org/10.1080/02791072.2000.10400221.

Arias-Reyes, C., Zubieta-DeUrioste, N., Poma-Machicao, L., Aliaga-Raduan, F., Carvajal-Rodriguez, F., Dutschmann, M., Schneider-Gasser, E.M., Zubieta-Calleja, G., Soliz, J., 2020. Does the pathogenesis of SARS-CoV-2 virus decrease at high-altitude? Respir. Physiol. Neurobiol. 277, 103443. https://doi.org/10.1016/j.resp.2020.103443. (Epub 2020 Apr 22).

Armistead-Jehle, P., Soble, J.R., Cooper, D.B., Belanger, H.G., 2017. Unique aspects of traumatic brain injury in military and veteran populations. Phys. Med. Rehabil. Clin. N. Am. 28 (2), 323–337. https://doi.org/10.1016/j.pmr.2016.12.008.

Aryal, N., Weatherall, M., Bhatta, Y.K.D., Mann, S., 2019. Blood pressure and hypertension in people living at high altitude in Nepal. Hypertens. Res. 42 (2), 284–291. https://doi.org/10.1038/s41440-018-0138-x. (Epub 2018 Nov 21).

Askew, E.W., 2002. Work at high altitude and oxidative stress: antioxidant nutrients. Toxicology 180 (2), 107–119. https://doi.org/10.1016/s0300-483x(02)00385-2.

Bahrke, M.S., Shukitt-Hale, B., 1993. Effects of altitude on mood, behaviour and cognitive functioning. A review. Sports Med. 16 (2), 97–125. https://doi.org/10.2165/00007256-199316020-00003.

Bai, Z., Voituron, N., Wuren, T., Jeton, F., Jin, G., Marchant, D., Richalet, J.P., Ge, R.L., Pichon, A.P., 2015. Role of glutamate and serotonin on the hypoxic ventilatory response in high-altitude-adapted plateau Pika. Respir. Physiol. Neurobiol. 212–214, 39–45. https://doi.org/10.1016/j.resp.2015.03.006. (Epub 2015 Apr).

Bailey, D.M., 2019. Oxygen, evolution and redox signalling in the human brain; quantum in the quotidian. J. Physiol. 597 (1), 15–28. https://doi.org/10.1113/JP276814. (Epub 2018 Nov 2).

Bailey, D.M., Roukens, R., Knauth, M., Kallenberg, K., Christ, S., Mohr, A., Genius, J., Storch-Hagenlocher, B., Meisel, F., McEneny, J., Young, I.S., Steiner, T., Hess, K., Bärtsch, P., 2006. Free radical-mediated damage to barrier function is not associated with altered brain morphology in high-altitude headache. J. Cereb. Blood Flow Metab. 26 (1), 99–111. https://doi.org/10.1038/sj.jcbfm.9600169.

Bakker-Dyos, J., Vanstone, S., Mellor, A.J., 2016. High altitude adaptation and illness: military implications. J. R. Nav. Med. Serv. 102 (1), 33–39.

Bakonyi, T., Radak, Z., 2004. High altitude and free radicals. J. Sports Sci. Med. 3 (2), 64–69. 24482580. PMC3899533.

Banchero, N., Sime, F., Peñaloza, D., Cruz, J., Gamboa, R., Marticorena, E., 1966. Pulmonary pressure, cardiac output, and arterial oxygen saturation during exercise at high altitude and at sea level. Circulation 33 (2), 249–262. https://doi.org/10.1161/01.cir.33.2.249.

Banderet, L.E., Burse, R.L., 1992. Effects of high terrestrial altitude on military performance. In: Gal, R., Mangelsdorff, D. (Eds.), Hand- Book of Military Psychology. Wiley & Sons, New York, pp. 1–45. ISBN: 978-0-471-92045-8 (April 1992, 812 pages Accession Nr. ADA209614).

Barati, M., Ahmadpanah, M., Soltanian, A.R., 2014. Prevalence and factors associated with methamphetamine use among adult substance abusers. J. Res. Health Sci. 14 (3), 221–226.

Bärtsch, P., 1999. High altitude pulmonary edema. Med. Sci. Sports Exerc. 31 (Suppl. 1), S23–S27. https://doi.org/10.1097/00005768-199901001-00004.

Basnyat, B., Wu, T., Gertsch, J.H., 2004. Neurological conditions at altitude that fall outside the usual definition of altitude sickness. High Alt. Med. Biol. 5 (2), 171–179. https://doi.org/10.1089/1527029041352126.

Bayazit, H., Dulgeroglu, D., Selek, S., 2020. Brain-derived neurotrophic factor and oxidative stress in cannabis dependence. Neuropsychobiology 79 (3), 186–190. https://doi.org/10.1159/000504626. (Epub 2019 Nov 28).

Bellis, F., Parris, R., Thake, D., Richards, P., 2005. Difficult decisions at altitude: the management of an acutely dyspneic porter at 5000 meters. Wilderness Environ. Med. 16 (4), 212–218. https://doi.org/10.1580/pr27-04.1.

Benz, F., Liebner, S., 2020. Structure and function of the blood-brain barrier (BBB). Handb. Exp. Pharmacol. https://doi.org/10.1007/164_2020_404 (Online ahead of print).

Berger, M.M., Macholz, F., Mairbäurl, H., Bärtsch, P., 2015. Remote ischemic preconditioning for prevention of high-altitude diseases: fact or fiction? J. Appl. Physiol. (1985) 119 (10), 1143–1151. https://doi.org/10.1152/japplphysiol.00156.2015. (Epub 2015 Jun 18).

Betz, M.E., Valley, M.A., Lowenstein, S.R., Hedegaard, H., Thomas, D., Stallones, L., Honigman, B., 2011. Elevated suicide rates at high altitude: sociodemographic and health issues may be to blame. Suicide Life Threat. Behav. 41 (5), 562–573. https://doi.org/10.1111/j.1943-278X.2011.00054.x. (Epub 2011).

Bhagi, S., Srivastava, S., Singh, S.B., 2014. High-altitude pulmonary edema: review. J. Occup. Health 56 (4), 235–243. https://doi.org/10.1539/joh.13-0256-ra. (Epub 2014 May 29).

Bhattrai, A., Irimia, A., Van Horn, J.D., 2019. Neuroimaging of traumatic brain injury in military personnel: an overview. J. Clin. Neurosci. 70, 1–10. https://doi.org/10.1016/j.jocn.2019.07.001. (Epub 2019 Jul 19).

Bilo, G., Caravita, S., Torlasco, C., Parati, G., 2019. Blood pressure at high altitude: physiology and clinical implications. Kardiol. Pol. 77 (6), 596–603. https://doi.org/10.33963/KP.14832. (Epub 2019 May 17).

Biondich, A.S., Joslin, J.D., 2015. Coca: high altitude remedy of the ancient Incas. Wilderness Environ. Med. 26 (4), 567–571. https://doi.org/10.1016/j.wem.2015.07.006. (Epub 2015 Oct 23).

Bordes, J., Joubert, C., Esnault, P., Montcriol, A., Nguyen, C., Meaudre, E., Dulou, R., Dagain, A., 2017. Coagulopathy and transfusion requirements in war related penetrating traumatic brain injury. A single centre study in a French role 3 medical treatment facility in Afghanistan. Injury 48 (5), 1047–1053. https://doi.org/10.1016/j.injury.2016.11.023. (Epub 2016 Nov 21).

Bowman, G.L., Dayon, L., Kirkland, R., Wojcik, J., Peyratout, G., Severin, I.C., Henry, H., Oikonomidi, A., Migliavacca, E., Bacher, M., Popp, J., 2018. Blood-brain barrier breakdown, neuroinflammation, and cognitive decline in older adults. Alzheimers Dement. 14 (12), 1640–1650. https://doi.org/10.1016/j.jalz.2018.06.2857. (Epub 2018 Aug).

Brühl, A.B., d'Angelo, C., Sahakian, B.J., 2019. Neuroethical issues in cognitive enhancement: modafinil as the example of a workplace drug? Brain Neurosci. Adv. 3. 2398212818816018. https://doi.org/10.1177/2398212818816018. (eCollection 2019 Jan-Dec).

Buguet, A., Moroz, D.E., Radomski, M.W., 2003. Modafinil—medical considerations for use in sustained operations. Aviat. Space Environ. Med. 74 (6 Pt. 1), 659–663.

Cadet, J.L., Krasnova, I.N., 2009. Molecular bases of methamphetamine-induced neurodegeneration. Int. Rev. Neurobiol. 88, 101–119. https://doi.org/10.1016/S0074-7742(09)88005-7.

Capizzi, A., Woo, J., Verduzco-Gutierrez, M., 2020. Traumatic brain injury: an overview of epidemiology, pathophysiology, and medical management. Med. Clin. North Am. 104 (2), 213–238. https://doi.org/10.1016/j.mcna.2019.11.001.

Carteron, L., Solari, D., Patet, C., Quintard, H., Miroz, J.P., Bloch, J., Daniel, R.T., Hirt, L., Eckert, P., Magistretti, P.J., Oddo, M., 2018. Hypertonic lactate to improve cerebral perfusion and glucose availability after acute brain injury. Crit. Care Med. 46 (10), 1649–1655. https://doi.org/10.1097/CCM.0000000000003274.

Centers for Disease Control and Prevention (CDC), 1995. Increasing morbidity and mortality associated with abuse of methamphetamine—United States, 1991-1994. MMWR Morb. Mortal. Wkly Rep. 44 (47), 882–886.

Chao, C.C., Chen, L.H., Lin, Y.C., Wang, S.H., Wu, S.H., Li, W.C., Huang, K.F., Chiu, T.F., Kuo, I.C., 2019. Impact of a 3-Day high-altitude trek on Xue Mountain (3886 m), Taiwan, on the emotional states of children: a prospective observational study. High Alt. Med. Biol. 20 (1), 28–34. https://doi.org/10.1089/ham.2018.0086. (Epub 2018 Dec 13).

Chaturvedi, H.K., Mahanta, J., 2004. Sociocultural diversity and substance use pattern in Arunachal Pradesh, India. Drug Alcohol Depend. 74 (1), 97–104. https://doi.org/10.1016/j.drugalcdep.2003.12.003.

Chen, T., Su, H., Zhong, N., Tan, H., Li, X., Meng, Y., Duan, C., Zhang, C., Bao, J., Xu, D., Song, W., Zou, J., Liu, T., Zhan, Q., Jiang, H., Zhao, M., 2020a. Disrupted brain network dynamics and cognitive functions in methamphetamine use disorder: insights from EEG microstates. BMC Psychiatry 20 (1), 334. https://doi.org/10.1186/s12888-020-02743-5.

Chen, Y., Chen, X., Tang, C., Liu, X., Luo, Y., 2020b. Modularization, standardization and actual combat' of high-altitude military medical geography in a comprehensive military medical exercise. BMJ Mil. Health 166 (3), 209. https://doi.org/10.1136/jramc-2019-001180. (Epub 2019 Mar 17).

Clark, J.M.R., Jak, A.J., Twamley, E.W., 2020. Cognition and functional capacity following traumatic brain injury in veterans. Rehabil. Psychol. 65 (1), 72–79. https://doi.org/10.1037/rep0000294. (Epub 2019 Oct 31).

Clauss, R., 2014. Disorders of consciousness and pharmaceuticals that act on oxygen based amino acid and monoamine neurotransmitter pathways of the brain. Curr. Pharm. Des. 20 (26), 4140–4153.

Clemente-Suárez, V.J., Robles-Pérez, J.J., Herrera-Mendoza, K., Herrera-Tapias, B., Fernández-Lucas, J., 2017. Psychophysiological response and fine motor skills in high-altitude parachute jumps. High Alt. Med. Biol. 18 (4), 392–399. https://doi.org/10.1089/ham.2017.0071. (Epub 2017 Oct 24).

Colangelo, C., Shichkova, P., Keller, D., Markram, H., Ramaswamy, S., 2019. Cellular, synaptic and network effects of acetylcholine in the neocortex. Front. Neural Circuits 13, 24. https://doi.org/10.3389/fncir.2019.00024 (eCollection 2019).

Connolly, D.M., Lee, V.M., Hodkinson, P.D., 2018. White matter status of participants in altitude chamber research and training. Aerosp. Med. Hum. Perform. 89 (9), 777–786. https://doi.org/10.3357/AMHP.5090.2018.

Cornelius, C., Crupi, R., Calabrese, V., Graziano, A., Milone, P., Pennisi, G., Radak, Z., Calabrese, E.J., Cuzzocrea, S., 2013. Traumatic brain injury: oxidative stress and neuroprotection. Antioxid. Redox Signal. 19 (8), 836–853. https://doi.org/10.1089/ars.2012.4981. (Epub 2013 May 10).

Costa, S.H., Yonamine, M., Ramos, A.L., Oliveira, F.G., Rodrigues, C.R., da Cunha, L.C., 2015. Prevalence of psychotropic drug use in military police units. Cien. Saude Colet. 20 (6), 1843–1849. https://doi.org/10.1590/1413-81232015206.00942014.

Cramer, N.P., Korotcov, A., Bosomtwi, A., Xu, X., Holman, D.R., Whiting, K., Jones, S., Hoy, A., Dardzinski, B.J., Galdzicki, Z., 2019. Neuronal and vascular deficits following chronic adaptation to high altitude. Exp. Neurol. 311, 293–304. https://doi.org/10.1016/j.expneurol.2018.10.007. (Epub 2018 Oct 13).

Critchley, H.D., Nicotra, A., Chiesa, P.A., Nagai, Y., Gray, M.A., Minati, L., Bernardi, L., 2015. Slow breathing and hypoxic challenge: cardiorespiratory consequences and their central neural substrates. PLoS One 10 (5), e0127082. https://doi.org/10.1371/journal.pone.0127082. (eCollection 2015).

Cymerman, A., Robinson, S.M., McCullough, D., 1972. Alteration of rat brain catecholamine metabolism during exposure to hypobaric hypoxia. Can. J. Physiol. Pharmacol. 50 (4), 321–327. https://doi.org/10.1139/y72-048.

Das, S.K., Dhar, P., Sharma, V.K., Barhwal, K., Hota, S.K., Norboo, T., Singh, S.B., 2018. High altitude with monotonous environment has significant impact on mood and cognitive performance of acclimatized lowlanders: possible role of altered serum BDNF and plasma homocysteine level. J. Affect. Disord. 237, 94–103. https://doi.org/10.1016/j.jad.2018.04.106. (Epub 2018 Apr 25).

Dawadi, S., Adhikari, S., 2019. Successful summit of two 8000 m peaks after recent high altitude pulmonary edema. Wilderness Environ. Med. 30 (2), 195–198. https://doi.org/10.1016/j.wem.2018.12.009. (Epub 2019 Mar 7).

de Aquino, L.V., Antunes, H.K., dos Santos, R.V., Lira, F.S., Tufik, S., de Mello, M.T., 2012. High altitude exposure impairs sleep patterns, mood, and cognitive functions. Psychophysiology 49 (9), 1298–1306. https://doi.org/10.1111/j.1469-8986.2012.01411.x. (Epub 2012 Jul 16).

Defalque, R.J., Wright, A.J., 2011. Methamphetamine for Hitler's Germany: 1937 to 1945. Bull. Anesth. Hist. 29 (2), 21–24. 32 https://doi.org/10.1016/s1522-8649(11)50016-2.

Degache, F., Serain, É., Roy, S., Faiss, R., Millet, G.P., 2020. The fatigue-induced alteration in postural control is larger in hypobaric than in normobaric hypoxia. Sci. Rep. 10 (1), 483. https://doi.org/10.1038/s41598-019-57166-4.

Dey, P.K., Sharma, H.S., 1984. Influence of ambient temperature and drug treatments on brain oedema induced by impact injury on skull in rats. Indian J. Physiol. Pharmacol. 28 (3), 177–186.

Donegani, E., Paal, P., Küpper, T., Hefti, U., Basnyat, B., Carceller, A., Bouzat, P., van der Spek, R., Hillebrandt, D., 2016. Drug use and misuse in the mountains: a UIAA MedCom consensus guide for medical professionals. High Alt. Med. Biol. 17 (3), 157–184. https://doi.org/10.1089/ham.2016.0080. (Epub 2016 Sep 1).

Dosek, A., Ohno, H., Acs, Z., Taylor, A.W., Radak, Z., 2007. High altitude and oxidative stress. Respir. Physiol. Neurobiol. 158 (2–3), 128–131. https://doi.org/10.1016/j.resp.2007.03.013. Epub 2007 Mar 31. PMID: 17482529.

Duan, X., Wen, Z., Shen, H., Shen, M., Chen, G., 2016. Intracerebral hemorrhage, oxidative stress, and antioxidant therapy. Oxid. Med. Cell. Longev. 2016, 1203285. https://doi.org/10.1155/2016/1203285. (Epub 2016 Apr 14).

Duong, J., Elia, C., Takayanagi, A., Lanzilotta, T., Ananda, A., Miulli, D., 2018. The impact of methamphetamines in patients with traumatic brain injury, a retrospective review. Clin. Neurol. Neurosurg. 170, 99–101. https://doi.org/10.1016/j.clineuro.2018.04.030. (Epub 2018 Apr 27).

Ebert-Santos, C., 2017. High-altitude pulmonary edema in mountain community residents. High Alt. Med. Biol. 18 (3), 278–284. https://doi.org/10.1089/ham.2016.0100. (Epub 2017 Aug 28).

Edeleano, L., 1887. Ueber einige Derivate der Phenylmethacrylsäure und der Phenylisobuttersäure. EuroJIC 12 (1), 616–622. First published: Januar–Juni Copyright © 1887 WILEY-VCH Verlag GmbH & co. KGaA, Weinheim; Version of Record online: 28 January 2006 https://doi.org/10.1002/cber.188702001142.

Edmundson, C., Heinsbroek, E., Glass, R., Hope, V., Mohammed, H., White, M., Desai, M., 2018. Sexualised drug use in the United Kingdom (UK): a review of the literature. Int. J. Drug Policy 55, 131–148. https://doi.org/10.1016/j.drugpo.2018.02.002. (Epub 2018 Apr 4).

Eghwrudjakpor, P.O., Allison, A.B., 2010. Oxidative stress following traumatic brain injury: enhancement of endogenous antioxidant defense systems and the promise of improved outcome. Niger. J. Med. 19 (1), 14–21. https://doi.org/10.4314/njm.v19i1.52466.

El Hayek, S., Allouch, F., Razafsha, M., Talih, F., Gold, M.S., Wang, K.K., Kobeissy, F., 2020. Traumatic brain injury and methamphetamine: a double-hit neurological insult. J. Neurol. Sci. 411, 116711. https://doi.org/10.1016/j.jns.2020.116711. (Epub 2020 Jan 30).

Elder, G.A., Cristian, A., 2009. Blast-related mild traumatic brain injury: mechanisms of injury and impact on clinical care. Mt. Sinai J. Med. 76 (2), 111–118. https://doi.org/10.1002/msj.20098.

Eliyahu, U., Berlin, S., Hadad, E., Heled, Y., Moran, D.S., 2007. Psychostimulants and military operations. Mil. Med. 172 (4), 383–387. https://doi.org/10.7205/milmed.172.4.383.

Elliott, K.A., Jasper, H., 1949. Measurement of experimentally induced brain swelling and shrinkage. Am. J. Physiol. 157 (1), 122–129. https://doi.org/10.1152/ajplegacy.1949.157.1.122. PMID: 18130228.

Elsworth, J.D., Groman, S.M., Jentsch, J.D., Valles, R., Shahid, M., Wong, E., Marston, H., Roth, R.H., 2012. Asenapine effects on cognitive and monoamine dysfunction elicited by subchronic phencyclidine administration. Neuropharmacology 62 (3), 1442–1452. https://doi.org/10.1016/j.neuropharm.2011.08.026. (Epub 2011).

Engwa, G.A., EnNwekegwa, F.N., Nkeh-Chungag, B.N., 2020. Free radicals, oxidative stress-related diseases and antioxidant supplementation. Altern. Ther. Health Med., AT6236 (Online ahead of print).

Evans, W.O., Witt, N.F., 1966. The interaction of high altitude and psychotropic drug action. Psychopharmacologia 10 (2), 184–188. https://doi.org/10.1007/BF00455979.

Fagenholz, P.J., Gutman, J.A., Murray, A.F., Noble, V.E., Camargo Jr., C.A., Harris, N.S., 2007. Evidence for increased intracranial pressure in high altitude pulmonary edema. High Alt. Med. Biol. 8 (4), 331–336. https://doi.org/10.1089/ham.2007.1019.

Farnia, V., Farshchian, F., Farshchian, N., Alikhani, M., Sadeghi Bahmani, D., Brand, S., 2020. Comparisons of voxel-based morphometric brain volumes of individuals with methamphetamine-induced psychotic disorder and schizophrenia spectrum disorder and healthy controls. Neuropsychobiology 79 (2), 170–178. https://doi.org/10.1159/000504576. (Epub 2019 Dec 3).

Fayed, N., Modrego, P.J., Morales, H., 2006. Evidence of brain damage after high-altitude climbing by means of magnetic resonance imaging. Am. J. Med. 119 (2), 168.e1–168.e6. https://doi.org/10.1016/j.amjmed.2005.07.062.

Fievisohn, E., Bailey, Z., Guettler, A., VandeVord, P., 2018. Primary blast brain injury mechanisms: current knowledge, limitations, and future directions. J. Biomech. Eng. 140 (2). https://doi.org/10.1115/1.4038710.

Finnoff, J.T., 2008. Environmental effects on brain function. Curr. Sports Med. Rep. 7 (1), 28–32. https://doi.org/10.1097/01.CSMR.0000308669.99816.71.

Fornasiero, A., Savoldelli, A., Stella, F., Callovini, A., Bortolan, L., Zignoli, A., Low, D.A., Mourot, L., Schena, F., Pellegrini, B., 2020. Shortening work-rest durations reduces physiological and perceptual load during uphill walking in simulated cold high-altitude conditions. High Alt. Med. Biol. 21 (3), 249–257. https://doi.org/10.1089/ham.2019.0136. (Epub 2020 May 15).

Forsling, M.L., Aziz, L.A., 1983. Release of vasopressin in response to hypoxia and the effect of aminergic and opioid antagonists. J. Endocrinol. 99 (1), 77–86. https://doi.org/10.1677/joe.0.0990077.

Friedrich, R.P., Pöttler, M., Cicha, I., Unterweger, H., Janko, C., Alexiou, C., 2019. Nanomedicine for neuroprotection. Nanomedicine (Lond.) 14 (2), 127–130. https://doi.org/10.2217/nnm-2018-0401. (Epub 2018 Dec 10).

Galvin, S.L., Ramage, M., Leiner, C., Sullivan, M.H., Fagan, E.B., 2020. A cohort comparison of differences between regional and buncombe county patients of a comprehensive perinatal substance use disorders program in Western North Carolina. N. C. Med. J. 81 (3), 157–165. https://doi.org/10.18043/ncm.81.3.157.

Gao, G., Wu, X., Feng, J., Hui, J., Mao, Q., Lecky, F., Lingsma, H., Maas, A.I.R., Jiang, J., 2020. China CENTER-TBI registry participants. Clinical characteristics and outcomes in patients with traumatic brain injury in China: a prospective, multicentre, longitudinal, observational study. Lancet Neurol. 19 (8), 670–677. https://doi.org/10.1016/S1474-4422(20)30182-4.

Gaur, P., Prasad, S., Kumar, B., Sharma, S.K., Vats, P., 2020. High-altitude hypoxia induced reactive oxygen species generation, signaling, and mitigation approaches. Int. J. Biometeorol. 65, 601–615. https://doi.org/10.1007/s00484-020-02037-1 (Online ahead of print).

Gibson, G.E., Peterson, C., Sansone, J., 1981a. Neurotransmitter and carbohydrate metabolism during aging and mild hypoxia. Neurobiol. Aging 2 (3), 165–172. https://doi.org/10.1016/0197-4580(81)90017-8.

Gibson, G.E., Pulsinelli, W., Blass, J.P., Duffy, T.E., 1981b. Brain dysfunction in mild to moderate hypoxia. Am. J. Med. 70 (6), 1247–1254. https://doi.org/10.1016/0002-9343(81)90834-2.

Giordano, F.J., 2005. Oxygen, oxidative stress, hypoxia, and heart failure. J. Clin. Invest. 115 (3), 500–508. https://doi.org/10.1172/JCI24408.

Gohy, B., Ali, E., Van den Bergh, R., Schillberg, E., Nasim, M., Naimi, M.M., Cheréstal, S., Falipou, P., Weerts, E., Skelton, P., Van Overloop, C., Trelles, M., 2016. Early physical and functional rehabilitation of trauma patients in the Médecins Sans Frontières trauma centre in Kunduz, Afghanistan: luxury or necessity? Int. Health 8 (6), 381–389. https://doi.org/10.1093/inthealth/ihw039. (Epub 2016 Oct 13).

Gold, M.S., Kobeissy, F.H., Wang, K.K., Merlo, L.J., Bruijnzeel, A.W., Krasnova, I.N., Cadet, J.L., 2009. Methamphetamine- and trauma-induced brain injuries: comparative cellular and molecular neurobiological substrates. Biol. Psychiatry 66 (2), 118–127. https://doi.org/10.1016/j.biopsych.2009.02.021. (Epub 2009 Apr 5).

Goldberg, S., Buhbut, E., Mimouni, F.B., Joseph, L., Picard, E., 2012. Effect of moderate elevation above sea level on blood oxygen saturation in healthy young adults. Respiration 84 (3), 207–211. https://doi.org/10.1159/000336554. (Epub 2012 Mar 21).

Hackett, P.H., 1999. High altitude cerebral edema and acute mountain sickness. A pathophysiology update. Adv. Exp. Med. Biol. 474, 23–45. https://doi.org/10.1007/978-1-4615-4711-2_2. PMID: 10634991.

Hackett, P., Rennie, D., 2002. High-altitude pulmonary edema. JAMA 287 (17), 2275–2278. https://doi.org/10.1001/jama.287.17.2275.

Hackett, P.H., Roach, R.C., 2004. High altitude cerebral edema. High Alt. Med. Biol. 5 (2), 136–146. https://doi.org/10.1089/1527029041352054.

Hackett, P.H., Yarnell, P.R., Weiland, D.A., Reynard, K.B., 2019. Acute and evolving MRI of high-altitude cerebral edema: microbleeds, edema, and pathophysiology. AJNR Am. J. Neuroradiol. 40 (3), 464–469. https://doi.org/10.3174/ajnr.A5897. (Epub 2019 Jan 24).

Halpin, L.E., Collins, S.A., Yamamoto, B.K., 2014. Neurotoxicity of methamphetamine and 3,4-methylenedioxymethamphetamine. Life Sci. 97 (1), 37–44. https://doi.org/10.1016/j.lfs.2013.07.014. (Epub 2013 Jul 24).

Hamilton, A.J., Cymmerman, A., Black, P.M., 1986. High altitude cerebral edema. Neurosurgery 19 (5), 841–849. https://doi.org/10.1227/00006123-198611000-00024.

Han, S., Zhao, L., Ma, S., Chen, Z., Wu, S., Shen, M., Xia, G., Jia, G., 2020. Alterations to cardiac morphology and function among high-altitude workers: a retrospective cohort study. Occup. Environ. Med. 77 (7), 447–453. https://doi.org/10.1136/oemed-2019-106108. (Epub 2020 Apr 8).

Hartmann, R., 2003. Japanische Studenten an der Berliner Universität 1920–1945. Kleine Reihe, 22, Mori-Ôgai-Gedenkstätte der Humboldt-Universität zu Berlin, Germany, pp. 1–102 (ISSN 1435-0351).

Hashmi, M.A., Rashid, M., Haleem, A., Bokhari, S.A., Hussain, T., 1998. Frostbite: epidemiology at high altitude in the Karakoram mountains. Ann. R. Coll. Surg. Engl. 80 (2), 91–95.

Hasoon, J., 2017. Blast-associated traumatic brain injury in the military as a potential trigger for dementia and chronic traumatic encephalopathy. U.S. Army Med. Dep. J. 1–17, 102–105.

Hassan, W.U., Syed, M.J., Alamgir, W., Awan, S., Bell, S.M., Majid, A., Wasay, M., 2019. Cerebral venous thrombosis at high altitude: analysis of 28 cases. Cerebrovasc. Dis. 48 (3–6), 184–192. https://doi.org/10.1159/000504504. (Epub 2019 Nov 27).

Heinrich, E.C., Djokic, M.A., Gilbertson, D., DeYoung, P.N., Bosompra, N.O., Wu, L., Anza-Ramirez, C., Orr, J.E., Powell, F.L., Malhotra, A., Simonson, T.S., 2019. Cognitive function and mood at high altitude following acclimatization and use of supplemental oxygen and adaptive servoventilation sleep treatments. PLoS One 14 (6), e0217089. https://doi.org/10.1371/journal.pone.0217089. (eCollection 2019).

Herst, P.M., Rowe, M.R., Carson, G.M., Berridge, M.V., 2017. Functional mitochondria in health and disease. Front. Endocrinol. (Lausanne) 8, 296. https://doi.org/10.3389/fendo.2017.00296 (eCollection 2017).

Hill, N.E., Stacey, M.J., Woods, D.R., 2011. Energy at high altitude. J. R. Army Med. Corps 157 (1), 43–48. https://doi.org/10.1136/jramc-157-01-08.

Hillstrom, K., 2015. Methamphetamine. Greenhaven Publishing LLC, Farmington Hills, MI, USA: New York, NY, USA, pp. 1–153. ISBN 978-1-4205-0872-7.

Hochachka, P.W., Clark, C.M., Matheson, G.O., Brown, W.D., Stone, C.K., Nickles, R.J., Holden, J.E., 1999. Effects on regional brain metabolism of high-altitude hypoxia: a study of six US marines. Am. J. Physiol. 277 (1), R314–R319. https://doi.org/10.1152/ajpregu.1999.277.1.R314.

Hoiland, R.L., Howe, C.A., Coombs, G.B., Ainslie, P.N., 2018. Ventilatory and cerebrovascular regulation and integration at high-altitude. Clin. Auton. Res. 28 (4), 423–435. https://doi.org/10.1007/s10286-018-0522-2. (Epub 2018 Mar 24).

Hoiland, R.L., Howe, C.A., Carter, H.H., Tremblay, J.C., Willie, C.K., Donnelly, J., MacLeod, D.B., Gasho, C., Stembridge, M., Boulet, L.M., Niroula, S., Ainslie, P.N., 2019. UBC-Nepal expedition: phenotypical evidence for evolutionary adaptation in the control of cerebral blood flow and oxygen delivery at high altitude. J. Physiol. 597 (12), 2993–3008. https://doi.org/10.1113/JP277596. (Epub 2019 May 13).

Holsinger, T., Steffens, D.C., Phillips, C., Helms, M.J., Havlik, R.J., Breitner, J.C., Guralnik, J.M., Plassman, B.L., 2002. Head injury in early adulthood and the lifetime risk of depression. Arch. Gen. Psychiatry 59 (1), 17–22. https://doi.org/10.1001/archpsyc.59.1.17.

Holtzer, S., Vigué, B., Ract, C., Samii, K., Escourrou, P., 2001. Hypoxia-hypotension decreases pressor responsiveness to exogenous catecholamines after severe traumatic brain injury in rats. Crit. Care Med. 29 (8), 1609–1614. https://doi.org/10.1097/00003246-200108000-00018.

Homer, B.D., Solomon, T.M., Moeller, R.W., Mascia, A., DeRaleau, L., Halkitis, P.N., 2008. Methamphetamine abuse and impairment of social functioning: a review of the underlying neurophysiological causes and behavioral implications. Psychol. Bull. 134 (2), 301–310. https://doi.org/10.1037/0033-2909.134.2.301.

Hornbein, T.F., 1992. Long term effects of high altitude on brain function. Int. J. Sports Med. 13 (Suppl. 1), S43–S45. https://doi.org/10.1055/s-2007-1024589.

Hornbein, T.F., 2001. The high-altitude brain. J. Exp. Biol. 204 (Pt. 18), 3129–3132. 11581326.

Hoyer, S., 1982. The young-adult and normally aged brain. Its blood flow and oxidative metabolism. A review—part I. Arch. Gerontol. Geriatr. 1 (2), 101–116. https://doi.org/10.1016/0167-4943(82)90010-3.

Hsieh, D.T., Warden, G.I., Butler, J.M., Nakanishi, E., Asano, Y., 2020. Multiple sclerosis exacerbation associated with high-altitude climbing exposure. Mil. Med. 185 (7–8), e1322–e1325. https://doi.org/10.1093/milmed/usz421.

Hu, S., Li, F., Luo, H., Xia, Y., Zhang, J., Hu, R., Cui, G., Meng, H., Feng, H., 2010. Amelioration of rCBF and PbtO2 following TBI at high altitude by hyperbaric oxygen preconditioning. Neurol. Res. 32 (2), 173–178. https://doi.org/10.1179/174313209X414524.

Huang, X., Chen, Y.Y., Shen, Y., Cao, X., Li, A., Liu, Q., Li, Z., Zhang, L.B., Dai, W., Tan, T., Arias-Carrion, O., Xue, Y.X., Su, H., Yuan, T.F., 2017. Methamphetamine abuse impairs motor cortical plasticity and function. Mol. Psychiatry 22 (9), 1274–1281. https://doi.org/10.1038/mp.2017.143. (Epub 2017 Jul 25).

Hübner, K., Cabochette, P., Diéguez-Hurtado, R., Wiesner, C., Wakayama, Y., Grassme, K.S., Hubert, M., Guenther, S., Belting, H.G., Affolter, M., Adams, R.H., Vanhollebeke, B., Herzog, W., 2018. Wnt/beta-catenin signaling regulates VE-cadherin-mediated anastomosis of brain capillaries by counteracting S1pr1 signaling. Nat. Commun. 9 (1), 4860. https://doi.org/10.1038/s41467-018-07302-x.

Hunt, D., Kuck, S., Truitt, L., 1996. Methamphetamine Use: Lessons Learned. National Institute of Justice; Office of Justice Programs acquisition Management Division, Abt Associate Inc, Washington DC, USA, Cambridge, MA, USA, pp. 1–66.

Imray, C., Wright, A., Subudhi, A., Roach, R., 2010. Acute mountain sickness: pathophysiology, prevention, and treatment. Prog. Cardiovasc. Dis. 52 (6), 467–484. https://doi.org/10.1016/j.pcad.2010.02.003.

Ishikawa, M., Yamanaka, G., Yamamoto, N., Nakaoka, T., Okumiya, K., Matsubayashi, K., Otsuka, K., Sakura, H., 2016. Depression and altitude: cross-sectional community-based study among elderly high-altitude residents in the himalayan regions. Cult. Med. Psychiatry 40 (1), 1–11. https://doi.org/10.1007/s11013-015-9462-7.

Ismailov, R.M., Lytle, J.M., 2016. Traumatic brain injury: its outcomes and high altitude. J. Spec. Oper. Med. 16 (1), 67–69.

Jalloh, I., Carpenter, K.L., Helmy, A., Carpenter, T.A., Menon, D.K., Hutchinson, P.J., 2015. Glucose metabolism following human traumatic brain injury: methods of assessment and pathophysiological findings. Metab. Brain Dis. 30 (3), 615–632. https://doi.org/10.1007/s11011-014-9628-y. (Epub 2014 Nov 21).

Jang, E.Y., Yang, C.H., Hedges, D.M., Kim, S.P., Lee, J.Y., Ekins, T.G., Garcia, B.T., Kim, H.Y., Nelson, A.C., Kim, N.J., Steffensen, S.C., 2017. The role of reactive oxygen species in methamphetamine self-administration and dopamine release in the nucleus accumbens. Addict. Biol. 22 (5), 1304–1315. https://doi.org/10.1111/adb.12419. (Epub 2016 Jul 14).

Janocha, A.J., Comhair, S.A.A., Basnyat, B., Neupane, M., Gebremedhin, A., Khan, A., Ricci, K.S., Zhang, R., Erzurum, S.C., Beall, C.M., 2017. Antioxidant defense and oxidative damage vary widely among high-altitude residents. Am. J. Hum. Biol. 29 (6), e23039. https://doi.org/10.1002/ajhb.23039. (Epub 2017 Jul 20).

Jansen, G.F., Krins, A., Basnyat, B., Bosch, A., Odoom, J.A., 2000. Cerebral autoregulation in subjects adapted and not adapted to high altitude. Stroke 31 (10), 2314–2318. https://doi.org/10.1161/01.str.31.10.2314.

Jason, G.W., Pajurkova, E.M., Lee, R.G., 1989. High-altitude mountaineering and brain function: neuropsychological testing of members of a Mount Everest expedition. Aviat. Space Environ. Med. 60 (2), 170–173.

Jefferson, J.A., Simoni, J., Escudero, E., Hurtado, M.E., Swenson, E.R., Wesson, D.E., Schreiner, G.F., Schoene, R.B., Johnson, R.J., Hurtado, A., 2004. Increased oxidative stress following acute and chronic high altitude exposure. High Alt. Med. Biol. 5 (1), 61–69. https://doi.org/10.1089/152702904322963690.

Jeng, W., Ramkissoon, A., Parman, T., Wells, P.G., 2006. Prostaglandin H synthase-catalyzed bioactivation of amphetamines to free radical intermediates that cause CNS regional DNA oxidation and nerve terminal degeneration. FASEB J. 20 (6), 638–650. https://doi.org/10.1096/fj.05-5271com.

Jensen, J.D., Vincent, A.L., 2020. High altitude pulmonary edema. In: StatPearls [Internet]. StatPearls Publishing, Treasure Island (FL) (2021 Jan).

Ji, W., Zhang, Y., Ge, R.L., Wan, Y., Liu, J., 2020. NMDA receptor-mediated excitotoxicity is involved in neuronal apoptosis and cognitive impairment induced by chronic hypobaric hypoxia exposure at high altitude. High Alt. Med. Biol. 22, 45–57. https://doi.org/10.1089/ham.2020.0127 (Online ahead of print).

Jia, L., Piña-Crespo, J., Li, Y., 2019. Restoring Wnt/beta-catenin signaling is a promising therapeutic strategy for Alzheimer's disease. Mol. Brain 12 (1), 104. https://doi.org/10.1186/s13041-019-0525-5.

Jørgensen, O.S., Hansen, L.I., Hoffman, S.W., Fülöp, Z., Stein, D.G., 1997. Synaptic remodeling and free radical formation after brain contusion injury in the rat. Exp. Neurol. 144 (2), 326–338. https://doi.org/10.1006/exnr.1996.6372.

Joyce, K.E., Lucas, S.J.E., Imray, C.H.E., Balanos, G.M., Wright, A.D., 2018. Advances in the available non-biological pharmacotherapy prevention and treatment of acute mountain sickness and high altitude cerebral and pulmonary oedema. Expert Opin. Pharmacother. 19 (17), 1891–1902. https://doi.org/10.1080/14656566.2018.1528228. (Epub 2018 Oct 11).

Julian, C.G., 2017. Epigenomics and human adaptation to high altitude. J. Appl. Physiol. (1985) 123 (5), 1362–1370. https://doi.org/10.1152/japplphysiol.00351.2017. (Epub 2017 Aug 17).

Julian, C.G., Moore, L.G., 2019. Human genetic adaptation to high altitude: evidence from the andes. Genes (Basel) 10 (2), 150. https://doi.org/10.3390/genes10020150.

Kamat, S.R., Banerji, B.C., 1972. Study of cardiopulmonary function on exposure to high altitude. I. Acute acclimatization to an altitude of 3,5000 to 4,000 meters in relation to altitude sickness and cardiopulmonary function. Am. Rev. Respir. Dis. 106 (3), 404–413. https://doi.org/10.1164/arrd.1972.106.3.404.

Kaplan, G.B., Leite-Morris, K.A., Wang, L., Rumbika, K.K., Heinrichs, S.C., Zeng, X., Wu, L., Arena, D.T., Teng, Y.D., 2018. Pathophysiological bases of comorbidity: traumatic brain injury and post-traumatic stress disorder. J. Neurotrauma 35 (2), 210–225. https://doi.org/10.1089/neu.2016.4953. (Epub 2017 Nov 3).

Kasper, C.E., 2015. Chapter 2 traumatic brain injury research in military populations. Annu. Rev. Nurs. Res. 33, 13–29. https://doi.org/10.1891/0739-6686.33.13.

Kauser, H., Sahu, S., Panjwani, U., 2016. Guanfacine promotes neuronal survival in medial prefrontal cortex under hypobaric hypoxia. Brain Res. 1636, 152–160. https://doi.org/10.1016/j.brainres.2016.01.053. (Epub 2016 Feb 5).

Kehagia, A.A., Murray, G.K., Robbins, T.W., 2010. Learning and cognitive flexibility: frontostriatal function and monoaminergic modulation. Curr. Opin. Neurobiol. 20 (2), 199–204. https://doi.org/10.1016/j.conb.2010.01.007. (Epub 2010 Feb 16).

Kennedy, J.E., Lu, L.H., Reid, M.W., Leal, F.O., Cooper, D.B., 2019. Correlates of depression in U.S. military service members with a history of mild traumatic brain injury. Mil. Med. 184 (Suppl. 1), 148–154. https://doi.org/10.1093/milmed/usy321.

Kerry, R., Goovaerts, P., Vowles, M., Ingram, B., 2016. Spatial analysis of drug poisoning deaths in the American West, particularly Utah. Int. J. Drug Policy 33, 44–55. https://doi.org/10.1016/j.drugpo.2016.05.004. (Epub 2016 May 16).

Khodaee, M., Grothe, H.L., Seyfert, J.H., VanBaak, K., 2016. Athletes at high altitude. Sports Health 8 (2), 126–132. https://doi.org/10.1177/1941738116630948.

Kim, T.S., Kondo, D.G., Kim, N., Renshaw, P.F., 2014. Altitude may contribute to regional variation in methamphetamine use in the United States: a population database study. Psychiatry Investig. 11 (4), 430–436. https://doi.org/10.4306/pi.2014.11.4.430. (Epub 2014 Oct 20) 25395974. PMCID: PMC4225207.

Kim, Y.K., Na, K.S., Myint, A.M., Leonard, B.E., 2016. The role of pro-inflammatory cytokines in neuroinflammation, neurogenesis and the neuroendocrine system in major depression. Prog. Neuropsychopharmacol. Biol. Psychiatry 64, 277–284. https://doi.org/10.1016/j.pnpbp.2015.06.008. (Epub 2015 Jun 23).

Kinch, K., Fullerton, J.L., Stewart, W., 2019. One hundred years (and counting) of blast-associated traumatic brain injury. J. R. Army Med. Corps 165 (3), 180–182. https://doi.org/10.1136/jramc-2017-000867. (Epub 2018 Jan 10).

Kious, B.M., Kondo, D.G., Renshaw, P.F., 2018. Living high and feeling low: altitude, suicide, and depression. Harv. Rev. Psychiatry 26 (2), 43–56. https://doi.org/10.1097/HRP.0000000000000158.

Kious, B.M., Bakian, A., Zhao, J., Mickey, B., Guille, C., Renshaw, P., Sen, S., 2019. Altitude and risk of depression and anxiety: findings from the intern health study. Int. Rev. Psychiatry 31 (7–8), 637–645. https://doi.org/10.1080/09540261.2019.1586324. (Epub 2019 May 14).

Kita, T., Miyazaki, I., Asanuma, M., Takeshima, M., Wagner, G.C., 2009. Dopamine-induced behavioral changes and oxidative stress in methamphetamine-induced neurotoxicity. Int. Rev. Neurobiol. 88, 43–64. https://doi.org/10.1016/S0074-7742(09)88003-3.

Kiyatkin, E.A., Sharma, H.S., 2009. Acute methamphetamine intoxication: brain hyperthermia, blood-brain barrier, brain edema, and morphological cell abnormalities. Int. Rev. Neurobiol. 88, 65–100. https://doi.org/10.1016/S0074-7742(09)88004-5.

Kiyatkin, E.A., Sharma, H.S., 2012. Environmental conditions modulate neurotoxic effects of psychomotor stimulant drugs of abuse. Int. Rev. Neurobiol. 102, 147–171. https://doi.org/10.1016/B978-0-12-386986-9.00006-5.

Kiyatkin, E.A., Sharma, H.S., 2015. Not just the brain: methamphetamine disrupts blood-spinal cord barrier and induces acute glial activation and structural damage of spinal cord cells. CNS Neurol. Disord. Drug Targets 14 (2), 282–294. https://doi.org/10.2174/1871527314666150217121354.

Kiyatkin, E.A., Sharma, H.S., 2016. Breakdown of blood-brain and blood-spinal cord barriers during acute methamphetamine intoxication: role of brain temperature. CNS Neurol. Disord. Drug Targets 15 (9), 1129–1138. https://doi.org/10.2174/1871527315666160920112445.

Kiyatkin, E.A., Sharma, H.S., 2019. Leakage of the blood-brain barrier followed by vasogenic edema as the ultimate cause of death induced by acute methamphetamine overdose. Int. Rev. Neurobiol. 146, 189–207. https://doi.org/10.1016/bs.irn.2019.06.010. (Epub 2019 Jul 9).

Kiyatkin, E.A., Brown, P.L., Sharma, H.S., 2007. Brain edema and breakdown of the blood-brain barrier during methamphetamine intoxication: critical role of brain hyperthermia. Eur. J. Neurosci. 26 (5), 1242–1253. https://doi.org/10.1111/j.1460-9568.2007.05741.x.

Klette, K.L., Kettle, A.R., Jamerson, M.H., 2006. Prevalence of use study for amphetamine (AMP), methamphetamine (MAMP), 3,4-methylenedioxy-amphetamine (MDA), 3,4-methylenedioxy-methamphetamine (MDMA), and 3,4-methylenedioxy-ethylamphetamine (MDEA) in military entrance processing stations (MEPS) specimens. J. Anal. Toxicol. 30 (5), 319–322. https://doi.org/10.1093/jat/30.5.319.

Knapik, J.J., Steelman, R., Hoedebecke, K., Klug, K.L., Rankin, S., Proctor, S., Graham, B., Jones, B.H., 2014. Risk factors for closed-head injuries during military airborne operations. Aviat. Space Environ. Med. 85 (2), 105–111. https://doi.org/10.3357/asem.3788.2014.

Kogachi, S., Chang, L., Alicata, D., Cunningham, E., Ernst, T., 2017. Sex differences in impulsivity and brain morphometry in methamphetamine users. Brain Struct. Funct. 222 (1), 215–227. https://doi.org/10.1007/s00429-016-1212-2. (Epub 2016 Apr 19).

Kogure, K., Scheinberg, P., Utsunomiya, Y., Kishikawa, H., Busto, R., 1977. Sequential cerebral biochemical and physiological events in controlled hypoxemia. Ann. Neurol. 2 (4), 304–310. https://doi.org/10.1002/ana.410020408.

Korzeniewski, K., Nitsch-Osuch, A., Chciałowski, A., Korsak, J., 2013. Environmental factors, immune changes and respiratory diseases in troops during military activities. Respir. Physiol. Neurobiol. 187 (1), 118–122. https://doi.org/10.1016/j.resp.2013.02.003. (Epub 2013 Feb 10).

Krasnova, I.N., Cadet, J.L., 2009. Methamphetamine toxicity and messengers of death. Brain Res. Rev. 60 (2), 379–407. https://doi.org/10.1016/j.brainresrev.2009.03.002. (Epub 2009 Mar 25).

Lacy, B.W., Ditzler, T.F., Wilson, R.S., Martin, T.M., Ochikubo, J.T., Roussel, R.R., Pizarro-Matos, J.M., Vazquez, R., 2008. Regional methamphetamine use among U.S. Army personnel stationed in the continental United States and Hawaii: a six-year retrospective study (2000-2005). Mil. Med. 173 (4), 353–358. https://doi.org/10.7205/milmed.173.4.353.

Lafuente, J.V., Bermudez, G., Camargo-Arce, L., Bulnes, S., 2016. Blood-brain barrier changes in high altitude. CNS Neurol. Disord. Drug Targets 15 (9), 1188–1197. https://doi.org/10.2174/1871527315666160920123911.

Lafuente, J.V., Sharma, A., Muresanu, D.F., Ozkizilcik, A., Tian, Z.R., Patnaik, R., Sharma, H.S., 2018. Repeated forced swim exacerbates methamphetamine-induced neurotoxicity: neuroprotective effects of nanowired delivery of 5-HT3-receptor antagonist ondansetron. Mol. Neurobiol. 55 (1), 322–334. https://doi.org/10.1007/s12035-017-0744-7.

Lahiri, S., DeLaney, R.G., Brody, J.S., Simpser, M., Velasquez, T., Motoyama, E.K., Polgar, C., 1976. Relative role of environmental and genetic factors in respiratory adaptation to high altitude. Nature 261 (5556), 133–135. https://doi.org/10.1038/261133a0.

Laksitorini, M.D., Yathindranath, V., Xiong, W., Hombach-Klonisch, S., Miller, D.W., 2019. Modulation of Wnt/beta-catenin signaling promotes blood-brain barrier phenotype in cultured brain endothelial cells. Sci. Rep. 9 (1), 19718. https://doi.org/10.1038/s41598-019-56075-w.

Lassen, N.A., 1992. Increase of cerebral blood flow at high altitude: its possible relation to AMS. Int. J. Sports Med. 13 (Suppl. 1), S47–S48. https://doi.org/10.1055/s-2007-1024591.

Lawley, J.S., Levine, B.D., Williams, M.A., Malm, J., Eklund, A., Polaner, D.M., Subudhi, A.W., Hackett, P.H., Roach, R.C., 2016. Cerebral spinal fluid dynamics: effect of hypoxia and implications for high-altitude illness. J. Appl. Physiol. (1985) 120 (2), 251–262. https://doi.org/10.1152/japplphysiol.00370.2015. (Epub 2015 Oct 22).

Le Berre, A.P., Fama, R., Sullivan, E.V., 2017. Executive functions, memory, and social cognitive deficits and recovery in chronic alcoholism: a critical review to inform future research. Alcohol. Clin. Exp. Res. 41 (8), 1432–1443. https://doi.org/10.1111/acer.13431. (Epub 2017 Jul 4).

Lee, M.R., 2011. The history of Ephedra (ma-huang). J. R. Coll. Physicians Edinb. 41 (1), 78–84. https://doi.org/10.4997/JRCPE.2011.116.

Lefferts, W.K., DeBlois, J.P., White, C.N., Day, T.A., Heffernan, K.S., Brutsaert, T.D., 2019. Changes in cognitive function and latent processes of decision-making during incremental ascent to high altitude. Physiol. Behav. 201, 139–145. https://doi.org/10.1016/j.physbeh.2019.01.002. (Epub 2019 Jan 3).

Leissner, K.B., Mahmood, F.U., 2009. Physiology and pathophysiology at high altitude: considerations for the anesthesiologist. J. Anesth. 23 (4), 543–553. https://doi.org/10.1007/s00540-009-0787-7. (Epub 2009 Nov 18).

León-López, J., Calderón-Soto, C., Pérez-Sánchez, M., Feriche, B., Iglesias, X., Chaverri, D., Rodríguez, F.A., 2018. Oxidative stress in elite athletes training at moderate altitude and at sea level. Eur. J. Sport Sci. 18 (6), 832–841. https://doi.org/10.1080/17461391.2018.1453550. (Epub 2018 Mar 24).

Leu, T., Schützhold, V., Fandrey, J., Ferenz, K.B., 2019. When the brain yearns for oxygen. Neurosignals 27 (1), 50–61. https://doi.org/10.33594/000000199.

Lewis, N.C.S., Bain, A.R., Wildfong, K.W., Green, D.J., Ainslie, P.N., 2017. Acute hypoxaemia and vascular function in healthy humans. Exp. Physiol. 102 (12), 1635–1646. https://doi.org/10.1113/EP086532. (Epub 2017 Oct 25).

Li, J., Qi, Y., Liu, H., Cui, Y., Zhang, L., Gong, H., Li, Y., Li, L., Zhang, Y., 2013. Acute high-altitude hypoxic brain injury: identification of ten differential proteins. Neural Regen. Res. 8 (31), 2932–2941. https://doi.org/10.3969/j.issn.1673-5374.2013.31.006.

Li, W.H., Li, Y.X., Ren, J., 2019. High altitude hypoxia on brain ultrastructure of rats and Hsp70 expression changes. Br. J. Neurosurg. 33 (2), 192–195. https://doi.org/10.1080/02688697.2018.1519108. (Epub 2019 Jan 27).

Li, K., Peng, W., Zhou, Y., Ren, Y., Zhao, J., Fu, X., Nie, Y., 2020a. Host genetic and environmental factors shape the composition and function of gut microbiota in populations living at high altitude. Biomed. Res. Int. 2020, 1482109. https://doi.org/10.1155/2020/1482109 (eCollection 2020).

Li, S., Han, C., Asmaro, K., Quan, S., Li, M., Ren, C., Zhang, J., Zhao, W., Xu, J., Liu, Z., Zhang, P., Zhu, L., Ding, Y., Wang, K., Ji, X., Duan, L., 2020b. Remote ischemic conditioning improves attention network function and blood oxygen levels in unacclimatized adults exposed to high altitude. Aging Dis. 11 (4), 820–827. https://doi.org/10.14336/AD.2019.0605 (eCollection 2020 Jul).

Lichtenauer, M., Goebel, B., Fritzenwanger, M., Förster, M., Betge, S., Lauten, A., Figulla, H.-R., Jung, C., 2015. Simulated temporary hypoxia triggers the release of CD31+/Annexin+ endothelial microparticles: a prospective pilot study in humans. Clin. Hemorheol. Microcirc. 61 (1), 83–90. https://doi.org/10.3233/CH-141908.

Liebner, S., Corada, M., Bangsow, T., Babbage, J., Taddei, A., Czupalla, C.J., Reis, M., Felici, A., Wolburg, H., Fruttiger, M., Taketo, M.M., von Melchner, H., Plate, K.H., Gerhardt, H., Dejana, E., 2008. Wnt/beta-catenin signaling controls development of the blood-brain barrier. J. Cell Biol. 183 (3), 409–417. https://doi.org/10.1083/jcb.200806024.

Lim, Y.Y., Maruff, P., Schindler, R., Ott, B.R., Salloway, S., Yoo, D.C., Noto, R.B., Santos, C.Y., Snyder, P.J., 2015. Disruption of cholinergic neurotransmission exacerbates Abeta-related cognitive impairment in preclinical Alzheimer's disease. Neurobiol. Aging 36 (10), 2709–2715. https://doi.org/10.1016/j.neurobiolaging.2015.07.009. (Epub 2015).

Lin, H., Chang, C.P., Lin, H.J., Lin, M.T., Tsai, C.C., 2012. Attenuating brain edema, hippocampal oxidative stress, and cognitive dysfunction in rats using hyperbaric oxygen preconditioning during simulated high-altitude exposure. J. Trauma Acute Care Surg. 72 (5), 1220–1227. https://doi.org/10.1097/TA.0b013e318246ee70.

Ling, G., Ecklund, J.M., Bandak, F.A., 2015. Brain injury from explosive blast: description and clinical management. Handb. Clin. Neurol. 127, 173–180. https://doi.org/10.1016/B978-0-444-52892-6.00011-8.

Liu, X.S., Yang, X.R., Liu, L., Qin, X.K., Gao, Y.Q., 2018. A hypothesis study on a four-period prevention model for high altitude disease. Mil. Med. Res. 5 (1), 2. https://doi.org/10.1186/s40779-018-0150-0.

Lock, M.M., 1984. East Asian Medicine in Urban Japan: Varieties of Medical Experience. University of California Press, Oakland, CA, USA, pp. 1–336. (Reprint edition). ISBN 0-520-05231-5.

Loeffler, G., Hurst, D., Penn, A., Yung, K., 2012. Spice, bath salts, and the U.S. military: the emergence of synthetic cannabinoid receptor agonists and cathinones in the U.S. Armed Forces. Mil. Med. 177 (9), 1041–1048. https://doi.org/10.7205/milmed-d-12-00180.

Loftis, J.M., Taylor, J., Hudson, R., Firsick, E.J., 2019. Neuroinvasion and cognitive impairment in comorbid alcohol dependence and chronic viral infection: an initial investigation. J. Neuroimmunol. 335, 577006. https://doi.org/10.1016/j.jneuroim.2019.577006. (Epub 2019 Jul 10).

Lozano, A., Franchi, F., Seastres, R.J., Oddo, M., Lheureux, O., Badenes, R., Scolletta, S., Vincent, J.L., Creteur, J., Taccone, F.S., 2020. Glucose and lactate concentrations in cerebrospinal fluid after traumatic brain injury. J. Neurosurg. Anesthesiol. 32 (2), 162–169. https://doi.org/10.1097/ANA.0000000000000582.

Luan, F., Li, M., Han, K., Ma, Q., Wang, J., Qiu, Y., Yu, L., He, X., Liu, D., Lv, H., 2019. Phenylethanoid glycosides of Phlomis younghusbandii Mukerjee ameliorate acute hypobaric hypoxia-induced brain impairment in rats. Mol. Immunol. 108, 81–88. https://doi.org/10.1016/j.molimm.2019.02.002. PMID: 30784766. (Epub 2019 Feb 20).

Luo, X., Li, L., Zheng, W., Gu, L., Zhang, X., Li, Y., Xie, Z., Cheng, Y., 2020. HLY78 protects blood-brain barrier integrity through Wnt/beta-catenin signaling pathway following subarachnoid hemorrhage in rats. Brain Res. Bull. 162, 107–114. https://doi.org/10.1016/j.brainresbull.2020.06.003. (Epub 2020 Jun 18).

Ma, S.Q., Xu, X.X., He, Z.Z., Li, X.H., Luo, J.M., 2019. Dynamic changes in peripheral blood-targeted miRNA expression profiles in patients with severe traumatic brain injury at high altitude. Mil. Med. Res. 6 (1), 12. https://doi.org/10.1186/s40779-019-0203-z.

Ma, J., Wang, C., Sun, Y., Pang, L., Zhu, S., Liu, Y., Zhu, L., Zhang, S., Wang, L., Du, L., 2020. Comparative study of oral and intranasal puerarin for prevention of brain injury induced by acute high-altitude hypoxia. Int. J. Pharm. 591, 120002. https://doi.org/10.1016/j.ijpharm.2020.120002. (Epub 2020 Oct 22).

Maa, E.H., 2010. Hypobaric hypoxic cerebral insults: the neurological consequences of going higher. NeuroRehabilitation 26 (1), 73–84. https://doi.org/10.3233/NRE-2010-0537.

Magalhães, J., Ascensão, A., Marques, F., Soares, J.M., Ferreira, R., Neuparth, M.J., Duarte, J.-A., 2005. Effect of a high-altitude expedition to a Himalayan peak (Pumori, 7,161 m) on plasma and erythrocyte antioxidant profile. Eur. J. Appl. Physiol. 93 (5–6), 726–732. https://doi.org/10.1007/s00421-004-1222-2. (Epub 2004 Sep 29).

Maiti, P., Singh, S.B., Mallick, B., Muthuraju, S., Ilavazhagan, G., 2008. High altitude memory impairment is due to neuronal apoptosis in hippocampus, cortex and striatum. J. Chem. Neuroanat. 36 (3–4), 227–238. https://doi.org/10.1016/j.jchemneu.2008.07.003. (Epub 2008 Jul 15).

Marriott, B.M., Carlson, S.J. (Eds.), 1996. Nutritional Needs in Cold and in High-Altitude Environments: Applications for Military Personnel in Field Operations. Institute of Medicine (US) Committee on Military Nutrition Research; National Academies Press (US), Washington (DC).

Martinelli, I., Tomassoni, D., Moruzzi, M., Traini, E., Amenta, F., Tayebati, S.K., 2017. Obesity and metabolic syndrome affect the cholinergic transmission and cognitive functions. CNS Neurol. Disord. Drug Targets 16 (6), 664–676. https://doi.org/10.2174/1871527316666170428123853.

May, P.A., Keaster, C., Bozeman, R., Goodover, J., Blankenship, J., Kalberg, W.O., Buckley, D., Brooks, M., Hasken, J., Gossage, J.P., Robinson, L.K., Manning, M., Hoyme, H.E., 2015. Prevalence and characteristics of fetal alcohol syndrome and partial fetal alcohol syndrome in a Rocky Mountain Region City. Drug Alcohol Depend. 155, 118–127. https://doi.org/10.1016/j.drugalcdep.2015.08.006. (Epub 2015 Aug 14).

McKetin, R., Kelly, E., McLaren, J., 2006. The relationship between crystalline methamphetamine use and methamphetamine dependence. Drug Alcohol Depend. 85 (3), 198–204. https://doi.org/10.1016/j.drugalcdep.2006.04.007. (Epub).

McLaughlin, C.W., Skabelund, A.J., George, A.D., 2017. Impact of high altitude on military operations. Curr. Pulmonol. Rep. 6, 146–154. https://doi.org/10.1007/s13665-017-0181-0.

Mellor, A., Woods, D., 2014. Physiology studies at high altitude; why and how. J. R. Army Med. Corps 160 (2), 131–134. https://doi.org/10.1136/jramc-2013-000206. (Epub 2014 Jan 21).

Mellor, A., Boos, C., Holdsworth, D., Begley, J., Hall, D., Lumley, A., Burnett, A., Hawkins, A., O'Hara, J., Ball, S., Woods, D., 2014. Cardiac biomarkers at high altitude. High Alt. Med. Biol. 15 (4), 452–458. https://doi.org/10.1089/ham.2014.1035.

Menon, P.K., Muresanu, D.F., Sharma, A., Mössler, H., Sharma, H.S., 2012. Cerebrolysin, a mixture of neurotrophic factors induces marked neuroprotection in spinal cord injury following intoxication of engineered nanoparticles from metals. CNS Neurol. Disord. Drug Targets 11 (1), 40–49. https://doi.org/10.2174/187152712799960781.

Midla, G.S., 2004. Lessons learned: operation Anaconda. Mil. Med. 169 (10), 810–813. https://doi.org/10.7205/milmed.169.10.810.

Mishra, K.P., Ganju, L., 2010. Influence of high altitude exposure on the immune system: a review. Immunol. Invest. 39 (3), 219–234. https://doi.org/10.3109/08820131003681144.

Mizoguchi, H., Yamada, K., 2019. Methamphetamine use causes cognitive impairment and altered decision-making. Neurochem. Int. 124, 106–113. https://doi.org/10.1016/j.neuint.2018.12.019. (Epub 2019 Jan 3).

Moore, E., Indig, D., Haysom, L., 2014. Traumatic brain injury, mental health, substance use, and offending among incarcerated young people. J. Head Trauma Rehabil. 29 (3), 239–247. https://doi.org/10.1097/HTR.0b013e31828f9876.

Moore, B.A., Brock, M.S., Brager, A., Collen, J., LoPresti, M., Mysliwiec, V., 2020. Posttraumatic stress disorder, traumatic brain injury, sleep, and performance in military personnel. Sleep Med. Clin. 15 (1), 87–100. https://doi.org/10.1016/j.jsmc.2019.11.004. (Epub 2020 Jan 8).

Mori, T., Ito, S., Namiki, M., Suzuki, T., Kobayashi, S., Matsubayashi, K., Sawaguchi, T., 2007. Involvement of free radicals followed by the activation of phospholipase A2 in the mechanism that underlies the combined effects of methamphetamine and morphine on subacute toxicity or lethality in mice: comparison of the therapeutic potential of fullerene, mepacrine, and cooling. Toxicology 236 (3), 149–157. https://doi.org/10.1016/j.tox.2007.03.027. (Epub 2007 Apr 7).

Moszczynska, A., Callan, S.P., 2017. Molecular, behavioral, and physiological consequences of methamphetamine neurotoxicity: implications for treatment. J. Pharmacol. Exp. Ther. 362 (3), 474–488. https://doi.org/10.1124/jpet.116.238501. (Epub 2017 Jun 19).

Mrakic-Sposta, S., Gussoni, M., Dellanoce, C., Marzorati, M., Montorsi, M., Rasica, L., Pratali, L., D'Angelo, G., Martinelli, M., Bastiani, L., Di Natale, L., Vezzoli, A., 2021.

Effects of acute and sub-acute hypobaric hypoxia on oxidative stress: a field study in the Alps. Eur. J. Appl. Physiol. 121 (1), 297–306. https://doi.org/10.1007/s00421-020-04527-x. (Epub 2020 Oct 15).

Müller, C.P., Homberg, J.R., 2015. The role of serotonin in drug use and addiction. Behav. Brain Res. 277, 146–192. https://doi.org/10.1016/j.bbr.2014.04.007. (Epub 2014 Apr 25).

Muresanu, D.F., Sharma, A., Tian, Z.R., Smith, M.A., Sharma, H.S., 2012. Nanowired drug delivery of antioxidant compound H-290/51 enhances neuroprotection in hyperthermia-induced neurotoxicity. CNS Neurol. Disord. Drug Targets 11 (1), 50–64. https://doi.org/10.2174/187152712799960736.

Muresanu, D.F., Sharma, A., Sharma, H.S., Apr 2017. Pathophysiology of high-altitude traumatic brain edema. New roles of cererbrolysin and nanomedicine. In: Conference: 28th International Symposium on Cerebral Blood Flow, Metabolism and Function/13th International Conference on Quantification of Brain Function with PET Location: Int Soc Cerebral Blood Flow & Metab, Berlin. GERMANY Date: APR 01–04, 2017. JOURNAL OF CEREBRAL BLOOD FLOW AND METABOLISM Volume: 37 Supplement: 1 Pages: 208–209 Meeting Abstract: PS02–065 Published.

Muresanu, D.F., Sharma, A., Patnaik, R., Menon, P.K., Mössler, H., Sharma, H.S., 2019. Exacerbation of blood-brain barrier breakdown, edema formation, nitric oxide synthase upregulation and brain pathology after heat stroke in diabetic and hypertensive rats. Potential neuroprotection with cerebrolysin treatment. Int. Rev. Neurobiol. 146, 83–102. https://doi.org/10.1016/bs.irn.2019.06.007. (Epub 2019 Jul 18).

Murray, A.J., 2016. Energy metabolism and the high-altitude environment. Exp. Physiol. 101 (1), 23–27. https://doi.org/10.1113/EP085317. (Epub 2015 Sep 13).

Muza, S.R., 2007. Military applications of hypoxic training for high-altitude operations. Med. Sci. Sports Exerc. 39 (9), 1625–1631. https://doi.org/10.1249/mss.0b013e3180de49fe.

Naeije, R., 2010. Physiological adaptation of the cardiovascular system to high altitude. Prog. Cardiovasc. Dis. 52 (6), 456–466. https://doi.org/10.1016/j.pcad.2010.03.004.

Nation, D.A., Bondi, M.W., Gayles, E., Delis, D.C., 2017. Mechanisms of memory dysfunction during high altitude hypoxia training in military aircrew. J. Int. Neuropsychol. Soc. 23 (1), 1–10. https://doi.org/10.1017/S1355617716000965. (Epub 2016 Dec 7).

National Research Council, 2011. Guide for the Care and Use of Laboratory Animals, eighth ed. The National Academic Press, Washington DC. www.nap.edu.

Olson-Madden, J.H., Brenner, L.A., Corrigan, J.D., Emrick, C.D., Britton, P.C., 2012. Substance use and mild traumatic brain injury risk reduction and prevention: a novel model for treatment. Rehabil. Res. Pract. 2012, 174579. https://doi.org/10.1155/2012/174579. (Epub 2012 May 17).

Olsson, Y., Sharma, H.S., Pettersson, C.A., 1990. Effects of p-chlorophenylalanine on microvascular permeability changes in spinal cord trauma. An experimental study in the rat using 131I-sodium and lanthanum tracers. Acta Neuropathol. 79 (6), 595–603. https://doi.org/10.1007/BF00294236.

Oña, G., Bouso, J.C., 2021. Therapeutic potential of natural psychoactive drugs for central nervous system disorders: a perspective from polypharmacology. Curr. Med. Chem. 28 (1), 53–68. https://doi.org/10.2174/0929867326666191212103330.

Otani, S., Miyaoka, Y., Ikeda, A., Ohno, G., Imura, S., Watanabe, K., Kurozawa, Y., 2020. Evaluating health impact at high altitude in antarctica and effectiveness of monitoring oxygen saturation. Yonago Acta Med. 63 (3), 163–172. https://doi.org/10.33160/yam.2020.08.004 (eCollection 2020 Aug).

Otero-Losada, M., Capani, F., 2019. Oxidative medicine in brain injury. Curr. Pharm. Des. 25 (45), 4735–4736. https://doi.org/10.2174/138161282545200110113639.

Ott, E.P., Jacob, D.W., Baker, S.E., Holbein, W.W., Scruggs, Z.M., Shoemaker, J.K., Limberg, J.K., 2020. Sympathetic neural recruitment strategies following acute intermittent hypoxia in humans. Am. J. Physiol. Regul. Integr. Comp. Physiol. 318 (5), R961–R971. https://doi.org/10.1152/ajpregu.00004.2020. (Epub 2020 Apr 8).

Palmer, G.C., 1983. Effects of psychoactive drugs on cyclic nucleotides in the central nervous system. Prog. Neurobiol. 21 (1–2), 1–133. https://doi.org/10.1016/0301-0082(83)90018-7.

Parati, G., Ochoa, J.E., Torlasco, C., Salvi, P., Lombardi, C., Bilo, G., 2015. Aging, high altitude, and blood pressure: a complex relationship. High Alt. Med. Biol. 16 (2), 97–109. https://doi.org/10.1089/ham.2015.0010. (Epub 2015 May 26).

Patel, P., Taylor, D., Park, M.S., 2019. Characteristics of traumatic brain injury during Operation Enduring Freedom-Afghanistan: a retrospective case series. Neurosurg. Focus 47 (5), E13. https://doi.org/10.3171/2019.8.FOCUS19493.

Paul, S., Gangwar, A., Bhargava, K., Khurana, P., Ahmad, Y., 2018. Diagnosis and prophylaxis for high-altitude acclimatization: adherence to molecular rationale to evade high-altitude illnesses. Life Sci. 203, 171–176. https://doi.org/10.1016/j.lfs.2018.04.040. (Epub 2018 Apr 23).

Paulus, M.P., Stewart, J.L., 2020. Neurobiology, clinical presentation, and treatment of methamphetamine use disorder: a review. JAMA Psychiat. 77 (9), 959–966. https://doi.org/10.1001/jamapsychiatry.2020.0246.

Pavlovic, R., Panseri, S., Giupponi, L., Leoni, V., Citti, C., Cattaneo, C., Cavaletto, M., Giorgi, A., 2019. Phytochemical and ecological analysis of two varieties of hemp (Cannabis sativa L.) grown in a mountain environment of Italian Alps. Front. Plant Sci. 10, 1265. https://doi.org/10.3389/fpls.2019.01265 (eCollection 2019).

Pehrson, A.L., Sanchez, C., 2014. Serotonergic modulation of glutamate neurotransmission as a strategy for treating depression and cognitive dysfunction. CNS Spectr. 19 (2), 121–133. https://doi.org/10.1017/S1092852913000540. (Epub 2013 Aug 1).

Pfeiffer, J.M., Askew, E.W., Roberts, D.E., Wood, S.M., Benson, J.E., Johnson, S.C., Freedman, M.S., 1999. Effect of antioxidant supplementation on urine and blood markers of oxidative stress during extended moderate-altitude training. Wilderness Environ. Med. 10 (2), 66–74. https://doi.org/10.1580/1080-6032(1999)010[0066,eoasou]2.3.co;2.

Pitman, J.T., Harris, N.S., 2012. Possible association with amphetamine usage and development of high altitude pulmonary edema. Wilderness Environ. Med. 23 (4), 374–376. https://doi.org/10.1016/j.wem.2012.06.006. (Epub 2012 Sep 28).

Platteborze, P.L., Kippenberger, D.J., Martin, T.M., 2013. Drug positive rates for the Army, Army Reserve, and Army National Guard from fiscal year 2001 through 2011. Mil. Med. 178 (10), 1078–1084. https://doi.org/10.7205/MILMED-D-13-00193.

Potvin, S., Pelletier, J., Grot, S., Hébert, C., Barr, A.M., Lecomte, T., 2018. Cognitive deficits in individuals with methamphetamine use disorder: a meta-analysis. Addict. Behav. 80, 154–160. https://doi.org/10.1016/j.addbeh.2018.01.021. (Epub 2018 Jan 31).

Pubill, D., Chipana, C., Camins, A., Pallàs, M., Camarasa, J., Escubedo, E., 2005. Free radical production induced by methamphetamine in rat striatal synaptosomes. Toxicol. Appl. Pharmacol. 204 (1), 57–68. https://doi.org/10.1016/j.taap.2004.08.008.

Pugh, M.J., Swan, A.A., Amuan, M.E., Eapen, B.C., Jaramillo, C.A., Delgado, R., Tate, D.F., Yaffe, K., Wang, C.P., 2019. Deployment, suicide, and overdose among comorbidity phenotypes following mild traumatic brain injury: a retrospective cohort study from the

Chronic Effects of Neurotrauma Consortium. PLoS One 14 (9). https://doi.org/10.1371/journal.pone.0222674, e0222674 (eCollection 2019).

Pun, M., Guadagni, V., Drogos, L.L., Pon, C., Hartmann, S.E., Furian, M., Lichtblau, M., Muralt, L., Bader, P.R., Moraga, F.A., Soza, D., Lopez, I., Rawling, J.M., Ulrich, S., Bloch, K.E., Giesbrecht, B., Poulin, M.J., 2019. Cognitive effects of repeated acute exposure to very high altitude among altitude-experienced workers at 5050 m. High Alt. Med. Biol. 20 (4), 361–374. https://doi.org/10.1089/ham.2019.0012. (Epub 2019 Oct 25).

Quindry, J., Dumke, C., Slivka, D., Ruby, B., 2016. Impact of extreme exercise at high altitude on oxidative stress in humans. J. Physiol. 594 (18), 5093–5104. https://doi.org/10.1113/JP270651. (Epub 2015 Dec 7).

Quiñones-Laveriano, D.M., Espinoza-Chiong, C., Scarsi-Mejia, O., Rojas-Camayo, J., Mejia, C.R., 2016. Geographic altitude of residence and alcohol dependence in a Peruvian population. Rev. Colomb. Psiquiatr. 45 (3), 178–185. https://doi.org/10.1016/j.rcp.2015.11.002. (Epub 2015 Dec 17).

Rasmussen, N., 2006. Making the first anti-depressant: amphetamine in American medicine, 1929-1950. J. Hist. Med. Allied Sci. 61 (3), 288–323. https://doi.org/10.1093/jhmas/jrj039. (Epub 2006 Feb 21).

Rasmussen, N., 2011. Medical science and the military: the Allies' use of amphetamine during World War II. J. Interdiscip. Hist. 42 (2), 205–233. https://doi.org/10.1162/jinh_a_00212.

Rasmussen, N., 2015. Amphetamine-type stimulants: the early history of their medical and non-medical uses. Int. Rev. Neurobiol. 120, 9–25. https://doi.org/10.1016/bs.irn.2015.02.001. (Epub 2015 Mar 17).

Ray, K., Dutta, A., Panjwani, U., Thakur, L., Anand, J.P., Kumar, S., 2011. Hypobaric hypoxia modulates brain biogenic amines and disturbs sleep architecture. Neurochem. Int. 58 (1), 112–118. https://doi.org/10.1016/j.neuint.2010.11.003. (Epub 2010 Nov 12). PMID: 21075155.

Razavi, Y., Shabani, R., Mehdizadeh, M., Haghparast, A., 2020. Neuroprotective effect of chronic administration of cannabidiol during the abstinence period on methamphetamine-induced impairment of recognition memory in the rats. Behav. Pharmacol. 31 (4), 385–396. https://doi.org/10.1097/FBP.0000000000000544.

Reno, E., Brown, T.L., Betz, M.E., Allen, M.H., Hoffecker, L., Reitinger, J., Roach, R., Honigman, B., 2018. Suicide and high altitude: an integrative review. High Alt. Med. Biol. 19 (2), 99–108. https://doi.org/10.1089/ham.2016.0131. (Epub 2017 Nov 21).

Richalet, J.P., 2020. CrossTalk opposing view: barometric pressure, independent of PO2, is not the forgotten parameter in altitude physiology and mountain medicine. J. Physiol. 598 (5), 897–899. https://doi.org/10.1113/JP279160. (Epub 2020 Feb 13).

Richalet, J.P., Larmignat, P., Poignard, P., 2020. Transient cerebral ischemia at high altitude and hyper-responsiveness to hypoxia. High Alt. Med. Biol. 21 (1), 105–108. https://doi.org/10.1089/ham.2019.0100. (Epub 2020 Jan 22).

Richards, J.R., Laurin, E.G., 2020 Nov 20. Methamphetamine toxicity. In: StatPearls [Internet]. StatPearls Publishing, Treasure Island (FL) (2021 Jan).

Riech, S., Kallenberg, K., Moerer, O., Hellen, P., Bärtsch, P., Quintel, M., Knauth, M., 2015. The pattern of brain microhemorrhages after severe lung failure resembles the one seen in high-altitude cerebral edema. Crit. Care Med. 43 (9), e386–e389. https://doi.org/10.1097/CCM.0000000000001150.

Roberts, C.A., Jones, A., Sumnall, H., Gage, S.H., Montgomery, C., 2020. How effective are pharmaceuticals for cognitive enhancement in healthy adults? A series of meta-analyses of cognitive performance during acute administration of modafinil, methylphenidate and D-amphetamine. Eur. Neuropsychopharmacol. 38, 40–62. https://doi.org/10.1016/j.euroneuro.2020.07.002. (Epub 2020).

Rodway, G.W., Muza, S.R., 2011. Fighting in thin air: operational wilderness medicine in high Asia. Wilderness Environ. Med. 22 (4), 297–303. https://doi.org/10.1016/j.wem.2011.08.009. PMID: 22137862.

Rosen, L.N., Knudson, K.H., Brannen, S.J., Fancher, P., Killgore, T.E., Barasich, G.G., 2002. Intimate partner violence among U.S. Army soldiers in Alaska: a comparison of reported rates and survey results. Mil. Med. 167 (8), 688–691.

Roy, K., Chauhan, G., Kumari, P., Wadhwa, M., Alam, S., Ray, K., Panjwani, U., Kishore, K., 2018. Phosphorylated delta sleep inducing peptide restores spatial memory and p-CREB expression by improving sleep architecture at high altitude. Life Sci. 209, 282–290. https://doi.org/10.1016/j.lfs.2018.08.026. (Epub 2018 Aug 11).

Sabic, H., Kious, B., Boxer, D., Fitzgerald, C., Riley, C., Scholl, L., McGlade, E., Yurgelun-Todd, D., Renshaw, P.F., Kondo, D.G., 2019. Effect of altitude on veteran suicide rates. High Alt. Med. Biol. 20 (2), 171–177. https://doi.org/10.1089/ham.2018.0130. (Epub 2019 May 2).

Sabrini, S., Wang, G.Y., Lin, J.C., Ian, J.K., Curley, L.E., 2019. Methamphetamine use and cognitive function: a systematic review of neuroimaging research. Drug Alcohol Depend. 194, 75–87. https://doi.org/10.1016/j.drugalcdep.2018.08.041. (Epub 2018 Oct 24).

Sagoo, R.S., Hutchinson, C.E., Wright, A., Handford, C., Parsons, H., Sherwood, V., Wayte, S., Nagaraja, S., Ng'Andwe, E., Wilson, M.H., Imray, C.H., 2017. Birmingham medical research and expedition society. Magnetic Resonance investigation into the mechanisms involved in the development of high-altitude cerebral edema. J. Cereb. Blood Flow Metab. 37 (1), 319–331. https://doi.org/10.1177/0271678X15625350. (Epub 2016 Jan 8).

San, T., Polat, S., Cingi, C., Eskiizmir, G., Oghan, F., Cakir, B., 2013. Effects of high altitude on sleep and respiratory system and theirs adaptations. ScientificWorldJournal 2013, 241569. https://doi.org/10.1155/2013/241569. Print 2013.

Sareban, M., Berger, M.M., Pinter, D., Buchmann, A., Macholz, F., Schmidt, P., Schiefer, L., Schimke, M., Niebauer, J., Steinacker, J.M., Treff, G., Khalil, M., 2021. Serum neurofilament level increases after ascent to 4559 m but is not related to acute mountain sickness. Eur. J. Neurol. 28 (3), 1004–1008. https://doi.org/10.1111/ene.14606. (Epub 2020 Nov 26).

Schinder, E.O., Ruder, A.M., 1989. Epidemiology of coca and alcohol use among high-altitude miners in Argentina. Am. J. Ind. Med. 15 (5), 579–587. https://doi.org/10.1002/ajim.4700150510.

Schrot, R.J., Hubbard, J.R., 2016. Cannabinoids: medical implications. Ann. Med. 48 (3), 128–141. https://doi.org/10.3109/07853890.2016.1145794. (Epub 2016 Feb 25).

Shaerzadeh, F., Streit, W.J., Heysieattalab, S., Khoshbouei, H., 2018. Methamphetamine neurotoxicity, microglia, and neuroinflammation. J. Neuroinflammation 15 (1), 341. https://doi.org/10.1186/s12974-018-1385-0.

Sharma, H.S., 1987. Effect of captopril (a converting enzyme inhibitor) on blood-brain barrier permeability and cerebral blood flow in normotensive rats. Neuropharmacology 26 (1), 85–92. https://doi.org/10.1016/0028-3908(87)90049-9.

Sharma, H.S., 2007. Nanoneuroscience: emerging concepts on nanoneurotoxicity and nanoneuroprotection. Nanomedicine (Lond.) 2 (6), 753–758. https://doi.org/10.2217/17435889.2.6.753.

Sharma, H.S., 2011. Early microvascular reactions and blood-spinal cord barrier disruption are instrumental in pathophysiology of spinal cord injury and repair: novel therapeutic strategies including nanowired drug delivery to enhance neuroprotection. J. Neural Transm. (Vienna) 118 (1), 155–176. https://doi.org/10.1007/s00702-010-0514-4. (Epub 2010 Dec 16).

Sharma, H.S., Ali, S.F., 2006. Alterations in blood-brain barrier function by morphine and methamphetamine. Ann. N. Y. Acad. Sci. 1074, 198–224. https://doi.org/10.1196/annals.1369.020.

Sharma, H.S., Ali, S.F., 2008. Acute administration of 3,4-methylenedioxymethamphetamine induces profound hyperthermia, blood-brain barrier disruption, brain edema formation, and cell injury. Ann. N. Y. Acad. Sci. 1139, 242–258. https://doi.org/10.1196/annals.1432.052.

Sharma, H.S., Cervós-Navarro, J., 1990. Brain oedema and cellular changes induced by acute heat stress in young rats. Acta Neurochir. Suppl. (Wien) 51, 383–386. https://doi.org/10.1007/978-3-7091-9115-6_129.

Sharma, H.S., Dey, P.K., 1981. Impairment of blood-brain barrier (BBB) in rat by immobilization stress: role of serotonin (5-HT). Indian J. Physiol. Pharmacol. 25 (2), 111–122.

Sharma, H.S., Dey, P.K., 1984. Role of 5-HT on increased permeability of blood-brain barrier under heat stress. Indian J. Physiol. Pharmacol. 28 (4), 259–267.

Sharma, H.S., Dey, P.K., 1986a. Influence of long-term immobilization stress on regional blood-brain barrier permeability, cerebral blood flow and 5-HT level in conscious normotensive young rats. J. Neurol. Sci. 72 (1), 61–76. https://doi.org/10.1016/0022-510x(86)90036-5.

Sharma, H.S., Dey, P.K., 1986b. Probable involvement of 5-hydroxytryptamine in increased permeability of blood-brain barrier under heat stress in young rats. Neuropharmacology 25 (2), 161–167. https://doi.org/10.1016/0028-3908(86)90037-7.

Sharma, H.S., Dey, P.K., 1987. Influence of long-term acute heat exposure on regional blood-brain barrier permeability, cerebral blood flow and 5-HT level in conscious normotensive young rats. Brain Res. 424 (1), 153–162. https://doi.org/10.1016/0006-8993(87)91205-4.

Sharma, H.S., Kiyatkin, E.A., 2009. Rapid morphological brain abnormalities during acute methamphetamine intoxication in the rat: an experimental study using light and electron microscopy. J. Chem. Neuroanat. 37 (1), 18–32. https://doi.org/10.1016/j.jchemneu.2008.08.002. (Epub 2008 Aug 19).

Sharma, H.S., Olsson, Y., 1990. Edema formation and cellular alterations following spinal cord injury in the rat and their modification with p-chlorophenylalanine. Acta Neuropathol. 79 (6), 604–610. https://doi.org/10.1007/BF00294237.

Sharma, H.S., Sharma, A., 2012. Nanowired drug delivery for neuroprotection in central nervous system injuries: modulation by environmental temperature, intoxication of nanoparticles, and comorbidity factors. Wiley Interdiscip. Rev. Nanomed. Nanobiotechnol. 4 (2), 184–203.

Sharma, H.S., Sharma, A., 2013. New perspectives of nanoneuroprotection, nanoneuropharmacology and nanoneurotoxicity: modulatory role of amino acid neurotransmitters, stress, trauma, and co-morbidity factors in nanomedicine. Amino Acids 45 (5), 1055–1071. https://doi.org/10.1007/s00726-013-1584-z.

Sharma, H.S., Sjöquist, P.O., 2002. A new antioxidant compound H-290/51 modulates glutamate and GABA immunoreactivity in the rat spinal cord following trauma. Amino Acids 23 (1–3), 261–272. https://doi.org/10.1007/s00726-001-0137-z.

Sharma, H.S., Olsson, Y., Dey, P.K., 1990a. Changes in blood-brain barrier and cerebral blood flow following elevation of circulating serotonin level in anesthetized rats. Brain Res. 517 (1–2), 215–223. https://doi.org/10.1016/0006-8993(90)91029-g.

Sharma, H.S., Olsson, Y., Dey, P.K., 1990b. Early accumulation of serotonin in rat spinal cord subjected to traumatic injury. Relation to edema and blood flow changes. Neuroscience 36 (3), 725–730. https://doi.org/10.1016/0306-4522(90)90014-u.

Sharma, H.S., Cervós-Navarro, J., Dey, P.K., 1991. Acute heat exposure causes cellular alteration in cerebral cortex of young rats. Neuroreport 2 (3), 155–158. https://doi.org/10.1097/00001756-199103000-00012.

Sharma, H.S., Kretzschmar, R., Cervós-Navarro, J., Ermisch, A., Rühle, H.J., Dey, P.K., 1992a. Age-related pathophysiology of the blood-brain barrier in heat stress. Prog. Brain Res. 91, 189–196. https://doi.org/10.1016/s0079-6123(08)62334-1.

Sharma, H.S., Zimmer, C., Westman, J., Cervós-Navarro, J., 1992b. Acute systemic heat stress increases glial fibrillary acidic protein immunoreactivity in brain: experimental observations in conscious normotensive young rats. Neuroscience 48 (4), 889–901. https://doi.org/10.1016/0306-4522(92)90277-9.

Sharma, H.S., Olsson, Y., Cervós-Navarro, J., 1993a. Early perifocal cell changes and edema in traumatic injury of the spinal cord are reduced by indomethacin, an inhibitor of prostaglandin synthesis. Experimental study in the rat. Acta Neuropathol. 85 (2), 145–153. https://doi.org/10.1007/BF00227761.

Sharma, H.S., Olsson, Y., Nyberg, F., Dey, P.K., 1993b. Prostaglandins modulate alterations of microvascular permeability, blood flow, edema and serotonin levels following spinal cord injury: an experimental study in the rat. Neuroscience 57 (2), 443–449. https://doi.org/10.1016/0306-4522(93)90076-r.

Sharma, H.S., Olsson, Y., Persson, S., Nyberg, F., 1995. Trauma-induced opening of the blood-spinal cord barrier is reduced by indomethacin, an inhibitor of prostaglandin biosynthesis. Experimental observations in the rat using [131I]-sodium, Evans blue and lanthanum as tracers. Restor. Neurol. Neurosci. 7 (4), 207–215. https://doi.org/10.3233/RNN-1994-7403.

Sharma, H.S., Westman, J., Olsson, Y., Alm, P., 1996. Involvement of nitric oxide in acute spinal cord injury: an immunocytochemical study using light and electron microscopy in the rat. Neurosci. Res. 24 (4), 373–384. https://doi.org/10.1016/0168-0102(95)01015-7.

Sharma, H.S., Alm, P., Westman, J., 1998a. Nitric oxide and carbon monoxide in the brain pathology of heat stress. Prog. Brain Res. 115, 297–333. https://doi.org/10.1016/s0079-6123(08)62041-5.

Sharma, H.S., Westman, J., Nyberg, F., 1998b. Pathophysiology of brain edema and cell changes following hyperthermic brain injury. Prog. Brain Res. 115, 351–412. https://doi.org/10.1016/s0079-6123(08)62043-9.

Sharma, H.S., Alm, P., Sjöquist, P.O., Westman, J., 2000a. A new antioxidant compound H-290/51 attenuates upregulation of constitutive isoform of heme oxygenase (HO-2) following trauma to the rat spinal cord. Acta Neurochir. Suppl. 76, 153–157. https://doi.org/10.1007/978-3-7091-6346-7_31.

Sharma, H.S., Westman, J., Gordh, T., Nyberg, F., 2000b. Spinal cord injury induced c-fos expression is reduced by p-CPA, a serotonin synthesis inhibitor. An experimental study using immunohistochemistry in the rat. Acta Neurochir. Suppl. 76, 297–301. https://doi.org/10.1007/978-3-7091-6346-7_61.

Sharma, H.S., Sjöquist, P.-O., Westman, J., 2001. Pathophysiology of the blood-spinal cord barrier in spinal cord injury. Influence of a new antioxidant compound H-290/51. In: Kobiler, D., Lustig, S., Shapra, S. (Eds.), Blood-Brain Barrier. Drug Delivery and Brain Pathology. Kluwer Academic/Plenum Publishers, Springer, Boston, MA, New York, pp. 401–416. https://doi.org/10.1007/978-1-4615-0579-2_32.

Sharma, H.S., Sjöquist, P.O., Alm, P., 2003. A new antioxidant compound H-290151 attenuates spinal cord injury induced expression of constitutive and inducible isoforms of nitric oxide synthase and edema formation in the rat. Acta Neurochir. Suppl. 86, 415–420. https://doi.org/10.1007/978-3-7091-0651-8_86.

Sharma, H.S., Patnaik, R., Ray, A.K., Dey, P.K., 2004. Blood-central nervous system barriers in morphine dependence and withdrawal. In: Sharma, H.S., Westman, J. (Eds.), The Blood-Spinal Cord and Brain Barriers in Health and Disease. Elsevier Academic Press, San Diego, pp. 299–328.

Sharma, H.S., Gordh, T., Wiklund, L., Mohanty, S., Sjöquist, P.O., 2006a. Spinal cord injury induced heat shock protein expression is reduced by an antioxidant compound H-290/51. An experimental study using light and electron microscopy in the rat. J. Neural Transm. (Vienna) 113 (4), 521–536. https://doi.org/10.1007/s00702-005-0405-2.

Sharma, H.S., Lundstedt, T., Boman, A., Lek, P., Seifert, E., Wiklund, L., Ali, S.F., 2006b. A potent serotonin-modulating compound AP-267 attenuates morphine withdrawal-induced blood-brain barrier dysfunction in rats. Ann. N. Y. Acad. Sci. 1074, 482–496. https://doi.org/10.1196/annals.1369.049.

Sharma, H.S., Sjöquist, P.O., Mohanty, S., Wiklund, L., 2006c. Post-injury treatment with a new antioxidant compound H-290/51 attenuates spinal cord trauma-induced c-fos expression, motor dysfunction, edema formation, and cell injury in the rat. Acta Neurochir. Suppl. 96, 322–328. https://doi.org/10.1007/3-211-30714-1_68.

Sharma, H.S., Patnaik, R., Patnaik, S., Mohanty, S., Sharma, A., Vannemreddy, P., 2007a. Antibodies to serotonin attenuate closed head injury induced blood brain barrier disruption and brain pathology. Ann. N. Y. Acad. Sci. 1122, 295–312. https://doi.org/10.1196/annals.1403.022.

Sharma, H.S., Sjöquist, P.O., Ali, S.F., 2007b. Drugs of abuse-induced hyperthermia, blood-brain barrier dysfunction and neurotoxicity: neuroprotective effects of a new antioxidant compound H-290/51. Curr. Pharm. Des. 13 (18), 1903–1923. https://doi.org/10.2174/138161207780858375.

Sharma, H.S., Ali, S.F., Tian, Z.R., Hussain, S.M., Schlager, J.J., Sjöquist, P.O., Sharma, A., Muresanu, D.F., 2009a. Chronic treatment with nanoparticles exacerbate hyperthermia induced blood-brain barrier breakdown, cognitive dysfunction and brain pathology in the rat. Neuroprotective effects of nanowired-antioxidant compound H-290/51. J. Nanosci. Nanotechnol. 9 (8), 5073–5090. https://doi.org/10.1166/jnn.2009.gr10.

Sharma, H.S., Muresanu, D., Sharma, A., Patnaik, R., 2009b. Cocaine-induced breakdown of the blood-brain barrier and neurotoxicity. Int. Rev. Neurobiol. 88, 297–334. https://doi.org/10.1016/S0074-7742(09)88011-2.

Sharma, H.S., Patnaik, R., Sharma, A., Sjöquist, P.O., Lafuente, J.V., 2009c. Silicon dioxide nanoparticles (SiO2, 40-50 nm) exacerbate pathophysiology of traumatic spinal cord injury and deteriorate functional outcome in the rat. An experimental study using pharmacological and morphological approaches. J. Nanosci. Nanotechnol. 9 (8), 4970–4980. https://doi.org/10.1166/jnn.2009.1717.

Sharma, H.S., Sjöquist, P.O., Ali, S.F., 2010. Alterations in blood-brain barrier function and brain pathology by morphine in the rat. Neuroprotective effects of antioxidant H-290/51. Acta Neurochir. Suppl. 106, 61–66. https://doi.org/10.1007/978-3-211-98811-4_10.

Sharma, H.S., Ali, S.F., Patnaik, R., Zimmermann-Meinzingen, S., Sharma, A., Muresanu, D.F., 2011a. Cerebrolysin attenuates heat shock protein (HSP 72 KD) expression in the rat spinal cord following morphine dependence and withdrawal: possible new therapy for pain management. Curr. Neuropharmacol. 9 (1), 223–235. https://doi.org/10.2174/157015911795017100.

Sharma, H.S., Miclescu, A., Wiklund, L., 2011b. Cardiac arrest-induced regional blood-brain barrier breakdown, edema formation and brain pathology: a light and electron microscopic study on a new model for neurodegeneration and neuroprotection in porcine brain. J. Neural Transm. (Vienna) 118 (1), 87–114. https://doi.org/10.1007/s00702-010-0486-4. (Epub 2010 Oct 21).

Sharma, H.S., Muresanu, D.F., Patnaik, R., Stan, A.D., Vacaras, V., Perju-Dumbrav, L., Alexandru, B., Buzoianu, A., Opincariu, I., Menon, P.K., Sharma, A., 2011c. Superior neuroprotective effects of cerebrolysin in heat stroke following chronic intoxication of Cu or Ag engineered nanoparticles. A comparative study with other neuroprotective agents using biochemical and morphological approaches in the rat. J. Nanosci. Nanotechnol. 11 (9), 7549–7569. https://doi.org/10.1166/jnn.2011.5114.

Sharma, A., Muresanu, D.F., Mössler, H., Sharma, H.S., 2012a. Superior neuroprotective effects of cerebrolysin in nanoparticle-induced exacerbation of hyperthermia-induced brain pathology. CNS Neurol. Disord. Drug Targets 11 (1), 7–25. https://doi.org/10.2174/187152712799960790.

Sharma, H.S., Sharma, A., Mössler, H., Muresanu, D.F., 2012b. Neuroprotective effects of cerebrolysin, a combination of different active fragments of neurotrophic factors and peptides on the whole body hyperthermia-induced neurotoxicity: modulatory roles of co-morbidity factors and nanoparticle intoxication. Int. Rev. Neurobiol. 102, 249–276. https://doi.org/10.1016/B978-0-12-386986-9.00010-7.

Sharma, H.S., Menon, P., Lafuente, J.V., Muresanu, D.F., Tian, Z.R., Patnaik, R., Sharma, A., 2014. Development of in vivo drug-induced neurotoxicity models. Expert Opin. Drug Metab. Toxicol. 10 (12), 1637–1661. https://doi.org/10.1517/17425255.2014.970168.

Sharma, H.S., Kiyatkin, E.A., Patnaik, R., Lafuente, J.V., Muresanu, D.F., Sjöquist, P.O., Sharma, A., 2015a. Exacerbation of methamphetamine neurotoxicity in cold and hot environments: neuroprotective effects of an antioxidant compound H-290/51. Mol. Neurobiol. 52 (2), 1023–1033. https://doi.org/10.1007/s12035-015-9252-9. (Epub 2015 Jun 26).

Sharma, H.S., Muresanu, D.F., Lafuente, J.V., Sjöquist, P.O., Patnaik, R., Sharma, A., 2015b. Nanoparticles exacerbate both ubiquitin and heat shock protein expressions in spinal cord injury: neuroprotective effects of the proteasome inhibitor carfilzomib and the antioxidant compound H-290/51. Mol. Neurobiol. 52 (2), 882–898. https://doi.org/10.1007/s12035-015-9297-9. (Epub 2015 Jul 1).

Sharma, H.S., Muresanu, D.F., Lafuente, J.V., Sjöquist, P.O., Patnaik, R., Sharma, A., 2015c. Cord injury: neuroprotective effects of the proteasome inhibitor carfilzomib and the antioxidant compound H-290/51. Mol. Neurobiol. 52 (2), 882–898. https://doi.org/10.1007/s12035-015-9297-9. (Epub 2015 Jul 1).

Sharma, A., Menon, P., Muresanu, D.F., Ozkizilcik, A., Tian, Z.R., Lafuente, J.V., Sharma, H.-S., 2016a. Nanowired drug delivery across the blood-brain barrier in central nervous system injury and repair. CNS Neurol. Disord. Drug Targets 15 (9), 1092–1117. https://doi.org/10.2174/1871527315666160819123059.

Sharma, H.S., Muresanu, D.F., Lafuente, J.V., Nozari, A., Patnaik, R., Skaper, S.D., Sharma, A., 2016b. Pathophysiology of blood-brain barrier in brain injury in cold and hot environments: novel drug targets for neuroprotection. CNS Neurol. Disord. Drug Targets 15 (9), 1045–1071. https://doi.org/10.2174/1871527315666160902145145.

Sharma, A., Muresanu, D.F., Lafuente, J.V., Sjöquist, P.O., Patnaik, R., Ryan Tian, Z., Ozkizilcik, A., Sharma, H.S., 2018a. Cold environment exacerbates brain pathology and oxidative stress following traumatic brain injuries: potential therapeutic effects of nanowired antioxidant compound H-290/51. Mol. Neurobiol. 55 (1), 276–285. https://doi.org/10.1007/s12035-017-0740-y.

Sharma, H.S., Muresanu, D.F., Lafuente, J.V., Patnaik, R., Tian, Z.R., Ozkizilcik, A., Castellani, R.J., Mössler, H., Sharma, A., 2018b. Co-administration of TiO2 nanowired mesenchymal stem cells with cerebrolysin potentiates neprilysin level and reduces brain pathology in Alzheimer's Disease. Mol. Neurobiol. 55 (1), 300–311. https://doi.org/10.1007/s12035-017-0742-9.

Sharma, H., Muresanu, D., Lafuente, J., Nozari, A., Ozkizilcik, A., Tian, Z.R., Sharma, A., 2018c. Pathophysiology of concussive head injury is exacerbated high altitude. neuroprotective effects of TiO2 nanodelivery of cerebrolysin. J. Head Trauma Rehabil. 33 (3), E72 (Meeting Abstract: 0011 Published).

Sharma, A., Patnaik, R., Sharma, H.S., 2019a. Neuroprotective effects of 5-HT(3) receptor antagonist ondansetron on morphine withdrawal induced brain edema formation, blood-brain barrier dysfunction, neuronal injuries, glial activation and heat shock protein upregulation in the brain. Int. Rev. Neurobiol. 146, 209–228. https://doi.org/10.1016/bs.irn.2019.06.011. (Epub 2019 Jul 8).

Sharma, H.S., Muresanu, D.F., Nozari, A., Castellani, R.J., Dey, P.K., Wiklund, L., Sharma, A., 2019b. Anesthetics influence concussive head injury induced blood-brain barrier breakdown, brain edema formation, cerebral blood flow, serotonin levels, brain pathology and functional outcome. Int. Rev. Neurobiol. 146, 45–81. https://doi.org/10.1016/bs.irn.2019.06.006. (Epub 2019 Jul 8).

Sharma, R., Cramer, N.P., Perry, B., Adahman, Z., Murphy, E.K., Xu, X., Dardzinski, B.J., Galdzicki, Z., Perl, D.P., Dickstein, D.L., Iacono, D., 2019c. Chronic exposure to high altitude: synaptic, astroglial and memory changes. Sci. Rep. 9 (1), 16406. https://doi.org/10.1038/s41598-019-52563-1.

Sharma, H.S., Muresanu, D.F., Castellani, R.J., Nozari, A., Lafuente, J.V., Tian, Z.R., Sahib, S., Bryukhovetskiy, I., Bryukhovetskiy, A., Buzoianu, A.D., Patnaik, R., Wiklund, L., Sharma, A., 2020a. Pathophysiology of blood-brain barrier in brain tumor. Novel therapeutic advances using nanomedicine. Int. Rev. Neurobiol. 151, 1–66. https://doi.org/10.1016/bs.irn.2020.03.001. (Epub 2020 May 13).

Sharma, H.S., Sahib, S., Tian, Z.R., Muresanu, D.F., Nozari, A., Castellani, R.J., Lafuente, J.V., Wiklund, L., Sharma, A., 2020b. Protein kinase inhibitors in traumatic brain injury and repair: new roles of nanomedicine. Prog. Brain Res. 258, 233–283. https://doi.org/10.1016/bs.pbr.2020.09.009. (Epub 2020 Oct 24).

Shin, E.J., Tran, H.Q., Nguyen, P.T., Jeong, J.H., Nah, S.Y., Jang, C.G., Nabeshima, T., Kim, H.C., 2018. Role of mitochondria in methamphetamine-induced dopaminergic neurotoxicity: involvement in oxidative stress, neuroinflammation, and pro-apoptosis—a review. Neurochem. Res. 43 (1), 66–78. https://doi.org/10.1007/s11064-017-2318-5. (Epub 2017 Jun 7).

Shukitt, B.L., Banderet, L.E., 1988. Mood states at 1600 and 4300 meters terrestrial altitude. Aviat. Space Environ. Med. 59 (6), 530–532. 3390110.

Shukitt-Hale, B., Liebermann, H., 1996. Effects of altitude on cognitive performances and mood states, chapter 22. In: Marriott, B.M., Carlson, S.J. (Eds.), Nutritional Needs in Cold and in High Altitude Environments. Applications for Military Personnel in Field Operatins. National Axademy Press, Wshington DC, pp. 435–452. http://www.nap.edu/catalog/5197.html.

Shukitt-Hale, B., Rauch, T.M., Foutch, R., 1990. Altitude symptomatology and mood states during a climb to 3,630 meters. Aviat. Space Environ. Med. 61 (3), 225–228. PMID: 2317176.

Shukitt-Hale, B., Banderet, L.E., Lieberman, H.R., 1991. Relationships between symptoms, moods, performance, and acute mountain sickness at 4,700 meters. Aviat. Space Environ. Med. 62 (9 Pt. 1), 865–869.

Shukitt-Hale, B., Stillman, M.J., Welch, D.I., Levy, A., Devine, J.A., Lieberman, H.R., 1994. Hypobaric hypoxia impairs spatial memory in an elevation-dependent fashion. Behav. Neural Biol. 62, 244–252.

Shukitt-Hale, B., Kadar, T., Marlowe, B.E., Stillman, M.J., Galli, R.L., Levy, A., Devine, J.A., Lieberman, H.R., 1996. Morphological alterations in the hippocampus following hypobaric hypoxia. Hum. Exp. Toxicol. 15 (4), 312–319. https://doi.org/10.1177/096032719601500407.

Shukitt-Hale, B., Banderet, L.E., Lieberman, H.R., 1998. Elevation-dependent symptom, mood, and performance changes produced by exposure to hypobaric hypoxia. Int. J. Aviat. Psychol. 8 (4), 319–334. https://doi.org/10.1207/s15327108ijap0804_1.

Sikri, G., Kotwal, A., Singh, S.P., Bhattachar, S., Bhatia, S.S., Dutt, M., Srinath, N., 2019. Is it time to revise the acclimatization schedule at high altitude? Evidence from a field trial in Western Himalayas. Med. J. Armed Forces India 75 (3), 251–258. https://doi.org/10.1016/j.mjafi.2018.01.001. (Epub 2018 Aug).

Silverberg, N.D., Iaccarino, M.A., Panenka, W.J., Iverson, G.L., McCulloch, K.L., Dams-O'Connor, K., Reed, N., McCrea, M., 2020. American congress of rehabilitation medicine brain injury interdisciplinary special interest group mild TBI task force. Management of concussion and mild traumatic brain injury: a synthesis of practice guidelines. Arch. Phys. Med. Rehabil. 101 (2), 382–393. https://doi.org/10.1016/j.apmr.2019.10.179. (Epub 2019 Oct 23).

Simonson, T.S., Wei, G., Wagner, H.E., Wuren, T., Bui, A., Fine, J.M., Qin, G., Beltrami, F.G., Yan, M., Wagner, P.D., Ge, R.L., 2014. Increased blood-oxygen binding affinity in Tibetan and Han Chinese residents at 4200 m. Exp. Physiol. 99 (12), 1624–1635. https://doi.org/10.1113/expphysiol.2014.080820. (Epub 2014 Aug 28). PMID: 25172885.

Smith, D.W., Myer, G.D., Currie, D.W., Comstock, R.D., Clark, J.F., Bailes, J.E., 2013. Altitude modulates concussion incidence: implications for optimizing brain compliance to prevent brain injury in athletes. Orthop. J. Sports Med. 1 (6), 2325967113511588. https://doi.org/10.1177/2325967113511588. (eCollection 2013 Nov).

Solomon, T.M., Kiang, M.V., Halkitis, P.N., Moeller, R.W., Pappas, M.K., 2010. Personality traits and mental health states of methamphetamine-dependent and methamphetamine non-using MSM. Addict. Behav. 35 (2), 161–163. https://doi.org/10.1016/j.addbeh.2009.09.002. (Epub 2009 Sep 9).

Speidel, V., Purrucker, J.C., Klobučníková, K., 2021. Manifestation of intracranial lesions at high altitude: case report and review of the literature. High Alt. Med. Biol. 22, 87–89. https://doi.org/10.1089/ham.2020.0223 (Online ahead of print).

Spielberg, J.M., McGlinchey, R.E., Milberg, W.P., Salat, D.H., 2015. Brain network disturbance related to posttraumatic stress and traumatic brain injury in veterans. Biol. Psychiatry 78 (3), 210–216. https://doi.org/10.1016/j.biopsych.2015.02.013. (Epub 2015 Feb 18).

Stepanova, A., Konrad, C., Manfredi, G., Springett, R., Ten, V., Galkin, A., 2019. The dependence of brain mitochondria reactive oxygen species production on oxygen level is linear, except when inhibited by antimycin A. J. Neurochem. 148 (6), 731–745. https://doi.org/10.1111/jnc.14654. (Epub 2019 Jan 24).

Stewart, S.K., Pearce, A.P., Clasper, J.C., 2019. Fatal head and neck injuries in military underbody blast casualties. J. R. Army Med. Corps 165 (1), 18–21. https://doi.org/10.1136/jramc-2018-000942. (Epub 2018 Apr 21).

Swanson, T.M., Isaacson, B.M., Cyborski, C.M., French, L.M., Tsao, J.W., Pasquina, P.F., 2017. Traumatic brain injury incidence, clinical overview, and policies in the US military health system since 2000. Public Health Rep. 132 (2), 251–259. https://doi.org/10.1177/0033354916687748. (Epub 2017 Jan 30).

Tamura, M., United Nations Office on Drugs and Crime, 1989. Japan: stimulant epidemics past and present. Bull. Narc. 41 (1–2), 83–93. PMID: 2765722.

Thörnwall, M., Sharma, H.S., Gordh, T., Sjöquist, P.O., Nyberg, F., 1997. Substance P endopeptidase activity in the rat spinal cord following injury: influence of the new anti-oxidant compound H 290/51. Acta Neurochir. Suppl. 70, 212–215. https://doi.org/10.1007/978-3-7091-6837-0_65.

Tian, J., Tai, Y., Shi, M., Zhao, C., Xu, W., Ge, X., Zhu, G., 2020. Atorvastatin relieves cognitive disorder after sepsis through reverting inflammatory cytokines, oxidative stress, and neuronal apoptosis in hippocampus. Cell. Mol. Neurobiol. 40 (4), 521–530. https://doi.org/10.1007/s10571-019-00750-z. (Epub 2019 Nov 6).

Tran, K.A., Zhang, X., Predescu, D., Huang, X., Machado, R.F., Göthert, J.R., Malik, A.B., Valyi-Nagy, T., Zhao, Y.Y., 2016. Endothelial beta-catenin signaling is required for maintaining adult blood-brain barrier integrity and central nervous system homeostasis. Circulation 133 (2), 177–186. https://doi.org/10.1161/CIRCULATIONAHA.115.015982. (Epub 2015 Nov 4).

Trinity, J.D., Broxterman, R.M., Richardson, R.S., 2016. Regulation of exercise blood flow: role of free radicals. Free Radic. Biol. Med. 98, 90–102. https://doi.org/10.1016/j.freeradbiomed.2016.01.017. (Epub 2016 Feb 10).

Truesdell, A.G., Wilson, R.L., 2006. Training for medical support of mountain operations. Mil. Med. 171 (6), 463–467. https://doi.org/10.7205/milmed.171.6.463.

United Nations Office on Drugs and Crime, 2007. World Drug Report Volume 1. Analysis. United Nations Office on Drugs and Crime, Vienna.

Van Osta, A., Moraine, J.J., Mélot, C., Mairbäurl, H., Maggiorini, M., Naeije, R., 2005. Effects of high altitude exposure on cerebral hemodynamics in normal subjects. Stroke 36 (3), 557–560. https://doi.org/10.1161/01.STR.0000155735.85888.13. (Epub 2005 Feb 3).

Vandenbark, A.A., Meza-Romero, R., Benedek, G., Offner, H., 2019. A novel neurotherapeutic for multiple sclerosis, ischemic injury, methamphetamine addiction, and traumatic brain injury. J. Neuroinflammation 16 (1), 14. https://doi.org/10.1186/s12974-018-1393-0.

Vearrier, D., Greenberg, M.I., Miller, S.N., Okaneku, J.T., Haggerty, D.A., 2012. Methamphetamine: history, pathophysiology, adverse health effects, current trends, and hazards associated with the clandestine manufacture of methamphetamine. Dis. Mon. 58 (2), 38–89. https://doi.org/10.1016/j.disamonth.2011.09.004.

Vella, M.A., Crandall, M.L., Patel, M.B., 2017. Acute management of traumatic brain injury. Surg. Clin. North Am. 97 (5), 1015–1030. https://doi.org/10.1016/j.suc.2017.06.003.

Vij, A.G., Dutta, R., Satija, N.K., 2005. Acclimatization to oxidative stress at high altitude. High Alt. Med. Biol. 6 (4), 301–310. https://doi.org/10.1089/ham.2005.6.301.

Villafuerte, F.C., Corante, N., 2016. Chronic mountain sickness: clinical aspects, etiology, management, and treatment. High Alt. Med. Biol. 17 (2), 61–69. https://doi.org/10.1089/ham.2016.0031. (Epub 2016 May 24).

Virués-Ortega, J., Garrido, E., Javierre, C., Kloezeman, K.C., 2006. Human behaviour and development under high-altitude conditions. Dev. Sci. 9 (4), 400–410. https://doi.org/10.1111/j.1467-7687.2006.00505.x.

Volkow, N.D., Fowler, J.S., Wang, G.J., Shumay, E., Telang, F., Thanos, P.K., Alexoff, D., 2010. Distribution and pharmacokinetics of methamphetamine in the human body: clinical implications. PLoS One 5 (12), e15269. https://doi.org/10.1371/journal.pone.0015269.

Wang, J., Ke, T., Zhang, X., Chen, Y., Liu, M., Chen, J., Luo, W., 2013. Effects of acetazolamide on cognitive performance during high-altitude exposure. Neurotoxicol. Teratol. 35, 28–33. https://doi.org/10.1016/j.ntt.2012.12.003. (Epub 2012 Dec 30).

Wang, K., Zhang, M., Li, Y., Pu, W., Ma, Y., Wang, Y., Liu, X., Kang, L., Wang, X., Wang, J., Qiao, B., Jin, L., 2018. Physiological, hematological and biochemical factors associated with high-altitude headache in young Chinese males following acute exposure at 3700 m. J. Headache Pain 19 (1), 59. https://doi.org/10.1186/s10194-018-0878-7.

Wang, H., Zhu, X., Xiang, H., Liao, Z., Gao, M., Luo, Y., Wu, P., Zhang, Y., Ren, M., Zhao, H., Xu, M., 2019. Effects of altitude changes on mild-to-moderate closed-head injury in rats following acute high-altitude exposure. Exp. Ther. Med. 17 (1), 847–856. https://doi.org/10.3892/etm.2018.7020. (Epub 2018 Nov 27).

Wang, X.M., Liu, H., Li, J.Y., Wei, J.X., Li, X., Zhang, Y.L., Li, L.Z., Zhang, X.Z., 2020. Rosamultin attenuates acute hypobaric hypoxia-induced bone injuries by regulation of sclerostin and its downstream signals. High Alt. Med. Biol. 21 (3), 273–286. https://doi.org/10.1089/ham.2019.0113. (Epub 2020 Jun 25).

Warren, M.W., Kobeissy, F.H., Liu, M.C., Hayes, R.L., Gold, M.S., Wang, K.K., 2006. Ecstasy toxicity: a comparison to methamphetamine and traumatic brain injury. J. Addict. Dis. 25 (4), 115–123. https://doi.org/10.1300/J069v25n04_11.

Wei, L., Zhang, J., Zhang, B., Geng, J., Tan, Q., Wang, L., Chen, Z., Feng, H., Zhu, G., 2020. Complement C3 participates in the function and mechanism of traumatic brain injury at simulated high altitude. Brain Res. 1726, 146423. https://doi.org/10.1016/j.brainres.2019.146423. (Epub 2019 Oct 22).

Weizman, A., Weizman, R., 2000. Serotonin transporter polymorphism and response to SSRIs in major depression and relevance to anxiety disorders and substance abuse. Pharmacogenomics 1 (3), 335–341. https://doi.org/10.1517/14622416.1.3.335.

West, J.B., 2016a. Barcroft's bold assertion: all dwellers at high altitudes are persons of impaired physical and mental powers. J. Physiol. 594 (5), 1127–1134. https://doi.org/10.1113/JP270284. (Epub 2015 Jun 26).

West, J.B., 2016b. Oxygen conditioning: a new technique for improving living and working at high altitude. Physiology (Bethesda) 31 (3), 216–222. https://doi.org/10.1152/physiol.00057.2015.

Westerlund, C., Ostlund-Lindqvist, A.M., Sainsbury, M., Shertzer, H.G., Sjöquist, P.O., 1996. Characterization of novel indenoindoles. Part I. Structure-activity relationships in different model systems of lipid peroxidation. Biochem. Pharmacol. 51 (10), 1397–1402. https://doi.org/10.1016/0006-2952(96)00080-9.

Wettervik, T.S., Engquist, H., Howells, T., Lenell, S., Rostami, E., Hillered, L., Enblad, P., Lewén, A., 2020. Arterial oxygenation in traumatic brain injury-relation to cerebral energy metabolism, autoregulation, and clinical outcome. Intensive Care Med. 27. https://doi.org/10.1177/0885066620944097. Online ahead of print, 885066620944097.

Whayne Jr., T.F., 2014. Altitude and cold weather: are they vascular risks? Curr. Opin. Cardiol. 29 (4), 396–402. https://doi.org/10.1097/HCO.0000000000000064.

Wiegmann, D.A., Stanny, R.R., McKay, D.L., Neri, D.F., McCardie, A.H., 1996. Methamphetamine effects on cognitive processing during extended wakefulness. Int. J. Aviat. Psychol. 6 (4), 379–397. https://doi.org/10.1207/s15327108ijap0604_5.

Wilson, M.H., Newman, S., Imray, C.H., 2009. The cerebral effects of ascent to high altitudes. Lancet Neurol. 8 (2), 175–191. https://doi.org/10.1016/S1474-4422(09)70014-6.

Wilson, M.H., Wright, A., Imray, C.H., 2014. Intracranial pressure at altitude. High Alt. Med. Biol. 15 (2), 123–132. https://doi.org/10.1089/ham.2013.1151. PMID: 24971766.

Winter, C.D., Whyte, T., Cardinal, J., Kenny, R., Ballard, E., 2016. Re-exposure to the hypobaric hypoxic brain injury of high altitude: plasma S100B levels and the possible effect of acclimatisation on blood-brain barrier dysfunction. Neurol. Sci. 37 (4), 533–539. https://doi.org/10.1007/s10072-016-2521-1. (Epub 2016 Feb 29).

Wozniak, A., Drewa, G., Chesy, G., Rakowski, A., Rozwodowska, M., Olszewska, D., 2001. Effect of altitude training on the peroxidation and antioxidant enzymes in sportsmen. Med. Sci. Sports Exerc. 33 (7), 1109–1113. https://doi.org/10.1097/00005768-200107000-00007.

Wright, A.D., Imray, C.H., Morrissey, M.S., Marchbanks, R.J., Bradwell, A.R., 1995. Intracranial pressure at high altitude and acute mountain sickness. Clin. Sci. (Lond.) 89 (2), 201–204. https://doi.org/10.1042/cs0890201.

Wu, T., Ding, S., Liu, J., Jia, J., Dai, R., Liang, B., Zhao, J., Qi, D., 2006. Ataxia: an early indicator in high altitude cerebral edema. High Alt. Med. Biol. 7 (4), 275–280. https://doi.org/10.1089/ham.2006.7.275.

Wustmann, C., Fischer, H.D., Schmidt, J., 1982. Inhibitory effects of hypoxia on dopamine release. Acta Biol. Med. Ger. 41 (6), 571–574.

Xie, Y., Qin, S., Zhang, R., Wu, H., Sun, G., Liu, L., Hou, X., 2020. The effects of high-altitude environment on brain function in a seizure model of young-aged rats. Front. Pediatr. 8, 561. https://doi.org/10.3389/fped.2020.00561 (eCollection 2020).

Yamamoto, S., DeWitt, D.S., Prough, D.S., 2018. Impact & blast traumatic brain injury: implications for therapy. Molecules 23 (2), 245. https://doi.org/10.3390/molecules23020245.

Yanamandra, U., Gupta, A., Patyal, S., Varma, P.P., 2014. High-altitude cerebral oedema mimicking stroke. BMJ Case Rep. 2014. https://doi.org/10.1136/bcr-2013-201897, bcr2013201897.

Yang, J., Li, W., Liu, S., Yuan, D., Guo, Y., Jia, C., Song, T., Huang, C., 2016. Identification of novel serum peptide biomarkers for high-altitude adaptation: a comparative approach. Sci. Rep. 6, 25489. https://doi.org/10.1038/srep25489.

Yasaei, R., Saadabadi, A., 2020. Methamphetamine. In: StatPearls [Internet]. StatPearls Publishing, Treasure Island (FL) (2021 Jan).

Zafren, K., Pun, M., Regmi, N., Bashyal, G., Acharya, B., Gautam, S., Jamarkattel, S., Lamichhane, S.R., Acharya, S., Basnyat, B., 2017. High altitude illness in pilgrims after rapid ascent to 4380 M. Travel Med. Infect. Dis. 16, 31–34. https://doi.org/10.1016/j.tmaid.2017.03.002. (Epub 2017 Mar 9).

Zanella, S., Doi, A., Garcia 3rd, A.J., Elsen, F., Kirsch, S., Wei, A.D., Ramirez, J.M., 2014. When norepinephrine becomes a driver of breathing irregularities: how intermittent hypoxia fundamentally alters the modulatory response of the respiratory network. J. Neurosci. 34 (1), 36–50. https://doi.org/10.1523/JNEUROSCI.3644-12.2014.

Zhang, L.L., Liu, H.Q., Yu, X.H., Zhang, Y., Tian, J.S., Song, X.R., Han, B., Liu, A.J., 2016. The combination of scopolamine and psychostimulants for the prevention of severe motion sickness. CNS Neurosci. Ther. 22 (8), 715–722. https://doi.org/10.1111/cns.12566. (Epub 2016 May 9).

Zhang, C.J., Jiang, M., Zhou, H., Liu, W., Wang, C., Kang, Z., Han, B., Zhang, Q., Chen, X., Xiao, J., Fisher, A., Kaiser, W.J., Murayama, M.A., Iwakura, Y., Gao, J., Carman, J., Dongre, A., Dubyak, G., Abbott, D.W., Shi, F.D., Ransohoff, R.M., Li, X., 2018. TLR-stimulated IRAKM activates caspase-8 inflammasome in microglia and promotes neuroinflammation. J. Clin. Invest. 128 (12), 5399–5412. https://doi.org/10.1172/JCI121901. (Epub 2018 Oct 29).

Zhang, Y., Liu, L., Liang, C., Zhou, L., Tan, L., Zong, Y., Wu, L., Liu, T., 2020. Expression profiles of long noncoding RNAs in mice with high-altitude hypoxia-induced brain injury treated with Gymnadenia conopsea (L.) R. Br. Neuropsychiatr. Dis. Treat. 16, 1239–1248. https://doi.org/10.2147/NDT.S246504 (eCollection 2020).

Zhong, Z., Zhou, S., Xiang, B., Wu, Y., Xie, J., Li, P., 2021. Association of peripheral plasma neurotransmitters with cognitive performance in chronic high-altitude exposure. Neuroscience 463, 97–107. https://doi.org/10.1016/j.neuroscience.2021.01.031. S0306-4522(21)00047-6 Online ahead of print.

Zhou, Y., Huang, X., Zhao, T., Qiao, M., Zhao, X., Zhao, M., Xu, L., Zhao, Y., Wu, L., Wu, K., Chen, R., Fan, M., Zhu, L., 2017. Hypoxia augments LPS-induced inflammation and triggers high altitude cerebral edema in mice. Brain Behav. Immun. 64, 266–275. https://doi.org/10.1016/j.bbi.2017.04.013. (Epub 2017 Apr 20).

Zhou, X., Nian, Y., Qiao, Y., Yang, M., Xin, Y., Li, X., 2018. Hypoxia plays a key role in the pharmacokinetic changes of drugs at high altitude. Curr. Drug Metab. 19 (11), 960–969. https://doi.org/10.2174/1389200219666180529112913.

CHAPTER

5

Effectiveness of bortezomib and temozolomide for eradication of recurrent human glioblastoma cells, resistant to radiation

Oleg Pak[a], Sergei Zaitsev[a], Valery Shevchenko[b], Aruna Sharma[c], Hari Shanker Sharma[c,*], and Igor Bryukhovetskiy[a,*]

[a]*School of Biomedicine, Medical Center, Far Eastern Federal University, Vladivostok, Russia*

[b]*Institute of Carcinogenesis, N.N. Blokhin National Medical Research Center of Oncology, Moscow, Russia*

[c]*International Experimental Central Nervous System Injury & Repair (IECNSIR), Department of Surgical Sciences, Anesthesiology & Intensive Care Medicine, Uppsala University Hospital, Uppsala University, Uppsala, Sweden*

**Corresponding authors: Tel.: +46-70-20111801 (Hari Shanker Sharma); Tel.: +7-914-723-05-03 (Igor Bryukhovetskiy), e-mail address: sharma@surgsci.uu.se; igbryukhovetskiy@gmail.com*

Abstract

Background: Glioblastoma multiforme (GBM) is a primary human brain tumor with the highest mortality rate. The prognosis for such patients is unfavorable, since the tumor is highly resistant to treatment, and the median survival of patients is 13 months. Chemotherapy might extend patients' life, but a tumor, that reappears after chemoradiotherapy, is resistant to temozolomide (TMZ). Using postgenome technologies in clinical practice might have a positive effect on the treatment of a recurrent GBM.

Methods: T98G cells of human GBM have been used. Radiation treatment was performed with Rokus-M gamma-therapeutic system, using ^{60}Co as a source of radionuclide emissions. High-performance liquid chromatography-mass spectrometry was used for proteome analysis. Mass spectrometry data were processed with MaxQuant (version 1.6.1.0) and Perseus (version 1.6.1) software, Max Planck Institute of Biochemistry (Germany). Biological processes, molecular functions, cells locations and protein pathways were annotated with a help of PubMed, PANTHER, Gene Ontology and KEGG and STRING v10 databases. Pharmaceutical testing was performed in vitro with a panel of traditional chemotherapeutic agents.

Results: GBM cells proliferation speed is inversely proportional to the irradiation dose and recedes when the dosage is increased, as expected. Synthesis of ERC1, NARG1L, PLCD3,

Progress in Brain Research, Volume 266, ISSN 0079-6123, https://doi.org/10.1016/bs.pbr.2021.06.010
Copyright © 2021 Elsevier B.V. All rights reserved.

ROCK2, SARNP, TMSB4X and YTHDF2 in GBM cells, treated with 60 Gy of radiation, shows more than a fourfold increase, while the synthesis level of PSMA2, PSMA3, PSMA4, PSMB2, PSMB3, PSMB7, PSMC3, PSMD1, PSMD3 proteins increases significantly. Traditional chemotherapeutic agents are not very effective against cancer cells of the recurrent GBM. Combination of TMZ and CCNU with a proteasome inhibitor—bortezomib—significantly increases their ability to eradicate cells of a radioresistant GBM.

Conclusions: Bortezomib and temozolomide effectively destroy cells of a radioresistant recurrent human glioblastoma; proteome mapping of the recurrent GBM cancer cells allows to identify new targets for therapy to improve the treatment results.

Keywords

Glioblastoma, Radiation therapy, Temozolomide, Chemoresistance, Proteome mapping, Targeted therapy

1 Introduction

Glioblastoma multiforme (GBM) is one of the most aggressive human brain tumors. In Europe and in the United States its occurrence is 3–5 cases per 100,000 people, it can appear at any age, but the majority of patients are 50–60 years old (Nam and de Groot, 2017). Treatment for all GBM patients involves surgery, irradiation and chemotherapy (Sulman et al., 2017). Their life expectancy correlates with the received dosage of irradiation, reaching 60 Gy. Subsequent chemotherapy with TMZ might extend the relapse-free period. Basic drug (Mathen et al., 2020; Thomas et al., 2017) of the modern protocol for complex treatment is temozolomide (TMZ). Lomustine, carboplatin (Roberts et al., 2016) and some other cytotoxic agents are used as second- and third-line chemotherapeutic agents. The prognosis is unfavorable, and, despite all efforts of medical expects, the median survival of patients, who have GBM with IDH-mutation, is 23 months, 49% of people manage to live up to 2 years. The median survival of patients with IDH-wild-type GBM is 13 months (Stupp et al., 2017), and only 12% of them manage to live for 2 years after being diagnosed.

Low treatment effectiveness is usually associated with (Weller et al., 2019) selective penetrability of the blood–brain barrier for anti-tumor drugs, heterogeneous nature of GBM cell population and high plasticity of cancer cells (CCs), allowing them to immediately trigger DNA repair mechanisms, survive and actively proliferate, causing tumor relapse and death of a patient. Combination of chemotherapeutic agents with various targeted medication (Touat et al., 2017) is one of the most promising methods of prolonging GBM patients' life expectancy, but no significant success has been achieved in this area yet, thus, requiring development of more personally tailored pharmacotherapeutic strategies. The present study uses postgenome technologies of multidimensional biology to analyze metabolic (proteome) maps of GBM cells that received high irradiation dosage, as well as to discover and interact with new molecular targets in the cells of a recurrent GBM together with using traditional cytotoxic agents in the experimental in vitro models.

2 Materials and methods

2.1 Human glioblastoma cells

T98-G GBM cells were obtained from the American Type Culture Collection (ATCC® CRL-1690™). This cell line represents IDH-wild-type glioblastoma and is poorly responsive to TMZ (Paul-Samojedny et al., 2016) that is mainly associated with a low level of MGMT methylation and high content of O6-methylguanin-DNA methyltransferase, allowing to recreate the tumor condition right after its complex treatment with radiation and chemotherapy. T98-G GBM cells were tested for mycoplasma contamination with the Universal Mycoplasma Detection Kit (ATCC® 30-1012K™). All cells were cultured in 6-well plates with DMEM—Dulbecco's Modified Eagle Medium (DMEM) with 10% fetal bovine serum (FBS) with penicillin/streptomycin (100 U/mL, Antibiotic-Antimitotic 100× (cat. no. 15240062, Gibco, US) at 37 °C (5% CO_2). All chemicals were obtained from Gibco (Thermo, Inc.) Adhesive cells were cultured until 80% confluent and passaged at a 1:3 ratio. Cells were used in experiments after the third passage since the moment of being obtained from the manufacturer.

2.2 Recurrent GBM model

CCs were exposed to ionizing radiation. Rokus-M gamma-therapeutic system (Russia, St. Petersburg) was used for the procedure with ^{60}Co as a source of radionuclide emissions. Single fractions of 2, 4 or 6 Gy were used, and 10 fractions with 72-h interval were administered to reach the total dose of 20, 40 and 60 Gy. Culture viability was assessed with 0.4% trypan blue staining (Merk, Germany, cat. no. T8154) in hemocytometer. Cells were transplanted in an hour after their radiation treatment.

2.3 Proteomics-based comparative mapping of GBM cells

A combination of high-performance liquid chromatography and mass spectrometry (HPLC-MS) was used. The label-free method was used to evaluate the changes in proteins expression level. Two cell samples—irradiated cells of GBM and non-irradiated cells of GBM—were lysed with Mammalian Cell Lysis Kit (Merck, Germany), low-molecular compounds were removed, then enzymatic cleavage was performed, and each 4 μL of solution was analyzed, using mass spectrometry for controlled trypsinolysis. The samples were incubated at 30 °C in a Labconco CentriVap centrifugal concentrator to remove ammonium bicarbonate. Tryptic peptides were diluted during the mobile stage (30% acetonitrile, 70% water and 0.1% formic acid, pH 2.7) and divided into 24 fractions, using a Dionex UltiMate 3000 chromatograph (Dionex, Sunnyvale, CA, US), equipped with a fraction collector and cation exchange column MIC-10-CP (Puros 10S, 1 mm × 10 cm, Thermo, US). The obtained fractions were concentrated at 30 °C in the centrifugal concentrator and diluted again with 100 μL of 0.1% formic acid. Tryptic peptides were analyzed with Dionex Ultimate

3000 (Dionex, Netherlands) and LTQ Orbitrap XL mass-spectrometer (Thermo) with NSI ionization. Peptide division was performed with Acclaim C18 PepMap100 column (75 μm × 150 mm, grit size—3 μm, Dionex), equipped with a precolumn. HPLC-MS data were processed with MaxQuant (version 1.6.1.0) and Perseus software (version 1.6.1), Max Planck Institute of Biochemistry (Germany). Biological processes, molecular functions, cell location and protein signaling pathways were annotated, using such databases as PubMed (http://www.ncbi.nlm.nih.gov/pubmed), PANTHER (http://www.pantherdb.org), Gene Ontology (http://www.geneontology.org) and KEGG (http://www.genome.jp/kegg/) and STRING v10 (https://string-db.org/).

2.4 Used chemicals

Acetonitrile (ACN) HPLC gradient grade was purchased from Prolabo (VWR BDH Prolabo® USA). Formic acid (F0507), urea for molecular biology (U5378), ammonium bicarbonate (BioUltra, 09830), dithiothreitol for molecular biology (BioUltra, 43,815), iodoacetamide (IAA, I1149), 2.2.2-trifluoroethanol (T63002, Reagent Plus), acetic acid (HAc, A6283 Reagent Plus), protease inhibitor cocktail (P8340), mammalian cell lysis kit (MCL1), phosphate buffered saline (PBS, P5493), trypsin (modified, proteomics grade, T7575) ammonium bicarbonate NH_4HCO_3, (BioUltra, 09830), trichloroethylphosphate (TCEP, 07296), trifluoroacetic acids for protein sequence analysis (TFA, 299537), hydrochloric acid for molecular biology (BioReagent, H1758) were obtained from Merck (Germany). Ultrapure water was prepared by Milli-Q water purification system Milli-Q® CLX 7000 (Millipore, Bedford, MA, US).

2.5 Pharmaceutical agents

Temozolomide (T2577), CCNU (L0745000), carboplantin, (C2538) and bortezomib (504314), manufactured by Merck (Germany), were used in the study. Bortezomib was included into the list of tested agents based on the data of the bioinformation analysis of proteome profiles.

2.6 Method of in vitro testing

To assess the cytotoxic effect of the tested compounds the cells were seeded into 6-well plates with 6×10^4 cells/well. Then 500 μmol/L of the tested cytotoxic agents and 10 nmol/L of bortezomib (Tang et al., 2019) were added and incubated in standard conditions (T 37 °C, 5%, CO_2). Cytotoxic effect of the tested substances on cancer cells was assessed with flow cytometry.

2.7 Flow cytometry

Aliquots of cell suspension (100 μL) were incubated with a 20-fold dilution of TMRM (T668, Invitrogen, US) with the final concentration of 150 nmol, for 20 min at 37°C in 5% CO_2 and in a light-proof place. After incubation the samples were washed with excess PBS, containing 2% FBS (Merk, Germany). Then 5 μL of DRAQ7 solution (ab109202, Abcam, US) were introduced into the obtained cell suspension, and the samples were incubated. After the incubation, 100 μL of PBS was added to the samples, and each of them was analyzed on at least 15,000 events with BD Accuri C6 flow cytometer (BD Biosciences). To assess the mitochondrial membrane potential, the 20-fold dilution of DiOC6(3), prepared according to the manufacturer's instructions (D273, Invitrogen, US), was added to 100 μL of cell suspension (3×10^6 cells/mL). The samples with dye were incubated for 15–20 min at room temperature in a dark room. After incubation, 200 μL of PBS were added to the samples, and they were analyzed with the Navios flow cytometer (Beckman Coulter). At least 10,000 events have been analyzed for each sample. To distinguish between single cells and aggregates and eliminate them from the analysis, the combination of peak and integral DAPI fluorescence were used. Cytometric analysis data were processed with Kaluza Software, 10 UserNetworkPack (FullVersion) and ModFit TL (Verity Software House).

3 Results

3.1 Characteristics of T98G line of GBM cells after radiation treatment

The analysis of variance showed the significant differences in proliferation speed of cancer cells, treated with various radiation doses (Fig. 1A). The subsequent Student's *t*-test (a posteriori) showed that proliferation speed of cancer cells that received different radiation doses, was significantly diverged ($\alpha = 0.05$) from the control group, and the lowest speed of proliferation was detected for the cancer cells that had been treated with a total radiation dose of 60 Gy (Fig. 1B). It was that culture of stem cells that was selected for proteomic analysis.

3.2 Proteomics-based comparative mapping of T98G line cell of GBM after radiation treatment

Proteomic analysis revealed 978 proteins. The control sample of cancer cells (control cells, CC) contained 957 proteins based on 4664 peptides. The sample of cancer cells, treated with radiation of 60 Gy (irradiated cells, IrC), had 945 proteins based on 4652 peptides. Identified proteins exhibited a very high percentage of overlap for the two cell populations: 924 proteins (94% of 978 proteins) were discovered in all cell lysates, and 68 of those proteins had significant changes ($P < 0.05$) in expression of IrC cell line, if compared to CCs: the expression of 23 proteins was higher, while the expression of 45 proteins was lower. Bioinformation analysis revealed seven proteins with a more than fourfold increase in their synthesis, if compared to CCs (Table 1). Comprehensive bioinformation analysis revealed that nine differentially expressed proteins, identified in IrCs, pertain to a signaling cascade, associated with

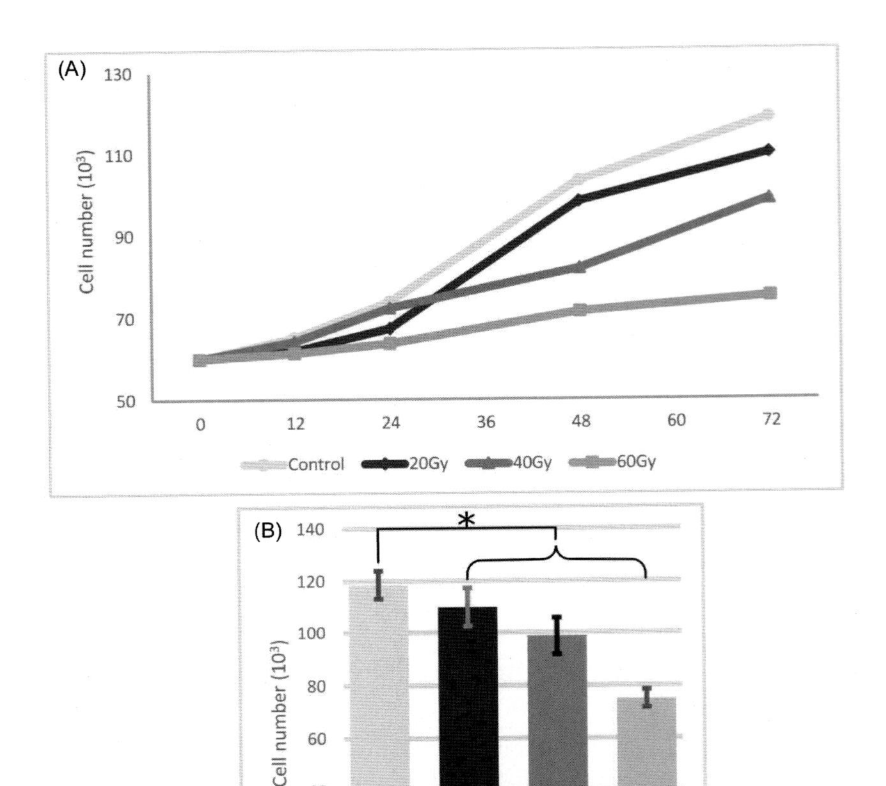

FIG. 1

Proliferation speed of T98G line of GBM cells that received a different dosage of irradiation. (A) Growth dynamics of cancer cells number in the culture medium. (B) Cancer cells number in the culture after 72 h from the beginning of the experiment. Analysis of variance with subsequent Student's *t*-test (a posteriori). Significance level is $\alpha = 0.05$.

proteasomes (Table 2). Based on the obtained data from bioinformation analysis, the further in vitro drug testing included bortezomib—a highly selective reversible inhibitor of 26S proteasome.

3.3 Testing the effect of cytotoxic and targeted agents on an in vitro model of recurrent GBM

All tested agents were able to destroy cancer cells of GBM, not treated with radiation. Bortezomib had the weakest cytotoxic effect, while other agents had a more visible results and eliminated more than a half of the cancer cells population during the 72-h

Table 1 Differentially expressed proteins in GBM cells that received 60 Gy of irradiation with a more than fourfold increase in the level of expression, when compared to the control cells.

Gene index	Protein name	CC/IrC
ERC1	ELKS/RAB6-interacting/CAST family member 1	9.5
NARG1L	NMDA receptor regulated 1-like; N-alpha-acetyltransferase 15	6.3
PLCD3	Phospholipase C, delta 3	5.7
ROCK2	Rho-associated, coiled-coil containing protein kinase 2	4.5
SARNP	SAP domain containing ribonucleoprotein	4.2
TMSB4X	Thymosin-like 2 (pseudogene); thymosin-like 1 (pseudogene); thymosin beta 4, X-linked	4.1
YTHDF2	YTH domain family, member 2; YTH domain family, member 3	6.8

Table 2 Determining factors of proteasomes with a significant ($P < 0.05$) difference in expression levels between the CC (control cells) and the IrC (irradiated cells) of T98G cell line of GBM.

Gene index	Protein name	CC\IrC
PSMA2	Proteasome (prosome, macropain) subunit, alpha type, 2	3.2
PSMA3	Proteasome (prosome, macropain) subunit, alpha type, 3	2.3
PSMA4	Proteasome (prosome, macropain) subunit, alpha type, 4	2.2
PSMB2	Proteasome (prosome, macropain) subunit, beta type, 2	2.3
PSMB3	Proteasome (prosome, macropain) subunit, beta type, 3	2.9
PSMB7	Proteasome (prosome, macropain) subunit, beta type, 7	3.5
PSMC3	Proteasome (prosome, macropain) 26S subunit, ATPase, 3	3.3
PSMD1	Proteasome (prosome, macropain) 26S subunit, non-ATPase, 1	2.4
PSMD3	Proteasome (prosome, macropain) 26S subunit, non-ATPase, 3	3.9

observation period (Fig. 2A). TMZ and carboplatin exhibited the best cytotoxic effect, and there were no significant differences between the numbers of surviving cancer cells in the culture media with these substances.

When the tested agents were added to the GBM cells, that had been treated with radiation of 60 Gy, the situation changed dramatically (Fig. 2B). Cytotoxic effect of bortezomib increased, while temozolomide and CCNB were less efficient. Carboplatin exhibited the best ability to destroy cancer cells of radiation-treated GBM. It is worth mentioning that, when combined with bortezomib, the effect of the tested agents was significantly higher, that was mostly evident in the combination of bortezomib and TMZ (Fig. 2C). In turn, TMZ and bortezomib combination (Fig. 2D) with other tested substances was accompanied by the decrease in the number of surviving cancer cells in the culture medium.

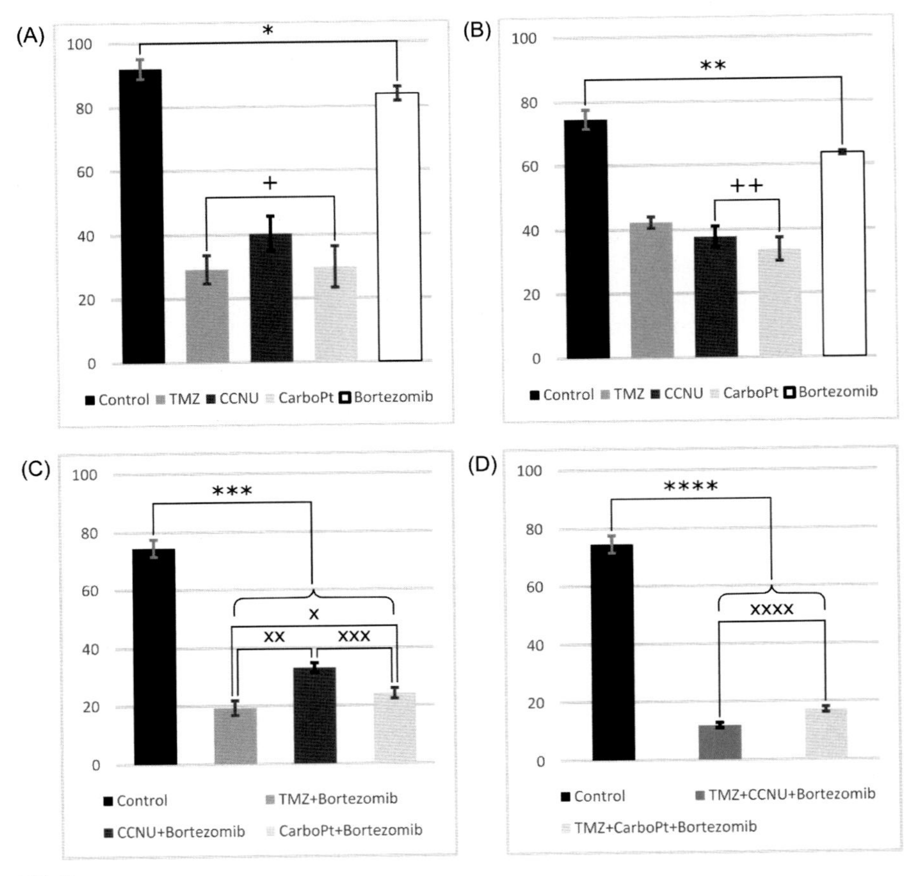

FIG. 2

The result of pharmaceutical in vitro testing of cytotoxic effect of the compared substances, 72 h. Analysis of variance with subsequent Student's *t*-test (a posteriori), significance level is $\alpha = 0.05$. (A) When the nonirradiated GBM cells came in contact with the tested substances, proliferation was inhibited in the cultures with TMZ and carboplatin, and cytotoxic effect of these substances was similar (+). Bortezomib was less effective in its interaction with the nonirradiated cancer cells, but this culture had significantly smaller (*) amount of viable GBM cells, if compared to the control group. (B) The culture with the irradiated cancer cells had significantly smaller amount of viable cancer cells, if compared to the control group (**). CCNU and CarboPt were leaders in inhibiting the irradiated cells and had similar effect (++), and their results were different from other substances and the control group. (C) Combination of TMZ + Bortezomib had the best effect on the irradiated cells. (D) Comparison of the tested substances and their effect on the irradiated cells showed that the combination of TMZ + CCNU + Bortezomib was the most successful. *** and **** mark the combination of drugs that significantly decreased the amount of viable cancer cells in the culture, if compared to the control group, while the cytotoxic effect of the compared substances was different (x, xx, xxx, xxxx).

4 Discussion

Radiation treatment is an effective adjuvant therapy of GBM, and some large-scale randomized studies under the auspices of the American Society of Clinical Oncology (Sulman et al., 2017) and European Association for Neuro-Oncology (Weller et al., 2017) showed better survival rates of GBM patients, treated with a combination of radiation and chemotherapy, if compared to only a surgery or only chemotherapy. Usually a patient is prescribed γ-radiation of 60 Gy: 2 Gy daily and overall 30 fractions for 6 weeks; exceptions are made for senior patients who should have only 40 Gy (Braun and Ahluwalia, 2017), but if their Karnofsky score is acceptable (>70), a standard treatment method is recommended. The results of our experiment definitely justify such aggressive treatment method. GBM cells proliferation speed is inversely proportional to the irradiation dose and recedes when the dosage is increased, while the minimal speed of proliferation was observed in cancer cells treated with radiation of 60 Gy. That is why maximally possible doses of radiation shall be recommended for all patients, but extreme dosage is not a solution as well, since it results in radiation necrosis that could cause severe neurological damage or a patient's death.

In their turn, cancer cells undergo a transformation, while adapting to systematic exposure to radiation. The proteome-based analysis shows that GBM cells, treated with 60 Gy radiation, have a higher content of some vital proteins, especially ERC1, NARG1L, PLCD3, ROCK2, SARNP, TMSB4X and YTHDF2 that had more than a fourfold increase in synthesis.

ERC1 or ELKS/RAB6—CAST interaction protein—is encoded by the *ERC1* gene and shows a 9.5-fold increase in synthesis due to radiation. The protein is a part of the RIM-binding protein group that regulates the presynaptic Ca^{+2} channel function (Kaeser et al., 2011); in mammalian cells it directs RAB6A to melanosome, takes part in transporting products of intracellular synthesis from the Golgi apparatus to the plasma membrane via the network of microtubes. According to the STRING protein–protein interaction database, the ERC1 protein interacts with the transcription factor NF-κβ that controls expression of immune response genes, apoptosis, cell cycle and DNA repair. This protein is an element of ubiquitin-dependent intracellular proteolysis and is closely related with stress response and DNA repair. There are some data (Astro et al., 2016) on this protein being involved in regulation of focal adhesion and micromechanical properties of cancer cells, activation of TGF-β and MAPK signaling. Critically high content of ERC1 in radiation-treated cell is likely to indicate activation of these processes.

NMDA (*N-methyl-D-aspartate* receptor)—encoded by NARG1L—shows 6.3-fold synthesis increase after radiation treatment. This protein is known to be a gastric cancer antigen. According to the STRING protein–protein interaction database, NARG1L is involved into vascularisation, hematopoiesis, neuron growth and development. There are some data (Sugiura et al., 2001) on this protein being involved with transmission of signals between neurons and differentiation of normal neural stem cells into neurons.

PLC delta 3 is encoded by *PLCD3* gene. The concentration of this protein in GBM cells after experimental radiation treatment increased 5.7-fold. This gene belongs to the phospholipase C family that catalyze hydrolysis of phosphatidylinositol 4,5-bisphosphate with creation of second messengers—diacylglycerol and inositol 1,4,5-triphosphate that are responsible for different cell responses to extracellular stimuli, inducing protein kinase C and increasing the concentration of cytosolic Ca^{2+}. The protein is involved in proliferation, migration and invasion of nasopharyngeal carcinoma cells (Liu et al., 2018), and PLCD3 expression is associated with low survival rates of patients with pancreatic cancer (Zhou et al., 2020).

Synthesis of Rho-associated protein kinase 2—encoded by ROCK2—shows 4.5-fold increase in GBM cells. According to Gene Cards database, this protein is a key regulator of actin cytoskeleton and cell polarity, is directly involved in focal adhesion to extracellular matrix, modulation of circadian rhythms and Ca^{+2} signaling. According to the KEGG database, Rho-associated protein kinase 2 is a signaling pathway component of tight intercellular junctions, Wnt, thrombocytes activation, regulation of actin cytoskeleton, focal adhesion and some other intracellular mechanisms responsible for GBM pathogenesis.

SAP-domain-containing ribonucleoprotein is encoded by SARNP and after radiation treatment of GBM cell shows 4.2-fold increase in synthesis. According to the STRING protein–protein interaction database, SARNP protein connects single-stranded and double-stranded DNA, is a component of TREX (TRanscription-EXport) complex, takes part in transcription or translation control of cell growth, metabolism and carcinogenesis. It negatively regulates expression of adhesive E-cadherin (Kang et al., 2020) and stimulates invasive processes.

Thymosin β4—a protein encoded by *TMSB4X* gene—has a 4.1-fold increase in synthesis in radiation-treated GBM cells. This protein is important for cytoskeleton structure, it binds and isolates actin monomers and inhibits actin polymerization. Injection of thymosin β4 promotes cells migration, blood vessels formation, stem cells development, survival of different cell types and decrease of proinflammatory cytokines production (Philp and Kleinman, 2010), creating vital conditions for GBM survival. This protein has strong anti-inflammatory effect on neutrophils, modulates biological activity of corticosteroids in cancer cells, is involved in focal adhesion, migration and invasion of human melanoma cells (Makowiecka et al., 2019). Moreover, thymosin β4 and TGF-β are considered (Morita and Hayashi, 2018) to be the most important factors, triggering epithelial-mesenchymal transition in all types of cancer cells.

YTH N6-methyladenosin RNA-binding protein 2, encoded by *YTHDF2* gene, has a 6.8-fold increase in synthesis in radiation-treated GBM cells. This gene is a part of YTH family. YTH protein domain is conservative for all eukaryotes, and its conservative C-terminal area is critical for transmitting cytoplasmic Ca-signals to the nucleus and regulation of gene expression. But *YTHDF2* functions are much more extensive. It controls the mRNA stability of proinflammatory genes, *YTHDF2* knockdown (Yu et al., 2019), increases expression of TNF-α, IL-1β and other proinflammatory cytokines.

Therefore, the proteins, that have significantly increased expression after treatment of GBM cells with radiation, are directly connected with calcium signaling, increasing cancer cells mobility, intensifying interaction with extracellular matrix and suppressing inflammatory stimuli of the microsurroundings. These are cells with complex mechanisms of active immunosuppressive modification of microsurroundings, cells of migration and fight for survival.

Significant proteasome upregulation of nine proteins requires special attention. Ubiquitin-dependent degradation of protein in proteasomes is energy-consuming and needs ATP energy that is relatively scarce in cancer cells, damaged by radiation and oxidative stress. But GBM cells increase the synthesis of these proteins, meaning that protein degradation in proteasomes is crucial for cancer cells survival. Proteasomes eliminate intracellular cyclins and other proteins, determining cellular movement through the key points of the cell cycle, inhibiting proliferation and helping cancer cells to survive the exposure to radiation. Proteasomes destroy oxidized proteins; for instance, proteasomes of a cell nucleus are controlled by poly(ADP-ribose) polymerases (PARP) and destroy oxidized histones (Bader and Grune, 2006) that are necessary for effective DNA repair. In the light of the above mentioned, targeting proteasomes is a very promising way of increasing the effect of genotoxic therapy.

Damage to genome causes active synthesis of p53, nicknamed as "the Guardian of the Genome," that triggers apoptosis. Theoretically speaking, proteasome activation is crucial for p53 destruction and preservation of cancer cells. Proteasomes control and dispose of the NF-κB transcription factor (Adams et al., 2003), that activates synthesis of proinflammatory cytokines and immune response, as well as suppresses leukocytes proliferation during inflammation (Ben-Neriah, 2002). In the majority of cells proteasomes inhibition results in apoptosis, and in that respect both well-known drugs—*lactacystin*, bortezomib, ritonavir—and a relatively new one—marizomib (Manton et al., 2016)—have promising prospects of application in therapy. Hypothetically speaking, proteasome inhibition might significantly increase oxidative stress, hinder DNA repair, inducing genome instability of cancer cells that has been demonstrated in our research.

The choice of bortezomib for our research was determined by the existing practice of using it for GBM treatment (Rahman et al., 2019), as well as the results of the second stage of studies, involving combination of bortezomib with TMZ and regional radiation therapy for preliminary treatment of patients with a first-time diagnosed GBM (Kong et al., 2018). Apart from highly selective reversible inhibition of 26S proteasome, this drug suppresses PTEN/PI3K/AKT/mTOR signaling pathway in cancer cells (Wang et al., 2020) and mitochondrial pathway of apoptosis (Gras Navarro et al., 2019).

Bortezomib has the strongest cytotoxic effect on GBM cells, treated with radiation of 60 Gy (Fig. 2B) that is likely to be due to the high concentration of proteasome proteins in these cells (Table 2). It worth emphasizing that cytotoxic effect of TMZ on irradiated GBM cells is significantly lower than on non-irradiated culture. The combination of bortezomib with TMZ (Fig. 2C) or TMZ+CCNU+Bortezomib significantly decreased the amount of viable cancer cells in the culture. In reality this result might signify the extension of a relapse-free period and longer life expectance of patients with a recurrent GBM.

This study indicates the need for selecting the proper strategy of cytotoxic chemotherapy for a patient with GBM relapse. The overwhelming majority of GBM patients experience a relapse 3–6 months after the surgery, meaning that the patient's life is still at risk. Our study shows that the issue cannot be resolved with a simple and traditional prescription of several rounds of TMZ monotherapy or together with CCNU. Personalized approach is required, and using omix technologies is essential for this. However, there are certain limitations. Young patients with a good functional status have a chance of another surgery, making it possible to obtain materials for proteome mapping and bioinformation analysis. Despite a very busy neurosurgical schedule of the Medical Centre of the Far Eastern Federal University, only every fifth patient with a GBM relapse gets a reoperation. These statistics are typical for the majority of neurosurgical hospitals (Lu et al., 2019; Woodroffe et al., 2020).

Nevertheless, proteome mapping is an additional chance for GBM patients that increases the effect of cytotoxic chemotherapy and extends life expectancy. This issue requires further research, but it is already obvious that its solution lies with nanotechnologies.

Therefore, the study leads us to the following conclusions. Experimental model of a recurrent GBM has been developed. Speed of GBM cells proliferation is inversely proportional to the irradiation dose and recedes when the dosage is increased, while the minimal speed of proliferation was observed in cancer cells, treated with radiation of 60 Gy. Irradiation causes more than a fourfold increase in synthesis of ERC1, NARG1L, PLCD3, ROCK2, SARNP, TMSB4X and YTHDF2 in GBM cells, while the synthesis of PSMA2, PSMA3, PSMA4, PSMB2, PSMB3, PSMB7, PSMC3, PSMD1, PSMD3 significantly intensifies. Traditional cytotoxic agents are much less effective for recurrent GBM cells. Combination of TMZ and CCNU with bortezomib—a highly selective inhibitor of 26S—improves the ability of these agents to destroy cancer cells of the recurrent GBM that is resistant to radiation treatment.

Funding

The present study was funded by Russian Foundation of Basic Research, project 19-315-90077 *Availability of data and materials.*

The datasets used and/or analyzed during the current study are available from the corresponding author on reasonable request.

Ethics approval and consent to participate

Informed consent was obtained from all individual participants included in the study and all procedures performed in studies involving human participants were in accordance with the ethical standards of the Far Eastern Federal University and University of Uppsala.

Authors' contributions

V.S. prepared and analyzed the samples, as well as performed the cell lysis, chromatography and mass spectrometry, and contributed to the bioinformatics analysis. S.Z. provided and performed the statistical analysis and was responsible for the mathematical process of the results. A.S. and H.S. discussed, analyzed and interpreted the results of the study, and also worked on the manuscript. O.P. and I.B. wrote the manuscript, proposed the study idea, designed the study, offered support with the experiments, organized the scientific team, cultured the cancer cells and isolated the cancer cells of glioblastoma for the experiment, provided scientific guidance and contributed to the bioinformatics analysis. All authors read and approved the manuscript and agree to be accountable for all aspects of the research in ensuring that the accuracy or integrity of any part of the work are appropriately investigated and resolved.

Informed consent for publication was obtained from all individual participants included in the study.

Conflict of interest

The authors declare there are no conflicts of interests.

References

Adams, V., Späte, U., Kränkel, N., Schulze, P.C., Linke, A., Schuler, G., Hambrecht, R., 2003. Nuclear factor-kappa B activation in skeletal muscle of patients with chronic heart failure: correlation with the expression of inducible nitric oxide synthase. Eur. J. Cardiovasc. Prev. Rehabil. 10 (4), 273–277. https://doi.org/10.1097/00149831-200308000-00009.

Astro, V., Tonoli, D., Chiaretti, S., Badanai, S., Sala, K., Zerial, M., de Curtis, I., 2016. Liprin-α1 and ERC1 control cell edge dynamics by promoting focal adhesion turnover. Sci. Rep. 6, 33653. https://doi.org/10.1038/srep33653.

Bader, N., Grune, T., 2006. Protein oxidation and proteolysis. Biol. Chem. 387 (10 – 11), 1351–1355. https://doi.org/10.1515/BC.2006.169.

Ben-Neriah, Y., 2002. Regulatory functions of ubiquitination in the immune system. Nat. Immunol. 3 (1), 20–26. https://doi.org/10.1038/ni0102-20.

Braun, K., Ahluwalia, M.S., 2017. Treatment of glioblastoma in older adults. Curr. Oncol. Rep. 19 (12), 81. https://doi.org/10.1007/s11912-017-0644-z.

Gras Navarro, A., Espedal, H., Joseph, J.V., Trachsel-Moncho, L., Bahador, M., Gjertsen, B.T., Kristoffersen, E.K., Simonsen, A., Miletic, H., Enger, P.O., Rahman, M.A., Chekenya, M., 2019. Pretreatment of glioblastoma with bortezomib potentiates natural killer cell cytotoxicity through TRAIL/DR5 mediated apoptosis and prolongs animal survival. Cancers (Basel) 11 (7), 996. https://doi.org/10.3390/cancers11070996.

Kaeser, P.S., Deng, L., Wang, Y., Dulubova, I., Liu, X., Rizo, J., Südhof, T.C., 2011. RIM proteins tether Ca2+ channels to presynaptic active zones via a direct PDZ-domain interaction. Cell 144 (2), 282–295. https://doi.org/10.1016/j.cell.2010.12.029.

Kang, G.J., Park, M.K., Byun, H.J., Kim, H.J., Kim, E.J., Yu, L., Kim, B., Shim, J.G., Lee, H., Lee, C.H., 2020. SARNP, a participant in mRNA splicing and export, negatively regulates E-cadherin expression via interaction with pinin. J. Cell. Physiol. 235 (2), 1543–1555. https://doi.org/10.1002/jcp.29073.

Kong, X.T., Nguyen, N.T., Choi, Y.J., Zhang, G., Nguyen, H.N., Filka, E., Green, S., Yong, W.H., Liau, L.M., Green, R.M., Kaprealian, T., Pope, W.B., Nghiemphu, P.L., Cloughesy, T., Lassman, A., Lai, A., 2018. Phase 2 study of bortezomib combined with temozolomide and regional radiation therapy for upfront treatment of patients with newly diagnosed glioblastoma multiforme: safety and efficacy assessment. Int. J. Radiat. Oncol. Biol. Phys. 100 (5), 1195–1203. https://doi.org/10.1016/j.ijrobp.2018.01.001.

Liu, W., Liu, X., Wang, L., Zhu, B., Zhang, C., Jia, W., Zhu, H., Liu, X., Zhong, M., Xie, D., Liu, Y., Li, S., Shi, J., Lin, J., Xia, X., Jiang, X., Ren, C., 2018. PLCD3, a flotillin2-interacting protein, is involved in proliferation, migration and invasion of nasopharyngeal carcinoma cells. Oncol. Rep. 39 (1), 45–52. https://doi.org/10.3892/or.2017.6080.

Lu, V.M., Goyal, A., Graffeo, C.S., Perry, A., Burns, T.C., Parney, I.F., Quinones-Hinojosa, A., Chaichana, K.L., 2019. Survival benefit of maximal resection for glioblastoma reoperation in the temozolomide era: a meta-analysis. World Neurosurg. 127, 31–37. https://doi.org/10.1016/j.wneu.2019.03.250.

Makowiecka, A., Malek, N., Mazurkiewicz, E., Mrówczyńska, E., Nowak, D., Mazur, A.J., 2019. Thymosin β4 regulates focal adhesion formation in human melanoma cells and affects their migration and invasion. Front. Cell Dev. Biol. 7, 304. https://doi.org/10.3389/fcell.2019.00304.

Manton, C.A., Johnson, B., Singh, M., Bailey, C.P., Bouchier-Hayes, L., Chandra, J., 2016. Induction of cell death by the novel proteasome inhibitor marizomib in glioblastoma in vitro and in vivo. Sci. Rep. 6, 18953. https://doi.org/10.1038/srep18953.

Mathen, P., Rowe, L., Mackey, M., Smart, D., Tofilon, P., Camphausen, K., 2020. Radiosensitizers in the temozolomide era for newly diagnosed glioblastoma. Neurooncol. Pract. 7 (3), 268–276. https://doi.org/10.1093/nop/npz057.

Morita, T., Hayashi, K., 2018. Tumor progression is mediated by Thymosin-β4 through a TGFβ/MRTF signaling axis. Mol. Cancer Res. 16 (5), 880–893. https://doi.org/10.1158/1541-7786.MCR-17-0715.

Nam, J.Y., de Groot, J.F., 2017. Treatment of glioblastoma. J. Oncol. Pract. 13 (10), 629–638. https://doi.org/10.1200/JOP.2017.025536.

Paul-Samojedny, M., Łasut, B., Pudełko, A., Fila-Daniłow, A., Kowalczyk, M., Suchanek-Raif, R., Zieliński, M., Borkowska, P., Kowalski, J., 2016. Methylglyoxal (MGO) inhibits proliferation and induces cell death of human glioblastoma multiforme T98G and U87MG cells. Biomed. Pharmacother. 80, 236–243. https://doi.org/10.1016/j.biopha.2016.03.021.

Philp, D., Kleinman, H.K., 2010. Animal studies with thymosin beta, a multifunctional tissue repair and regeneration peptide. Ann. N. Y. Acad. Sci. 1194, 81–86. https://doi.org/10.1111/j.1749-6632.2010.05479.x.

Rahman, M.A., Gras Navarro, A., Brekke, J., Engelsen, A., Bindesboll, C., Sarowar, S., Bahador, M., Bifulco, E., Goplen, D., Waha, A., Lie, S.A., Gjertsen, B.T., Selheim, F., Enger, P.O., Simonsen, A., Chekenya, M., 2019. Bortezomib administered prior to temozolomide depletes MGMT, chemosensitizes glioblastoma with unmethylated MGMT promoter and prolongs animal survival. Br. J. Cancer 121 (7), 545–555. https://doi.org/10.1038/s41416-019-0551-1.

Roberts, N.B., Wadajkar, A.S., Winkles, J.A., Davila, E., Kim, A.J., Woodworth, G.F., 2016. Repurposing platinum-based chemotherapies for multi-modal treatment of glioblastoma. Onco. Targets. Ther. 5 (9), e1208876. https://doi.org/10.1080/2162402X.2016.1208876.

Stupp, R., Taillibert, S., Kanner, A., Read, W., Steinberg, D., Lhermitte, B., Toms, S., Idbaih, A., Ahluwalia, M.S., Fink, K., Di Meco, F., Lieberman, F., Zhu, J.J., Stragliotto, G., Tran, D., Brem, S., Hottinger, A., Kirson, E.D., Lavy-Shahaf, G.,

Weinberg, U., Kim, C.Y., Paek, S.H., Nicholas, G., Bruna, J., Hirte, H., Weller, M., Palti, Y., Hegi, M.E., Ram, Z., 2017. Effect of tumor-treating fields plus maintenance temozolomide vs maintenance temozolomide alone on survival in patients with glioblastoma: a randomized clinical trial. JAMA 318 (23), 2306–2316. https://doi.org/10.1001/jama.2017.18718.

Sugiura, N., Patel, R.G., Corriveau, R.A., 2001. N-methyl-D-aspartate receptors regulate a group of transiently expressed genes in the developing brain. J. Biol. Chem. 276 (17), 14257–14263. https://doi.org/10.1074/jbc.M100011200.

Sulman, E.P., Ismaila, N., Armstrong, T.S., Tsien, C., Batchelor, T.T., Cloughesy, T., Galanis, E., Gilbert, M., Gondi, V., Lovely, M., Mehta, M., Mumber, M.P., Sloan, A., Chang, S.M., 2017. Radiation therapy for glioblastoma: American society of clinical oncology clinical practice guideline endorsement of the American society for radiation oncology guideline. J. Clin. Oncol. 35 (3), 361–369. https://doi.org/10.1200/JCO.2016.70.7562.

Tang, J.H., Yang, L., Chen, J.X., Li, Q.R., Zhu, L.R., Xu, Q.F., Huang, G.H., Zhang, Z.X., Xiang, Y., Du, L., Zhou, Z., Lv, S.Q., 2019. Bortezomib inhibits growth and sensitizes glioma to temozolomide (TMZ) via down-regulating the FOXM1-Survivin axis. Cancer Commun. (Lond.) 39 (1), 81. https://doi.org/10.1186/s40880-019-0424-2.

Thomas, A., Tanaka, M., Trepel, J., Reinhold, W.C., Rajapakse, V.N., Pommier, Y., 2017. Temozolomide in the era of precision medicine. Cancer Res. 77 (4), 823–826. https://doi.org/10.1158/0008-5472.CAN-16-2983.

Touat, M., Idbaih, A., Sanson, M., Ligon, K.L., 2017. Glioblastoma targeted therapy: updated approaches from recent biological insights. Ann. Oncol. 28 (7), 1457–1472. https://doi.org/10.1093/annonc/mdx106.

Wang, J., Ren, D., Sun, Y., Xu, C., Wang, C., Cheng, R., Wang, L., Jia, G., Ren, J., Ma, J., Tu, Y., Ji, H., 2020. Inhibition of PLK4 might enhance the anti-tumour effect of bortezomib on glioblastoma via PTEN/PI3K/AKT/mTOR signalling pathway. J. Cell. Mol. Med. 24 (7), 3931–3947. https://doi.org/10.1111/jcmm.14996.

Weller, M., van den Bent, M., Tonn, J.C., Stupp, R., Preusser, M., Cohen-Jonathan-Moyal, E., Henriksson, R., Le Rhun, E., Balana, C., Chinot, O., Bendszus, M., Reijneveld, J.C., Dhermain, F., French, P., Marosi, C., Watts, C., Oberg, I., Pilkington, G., Baumert, B.G., Taphoorn, M.J.B., Hegi, M., Westphal, M., Reifenberger, G., Soffietti, R., Wick, W., 2017. European Association for Neuro-Oncology (EANO) guideline on the diagnosis and treatment of adult astrocytic and oligodendroglial gliomas. Lancet Oncol. 18 (6), e315–e329. https://doi.org/10.1016/S1470-2045(17)30194-8.

Weller, M., Le Rhun, E., Preusser, M., Tonn, J.C., Roth, P., 2019. How we treat glioblastoma. ESMO Open 4 (Suppl. 2), e000520. https://doi.org/10.1136/esmoopen-2019-000520.

Woodroffe, R.W., Zanaty, M., Soni, N., Mott, S.L., Helland, L.C., Pasha, A., Maley, J., Dhungana, N., Jones, K.A., Monga, V., Greenlee, J.D.W., 2020. Survival after reoperation for recurrent glioblastoma. J. Clin. Neurosci. 73, 118–124. https://doi.org/10.1016/j.jocn.2020.01.009.

Yu, R., Li, Q., Feng, Z., Cai, L., Xu, Q., 2019. m6A reader YTHDF2 regulates LPS-induced inflammatory response. Int. J. Mol. Sci. 20 (6), 1323. https://doi.org/10.3390/ijms20061323.

Zhou, X., Liao, X., Wang, X., Huang, K., Yang, C., Yu, T., Han, C., Zhu, G., Su, H., Han, Q., Chen, Z., Huang, J., Gong, Y., Ruan, G., Ye, X., Peng, T., 2020. Noteworthy prognostic value of phospholipase C delta genes in early stage pancreatic ductal adenocarcinoma patients after pancreaticoduodenectomy and potential molecular mechanisms. Cancer Med. 9 (3), 859–871. https://doi.org/10.1002/cam4.2699.

CHAPTER

Cerebrolysin restores balance between excitatory and inhibitory amino acids in brain following concussive head injury. Superior neuroprotective effects of TiO_2 nanowired drug delivery

Hari Shanker Sharma[a,*], Dafin F. Muresanu[c,d], Seaab Sahib[e], Z. Ryan Tian[e], José Vicente Lafuente[b], Anca D. Buzoianu[f], Rudy J. Castellani[g], Ala Nozari[h], Cong Li[i], Zhiquiang Zhang[i], Lars Wiklund[a], and Aruna Sharma[a,*]

[a]*International Experimental Central Nervous System Injury & Repair (IECNSIR), Department of Surgical Sciences, Anesthesiology & Intensive Care Medicine, Uppsala University Hospital, Uppsala University, Uppsala, Sweden*

[b]*LaNCE, Department of Neuroscience, University of the Basque Country (UPV/EHU), Leioa, Bizkaia, Spain*

[c]*Department of Clinical Neurosciences, University of Medicine & Pharmacy, Cluj-Napoca, Romania*

[d]*"RoNeuro" Institute for Neurological Research and Diagnostic, Cluj-Napoca, Romania*

[e]*Department of Chemistry & Biochemistry, University of Arkansas, Fayetteville, AR, United States*

[f]*Department of Clinical Pharmacology and Toxicology, "Iuliu Hatieganu" University of Medicine and Pharmacy, Cluj-Napoca, Romania*

[g]*Department of Pathology, University of Maryland, Baltimore, MD, United States*

[h]*Anesthesiology & Intensive Care, Massachusetts General Hospital, Boston, MA, United States*

[i]*Department of Neurosurgery, Chinese Medicine Hospital of Guangdong Province; The Second Affiliated Hospital, Guangzhou University of Chinese Medicine, Yuexiu District, Guangzhou, China*

[*]*Corresponding authors: Tel.: +46-70-2011-801; Fax: +46-18-24-38-99 (Hari Shanker Sharma); Tel.: +46-70-21-95-963; Fax: +46-18-24-38-99 (Aruna Sharma), e-mail address: sharma@surgsci.uu.se; aruna.sharma@surgsci.uu.se*

Progress in Brain Research, Volume 266, ISSN 0079-6123, https://doi.org/10.1016/bs.pbr.2021.06.016

Copyright © 2021 Elsevier B.V. All rights reserved.

Abstract

Concussive head injury (CHI) often associated with military personnel, soccer players and related sports personnel leads to serious clinical situation causing lifetime disabilities. About 3–4 k head injury per 100 k populations are recorded in the United States since 2000–2014. The annual incidence of concussion has now reached to 1.2% of population in recent years. Thus, CHI inflicts a huge financial burden on the society for rehabilitation. Thus, new efforts are needed to explore novel therapeutic strategies to treat CHI cases to enhance quality of life of the victims. CHI is well known to alter endogenous balance of excitatory and inhibitory amino acid neurotransmitters in the central nervous system (CNS) leading to brain pathology. Thus, a possibility exists that restoring the balance of amino acids in the CNS following CHI using therapeutic measures may benefit the victims in improving their quality of life. In this investigation, we used a multimodal drug Cerebrolysin (Ever NeuroPharma, Austria) that is a well-balanced composition of several neurotrophic factors and active peptide fragments in exploring its effects on CHI induced alterations in key excitatory (Glutamate, Aspartate) and inhibitory (GABA, Glycine) amino acids in the CNS in relation brain pathology in dose and time-dependent manner. CHI was produced in anesthetized rats by dropping a weight of 114.6 g over the right exposed parietal skull from a distance of 20 cm height (0.224 N impact) and blood–brain barrier (BBB), brain edema, neuronal injuries and behavioral dysfunctions were measured 8, 24, 48 and 72 h after injury. Cerebrolysin (CBL) was administered (2.5, 5 or 10 mL/kg, i.v.) after 4–72 h following injury. Our observations show that repeated CBL induced a dose-dependent neuroprotection in CHI (5–10 mL/kg) and also improved behavioral functions. Interestingly when CBL is delivered through TiO_2 nanowires superior neuroprotective effects were observed in CHI even at a lower doses (2.5–5 mL/kg). These observations are the first to demonstrate that CBL is effectively capable to attenuate CHI induced brain pathology and behavioral disturbances in a dose dependent manner, not reported earlier.

Keywords

Concussive head injury, Cerebrolysin, Head injury in military, Blood–brain barrier, Brain edema, Neuronal injury, Glutamate, Aspartate, Glycine, GABA, Neuroprotection, Nanowired delivery

1 Introduction

Concussive head injury (CHI) is a serious clinical problem affecting especially military and/or sports personnel quite frequently (Armistead-Jehle et al., 2017; Asken et al., 2018; Blennow et al., 2016; Bryan, 2013; Dixon, 2017; Silverberg et al., 2020). In recent years, CHI has reached to the highest levels of 1.2% of population in America (Langer et al., 2020). As a result, CHI could cause pronounced behavioral, cognitive and neuropathological consequences in >90% of cases (Broshek et al., 2015; Fehily and Fitzgerald, 2017; Harmon et al., 2013; Leddy et al., 2017; McInnes et al., 2017; Montenigro et al., 2017; Zetterberg and Blennow, 2018). Patients affected by CHI thus have deterioration in their quality of life and require rehabilitation and continuous support for their day-to-day activities (Hadanny and Efrati, 2016; Howell et al., 2019; Jones et al., 2019; Pervez et al., 2018; Troup et al., 2020). This would impose a heavy financial burden on our society (Gardner and Zafonte, 2016; Ommaya, 1995; Seabury et al., 2018; Taylor et al., 2017; Temple et al., 2016). Thus, efforts are needed to explore the possible mechanisms of CHI and to develop novel therapeutic strategies to minimize misery of CHI victims.

Military personnel are often suffering from a variety of concussive injuries that are either caused by roadside bomb explosion, missile explosion or blast injuries (Eierud et al., 2019; Elder et al., 2019; Goldstein et al., 2012; Hasoon, 2017; Mu et al., 2017). In addition during combat operations, as a direct insult on their head using blunt objects induces severe CHI (Agimi et al., 2019; Bigler and Tsao, 2017; Heltemes et al., 2011; Larkin et al., 2018; McKeon et al., 2019; Pilipenko et al., 2020). On the other hand, sports personnel can get concussive injury during soccer game when players use their head to receive the ball repeatedly causing mild to moderate degree of CHI (Hon et al., 2019; Hubertus et al., 2019; McAllister and McCrea, 2017; Mullally, 2017; Provencher et al., 2019; Sharp and Jenkins, 2015). Moreover in other sports where accidental blunt head injury may occur after fall or running accidents in the ground could also result in head trauma (Martin, 2016; Mätzsch and Karlsson, 1986). Recently, studies using biomechanics and pathophysiological aspects of CHI in military and sports personnel show that there are no major differences between blast injuries or sports related CHI (Gardner and Yaffe, 2015; Regasa et al., 2019). The other causes of CHI include motor vehicle accidents, street fighting or blunt head injury from any object either at home or at work place (Jagnoor and Cameron, 2015; Segev et al., 2016; Viano and Parenteau, 2015).

During CHI, the brain rotates faster within the cranium resulting in trauma to the most part of the brain caused by mechanical forces created by brain movement (Choe, 2016; Scorza and Cole, 2019; VanItallie, 2019). As a result, the symptoms of CHI may appear instantly or will develop sometimes after the episode (Dashnaw et al., 2012; Jackson and Starling, 2019; Lu et al., 2019). Depending on the severity of CHI, patients could develop life-threatening injuries within hours to days (Feddermann-Demont et al., 2020; Marshall et al., 2015; Mercier et al., 2020). Recent evidences point out that CHI could also lead to slowly developing neurodegenerative diseases including Alzheimer's or Parkinson's over time (Ascherio and Schwarzschild, 2016; Delic et al., 2020; Hook et al., 2020; Jenkins et al., 2020; Kokiko-Cochran and Godbout, 2018; Perry et al., 2016; Ramos-Cejudo et al., 2018; Wu et al., 2020). In extreme cases, the whole brain is swollen and softened like sponge due to excessive edema and volume swelling (Bigler et al., 2016; Ito et al., 1986; Kawamata et al., 2007). Since brain has very little space to expand in the rigid cranium, severe CHI results in instant death in more than 50% of the victims (Adams et al., 2018; Giza and Hovda, 2014; Mez et al., 2017). Those who survive, lifetime disabilities ensue following CHI (Bramley et al., 2016; Iverson et al., 2017; Ryan and Warden, 2003; van Ierssel et al., 2018; Yrondi et al., 2017). In spite of large number of cases occurring in the United States or Worldwide involving more than

400 to 800 thousand patients of CHI every year there is no suitable therapeutic measures are developed so far for effective treatment (Baldwin et al., 2018; Jinguji et al., 2011; Kazl and Torres, 2019; Voss et al., 2015). Thus, the need of the hour is to explore novel therapeutic strategies to treat cases of CHI in order to minimize the sufferings of patients and to enhance their quality of life.

Previous reports from our laboratory showed that hyperthermic brain injury (HBI) or spinal cord injury (SCI) induces imbalance among excitatory and inhibitory amino acids in the CNS (Sharma, 2004a,b,c, 2006, 2007a,b,c; Sharma and Sharma, 2012b, 2013). Thus, increased levels of glutamate or aspartate and a decreased concentration of GABA and glycine following HBI or SCI are associated with cell damage (Sharma, 2004a,b,c, 2006, 2007a,b). Drugs that restore these balance after injury results in neuroprotection (Sharma, 2006, 2007a,b,c). Since basic mechanisms of cellular injury could be similar in nature, it appears that CHI induced alterations in excitatory and inhibitory amino acid neurotransmitters balance will result in brain pathology. As a result neuronal, glial, endothelial and axonal degeneration occur following CHI (Sharma et al., 2016a,b). These biochemical changes are often associated with reduction in several neurotrophic factors, peptides and enzymes in CHI (Sharma et al., 2020a,b). Thus, exogenous supplement of neurotrophic factors may enhance neuroregeneration or neurorepair in CHI cases (Menon et al., 2012; Sharma, 2007b; Sharma et al., 2012).

In present investigation, we used Cerebrolysin, a balanced composition of several neurotrophic factors and active peptide fragments in a rat model of CHI in dose- and time-dependent manner. We measured alterations in two excitatory amino acids glutamate and aspartate in relation to two inhibitory amino acids GABA and glycine following cerebrolysin therapy in CHI in relation to brain pathology and neuroprotection.

2 Materials and methods

2.1 Animals

Experiments were carried out on male Sprague Dawley rats (250–300 g body weight) housed at controlled room temperature (21 ± 1 °C) with 12 h light and 12 h dark schedule. Rat food pellets and tap water were supplied ad libitum before experiments.

All experiments were conducted according to National Institute of Health (NIH) care and guidelines of experimental animals and approved by local Institutional Ethics Committee.

2.1.1 Concussive head injury (CHI)

Concussive head injury (CHI) was produced in rats using a weight drop method as described earlier (Ruozi et al., 2015; Sharma et al., 2007, 2010, 2019, 2020a, b; Vannemreddy et al., 2006). In brief, rats were anesthetized with Equithesin (3 mL/kg, i.p.) and their parietal skull was exposed aseptically after incision of the skin under stereotaxic guidance. The skull surface was cleaned with any remaining muscle and washed with 70% alcohol to make it dry. An iron cylindrical metal tapered at the one end (2 mm^2) weighing 114.6 g was then dropped on the right parietal skull bone from a height of 20 cm through an aluminum guide tube fixed over the skull using a clamp on stand. The rat head was fixed in the stereotaxic apparatus and a soft pad was placed under the belly and the head to avoid head movement. The tapering end of the metal bar was placed over stamp-pad blue ink before the impact in order to mark the area of injury clearly. No bounce-back of the metal was observed under this condition and the head movement was also not seen in this model. The skull surface remained intact and no fracture of the bone was observed. In less than 4% cases, skull fracture, if any was seen no data were collected from those rats and they are discarded from the experimental group.

This weight and height adjustment induces an impact pf 0.224 N on the skull surface that induced CHI as seen in clinical situations. The impact of injury is calculated according the formula:

$E = m.g.h,$
where:
$E =$ impact force (Newton)
$m =$ mass (kg)
$g =$ acceleration of gravitation (9.81 m/s^2)
$h =$ falling height (meters)
$m\ 0.1146\,kg \times g\ 9.81 \times h\ 0.2\ m = 0.2248\,N$ (see Fig. 1)

2.2 Control group

The control group was anesthetized identically and skull exposed, head is placed in the stereotaxic apparatus but no impact was made in these rats.

2.3 Survival periods

The control or injured rats were allowed to survive 4, 8, 12, 24 and 72 h after the skull exposure (control group) or CHI (experimental group). These rats were placed individually in cages and all efforts are made to see that these animals do not have pain as their skin was sutured and local anesthesia was applied over the incision. All these animals were cared and handled by expert handler every 6 h until the observations period of 72 h as per NIH guidelines and care for experimental animals.

2.4 Cerebrolysin treatment

Cerebrolysin (CBL, Ever NeuroPharma, Austria) was administered in control or CHI animals in a dose of 5 mL, or 10 mL/kg in separate groups 2 h after CHI in 4 h group (1 injection); 8 h CHI in 2 and 6 h after (2 injection), in 12 h CHI 2, 6 and 10 h after (3 times); in 24 h CHI, 4, 12 and 20 h (3 times); in 48 h CHI; 4, 12, 28, 36 and 44 h (6 injections) and in 72 h CHI; 4, 12, 20, 28, 36, 44, 52, 60 and 68 h (9 injections). In control Untraumatized group also identical administration of CBL was done

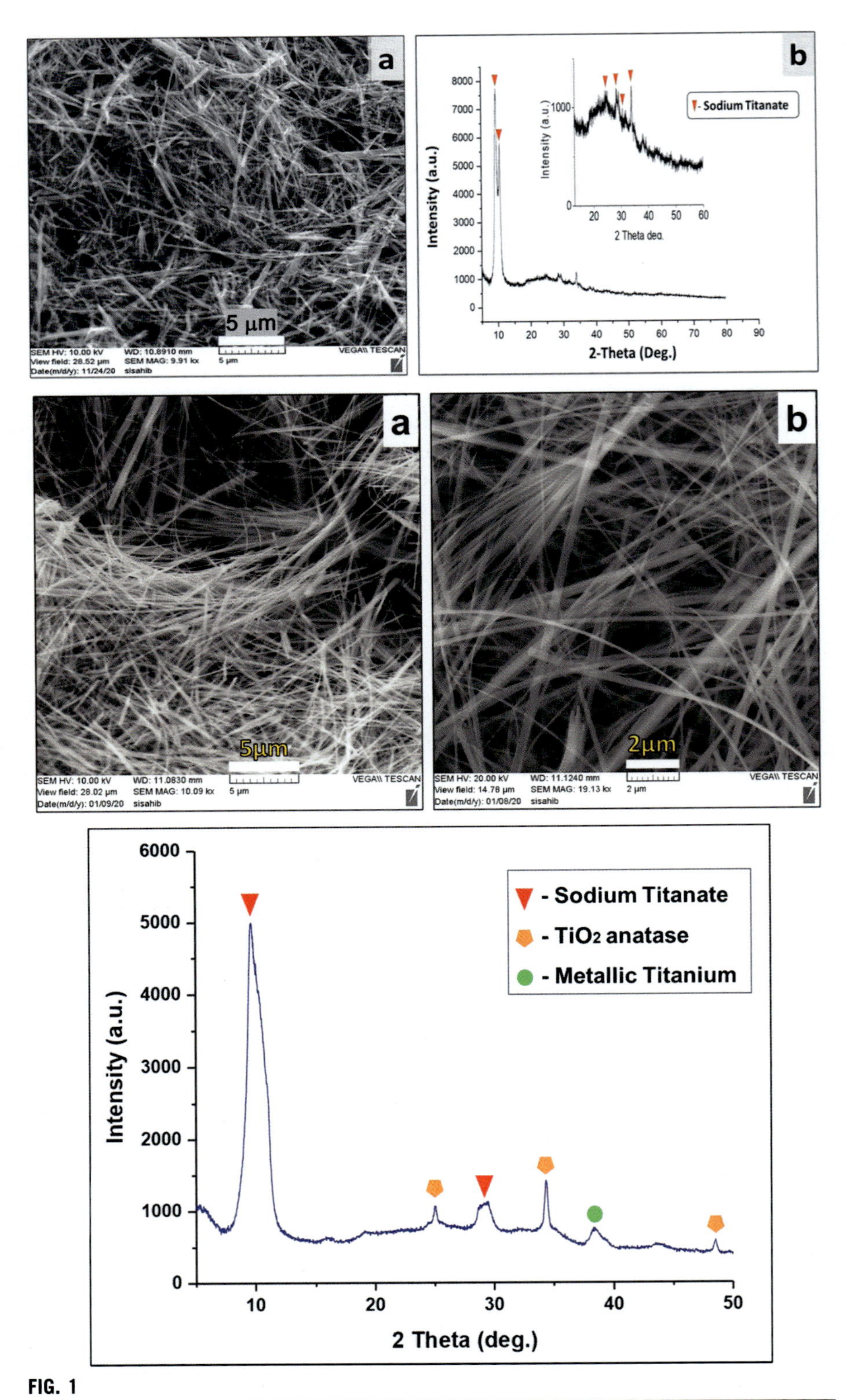

FIG. 1

SEM image of sodium titanate nanowires (A), and the XRD patterns of sodium titanate nanowires (B) fabricated via the hydrothermal method. (C) SEM images of the sodium titanate nanowires (Na-TiO2 NWs) (a) at low magnification 5μm, (b) at high magnification 2μm. (D) The XRD patterns for the Sodium titanate nanowires prepared by the hydrothermal method.

according to survival periods as mentioned above (Ruozi et al., 2015; Sharma et al., 2020a,b).

In CHI control group, saline (0.9%) instead of CBL in identical manner was administered for comparison.

2.5 TiO_2 nanowired cerebrolysin delivery

In separate groups, TiO_2 nanowired cerebrolysin (NWCBL) was administered in control and CHI rats to see whether NWCBL has superior effects on neuroprotection as compared to CBL alone (Sharma et al., 2020a,b). The TiO_2 NWCBL was prepared according to standard protocol as described earlier. In brief, we used a TiO_2-based (hydrogen titanate) single crystalline ceramic biomaterial with a typical diameter ranging from 50 to 60nm and a superb chemical stability that can be used to enhance drug delivery within the CNS, as described earlier. Standard procedures developed in our laboratory were used to synthesize typical TiO_2 nanowires (Fig. 1A). X-ray diffraction patterns (Fig. 1B) of the nanofibers confirm that the nanowires resemble the titanate in lattice structure. Energy dispersive X-ray elementary analysis of this film (Fig. 1D) clearly showed the existence of Ti, Na, and O in the nanowire membrane (Fig. 1C).

For CBL labeling, 0.20g of TiO_2 powder (Degussa P25) was introduced into 40mL of 10M alkali solution in a 150mL Teflon-lined autoclave container. After hydrothermal reaction in an oven for 1–15days at temperatures above 180°C, the white paper-like product was collected from the Teflon rod template and washed with distilled water. CBL was tagged separately to nanowires. In our study, the film was first sterilized in 70% ethanol, and then rinsed in sterile 0.9% saline. Subsequently, the membrane (1.0cm × 1.0cm) was soaked in a different strength of CBL solution at room temperature for 12h them washed with deionized water before use. The nanowired-CBL complex was administered separately intravenously after injury in an identical manner of normal CBL administration as described earlier (Sharma et al., 2017, 2018a,b).

2.6 Parameters measured

The following parameters were measured in control, CHI and CBL treated groups as follows:

2.7 Blood–brain barrier permeability

The blood–brain barrier (BBB) permeability was measured using two protein tracers Evans blue albumin (2% solution, 3mL/kg, i.v., pH 7.4) and radioiodine ([131]-Iodine, 100μCi/kg, i.v.) as described earlier (Sharma, 1987; Sharma and Dey, 1986a,b, 1987). These tracers when administered into the systemic circulation will bind to serum proteins with different capacities (largely albumin) and form serum protein complex. Thus, any leakage of these tracers represents extravasation of serum–protein complex in the brain. These tracers were administered into right jugular vein after experiment under Equithesin anesthesia (3 mL/kg, i.p.) and allowed for circulation for 10 min. After that the intravascular tracers were washed out by 0.9 & physiological saline *in situ* via cardiac puncture at 90 Torr perfusion pressure. Immediately before cardiac puncture, about 1 mL whole blood was withdrawn from the left ventricle to determine whole blood radioactivity.

After saline perfusion, the brains were immediately removed and placed on a cold saline wet filter paper (+4 °C). After that the spread of blue dye over the brain surfaces were examined and recorded. After that penetration of the blue dye was examined inside the brain ventricular spaces and internal brain structures after a midline sagittal section. After that various brain regions were dissected out quickly and weighed to obtain the wet weight of the samples. These samples were counted in a 3-in Gamma Counter (Packard, USA) for determination of brain radioactivity. The whole blood radioactivity was also determined in 1 mL blood sample in identical manner. After counting the radioactivity, the brain samples were homogenized in a mixture of 0.5% sodium sulfate and Acetone to extract Evans blue dye. After centrifugation (×900 g) the aliquot was taken out and measured for Evans blue dye concentration in spectrophotometer at 620 nm wavelength. The dye entered into the solution was calculated according to standard curve of Evans blue. The values of Evans blue entered into the brain was expressed as mg % as described earlier (Sharma, 1987; Sharma and Dey, 1986a,b, 1987).

Extravasation of radioactivity on the brain samples was calculated according to the formula:

Radioactivity in the brain sample (CPM/g brain) – Radioactivity in Blood (CPM blood 1 mL)/CPM whole blood × 100.

Extravasation of radioiodine was expressed as percentage of whole blood radioactivity (see Sharma, 1987).

2.8 Brain edema formation

Brain edema formation was measured using brain water content of the samples in six brain regions. After the end of the experiments the brain from control and experimental animals were removed after decapitation and placed on the cold saline wet filter papers. The large superficial blood vessels and blood clots if any were removed and discarded. The brains were bisected into two halves and each halves were divided into the six regions, namely, cerebral cortex, hippocampus, cerebellum, thalamus, hypothalamus and the brain stem and weighed immediately. The wet weight of the samples was between 70 and 150 mg approximately. Immediately after taking wet weight of the samples on preweighed filter papers (Whatman Filter Paper No. 1, GE Life Sciences, USA) the samples were placed in an incubator maintained at 90 °C for 48–72 h until the dry wet of the samples became constant on three subsequent determinations on a Electronic analytical balance (Mettler Toledo, ME104TE, Columbia MD, USA sensitivity 0.1 mg). The water content was calculated from the differences between wet and dry weight as described earlier (Sharma and Olsson, 1990).

The volume swelling ($\%f$) is calculated from the differences between control and experimental brain water changes according to the formula of Elliott and Jasper (1949) as described earlier (Sharma and Cervós-Navarro, 1990).

2.9 Neuropathology

Cell changes in the brain were examined using standard histopathological techniques under Light Microscopy. Some samples from also processed for examining under Transmission Electron Microscopy (see below).

2.9.1 Perfusion and fixation

After the end of the experiments animals were deeply anesthetized with Equithesin (3 mL/kg, i.p.) and the chest was opened rapidly. A butterfly cannula was inserted into the left cardiac ventricle and the right auricle was cut. The intravascular tracer as washed out with about 50 mL of cold Phosphate buffer saline (PBS, 0.1 M, pH 7.0) at 90 mm Torr. This was followed with perfusion of about 200 mL of cold 4% buffered paraformaldehyde or Somogyi Fixative with 0.2% picric acid (Sharma et al., 1990a,b, 1991a,b, 1992a,b; Somogyi and Takagi, 1982; Zimmer et al., 1991).

After perfusion *in situ*, the animals were wrapped in an aluminum foil and placed in a refrigerator at 4 °C for overnight. On the 2nd day, the brains were removed and coronal sections were cut and placed in the same fixative at 4 °C for 1 week.

2.9.2 Light microscopy

For light microscopy, the coronal sections passing through caudate nucleus, hippocampus, cerebellum and brain stem were embedded in paraffin using an automated tissue processor for histology (Tissue-Tek Sakura FineTek, Torrance, CA, USA). About 3 μm thick sections were cut on a manual microtome (Leica, Wetzlar, Germany). The sections were stained with Nissl or Hematoxylin & Eosin (H&E) to study neuronal injury using standard protocol (Sharma, 2000; Sharma and Cervós-Navarro, 1990).

To Sections were examined under an Inverted Bright Field Microscope (Carl-Zeiss Axio Inverted Microscope, Jena, Germany) and photographed using an attached digital camera (Olympus SC-180, Düsseldorf, Germany). The images are stored and processed using commercial software Adobe Photoshop extended Version 12.04 × 64 on an Apple Mac Book Pro system Os El Capitan version 10.11.6. All control, experimental and drug treated images are processed using identical filters and color balance (Sharma and Alm, 2002; Sharma and Sjöquist, 2002).

2.9.3 Transmission electron microscopy

For Transmission Electron Microscopy (TEM), small tissue pieces in Somogyi Fixative with picric acid from cerebral cortex, hippocampus and thalamus were cut and post fixed in Osmium Tetraoxide (OsO4) and embedded in Epon 812 (Sharma et al., 1993a,b, 1995a,b; Sharma and Olsson, 1990). About 1 μm thin thick sections were cut and stained with Toluidine blue and examined udder a Bench microscope to find the desired area for TEM analysis. The desired area identified was then trimmed in Epon blocks and ultrathin sections (50 nm) were cut on an Ultramicrotome (LKB, Sweden) using a diamond knife (Sharma and Olsson, 1990). The serial ultrathin sections were then collected on a one hole copper grid and counterstained with uranyl acetate and lead citrate. Some sections were examined unstained. These ultrathin sections were examined under a Philips 400 TEM and the images were captured by attached digital camera (Gatan Image System, Gatan Inc., Pleasanton, CA, USA). The images are acquired from TEM at 4000–12,000 magnification and processed using commercial software Photoshop extended Version 12.04 × 64 at Apple Mac Book Pro system Os El Capital Version 10.11.6 using identical color balance and filters from control, experimental and drug treated samples (Cervós-Navarro et al., 1998; Sharma et al., 1998a,b; Sharma and Westman, 1997; Westman and Sharma, 1998).

2.10 Measurement of amino acid neurotransmitters

In desired areas of the brain, amino acid neurotransmitters were measured using high-performance liquid chromatography (HPLC) and electrochemical detection according to standard protocol as described earlier (Sharma, 2006, 2007a,b). In brief, rats were killed by decapitation and the brains were immediately excised and frozen in dry ice. Then, the brains were weighed and homogenized in 5% trichloroacetic acid using 10 mL/g of tissue. A fraction of the homogenate was further deproteinized with sulfosalycilic acid. The proteins were removed by centrifugation 20 min at 3000 × *g*. The amino acids glutamate, aspartate, y-aminobutyric acid (GABA) and glycine were measured after conjugated with ortho-phthaldialdehyde and t-butyl thiol, amino acids separated by HPLC and detected electrochemically (Christensen and Fonnum, 1991; Sharma, 2006). The value of amino acids was expressed as μmol/g tissue.

2.11 Physiological variables

The mean arterial blood pressure (MABP), arterial pH and partial pressure of oxygen (PaO2) and carbon dioxide ($PaCO_2$) were measured using standard procedures as describe earlier (Sharma, 1987). In brief, the MABP was measured using an arterial catheter (PE10) inserted into the lift carotid artery retrogradely toward heart under ascetical conditions 1 week before the experiment (Sharma and Dey, 1986a,b). The arterial catheter was flushed daily with heparinized saline to keep it patency once a day till the time of the experiment. At the time of the experiment, the arterial catheter was connected to pressure transducer (Strain Gauge, Statham P23, USA) for blood pressure recording that was connected to a chart recorder (Electromed, UK). The heart rate (beats per min) and respiration rate (cycle per min) was also recorded (Dey et al., 1980; Sharma, 1987).

Before connecting the arterial catheter to pressure transducer the arterial blood sample was withdrawn for later determination of the Arterial pH, and blood gases using Radiometer Apparatus (Copenhagen, Denmark) (Sharma and Dey, 1986a,b).

2.12 Behavioral parameters

The following behavioral parameters for sensory motor and cognitive dysfunctions were examined in the control experimental and drug treated groups.

2.12.1 Rota-rod treadmill

The Rota-Rod Treadmill was used to evaluate motor and cognitive function of animals as described earlier (Sharma, 2006). Animals were placed on a Rota-Rod Treadmill set at 16 rotations per min (rpm) with cut off time of 120 s. The time stayed on the Rota-Rod treadmill by control, experimental and drug treated animals were counted manually (Sharma, 2006). The animals were trained to stay on the Rota Rod treadmill daily for 1 week before the experiment. Rats no falling off from the treadmill for 120 s in a session of 180 s are considered normal (Sharma, 2006).

2.12.2 Inclined plane-angle test

Animals were placed on an inclined plane angle test where the animal can stay on the platform for 5 s. Control animals could stay at the platform at an angle of 60° whereas traumatized animals require to raise the angle of platform or for 5 s that was recorded manually (Sharma, 2006). Animals were trained for 1 week on the inclined plane platform before the experiment (see Sharma, 2006).

2.12.3 Walking on mesh grid and placement error

A stainless steel mesh grid elevated at 30° platform with mesh size of 30 mm was used on which animals are allowed to walk for 60 s (Kanagal and Muir, 2008; Manser et al., 1995). The animals were trained to walk on mesh grid for 1 week before the experiment. The total nr of steps was counted manually for control, experimental and drug treated groups. During Mesh Grid walking the forepaw placement error was counted during a 60 s time in control, experimental and drug treated rats (Sharma, 2006, 2007a,b,c). The number of placement errors is recorded manually.

2.12.4 Foot print analysis

Foot print analysis is a measure of gait and motor function (Bain et al., 1989; Taoka et al., 1998). The animals gait was assessed using the stride length and hind feet distance was measured using standard procedures as described earlier (Sharma, 2006). For this purpose the hind paws were wetted and allowed to walk on a paper coated with bromophenol blue in acetone (Sharma, 2006). The foot imprints on the paper by each individual animal was then analyzed for the distance (in mm) between the hind paws from the base of the central pads and the stride length was calculated from the distance between the hind paws in two consecutive steps (Sharma, 2006).

3 Results

3.1 Physiological variables

Measurements of mean arterial blood pressure (MABP) exhibited significant reduction after 4h CHI (Fig. 2A) following 4h CHI that was further decreased at 8h CHI (Fig. 2A). The MABP was then slightly increased from 12h CHI that was more or less maintained until 48h CHI. At 72h CHI slight decrease in MABP occurred.

The arterial pH decreased after 4h CHI that decreased further after 8h CHI. After 12h CHI there was a gradual reduction in arterial pH that continued until 72h CHI (Fig. 2C).

The arterial PaO_2 showed a gradual decrease after CHI from 4h period to 72h CHI (Fig. 2D) whereas the arterial PaCO2 showed gradual increase from the control values after 4–24h CHI (Fig. 2D). After that 48 and 72h CHI exhibited slight decrease in PaCO2 values in CHI (Fig. 2D).

The heart rate in CHI showed significant decrease from the control values after 4h CHI (Fig. 2E) that exhibited gradual decrease after 8–72h CHI (Fig. 2E). On the other hand, the respiratory rat increased significantly after 4h CHI and continues to increase throughout until 72h of CHI as compared to the control group (Fig. 2F).

3.2 Behavioral symptoms

Evaluation of Rota-Rod performances after CHI showed a gradual decline in performances regarding staying capacity on the treadmill (Fig. 3A). Thus, there was a continuous decline on staying period on the Rota-Rod treadmill after 4h CHI to 72h CHI (Fig. 3A).

CHI has also reduced the ability to stay on inclined plane angle test (Fig. 3B). Thus, the angle on which CHI rats could stay for 5s requires to gradually lowering down so that the injured rats could stay for 5s on that plane (Fig. 3B). As evident from the figure rats after 4h CHI to 72h CHI the inclined plane angle was significantly reduced with advancing survival period after CHI (Fig. 3B).

Similar observations were seen regarding reduction in steps taken by CHI rats walking on mesh grid (Fig. 3C). Thus, there was a significant decline on number of steps taken on the mesh grid by rats after 4h CHI and this decrease in number of steps increased further with advancing survival periods of CHI from 4 to 72h periods (Fig. 3C).

CHI rats when walk on a mesh grid increases placement error that is gradually increased with survival period after injury (Fig. 3D). Thus, the placement error increased on walking on a mesh grid from 4h CHI to 72h CHI periods as compared to the controls (Fig. 3D).

Another test of gait regarding stride length (Fig. 3E) and distances between hind legs (Fig. 3F) was significantly affected by CHI survival periods. Thus, there was a significant decrease in stride length after 4h CHI and this decrease continued further with advancing survival periods of CHI up to 72h (Fig. 3E). On the other hand, the distances between feet increased significantly after 4h CHI and continued to increase with advancing survival periods up to 72h CHI (Fig. 3F).

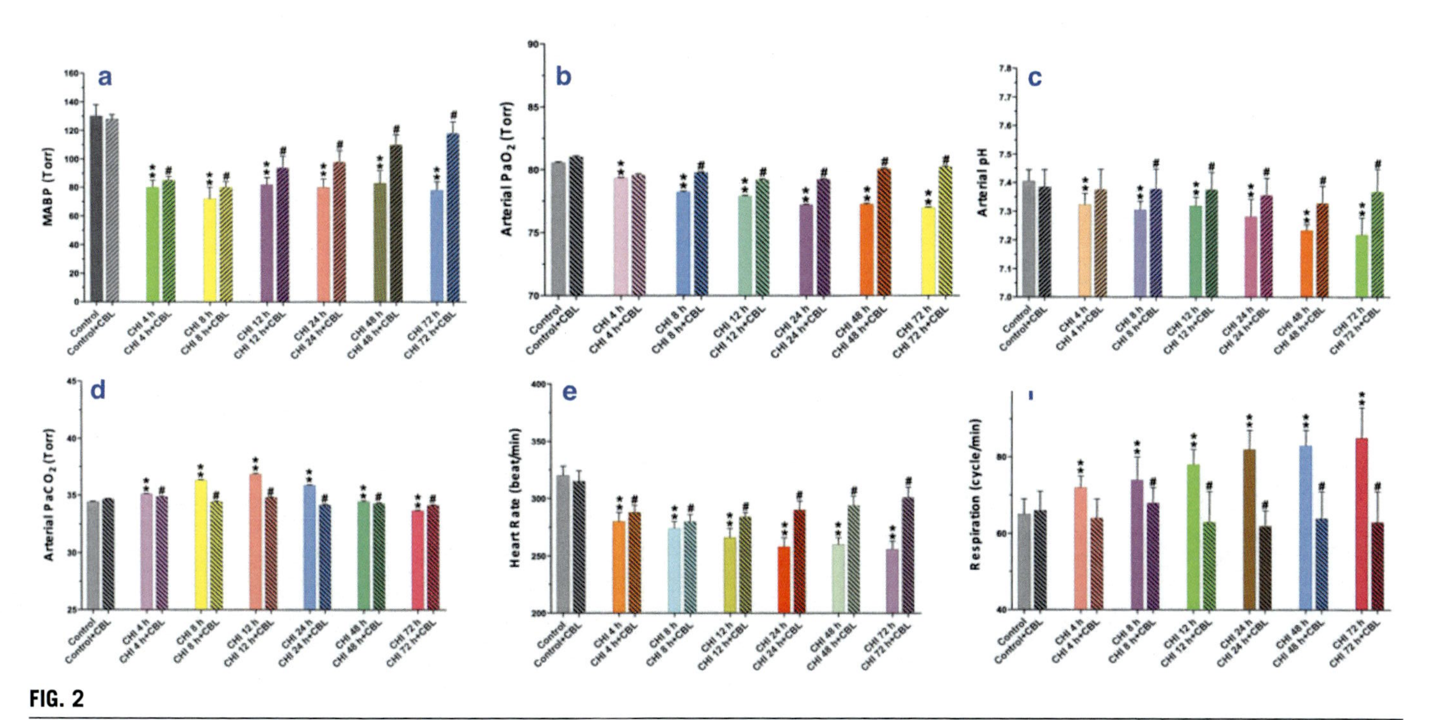

FIG. 2

Physiological variables in concussive head injury (CHI) and following cerebrolysin (CBL) treatment from CHI 4 to 72 h and their modification with CBL. (A) Mean arterial blood pressure (MABP); (B) Arterial PaO_2; (C) Arterial pH; (D) Arterial $PaCO_2$; (E) Heart rate; (F) Respiration rate. ANOVA followed by Dunnett's test for multiple group comparison from one control showing $^{**}P<0.01$ from control; $\# = P<0.05$ from control.

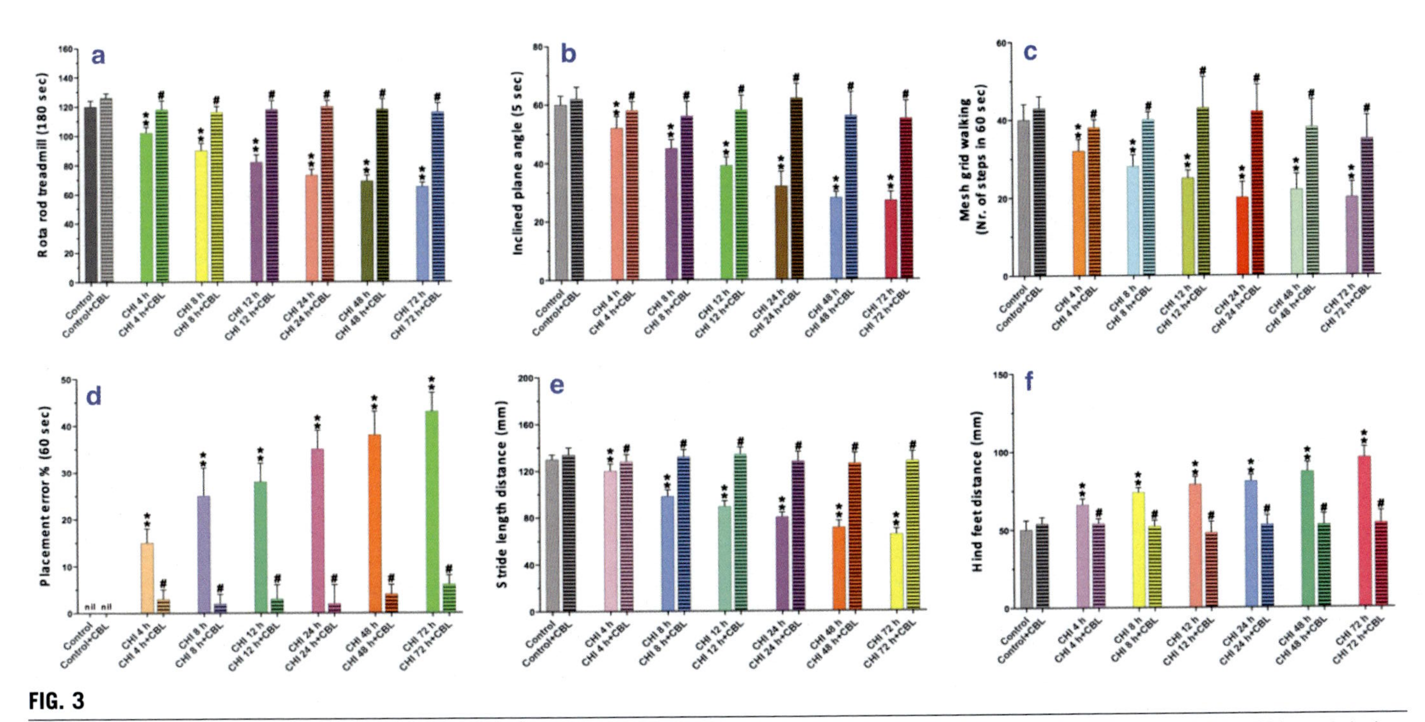

FIG. 3

Behavioral and cognitive functions in concussive head injury (CHI) and following cerebrolysin (CBL) treatment from CHI 4 to 72 h and their modification with CBL. (A) Rota-rod treadmill; (B) Inclined plane angle; (C) Mesh grid walking; (D) Placement error; (E) Stride length; (F) Hind feet distance. ANOVA followed by Dunnett's test for multiple group comparison from one control showing $^{**}P<0.01$ from control; $\# = P<0.05$ from control, nil = absent.

These behavioral symptoms suggest that CHI influenced motor, sensory and cognitive functions depending on the duration of injury (Fig. 3A–F).

3.3 Excitatory amino acids aspartate and glutamate

Measurement of excitatory amino acids aspartate and glutamate following CHI in several brain regions taken from cerebral cortex (Fig. 4A), hippocampus (Fig. 4B), cerebellum (Fig. 4C) shows significant gradual increase in the injured right hemisphere (RH) and left hemisphere (LH) from 4h CHI to 24h that was most marked in the uninjured half (LH). At 48h CHI the aspartate and glutamate values declined from the peak 24h CHI and continued to decrease at 72h CHI (Fig. 4A–C). A significant higher increase in aspartate and glutamate is seen in same pattern in the LH as compared to RH in all brain regions (Fig. 4A–C). However, the glutamate levels were in general are higher than aspartate following CHI in both RH and LH.

In thalamus (Fig. 5A), hypothalamus (Fig. 5B) and in brain stem (Fig. 5C) the aspartate and glutamate level also exhibited same pattern of increase from 4h CH and reached its peak level after 24h CHI and then declined from 48h CHI that continued until 72h CHI except in brain stem (Fig. 5A–C). In brain stem there was no detectable changes in aspartate or glutamate level after 72h CHI (Fig. 5C). The glutamate levels were in general are higher than aspirate following CHI in both RH and LH. However, LH has higher levels of aspartate and glutamate after CHI in all these regions (Fig. 5A–C).

These changes show that glutamate and aspartate levels increased globally after CHI depending on survival periods.

3.4 Inhibitory amino acids glycine and GABA

Glycine and GABA appears to counteract aspartate and glutamate action in the brain after CHI. Thus, glycine and GABA exhibited increased level after CHI from 4h and reached to peak levels after 24h period. Then glycine and GABA levels in CHI declined following 24, 48 and 72h survival periods in the brain (Figs. 6 and 7). In general, the GABA exhibited greater increase after CHI as compared to glycine at various survival periods after CHI in different brain regions. Also the uninjured half showed greater changes in glycine and GABA levels after CHI on RH (Figs. 6 and 7).

Cerebral cortex (Fig. 6A), hippocampus (Fig. 6B), cerebellum (Fig. 6C), thalamus (Fig. 7A), hypothalamus (Fig. 7B) and brain stem (Fig. 7C) exhibited the peak levels of glycine and GABA at after 24h CHI and then declined gradually over the 72h CHI. The LH always showed higher changes in glycine and GABA than RH after CHI at different survival periods.

These changes in glycine and GABA in CHI appears to balance the increase in aspartate and glutamate levels in CHI (Figs. 4–7).

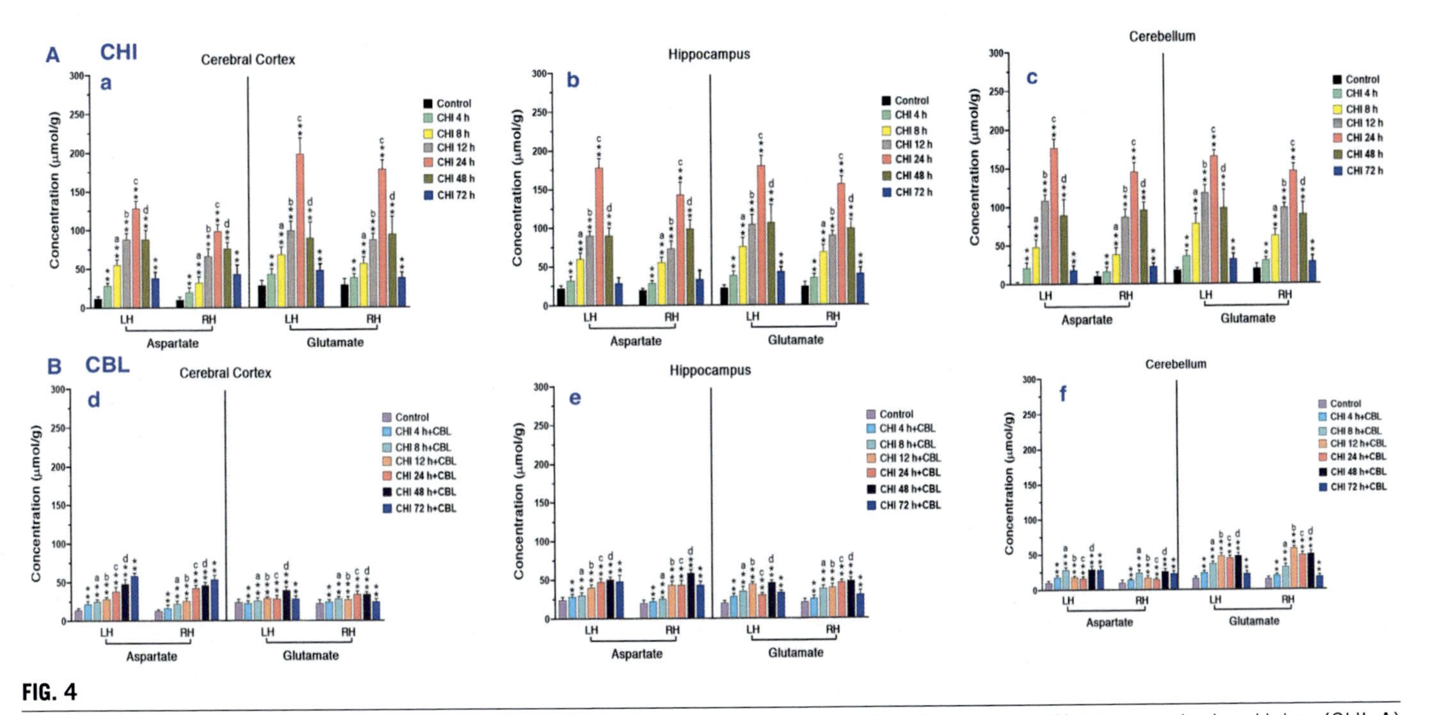

FIG. 4

Changes in Aspartate and Glutamate amino acids in cerebral cortex (A), hippocampus (B) and cerebellum (C) in concussive head injury (CHI, A) and following cerebrolysin (CBL, B) treatment with CHI in cerebral cortex (D), hippocampus (E) and cerebellum (F). ANOVA followed by Dunnett's test for multiple group comparison from one control showing $^{**}P<0.01$ from control; A = $P<0.05$ 8h from control; B = $P<0.05$ 12h from control; C = $P<0.01$ 24h from control; D = $P<0.01$ 48h from control. LH = Right injured half; LH = Left uninjured half.

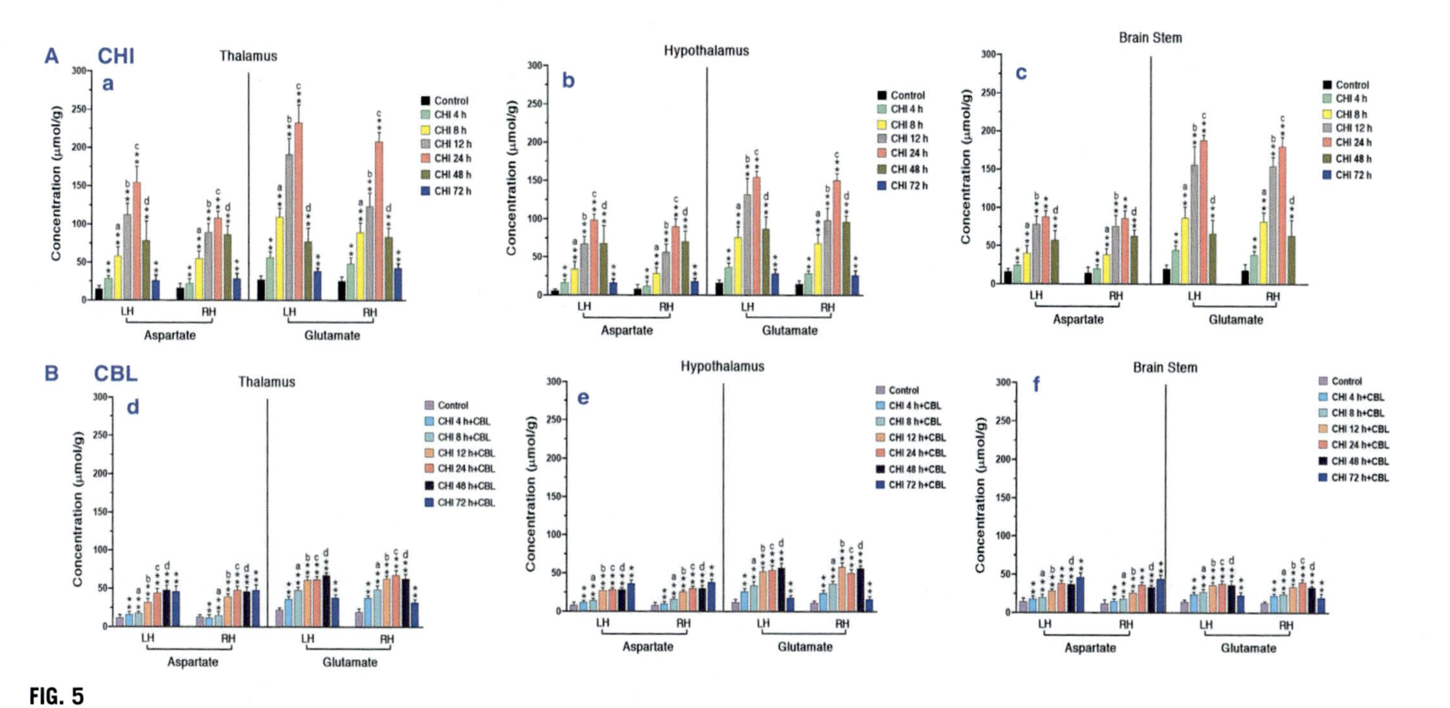

FIG. 5

Changes in Aspartate and Glutamate amino acids in thalamus (A), hypothalamus (B) and brain stem (C) in concussive head injury (CHI, A) and following cerebrolysin (CBL, B) treatment with CHI in thalamus (D), hypothalamus (E) and brain stem (F). ANOVA followed by Dunnett's test for multiple group comparison from one control showing $**P<0.01$ from control; A $= P<0.05$ 8h from control; B $= P<0.05$ 12h from control; C $= P<0.01$ 24h from control; D $= P<0.01$ 48h from control. LH = Right injured half; LH = Left uninjured half.

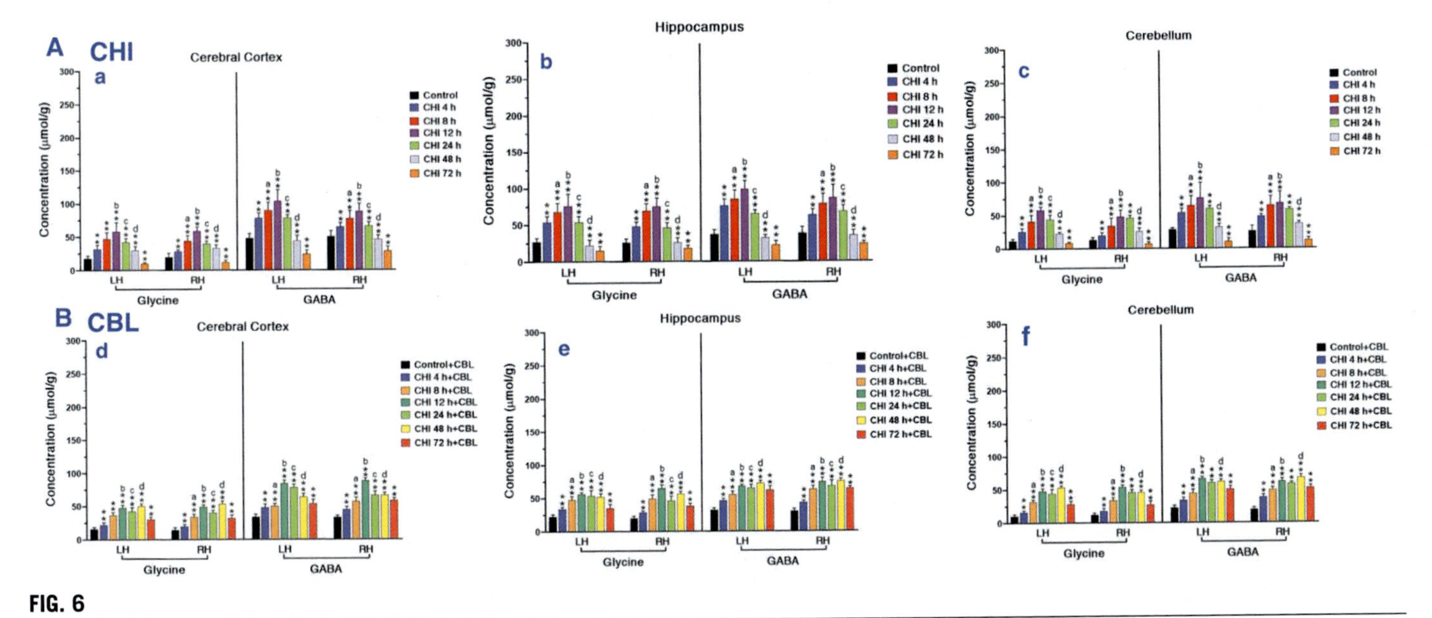

FIG. 6

Changes in Glycine and GABA amino acids in cerebral cortex (A), hippocampus (B) and cerebellum (C) in concussive head injury (CHI, A) and following cerebrolysin (CBL, B) treatment with CHI in cerebral cortex (D), hippocampus (E) and cerebellum (F). ANOVA followed by Dunnett's test for multiple group comparison from one control showing $^{**}P<0.01$ from control; A = $P<0.05$ 8 h from control; B = $P<0.05$ 12 h from control; C = $P<0.01$ 24 h from control; D = $P<0.01$ 48 h from control. LH = Right injured half; LH = Left uninjured half.

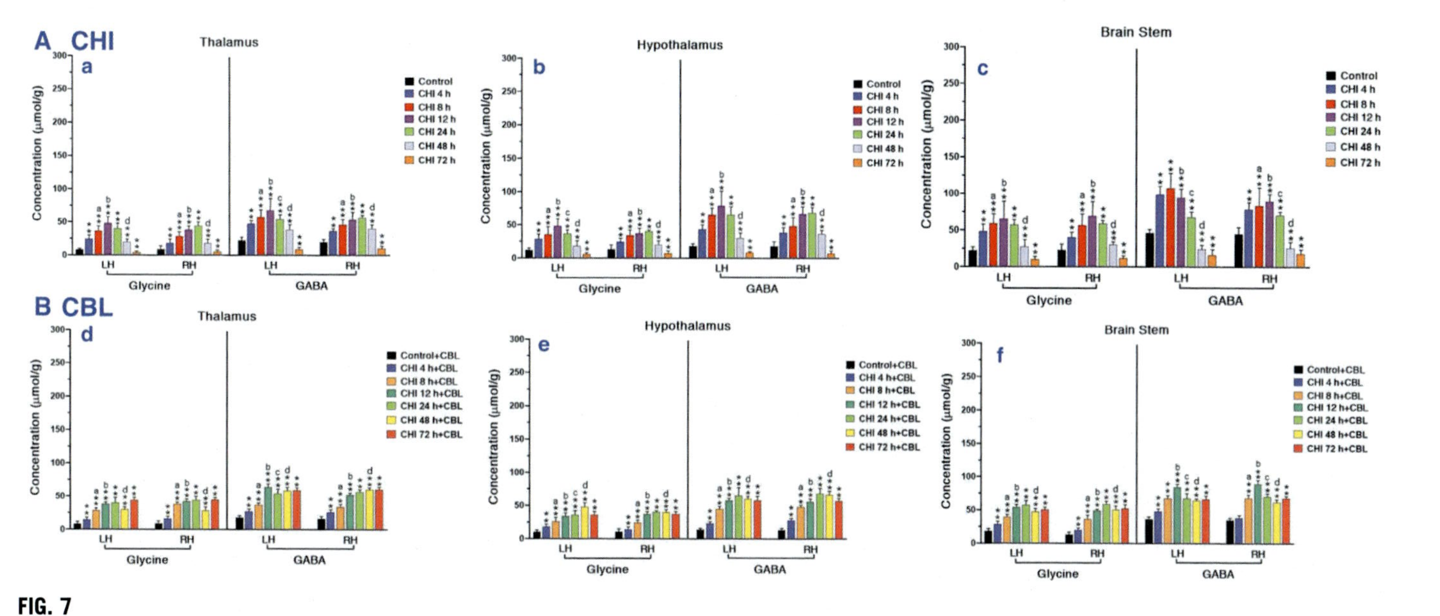

FIG. 7

Changes in Glycine and GABA amino acids in thalamus (A), hypothalamus (B) and brain stem (C) in concussive head injury (CHI, A) and following cerebrolysin (CBL, B) treatment with CHI in thalamus (D), hypothalamus (E) and brain stem (F). ANOVA followed by Dunnett's test for multiple group comparison from one control showing $^{**}P<0.01$ from control; A $= P<0.05$ 8h from control; B $= P<0.05$ 12h from control; C $= P<0.01$ 24h from control; D $= P<0.01$ 48h from control. LH = Right injured half; LH = Left uninjured half.

3.5 Blood–brain barrier permeability

Measurement of BBB breakdown in CHI was evaluated using EBA and radioiodine in several brain areas (Figs. 8–11). There was a significant increase in EBA leakage after 4h CHI that was progressive until 72h CHI in LH in the cerebral cortex (Fig. 8A). Whereas in hippocampus the BBB breakdown to EBA was progressive from 4h up to 24h CHI and then slightly declined in LH (Fig. 8B). In cerebellum (Fig. 8C) LH exhibited progressive increase in the EBA leakage from 4h CHI up to 48h CHI and then slightly declined. On the other hand EBA leakage in RH reached its peak progressively from 4h CHI to 24h CHI and then stayed at almost same level until 72h CHI in the cerebral cortex (Fig. 8A). In hippocampus (Fig. 8B) EBA reached its peak level in RH progressively after 4h CHI until 24h CHI and then gradually declined at 48–72h CHI. In cerebellum (Fig. 8C) EBA leakage showed a progressive increase in RH from 4h CHI to 48h CHI and then declined at 72h survival period (Fig. 8A).

In the thalamus (Fig. 9A), hypothalamus (Fig. 9B) and brain stem (Fig. 9C) showed EBA leakage in RH progressively from 4h CHI to 24h CHI and then slightly declined except in brain stem that exhibited continued progressive increase in BBB from 4 to 71h CHI (Fig. 9A). On the other hand RH showed similar progressive increase in BBB breakdown to EBA from 4h CHI to 24h CHI in thalamus (Fig. 9A) and hypothalamus (Fig. 9B) and then declined slightly until 72h CHI except in brain stem (Fig. 9C) where BBB breakdown to EBA continue until 72h CHI (Fig. 9A). In general LH exhibited higher leakage of EBA in all brain regions as compared to the RH from 4h CHI to 72h CHI period (Figs. 8A and 9A).

The leakage of radioiodine (Figs. 10A and 11A) in the cerebral cortex (Fig. 10A), hippocampus (Fig. 10B), cerebellum (Fig. 10C), thalamus (Fig. 11A), hypothalamus (Fig. 11B) and brain stem (Fig. 11C) showed almost similar increase in BBB with only minor variations in some brain areas after 4h to 72h CHI. In general the leakage of radioiodine in both LH and RH were slightly higher than EBA leakage in similar brain areas. However, leakage of radioiodine in CHI was higher in LH than in RH in all brain regions from 4 to 72h CHI groups (Figs. 10A and 11A).

3.6 Brain edema formation

Measurement of brain water content showed significant edema formation after CHI from 4 to 72h in cerebral cortex (Fig. 12A), hippocampus (Fig. 12B), cerebellum (Fig. 12C), thalamus (Fig. 13A), hypothalamus (Fig. 13B) and brain stem (Fig. 13C). In general LH showed greater accumulation of brain water as compared to the RH following CHI from 4 to 72h in all brain regions examined (Figs. 12A and 13A). This increase in brain water was progressive in both the LH and RH in most of the brain regions from 4 to 72h CHI (Figs. 12A and 13A).

Volume swelling (*%f*) calculated from differences in brain water content between control and after CHI revealed a progressive increase from 4 to 48h and then slightly declined after 72h in the cerebral cortex (Fig. 14A), hippocampus (Fig. 14B), cerebellum (Fig. 14C), thalamus (Fig. 15A), hypothalamus (Fig. 15B) and brain stem

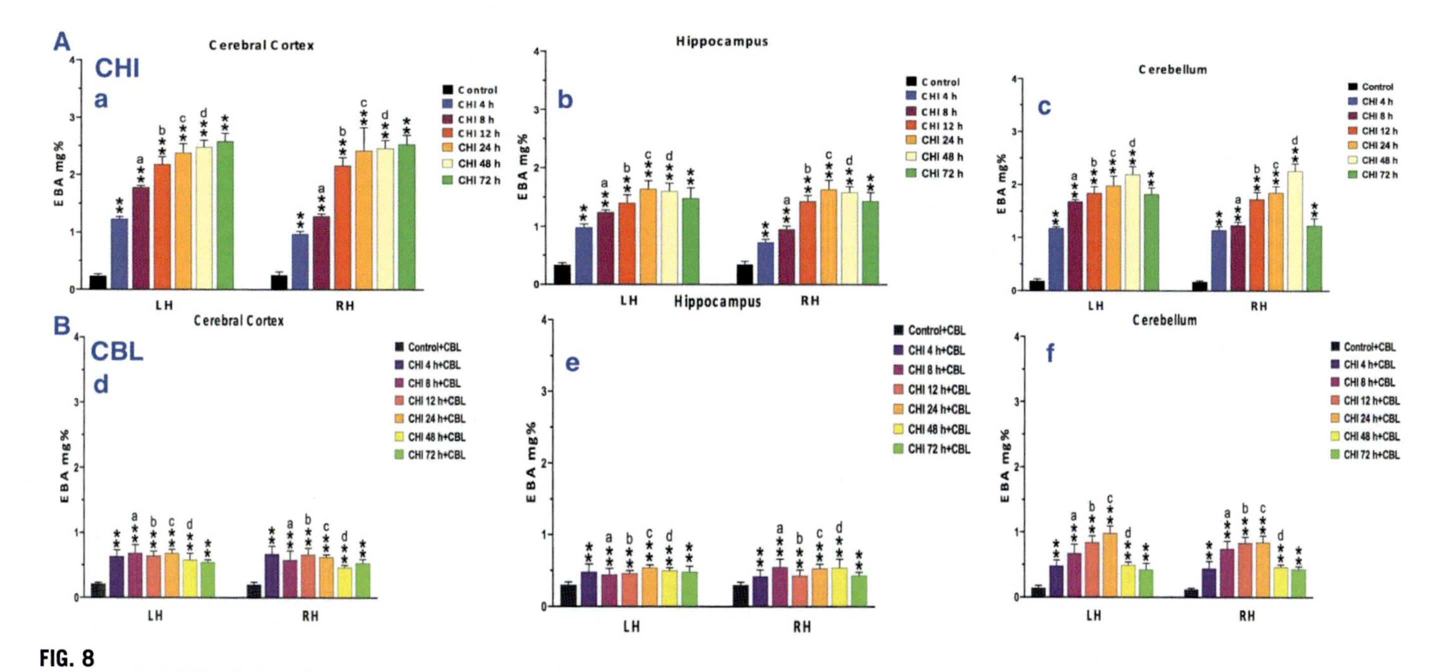

FIG. 8

Evans blue albumin (EBA) leakage across the blood-brain barrier (BBB) in cerebral cortex (A), hippocampus (B) and cerebellum (C) in concussive head injury (CHI, A) and following cerebrolysin (CBL, B) treatment with CHI in cerebral cortex (D), hippocampus (E) and cerebellum (F). ANOVA followed by Dunnett's test for multiple group comparison from one control showing $^{**}P<0.01$ from control; A $= P<0.05$ 8 h from control; B $= P<0.05$ 12 h from control; C $= P<0.01$ 24 h from control; D $= P<0.01$ 48 h from control. LH = Right injured half; LH = Left uninjured half.

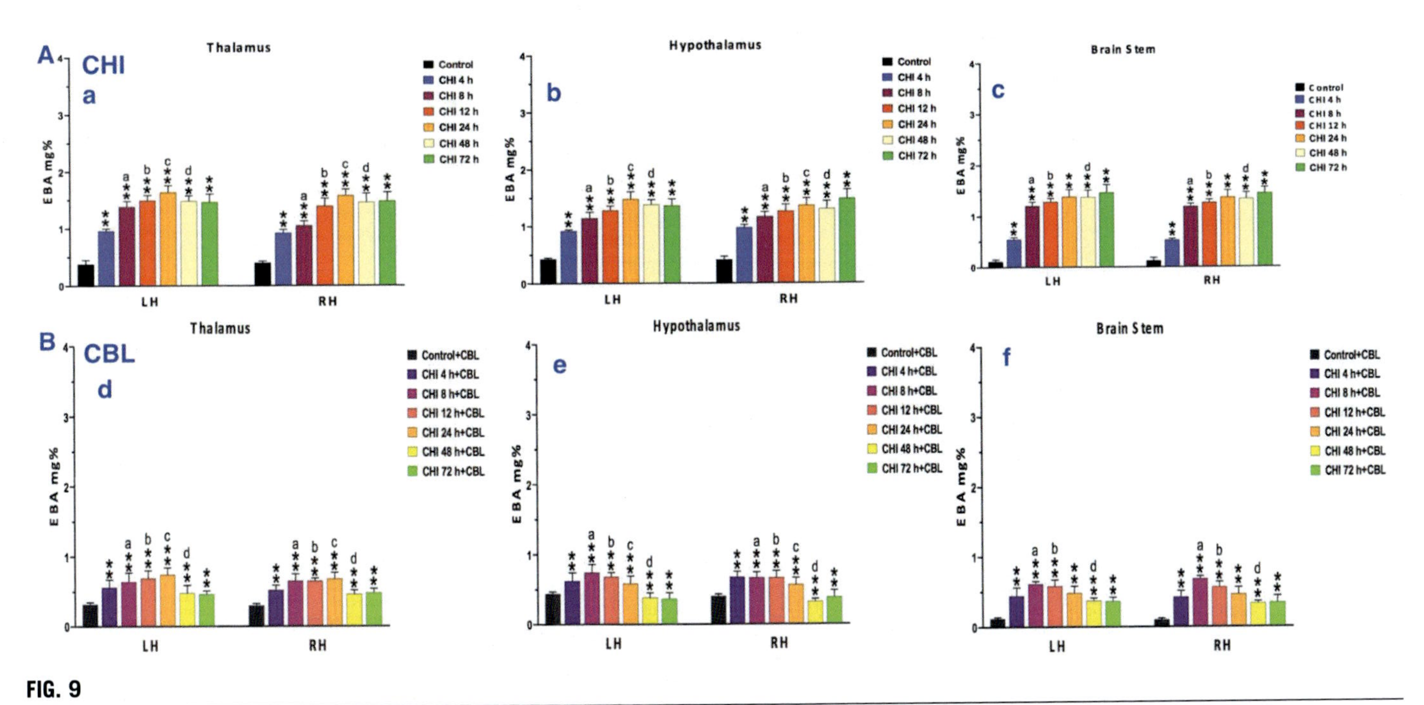

FIG. 9

Evans blue albumin (EBA) leakage across the blood-brain barrier (BBB) in thalamus (A), hypothalamus (B) and brain stem (C) in concussive head injury (CHI, A) and following cerebrolysin (CBL, B) treatment with CHI in thalamus (D), hypothalamus (E) and brain stem (F). ANOVA followed by Dunnett's test for multiple group comparison from one control showing $^{**}P<0.01$ from control; A $=P<0.05$ 8h from control; B $=P<0.05$ 12h from control; C $=P<0.01$ 24h from control; D $=P<0.01$ 48h from control. LH = Right injured half; LH = Left uninjured half.

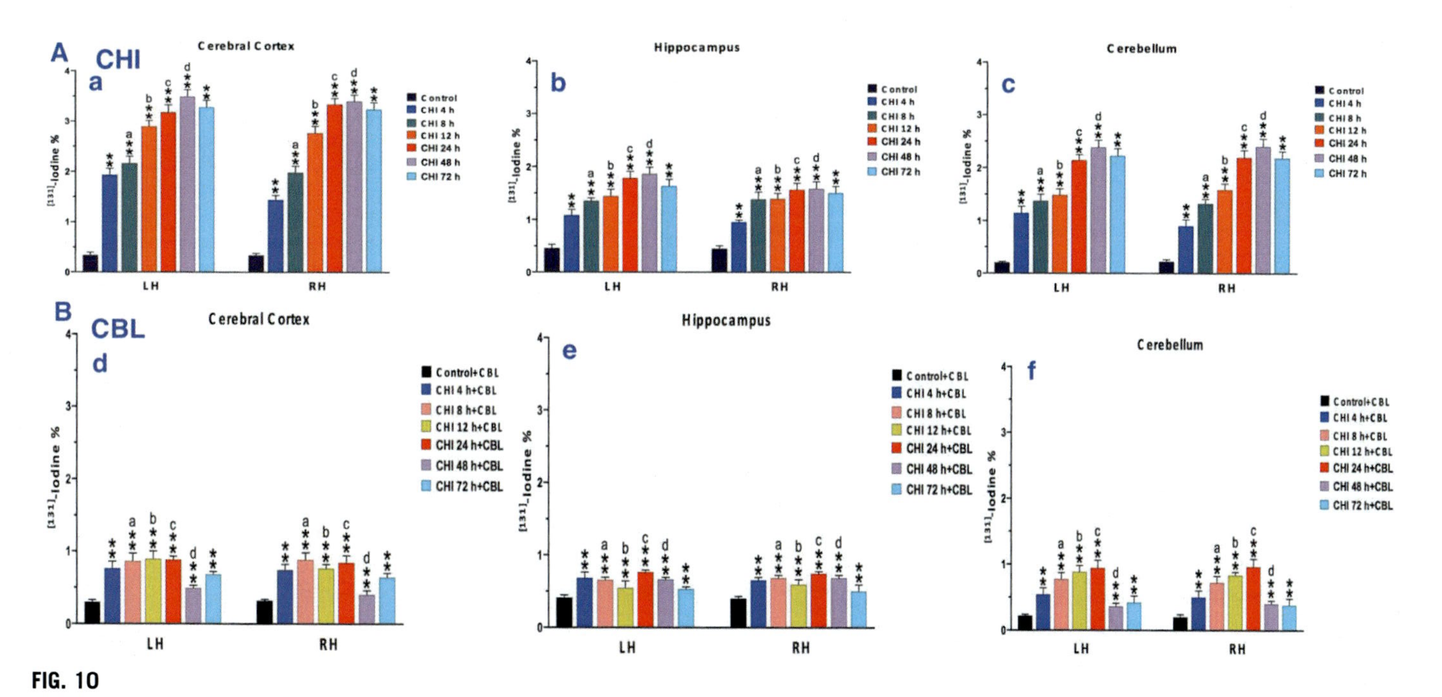

FIG. 10

Radioiodine ([131]-Iodine) leakage across the blood–brain barrier (BBB) in cerebral cortex (A), hippocampus (B) and cerebellum (C) in concussive head injury (CHI, A) and following cerebrolysin (CBL, B) treatment with CHI in cerebral cortex (D), hippocampus (E) and cerebellum (F). ANOVA followed by Dunnett's test for multiple group comparison from one control showing $^{**}P<0.01$ from control; A = $P<0.05$ 8 h from control; B = $P<0.05$ 12 h from control; C = $P<0.01$ 24 h from control; D = $P<0.01$ 48 h from control. LH = Right injured half; LH = Left uninjured half.

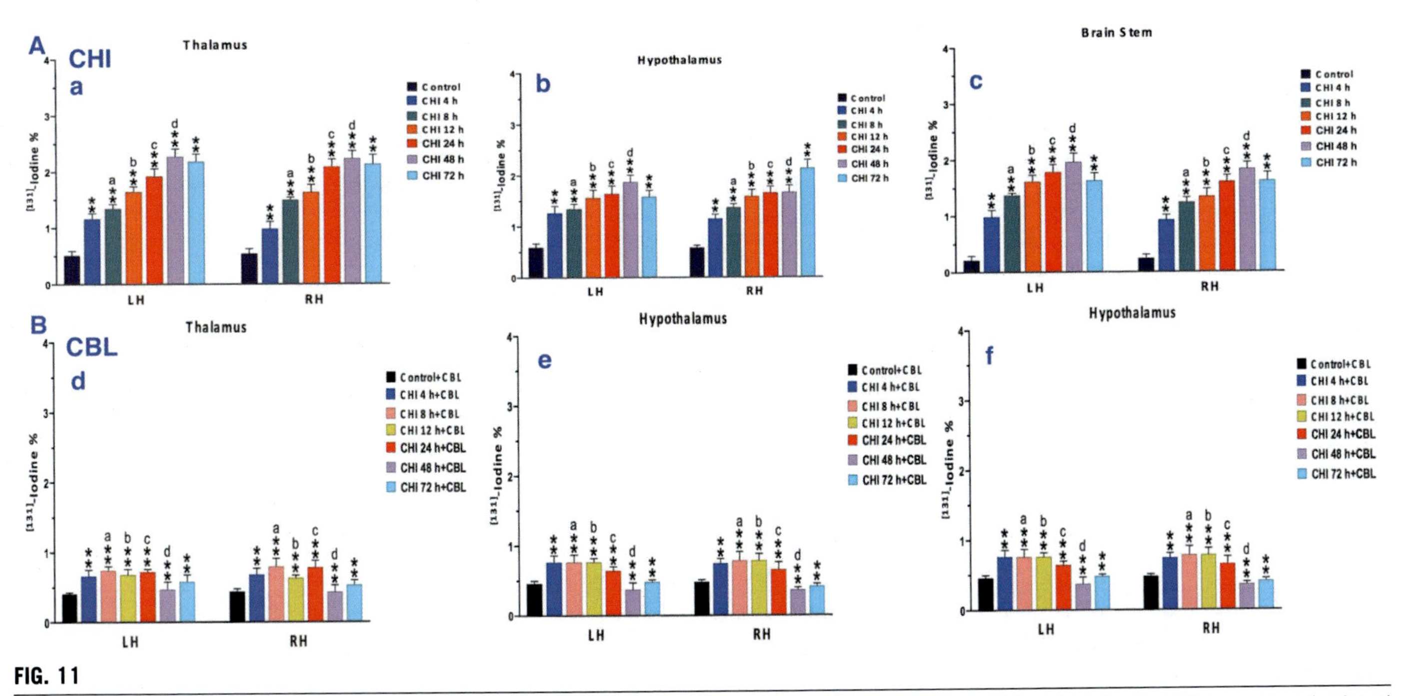

FIG. 11

Radioiodine ([131]-Iodine) leakage across the blood-brain barrier (BBB) in thalamus (A), hypothalamus (B) and brain stem (C) in concussive head injury (CHI, A) and following cerebrolysin (CBL, B) treatment with CHI in thalamus (D), hypothalamus (E) and brain stem (F). ANOVA followed by Dunnett's test for multiple group comparison from one control showing $**P<0.01$ from control; A = $P<0.05$ 8h from control; B = $P<0.05$ 12h from control; C = $P<0.01$ 24h from control; D = $P<0.01$ 48h from control. LH = Right injured half; LH = Left uninjured half.

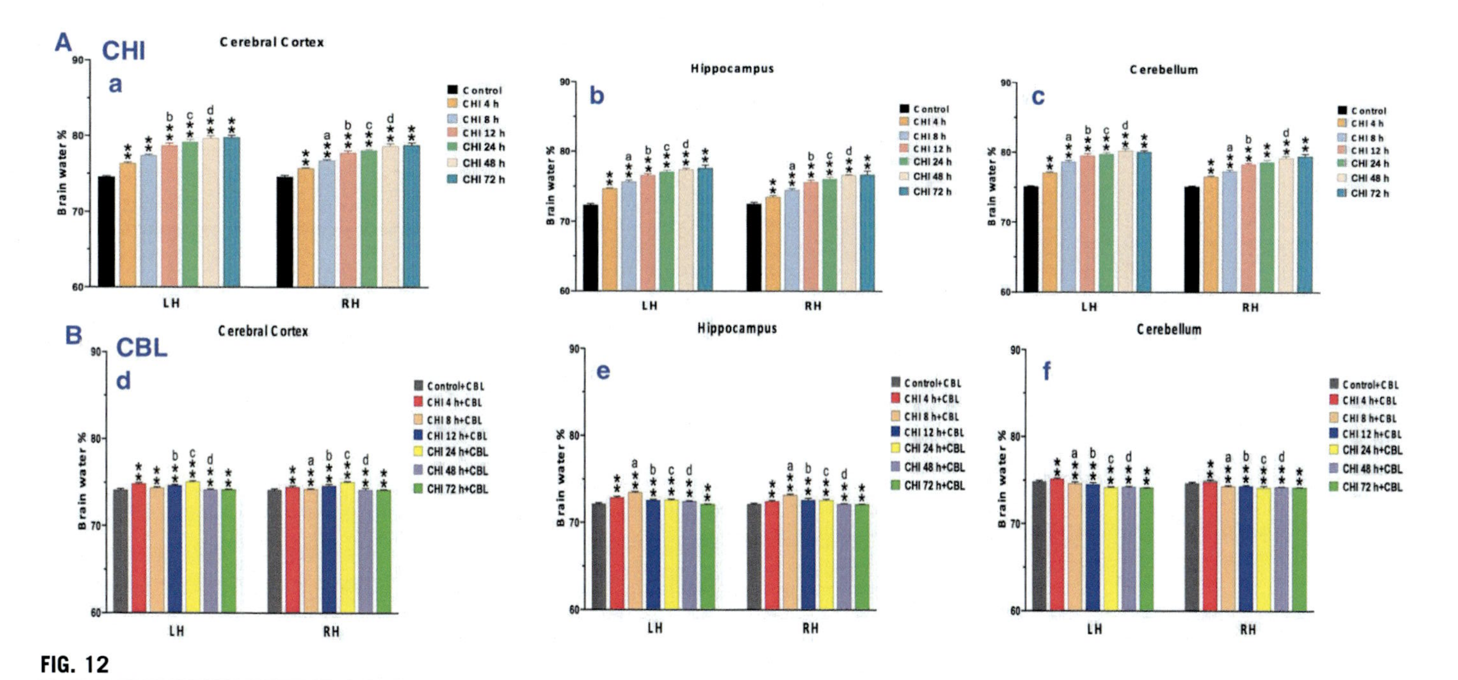

FIG. 12

Brain water content in cerebral cortex (A), hippocampus (B) and cerebellum (C) in concussive head injury (CHI, A) and following cerebrolysin (CBL, B) treatment with CHI in cerebral cortex (D), hippocampus (E) and cerebellum (F). ANOVA followed by Dunnett's test for multiple group comparison from one control showing $^{**}P<0.01$ from control; A = $P<0.05$ 8h from control; B = $P<0.05$ 12h from control; C = $P<0.01$ 24h from control; D = $P<0.01$ 48h from control. LH = Right injured half; LH = Left uninjured half.

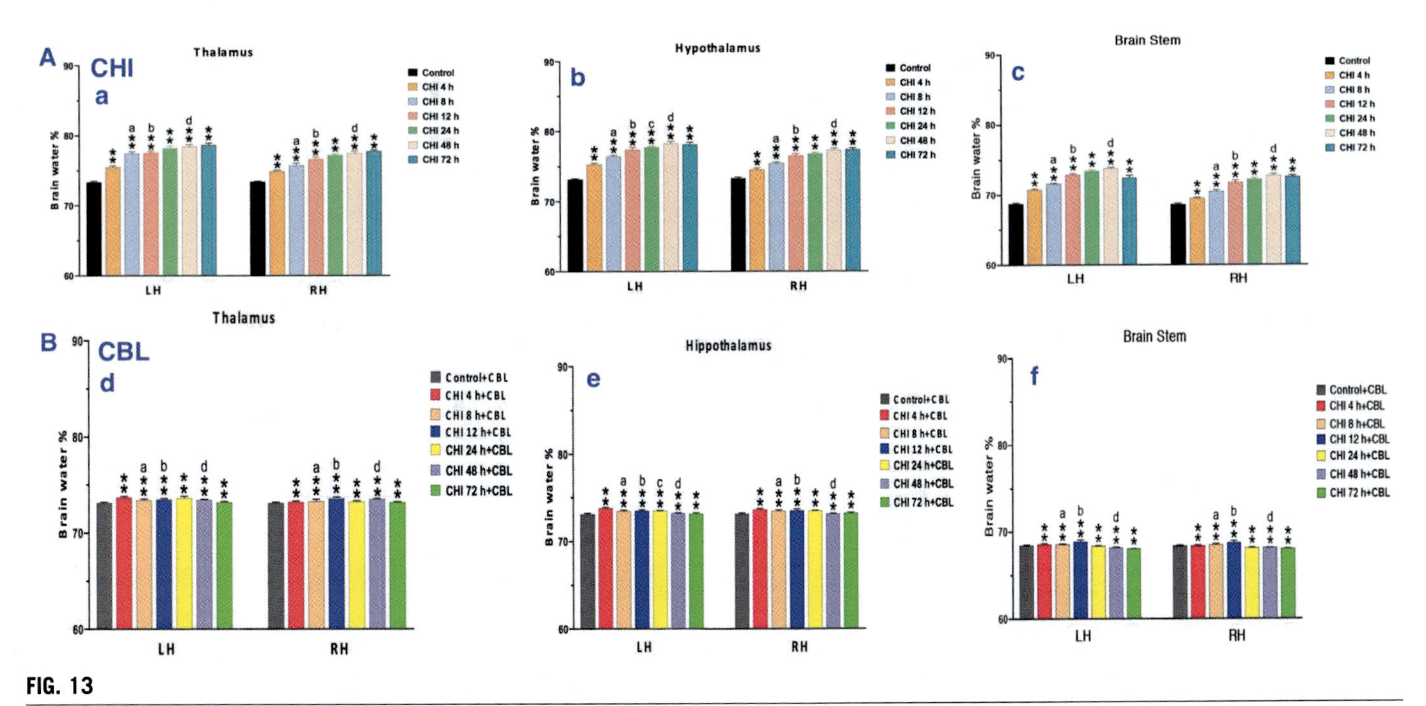

FIG. 13

Brain water content in thalamus (A), hypothalamus (B) and brain stem (C) in concussive head injury (CHI, A) and following cerebrolysin (CBL, B) treatment with CHI in thalamus (D), hypothalamus (E) and brain stem (F). ANOVA followed by Dunnett's test for multiple group comparison from one control showing **$P<0.01$ from control; A = $P<0.05$ 8h from control; B = $P<0.05$ 12h from control; C = $P<0.01$ 24h from control; D = $P<0.01$ 48h from control. LH = Right injured half; LH = Left uninjured half.

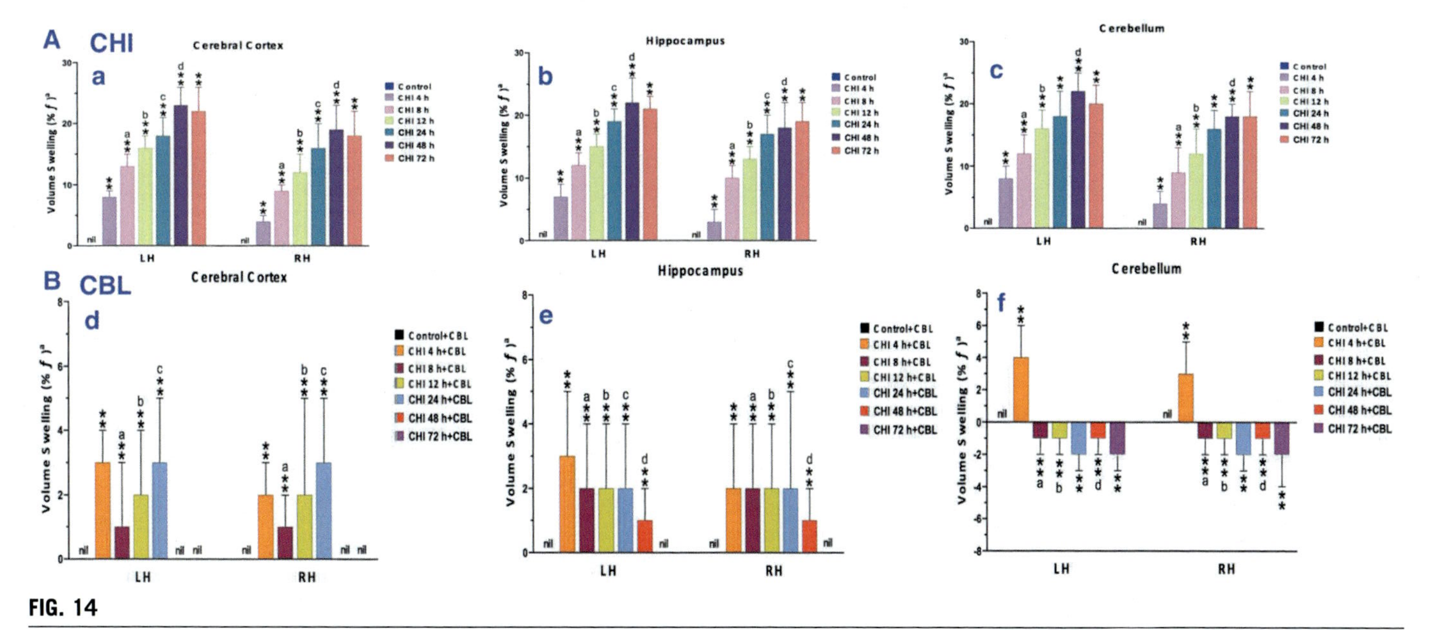

FIG. 14

Volume swelling calculated from changes in Brain water content in cerebral cortex (A), hippocampus (B) and cerebellum (C) in concussive head injury (CHI, A) and following cerebrolysin (CBL, B) treatment with CHI in cerebral cortex (D), hippocampus (E) and cerebellum (F). ANOVA followed by Dunnett's test for multiple group comparison from one control showing $^{**}P<0.01$ from control; A $= P<0.05$ 8h from control; B $= P<0.05$ 12h from control; C $= P<0.01$ 24h from control; D $= P<0.01$ 48h from control. LH = Right injured half; LH = Left uninjured half. Please note that the scale is different in cerebrolysin treated CHI to denote reduction in volume swelling at much lower scale as compared to the untreated CHI. For details see text.

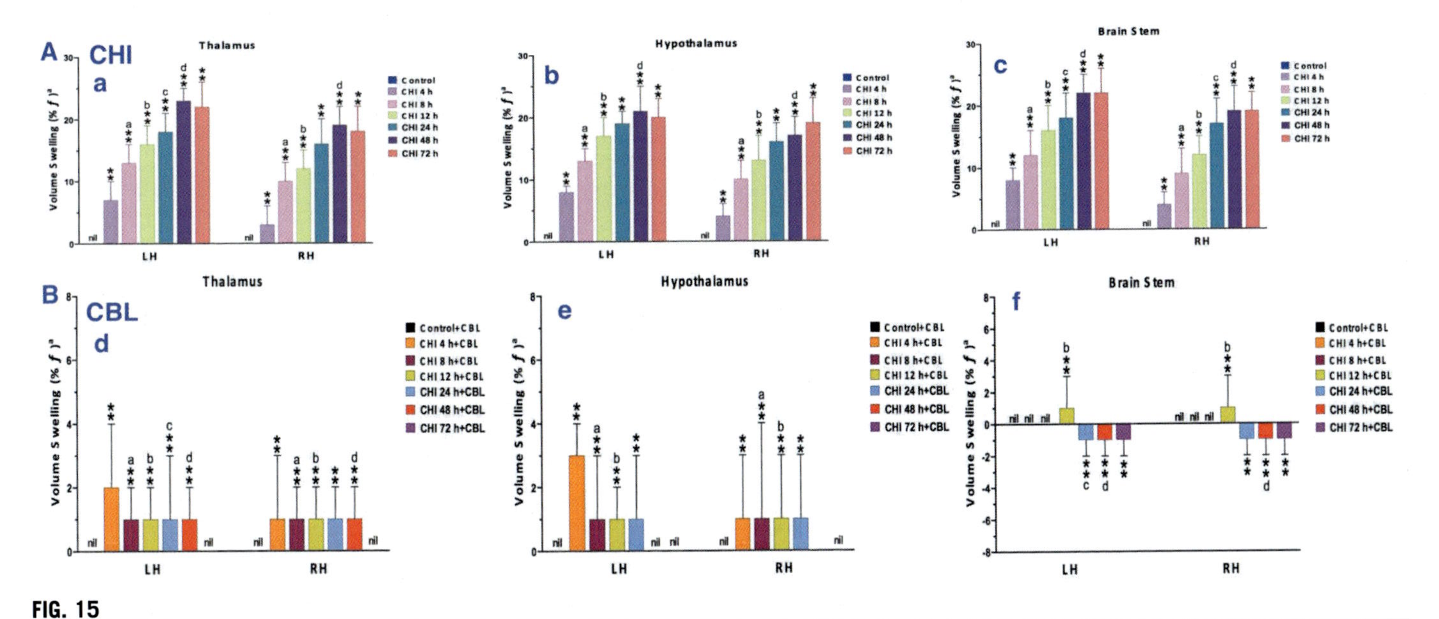

FIG. 15

Volume swelling calculated from changes in Brain water content in thalamus (A), hypothalamus (B) and brain stem (C) in concussive head injury (CHI, A) and following cerebrolysin (CBL, B) treatment with CHI in thalamus (D), hypothalamus (E) and brain stem (F). ANOVA followed by Dunnett's test for multiple group comparison from one control showing $^{**}P<0.01$ from control; A = $P<0.05$ 8 h from control; B = $P<0.05$ 12 h from control; C = $P<0.01$ 24 h from control; D = $P<0.01$ 48 h from control. LH = Right injured half; LH = Left uninjured half. Please note that the scale is different in cerebrolysin treated CHI to denote reduction in volume swelling at much lower scale as compared to the untreated CHI. For details see text.

(Fig. 15C) in both LH and in RH (Figs. 14A and 15A). In general the LH showed higher volume swelling (%f) after CHI as compared to the RH in identical brain regions and survival periods (Figs. 14A and 15A). Some minor variations are seen in some brain regions after CHI in brain edema or volume swelling (Figs. 12A–15A).

3.7 Brain pathology

Evaluation of brain pathology at light microscopy in CHI from 4 to 72h revealed progressive neuronal damages in the cerebral cortex, hippocampus, cerebellum, thalamus, hypothalamus and brain stem. Nerve cell damage showed shrunken and swollen neurons in the neuropil associated with perineuronal edema and sponginess of the neuropil. In general, the left hemisphere exhibited higher nerve cell injuries and brain damage as compared to the right hemisphere. A representative example of neuronal damages after 48h CHI is shown in the cerebral cortex in LH (Fig. 16A) and RH (Fig. 16B). Several neurons with dark appearance exhibiting profound perineuronal edema are clearly seen (Fig. 16A and B). Neuropil showed sponginess and edematous expansion in both halves of the brain. However, LH (Fig. 16A) exhibited higher neuronal injury and brain damage as compared to the LH (Fig. 16B) after 48 h of CHI on the RH. Similar changes were seen in hippocampus, cerebellum, thalamus, hypothalamus and brain stem (results not shown). In hippocampus dentate gyrus and CA1–CA3 exhibited most neuronal damages and expansion of the neuropil with sponginess that was more frequent in the LH than RH (H S Sharma unpublished observation). In the cerebellum both Purkinje cells and granule cells exhibited neuronal damage, swelling, degeneration in spongy neuropil with edematous expansion (results not shown). In brain stem reticular system several dark neurons were present in edematous neuropil (HS Sharma, unpublished observations). The neuronal damage could be seen from 4h CHI and progressively increased over 72h period of CHI (results not shown). The magnitude and intensity of neuronal damage after CHI was much more pronounced in the LH than in RH. This indicates that CHI induces global neuronal damages in the brain.

At transmission electron microscopy (TEM) these neuronal cells showed disintegration of karyoplasm, synaptic and dendritic cellular damage, membrane vacuolation and myelin vesiculation in the cerebral cortex, hippocampus, cerebellum, thalamus, hypothalamus and brain stem starting from 4h CHI and exhibited progressive cell damages up to 72h CHI observed in this study. A representative example of ultrastructural cell damage in the cerebral cortex after 48h CHI is shown in Fig. 17. In parietal cortex disintegration of neuronal karyoplasm, membrane damage and membrane vacuolation is clearly seen after 48h of CHI in LH (Fig. 17A). Myelin deformation and vesiculation is also prominent in in the neuropil of cerebral cortex after 48h CHI (Fig. 17B). Similar changes are observed in the hippocampus, cerebellum, thalamus, hypothalamus and brain stem in CHI (results not shown). These observations show that CHI is capable of ultrastructural cell damages in both LH and RH extensively.

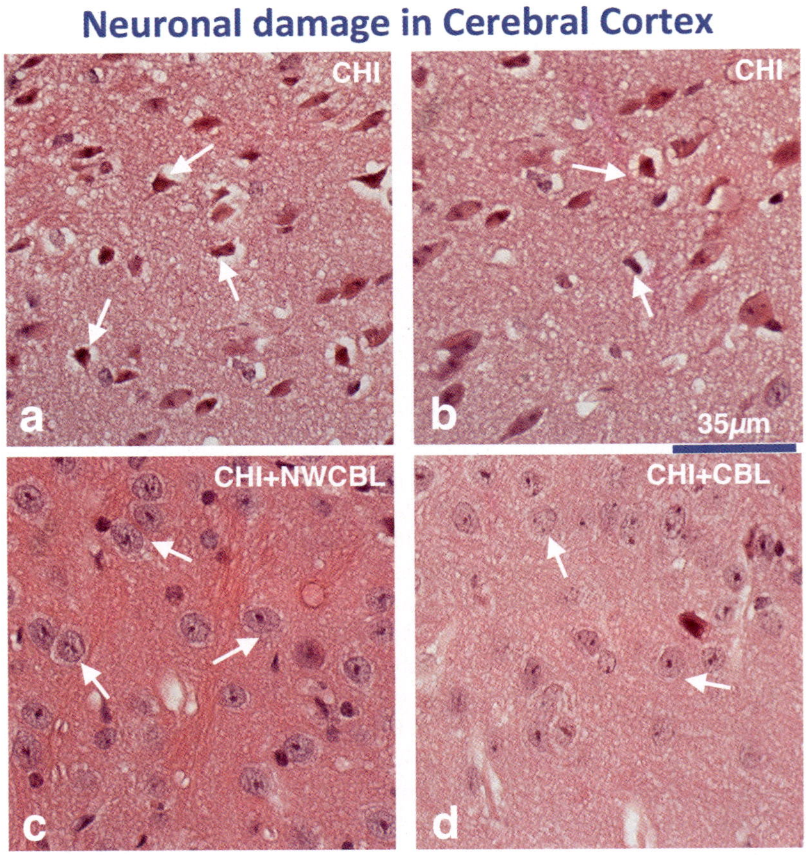

Neuronal damage in Cerebral Cortex

CHI CHI a b 35μm CHI+NWCBL CHI+CBL c d

FIG. 16

High power light micrograph of parietal cerebral cortex in concussive head injury (CHI) left half (A) and right half (B) and following nanowired cerebrolysin (NWCBL) treatment (C) or conventional cerebrolysin (CBL) treatment (D). In CHI several damaged and dark neurons (arrows) with perineuronal edema, sponginess and edematous expansion is seen in CHI that is more pronounced in the left half (A) as compared to the right half (B) after 48h injury. Treatment with CBL induced neuroprotection in CHI as seen by absence of neuronal damage (arrows) and expansion and sponginess of the neuropil is minimized (D). Whereas, NWCBL appears to induce superior neuroprotection in CHI as compared to conventional CBL (C). This is evident from appearance of much healthy nerve cells (arrows) and a better-preserved neuropil showing minimized sponginess and edematous expansion (C). H&E staining on paraffin section (3-μm), Bar = 35 μm.

3.8 Effect of cerebrolysin in concussive head injury

Treatment with cerebrolysin in CHI significantly improved behavioral, physiological, biochemical and pathological parameters in a dose-related manner. A brief description on neuroprotective effects of cerebrolysin is mentioned below.

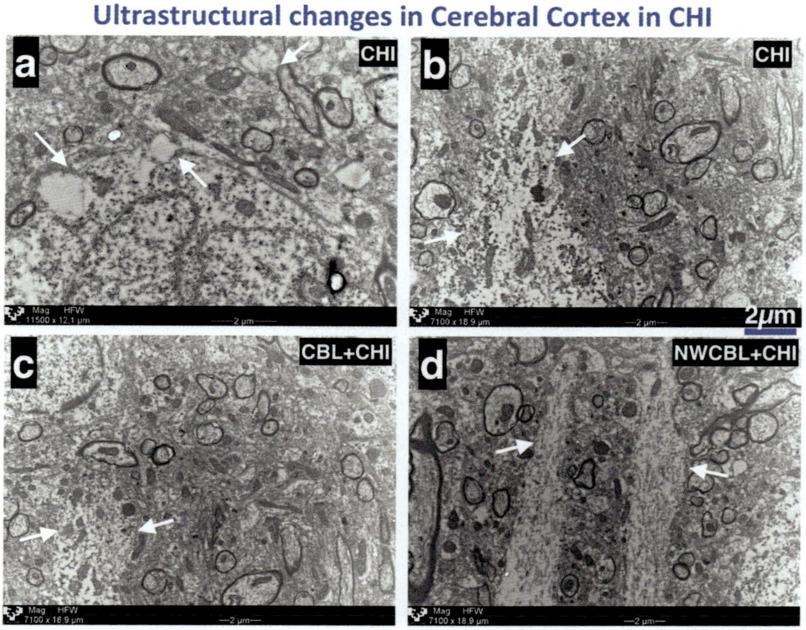

Ultrastructural changes in Cerebral Cortex in CHI

FIG. 17

High power Transmission Electron micrograph (TEM) showing axonal (arrow) and neuronal (arrow) damage (A) in the parietal cerebral cortex together with dendritic degeneration (arrows, B) after 48h concussive head injury. Treatment with conventional CBL resulted in compact neuropil with minimized damage (arrows) a dendrite (C). Well preserved two axonal structures (arrows) are seen in the cerebral cortex of NWCBL treated CHI group after 48h. (A) CHI, (B) CHI, (C) CHI+CBL, (D) CHI+NWCBL. CHI = concussive head injury; CBL = cerebrolysin; NWCBL = nanowired cerebrolysin. Bar=2μm.

3.9 Physiological variables

Cerebrolysin treatment in 5mL or 10mL/kg dose administered repeatedly after 4h CHI resulted in enhanced MABP following CHI after 4–72h (Fig. 2A). The arterial pH remained significantly higher in cerebrolysin treated CHI rats from 4 to 72h (Fig. 2C). The values of PaO_2 (Fig. 2B) and $PaCO_2$ (Fig. 2D) were aloe maintained at higher level in cerebrolysin treated CHI rats from 4 to 72h as compared to the untreated CHI group. Cerebrolysin treatment was able to maintain higher heart rate (Fig. 2E) and increased respiration cycle (Fig. 2F) in CHI following 4 to 72h CHI (Fig. 2).

3.10 Behavioral function

Cerebrolysin treatment significantly improved the behavioral functions after CHI in rats following 4–72h. Thus, staying time at Rota-Rod treadmill in cerebrolysin treated CHI rats increased significantly from 4 to 72h CHI (Fig. 3A). The angle of inclined plane on which cerebrolysin treated CHI rats could stay for 5 s has increased significantly (Fig. 3B). The number of steps taken by cerebrolysin treated CHI rats from 4 to 72 h has significantly increased (Fig. 3C) and the number of placement error on walking at mesh grid decreased significantly (Fig. 3D). In cerebrolysin treated CHI rats the stride length increased significantly in 4–72 h CHI group (Fig. 3E). Likewise, the distances between hind legs decreased in cerebrolysin treated CHI group from 4 to 72 h CHI (Fig. 3F).

3.11 Excitatory amino acids in brain

Treatment with cerebrolysin significantly reduced the increase of aspartate and glutamate in the cerebral cortex, hippocampus, cerebellum (Fig. 4B), thalamus, hypothalamus and brain stem (Fig. 5B) following 4–72 h CHI as compared to the untreated CHI group under identical conditions. This reduction was more pronounced in RH than the LH in cerebrolysin treated CHI group seen after 4–72 h CHI (Figs. 4B and 5B).

3.12 Inhibitory amino acids

Likewise cerebrolysin treatment in CHI resulted in significant alterations in glycine and GABA levels in LH and RH following 4–72 h periods in the cerebral cortex, hippocampus, cerebellum (Fig. 4B), thalamus, hypothalamus and brain stem (Fig. 5B) as compared to the untreated CHI group under identical conditions.

3.13 Blood–brain barrier

Cerebrolysin treatment in CHI significantly reduced BBB leakage of EBA in all brain regions examined (Fig. 8B and 9B) as compared to the untreated CHI group from 4 to 72 h CHI. Likewise, leakage of radioiodine in CHI was also reduced in cerebrolysin treated group following 4–72 h in the cerebral cortex, hippocampus, cerebellum (Fig. 10B), thalamus, hypothalamus and brain stem (Fig. 11B) as compared to the untreated CHI groups under identical conditions.

3.14 Brain edema

Cerebrolysin treatment in CHI resulted in significant reductions in brain water content following CHI after 4–72 h period in the cerebral cortex, hippocampus, cerebellum (Fig. 12B), thalamus, hypothalamus and brain stem (Fig. 13B) as compared to the untreated CHI group under identical conditions.

Volume swelling (%*f*) was also reduced in cerebrolysin treated CHI group after 4–72 h periods in the cerebral cortex, hippocampus, cerebellum (Fig. 14B), thalamus, hypothalamus and brain stem (Fig. 15B) as compared to the untreated CHI group under identical conditions.

3.15 Brain pathology

Cerebrolysin treatment in high doses was the most effective in reducing brain pathology after CHI from 24h survival to 72h CHI. In these cases neuropil was quite compact and edematous expansion was markedly reduced in CHI after cerebrolysin treatment. A representative example of 48h CHI with cerebrolysin treatment was shown in the cerebral cortex (Fig. 16D). As evident with the figure neuronal changes were markedly reduced by cerebrolysin treatment after 48h CHI. The nerve cells exhibit near normal appearances with a distinct cell nucleus and nucleolus is also clearly visible. Perineuronal edema is largely absent and sponginess of the neuropil and edematous expansion are much less evident (Fig. 16D). Similar protective change in the nerve cells was also seen in the hippocampus, cerebellum, thalamus, hypothalamus and brain stem (results not shown).

At TEM the neuronal structures, cytoplasm and karyoplasm appears quite normal and incidences of membrane vacuolation, synaptic damage, deformation of myelin and vesiculation were markedly reduced in the cerebral cortex, hippocampus, cerebellum, thalamus, hypothalamus and brain stem (results not shown). In a representative example of TEM from the cerebral cortex of 48h CHI is shown (Fig. 17C). As evident from the figure membrane vacuolation, myelin deformity, neuronal cell adages are much less evident and the cerebral cortex exhibited quite compact neuropil in cerebrolysin treated 48h CHI rat. These observations clearly show that high doses of cerebrolysin are quite effective in reducing CHI induced brain damage.

3.16 Nanodelivery of cerebrolysin

TiO_2 nanodelivery of cerebrolysin was done in identical manner in CHI using 5mL/kg following 24–72h CHI. We found that 5mL/kg dose was sufficient enough to reduce BBB breakdown to radioiodine, brain edema and volume swelling together with brain pathology (Figs. 16C and 17D). Treatment with nanowired cerebrolysin (NWCBL) following 48h or 72h CHI significantly reduced radioiodine extravasation in the cerebral cortex, cerebral edema formation and volume swelling as well as significant reductions in nerve cell damages.

A representative example of NWCBL treatment in 48h CHI showed marked neuroprotection in nerve cells in the cerebral cortex (Fig. 16C). Several nerve cells were round in appearance with distinct cell nucleus exhibiting clear nucleolus after 48h CHI in NWCBL treated group. The perineuronal edema, sponginess of the neuropil and edematous expansion were not evident in NWCBL treated CHI 48h rat (Fig. 16C). These neuroprotective effects appear to be superior in NWCBL treatment as compared to 10mL/kg conventional cerebrolysin (CBL) treatment in 48h CHI (Fig. 16D). Similar changes are seen after 72h CHI in NWCBL treated cerebral cortex, hippocampus and cerebellum (results not shown).

At the ultrastructural level NWCBL was able to protect nerve cell axons or dendrites within the neuropil (Fig. 17D) along with preservation of other cellular structures markedly in the cerebral cortex (Fig. 17C). Comparison of conventional cerebrolysin (CBL) treatment with 10 mL/kg, NWCBL treatment in 5 mL/kg appears to be superior in protecting ultrastructural changes at 48 h CHI (Fig. 17). Similar protection was also seen by NWCBL in the cerebral cortex after 48 h CHI (results not shown).

These observations suggest that NWCBL has superior neuroprotection abilities following 48–72 h of CHI.

4 Discussion

The salient new findings from this investigation clearly show that CHI for longer-term survival induces significant increase in excitotoxicity in the brain in a time-dependent manner. This is evident from the findings that the excitatory amino acids aspartate and glutamate show a progressive rise after CHI from 4 to 48 h survival and then decline slightly in the brain. It remains to be seen if the survival periods extend beyond 72 h after CHI whether these excitatory amino acids aspartate and glutamate could reach to base line before injury. This is a feature that requires additional investigation. These observations are in line with the idea that brain injury alters amino acids metabolism (Guerriero et al., 2015; Hajiaghamemar et al., 2020; Hayes et al., 1992; Sharma et al., 2018a,b; Wang et al., 2016).

It appears that to counteract excitotoxicity in CHI endogenous mechanisms of increase in inhibitory amino acids glycine and GABA also showed similar changes in the brain with peak rise occurs around 24 h after injury and then gradually declines over 72 h periods. This suggests that both excitatory and inhibitory amino acids increase over time to counterbalance cell injury after CHI, not reported earlier. These observations indicate that in CHI several endogenous neurodestructive elements like aspirate and glutamate are released together with several neuroprotective agents like glycine and GABA are aslo released in CHI to counteract the endogenous balance to for cell survival (Cai et al., 2019; Chen et al., 2019, 2020; Chitturi et al., 2018; Dempsey and Başkaya, 2000; Gu et al., 2016; Hosseini-Zare et al., 2017; Kierans et al., 2014; López-Picón et al., 2016; Shannon et al., 2016; Yasen et al., 2018; Zhang et al., 2020). Our studies are thus clearly suggests an active involvement of amino acids in the brain pathology of CHI.

Previous reports from our laboratory show that CHI in rats is associated with profound increase in brain serotonin, prostaglandin, neuropeptide and histamine causing neuroinflammation, breakdown of the blood–brain barrier (BBB), edema formation and cellular injuries (Dey and Sharma, 1983, 1984; Johanson et al., 2011; Mohanty et al., 1989; Olsson et al., 1992, 1995; Patnaik et al., 2000; Sharma, 2000, 2006, 2007a,b,c; Sharma and Sharma, 2012a,b; Sharma et al., 1993, 1997a,b, 2000a,b,c, 2017; Vannemreddy et al., 2006; Winkler et al., 1994). Thus, treatment with specific receptor or synthesis blocker of these neurotransmitters achieves neuroprotection in short survival periods of 5 h. However, large number of evidences also suggests an increase in excitotoxicity that is primarily responsible for neurotoxicity seen in traumatic brain injuries (TBI) in several preclinical and clinical conditions (Leist and

Nicotera, 1998; Woitzik et al., 2014; Nicotera et al., 1997; Matute, 2011). In ischemic brain injury models, an increase in glutamate and aspartate is observed. Whereas, enhanced GABA and glycine level is also seen in ischemia that is supposed to counteract the neurotoxicity caused by excitotoxicity (Khoshnam et al., 2017; Krzyżanowska et al., 2017; Martin et al., 1998; Vercelli et al., 2015).

Brain injury either caused by concussion (CHI) or perforating brain injury induces significant morbidity and mortality. Traumatic brain injury (TBI) leads to disruption of the BBB, brain edema formation and neuronal cell injury (Dey and Sharma, 1983, 1984; Mohanty et al., 1989; Sharma et al., 2016a,b). Likewise CHI depending on the severity of the cases results in excessive brain swelling and cellular injury that is affected in the whole brain within the closed cranium (Dey and Sharma, 1984; Sharma et al., 2020a,b). The pathological shift following CHI could results in excess excitability of the cerebral component and disturbs the balances between excitation and inhibition ratio.

Glutamate and GABA are the principal excitatory and inhibitory neurotransmitters in the brain (Fonnum, 1984; Fonnum and Storm-Mathisen, 1969; Fonnum and Walberg, 1973; Fykse and Fonnum, 1996; Gundersen et al., 2001). A balance between glutamate and GABA is crucial for physiological functions of the brain (Sharma, 2006, 2007a,b,c). Release of glutamate following CHI leads to excitotoxicity causing neuronal injuries, cell death and alterations in several neuronal functions (Cao et al., 2019; Petroff, 2002; Steel et al., 2020). Continued excitotoxicity after CHI disrupts the neuronal networks leading to cognitive and motor dysfunctions and alter neuronal plasticity responses (Cheriyan and Sheets, 2018; Rebec, 2018).

Glutamate is produced by the pyramidal neurons in the cerebral cortex and hippocampus as well as the nerve cells from cerebellum, thalamus, hypothalamus, and brain stem. Glutamate is the main excitatory signaling pathways in the brain (Agnesi et al., 2010; John et al., 2014; Marvin et al., 2013; Oja and Saransaari, 2009; Paukert et al., 2010; Su et al., 2019). On the other hand, GABA is produced within the interneurons and modulates cortical and thalamocortical circuits of inhibitory signals (Bryson et al., 2020; Herrera et al., 2016). Release of GABA could coordinate sensory-motor functions and is associated with memory and attention phenomena (Heaney and Kinney, 2016; Hoshino et al., 2018; Paine et al., 2020). Damage to GABA neurons alters the balance between excitatory and inhibitory functions resulting in neurodegeneration and other brain diseases (Leonelli et al., 2009). Glutamate excitotoxicity results in epilepsy, ischemia and other neurodegenerative diseases (Hanada, 2020; Lau and Tymianski, 2010). Imbalances between excitatory and inhibitory signals occur in TBI causing cerebral microhemorrhages and other vascular disorders, neuronal injuries, axonal damages and oxidative stress that all could result in profound brain pathology (Guerriero et al., 2015; Willis et al., 2004).

After a focal brain injury or CHI secondary injury cascade results in tissue damage, brain swelling, microvascular leakage and ischemia (Botteri et al., 2008; Jha et al., 2019; Webster et al., 2017). Glutamate induced excitotoxicity further spread the tissue damage in the surrounding tissue similar to penumbra in ischemic injuries

(Obrenovitch and Urenjak, 1997). Concussion further induces diffuse injury in the widespread areas in the brain including axonal injuries. At the molecular level, CHI causes membrane swelling and ion release into the extracellular space as well as Na+/K+ pump failure (Nathanson et al., 1995). Release of glutamate in CHI causes intracellular accumulation of Ca++ perpetuating cell injuries (Belov Kirdajova et al., 2020). This idea is further supported by clinical studies demonstrating extracellular glutamate increase following TBI that is well correlating with post-traumatic mortality (Yasen et al., 2020). At this time changes in GABA is closely related to the glutamate induced excitatory signals (Almeida-Suhett et al., 2015).

Apart from GABA, another inhibitory amino acid glycine receptors are present throughout all the regions in the hippocampus and play active roles in regulating excitability and plasticity (Chen et al., 2014; Lovett-Barron et al., 2014). Activation of glycine causes hyperpolarization due to increased Cl- flux and suppresses neuronal excitability regulating excitation-inhibition balance in the brain during TBI (Zhong et al., 2003). These studies suggest that glycine has neuroprotective ability. When concentrations of glycine increased in the brain it activates extrasynaptic glycine receptors to activate inhibitory effects to restore excitation-inhibition imbalance to achieve neuroprotection (Ye, 2008).

Aspartate is another excitatory amino acid that is involved in excitotoxicity together with glutamate (Cavallero et al., 2009; D'Aniello et al., 2011; Errico et al., 2009; Morland et al., 2012; Wenthold, 1981; Wiklund et al., 1982). Alterations in aspartate levels in brain injury in clinical and preclinical studies are well documented (Sager et al., 1997; Di Pietro et al., 2014; Madhavarao et al., 2004; Chakraborty et al., 2001; Moffett et al., 2007; Baslow, 2003; Tsai and Coyle, 1995; Belli et al., 2006; De Stefano et al., 1995; Al-Samsam et al., 2000; Shannon et al., 2016). Aspartate is needed for myelin function and has a modulatory role on glutamate induce excitotoxicity (de Rosa et al., 2019; Fannon et al., 2015; Fisher et al., 1986). In CHI we have seen both glutamate and aspartate increased from 4 to 72 h CHI. In some cases a decrease in aspartate is seen after brain injury in clinical settings (Shannon et al., 2016). Thus suggests that the magnitude and severity of brain injury is responsible for fluctuating aspartate levels in the brain.

Aspartate like glutamate is one of the major excitatory neurotransmitters in cortico-striatal pathways (Herrera-Marschitz et al., 1997; Meana et al., 1991). Like glutamate, aspartate is also released from nerve terminals in the brain under physiological or pathological conditions (Herrera-Marschitz et al., 1998; You et al., 1996). Released aspartate is taken up by surrounding neural and nonneuronal elements such as astrocytes and myelin (Herrera-Marschitz et al., 1998, 1997). Elevated levels of aspartate lead to overstimulation of NMDA-receptor subtypes to induce excitotoxicity (Krashia et al., 2016; Tranberg et al., 2004).

These observations support the idea that a balance between excitatory and inhibitory amino acids in the brain is the key for neuroprotection and imbalances among them leads to brain pathology and neurodegeneration (Sharma, 2006, 2007a,b,c). In TBI release of glutamate and GABA interacts with various other neurotransmitters and immune system. Thus, a potential strong interaction exists among amino acids neurotransmitters with serotonin, prostaglandins, neuropeptides and neurotrophins in the neuronal circuitry of the brain (Herrera-Marschitz et al., 1997; Lasoń, 1997; Randić et al., 1995; Shah and González-Maeso, 2019; van der Mast and Fekkes, 2000).

Several evidences suggest that neurotrophins like BDNF interact with glutamate to reduce neurotoxicity (Baker et al., 1991; Balazs, 2006; Gulyaeva, 2017; Kim et al., 2019; Prehn et al., 1993). Thus, treatment with neurotrophic factors reduces neurotoxicity after brain injury (Alberch et al., 2002; Mattson, 1990; Wiese et al., 1999). Moreover serotonin and glutamate interaction plays key role in sleep deprivation, ischemia, and TBI in brain pathology (Maloney et al., 1999, 2000; Sharma et al., 2016a,b). Treatment with glutamatergic receptor antagonists or serotonergic receptor antagonists is able to induce neuroprotection in various brain diseases including TBI (Sharma, 2004a,b,c). This suggests that alterations in amino acid metabolism in brain injury perturb various other neurotransmitters metabolism in brain.

We used cerebrolysin—a balance composition of several neurotrophic factors and active peptide fragments that are able to thwart brain pathology and alters glutamate, GABA, aspartate and glycine levels in CHI and induce neuroprotection. This effect of cerebrolysin is further enhanced with TiO_2 nanowired cerebrolysin administration in CHI. These observations indicate that neurotrophins induce neuroprotection after CHI and modulates amino acids metabolism in the brain. Potential superior neuroprotection caused by nanowired cerebrolysin even in low doses indicate that nanodelivery of drugs has the future clinical potentials in treating brain diseases. There are reasons to believe that when drugs are delivered through nanowired technology they could get quick and long term access into the brain fluid microenvironment to induce better beneficial effects (Sharma, 2006, 2011; Sharma and Sharma, 2012a,b, 2013; Sharma et al., 2016a,b,c). The nanowired could penetrate cell membranes by piercing them without making any injury and could release their content inside the cell or extracellular microenvironment to induce greater effects in the brain (Sharma et al., 2016a,b). Since drugs are bound to nanowires in the brain so that there metabolism by endogenous enzymes and adverse factors are slowed down considerably (Sharma and Sharma, 2012a,b, 2013; Sharma et al., 2016a,b). Thus, nanodelivery of drugs in the brain may be similar to slow released pumps that could last longer time to induce beneficial effects within the brain structures (Sharma et al., 2016a,b).

We found that treatment with conventional cerebrolysin or nanodelivered cerebrolysin is able to reduce the amino acids metabolism in the brain after CHI. Obviously, a reduction in amino acids in the brain could counteract the excitotoxicity and reduce cell injury. Also reduction in brain pathology affects improved functional behaviors as seen in our experiments after CHI (Sharma, 2011).

Amino acids induced excitotoxicity could affect microvasculature in the brain and alter BBB breakdown (Sharma, 2006, 2007a,b,c; Vazana et al., 2016). Significantly reduced BBB breakdown to large molecular tracers such as EBA or radioiodine reduces brain edema formation and volume swelling (Sharma et al., 1998a,b, 2000a,b,c). This is clearly evident in cerebrolysin treated group of CHI where brain edema and volume swellings are reduced in several brain areas examined.

When BBB breakdown is reduced and edema formation is thwarted in CHI it will lead to reduce brain pathology as seen in our investigations using light microscopy and transmission electron microscopy. As evident with our findings cerebrolysin treatment reduced nerve cell injury, edematous expansion and sponginess in the neuropil of several brain areas as seen in the cerebral cortex. At TEM axonal and dendritic structures appear well preserved after CHI in cerebrolysin treated group together with reductions in amino acid metabolism. These findings are inline with the idea that neurotrophins regulate amino acid metabolism and excitotoxicity in the brain following CHI.

Since serotonin, prostaglandin or neuropeptide also interacts with amino acid in the brain it would be interesting to see whether drugs affecting these neurotransmitters metabolism or their receptors could also affect amino acids levels in the brain and reduce brain pathology in CHI. This is a feature that require additional investigation.

In summary, our results are the first to show that cerebrolysin is able to restore balances between excitatory and inhibitory amino acids metabolism in CHI and reduce brain pathology. Further studies are needed to find out if blockade of glutamate or GABA receptors could affect amino acid metabolism and brain pathology. In addition whether aspartate and glycine receptors are equally involved in brain pathology of CHI, a feature that is currently being investigated in our laboratory.

Another important aspects in brain injury are the mode of trauma inflicted on the brain in relation to development of pathological reactions. Thus, it would be interesting to find out whether TBI caused by piercing lesion of the brain could also affect the amino acids metabolism and brain pathology in similar manner. Also, whether other neurodegenerative diseases like Alzheimer's and Parkinson's diseases, hyperthermic brain injury or sleep deprivation induces brain pathology are also associated with excitotoxicity. To answer these questions, further investigation on amino acids metabolism in brain pathology is needed.

Acknowledgment

This investigation is supported by grants from the Air Force Office of Scientific Research (EOARD, London, UK), and Air Force Material Command, USAF, under grant number FA8655-05-1-3065; Grants from the Alzheimer's Association (IIRG-09-132087), the National Institutes of Health (R01 AG028679) and the Dr. Robert M. Kohrman Memorial Fund (RJC); Swedish Medical Research Council (Nr 2710-HSS), the Ministry of Science & Technology, People Republic of China, Göran Gustafsson Foundation, Stockholm, Sweden (HSS), Astra Zeneca, Mölndal, Sweden (HSS/AS), The University Grants Commission, New Delhi, India (HSS/AS), Ministry of Science & Technology, Govt. of India (HSS/AS), Indian Medical Research Council, New Delhi, India (HSS/AS) and India-EU Co-operation Program (RP/AS/HSS) and IT-901/16 (JVL), Government of Basque Country and PPG 17/51 (JVL), J.V.L. thanks to the support of the University of the Basque Country (UPV/EHU) PPG 17/51 and 14/08, the Basque Government (IT-901/16 and CS-2203) Basque Country, Spain; and Foundation for Nanoneuroscience and Nanoneuroprotection (FSNN), Romania. Technical and human support provided by Dr. Ricardo Andrade from SGIker (UPV/EHU) is gratefully acknowledged. Dr. Seaab Sahib is supported by Research Fellowship at the University of Arkansas Fayetteville AR by Department of Community Health; Middle Technical University; Wassit; Iraq, and The Higher Committee for Education Development in Iraq; Baghdad; Iraq. We thank Suraj Sharma, Blekinge Inst. Technology, Karlskrona, Sweden and Dr. Saja Alshafeay, University of Arkansas Fayetteville, Fayetteville AR, USA for computer and graphic support. The U.S. Government is authorized to reproduce and distribute reprints for Government purpose notwithstanding any copyright notation thereon. The views and conclusions contained herein are those of the authors and should not be interpreted as necessarily representing the official policies or endorsements, either expressed or implied, of the Air Force Office of Scientific Research or the U.S. Government.

Conflict of interest

There is no conflict of interest between any entity and/or organization mentioned here.

References

Adams, W.M., Scarneo, S.E., Casa, D.J., 2018. Assessment of evidence-based health and safety policies on sudden death and concussion management in secondary school athletics: a benchmark study. J. Athl. Train. 53 (8), 756–767. https://doi.org/10.4085/1062-6050-220-17 (Epub 2018 Sep 13).

Agimi, Y., Regasa, L.E., Stout, K.C., 2019. Incidence of traumatic brain injury in the U.S. military, 2010-2014. Mil. Med. 184 (5–6), e233–e241. https://doi.org/10.1093/milmed/usy313.

Agnesi, F., Blaha, C.D., Lin, J., Lee, K.H., 2010. Local glutamate release in the rat ventral lateral thalamus evoked by high-frequency stimulation. J. Neural Eng. 7 (2), 26009. https://doi.org/10.1088/1741-2560/7/2/026009 (Epub 2010 Mar 23).

Alberch, J., Pérez-Navarro, E., Canals, J.M., 2002. Neuroprotection by neurotrophins and GDNF family members in the excitotoxic model of Huntington's disease. Brain Res. Bull. 57 (6), 817–822. https://doi.org/10.1016/s0361-9230(01)00775-4.

Almeida-Suhett, C.P., Prager, E.M., Pidoplichko, V., Figueiredo, T.H., Marini, A.M., Li, Z., Eiden, L.E., Braga, M.F., 2015. GABAergic interneuronal loss and reduced inhibitory synaptic transmission in the hippocampal CA1 region after mild traumatic brain injury. Exp. Neurol. 273, 11–23. https://doi.org/10.1016/j.expneurol.2015.07.028 (Epub 2015 Jul 31).

Al-Samsam, R.H., Alessandri, B., Bullock, R., 2000. Extracellular N-acetyl-aspartate as a biochemical marker of the severity of neuronal damage following experimental acute traumatic brain injury. J. Neurotrauma 17, 31–39.

Armistead-Jehle, P., Soble, J.R., Cooper, D.B., Belanger, H.G., 2017. Unique aspects of traumatic brain injury in military and veteran populations. Phys. Med. Rehabil. Clin. N. Am. 28 (2), 323–337. https://doi.org/10.1016/j.pmr.2016.12.008.

Ascherio, A., Schwarzschild, M.A., 2016. The epidemiology of Parkinson's disease: risk factors and prevention. Lancet Neurol. 15 (12), 1257–1272. https://doi.org/10.1016/S1474-4422(16)30230-7 (Epub 2016 Oct 11).

Asken, B.M., DeKosky, S.T., Clugston, J.R., Jaffee, M.S., Bauer, R.M., 2018. Diffusion tensor imaging (DTI) findings in adult civilian, military, and sport-related mild traumatic brain injury (mTBI): a systematic critical review. Brain Imaging Behav. 12 (2), 585–612. https://doi.org/10.1007/s11682-017-9708-9.

Bain, J.R., Mackinnon, S.E., Hunter, D.A., 1989. Functional evaluation of complete sciatic, peroneal, and posterior tibial nerve lesions in the rat. Plast. Reconstr. Surg. 83 (1), 129–138. https://doi.org/10.1097/00006534-198901000-00024.

Baker, A.J., Zornow, M.H., Scheller, M.S., Yaksh, T.L., Skilling, S.R., Smullin, D.H., Larson, A.A., Kuczenski, R., 1991. Changes in extracellular concentrations of glutamate, aspartate, glycine, dopamine, serotonin, and dopamine metabolites after transient global ischemia in the rabbit brain. J. Neurochem. 57 (4), 1370–1379. https://doi.org/10.1111/j.1471-4159.1991.tb08303.x.

Balazs, R., 2006. Trophic effect of glutamate. Curr. Top. Med. Chem. 6 (10), 961–968. https://doi.org/10.2174/156802606777323700.

Baldwin, G.T., Breiding, M.J., Dawn, C.R., 2018. Epidemiology of sports concussion in the United States. Handb. Clin. Neurol. 158, 63–74. https://doi.org/10.1016/B978-0-444-63954-7.00007-0.

Baslow, M.H., 2003. N-acetylaspartate in the vertebrate brain: metabolism and function. Neurochem. Res. 28, 941–953.

Belli, A., Sen, J., Petzold, A., Russo, S., Kitchen, N., Smith, M., Tavazzi, B., Vagnozzi, R., Signoretti, S., Amorini, A.M., 2006. Extracellular N-acetylaspartate depletion in traumatic brain injury. J. Neurochem. 96, 861–869.

Belov Kirdajova, D., Kriska, J., Tureckova, J., Anderova, M., 2020. Ischemia-triggered glutamate excitotoxicity from the perspective of glial cells. Front. Cell. Neurosci. 14, 51. https://doi.org/10.3389/fncel.2020.00051 (eCollection 2020).

Bigler, E.D., Tsao, J.W., 2017. Mild traumatic brain injury in soldiers returning from combat. Neurology 88 (16), 1490–1492. https://doi.org/10.1212/WNL.0000000000003852 (Epub 2017 Mar 17).

Bigler, E.D., Abildskov, T.J., Goodrich-Hunsaker, N.J., Black, G., Christensen, Z.P., Huff, T., Wood, D.M., Hesselink, J.R., Wilde, E.A., Max, J.E., 2016. Structural neuroimaging findings in mild traumatic brain injury. Sports Med. Arthrosc. Rev. 24 (3), e42–e52. https://doi.org/10.1097/JSA.0000000000000119.

Blennow, K., Brody, D.L., Kochanek, P.M., Levin, H., McKee, A., Ribbers, G.M., Yaffe, K., Zetterberg, H., 2016. Traumatic brain injuries. Nat. Rev. Dis. Primers. 2, 16084. https://doi.org/10.1038/nrdp.2016.84.

Botteri, M., Bandera, E., Minelli, C., Latronico, N., 2008. Cerebral blood flow thresholds for cerebral ischemia in traumatic brain injury. A systematic review. Crit. Care Med. 36 (11), 3089–3092. https://doi.org/10.1097/CCM.0b013e31818bd7df.

Bramley, H., Hong, J., Zacko, C., Royer, C., Silvis, M., 2016. Mild traumatic brain injury and post-concussion syndrome: treatment and related sequela for persistent symptomatic disease. Sports Med. Arthrosc. Rev. 24 (3), 123–129. https://doi.org/10.1097/JSA.0000000000000111.

Broshek, D.K., De Marco, A.P., Freeman, J.R., 2015. A review of post-concussion syndrome and psychological factors associated with concussion. Brain Inj. 29 (2), 228–237. https://doi.org/10.3109/02699052.2014.974674 (Epub 2014 Nov 10).

Bryan, C.J., 2013. Multiple traumatic brain injury and concussive symptoms among deployed military personnel. Brain Inj. 27 (12), 1333–1337. https://doi.org/10.3109/02699052.2013.823651 (Epub 2013 Sep 26).

Bryson, A., Hatch, R.J., Zandt, B.J., Rossert, C., Berkovic, S.F., Reid, C.A., Grayden, D.B., Hill, S.L., Petrou, S., 2020. GABA-mediated tonic inhibition differentially modulates gain in functional subtypes of cortical interneurons. Proc. Natl. Acad. Sci. U. S. A. 117 (6), 3192–3202. https://doi.org/10.1073/pnas.1906369117 (Epub 2020 Jan 23).

Cai, C.C., Zhu, J.H., Ye, L.X., Dai, Y.Y., Fang, M.C., Hu, Y.Y., Pan, S.L., Chen, S., Li, P.J., Fu, X.Q., Lin, Z.L., 2019. Glycine protects against hypoxic-ischemic brain injury by regulating mitochondria-mediated autophagy via the AMPK pathway. Oxid. Med. Cell. Longev. 2019, 4248529. https://doi.org/10.1155/2019/4248529 (eCollection 2019).

Cao, F.L., Xu, M., Gong, K., Wang, Y., Wang, R., Chen, X., Chen, J., 2019. Imbalance between excitatory and inhibitory synaptic transmission in the primary somatosensory cortex caused by persistent nociception in rats. J. Pain 20 (8), 917–931. https://doi.org/10.1016/j.jpain.2018.11.014 (Epub 2019 Feb 8).

Cavallero, A., Marte, A., Fedele, E., 2009. L-aspartate as an amino acid neurotransmitter: mechanisms of the depolarization-induced release from cerebrocortical synaptosomes. J. Neurochem. 110 (3), 924–934. https://doi.org/10.1111/j.1471-4159.2009.06187.x (Epub 2009 Jun 22).

Cervós-Navarro, J., Sharma, H.S., Westman, J., Bongcam-Rudloff, E., 1998. Glial reactions in the central nervous system following heat stress. Prog. Brain Res. 115, 241–274. https://doi.org/10.1016/s0079-6123(08)62039-7.

Chakraborty, G., Mekala, P., Yahya, D., Wu, G., Ledeen, R.W., 2001. Intraneuronal N-acetylaspartate supplies acetyl groups for myelin lipid synthesis: evidence for myelin-associated aspartoacylase. J. Neurochem. 78, 736–745.

Chen, R., Okabe, A., Sun, H., Sharopov, S., Hanganu-Opatz, I.L., Kolbaev, S.N., Fukuda, A., Luhmann, H.J., Kilb, W., 2014. Activation of glycine receptors modulates spontaneous epileptiform activity in the immature rat hippocampus. J. Physiol. 592 (10), 2153–2168. https://doi.org/10.1113/jphysiol.2014.271700 (Epub 2014 Mar 24).

Chen, C., Zhou, X., He, J., Xie, Z., Xia, S., Lu, G., 2019. The roles of GABA in ischemia-reperfusion injury in the central nervous system and peripheral organs. Oxid. Med. Cell. Longev. 2019, 4028394. https://doi.org/10.1155/2019/4028394 (eCollection 2019).

Chen, Z., Wang, X., Liao, H., Sheng, T., Chen, P., Zhou, H., Pan, Y., Liu, W., Yao, H., 2020. Glycine attenuates cerebrovascular remodeling via glycine receptor alpha 2 and vascular endothelial growth factor receptor 2 after stroke. Am. J. Transl. Res. 12 (10), 6895–6907 (eCollection 2020).

Cheriyan, J., Sheets, P.L., 2018. Altered excitability and local connectivity of mPFC-PAG neurons in a mouse model of neuropathic pain. J. Neurosci. 38 (20), 4829–4839. https://doi.org/10.1523/JNEUROSCI.2731-17.2018 (Epub 2018 Apr 25).

Chitturi, J., Li, Y., Santhakumar, V., Kannurpatti, S.S., 2018. Early behavioral and metabolomic change after mild to moderate traumatic brain injury in the developing brain. Neurochem. Int. 120, 75–86. https://doi.org/10.1016/j.neuint.2018.08.003 (Epub 2018 Aug 9).

Choe, M.C., 2016. The pathophysiology of concussion. Curr. Pain Headache Rep. 20 (6), 42. https://doi.org/10.1007/s11916-016-0573-9.

Christensen, H., Fonnum, F., 1991. Uptake of glycine, GABA and glutamate by synaptic vesicles isolated from different regions of rat CNS. Neurosci. Lett. 129 (2), 217–220. https://doi.org/10.1016/0304-3940(91)90465-6.

D'Aniello, S., Somorjai, I., Garcia-Fernàndez, J., Topo, E., D'Aniello, A., 2011. D-aspartic acid is a novel endogenous neurotransmitter. FASEB J. 25 (3), 1014–1027. https://doi.org/10.1096/fj.10-168492 (Epub 2010 Dec 16).

Dashnaw, M.L., Petraglia, A.L., Bailes, J.E., 2012. An overview of the basic science of concussion and subconcussion: where we are and where we are going. Neurosurg. Focus 33 (6), E5. 1–9 https://doi.org/10.3171/2012.10.FOCUS12284.

de Rosa, V., Secondo, A., Pannaccione, A., Ciccone, R., Formisano, L., Guida, N., Crispino, R., Fico, A., Polishchuk, R., D'Aniello, A., Annunziato, L., Boscia, F., 2019. D-Aspartate treatment attenuates myelin damage and stimulates myelin repair. EMBO Mol. Med. 11 (1), e9278. https://doi.org/10.15252/emmm.201809278.

De Stefano, N., Matthews, P.M., Arnold, D.L., 1995. Reversible decreases in N-acetylaspartate after acute brain injury. Magn. Reson. Med. 34, 721–727.

Delic, V., Beck, K.D., Pang, K.C.H., Citron, B.A., 2020. Biological links between traumatic brain injury and Parkinson's disease. Acta Neuropathol. Commun. 8 (1), 45. https://doi.org/10.1186/s40478-020-00924-7.

Dempsey, R.J., Başkaya, M.K., 2000. Doğan attenuation of brain edema, blood-brain barrier breakdown, and injury volume by ifenprodil, a polyamine-site N-methyl-D-aspartate receptor antagonist, after experimental traumatic brain injury in rats. Neurosurgery 47 (2), 399–404. discussion 404-6 https://doi.org/10.1097/00006123-200008000-00024.

Dey, P.K., Sharma, H.S., 1983. Ambient temperature and development of traumatic brain oedema in anaesthetized animals. Indian J. Med. Res. 77, 554–563.

Dey, P.K., Sharma, H.S., 1984. Influence of ambient temperature and drug treatments on brain oedema induced by impact injury on skull in rats. Indian J. Physiol. Pharmacol. 28 (3), 177–186.

Dey, P.K., Sharma, H.S., Rao, K.S., 1980. Effect of indomethacin (a prostaglandin synthetase inhibitor) on the permeability of blood-brain and blood-CSF barriers in rat. Indian J. Physiol. Pharmacol. 24 (1), 25–36.

Di Pietro, V., Amorini, A.M., Tavazzi, B., Vagnozzi, R., Logan, A., Lazzarino, G., Signoretti, S., Belli, A., 2014. The molecular mechanisms affecting N-acetylaspartate homeostasis following experimental graded traumatic brain injury. Mol. Med. 20, 147–157.

Dixon, K.J., 2017. Pathophysiology of traumatic brain injury. Phys. Med. Rehabil. Clin. N. Am. 28 (2), 215–225. https://doi.org/10.1016/j.pmr.2016.12.001 (Epub 2017 Mar 2).

Eierud, C., Nathan, D.E., Bonavia, G.H., Ollinger, J., Riedy, G., 2019. Cortical thinning in military blast compared to non-blast persistent mild traumatic brain injuries. Neuroimage Clin. 22, 101793. https://doi.org/10.1016/j.nicl.2019.101793 (Epub 2019 Mar 26).

Elder, G.A., Ehrlich, M.E., Gandy, S., 2019. Relationship of traumatic brain injury to chronic mental health problems and dementia in military veterans. Neurosci. Lett. 707, 134294. https://doi.org/10.1016/j.neulet.2019.134294 (Epub 2019 May 26).

Elliott, K.A., Jasper, H., 1949. Measurement of experimentally induced brain swelling and shrinkage. Am. J. Physiol. 157 (1), 122–129. https://doi.org/10.1152/ajplegacy.1949.157.1.122.

Errico, F., Napolitano, F., Nisticò, R., Centonze, D., Usiello, A., 2009. D-aspartate: an atypical amino acid with neuromodulatory activity in mammals. Rev. Neurosci. 20 (5–6), 429–440. https://doi.org/10.1515/revneuro.2009.20.5-6.429.

Fannon, J., Tarmier, W., Fulton, D., 2015. Neuronal activity and AMPA-type glutamate receptor activation regulates the morphological development of oligodendrocyte precursor cells. Glia 63, 1021–1035.

Feddermann-Demont, N., Chiampas, G., Cowie, C.M., Meyer, T., Nordström, A., Putukian, M., Straumann, D., Kramer, E., 2020. Recommendations for initial examination, differential diagnosis, and management of concussion and other head injuries in high-level football. Scand. J. Med. Sci. Sports 30 (10), 1846–1858. https://doi.org/10.1111/sms.13750 (Epub 2020 Jun 29).

Fehily, B., Fitzgerald, M., 2017. Repeated mild traumatic brain injury: potential mechanisms of damage. Cell Transplant. 26 (7), 1131–1155. https://doi.org/10.1177/0963689717714092.

Fisher, G.H., Garcia, N.M., Payan, I.L., Cadilla-Perezrios, R., Sheremata, W.A., Man, E.H., 1986. D-aspartic acid in purified myelin and myelin basic protein. Biochem. Biophys. Res. Commun. 135, 683–687.

Fonnum, F., 1984. Glutamate: a neurotransmitter in mammalian brain. J. Neurochem. 42 (1), 1–11. https://doi.org/10.1111/j.1471-4159.1984.tb09689.x.

Fonnum, F., Storm-Mathisen, J., 1969. GABA synthesis in rat hippocampus correlated to the distribution of inhibitory neurons. Acta Physiol. Scand. 76 (1), 35A–36A.

Fonnum, F., Walberg, F., 1973. The concentration of GABA within inhibitory nerve terminals. Brain Res. 62 (2), 577–579. https://doi.org/10.1016/0006-8993(73)90724-5.

Fykse, E.M., Fonnum, F., 1996. Amino acid neurotransmission: dynamics of vesicular uptake. Neurochem. Res. 21 (9), 1053–1060. https://doi.org/10.1007/BF02532415.

Gardner, R.C., Yaffe, K., 2015. Epidemiology of mild traumatic brain injury and neurodegenerative disease. Mol. Cell. Neurosci. 66 (Pt. B), 75–80. https://doi.org/10.1016/j.mcn.2015.03.001 (Epub 2015 Mar 5).

Gardner, A.J., Zafonte, R., 2016. Neuroepidemiology of traumatic brain injury. Handb. Clin. Neurol. 138, 207–223. https://doi.org/10.1016/B978-0-12-802973-2.00012-4.

Giza, C.C., Hovda, D.A., 2014. The new neurometabolic cascade of concussion. Neurosurgery 75 (Suppl. 4(0 4)), S24–S33. https://doi.org/10.1227/NEU.0000000000000505.

Goldstein, L.E., Fisher, A.M., Tagge, C.A., Zhang, X.L., Velisek, L., Sullivan, J.A., Upreti, C., Kracht, J.M., Ericsson, M., Wojnarowicz, M.W., Goletiani, C.J., Maglakelidze, G.M., Casey, N., Moncaster, J.A., Minaeva, O., Moir, R.D., Nowinski, C.J., Stern, R.A., Cantu, R.C., Geiling, J., Blusztajn, J.K., Wolozin, B.L., Ikezu, T., Stein, T.D., Budson, A.-E., Kowall, N.W., Chargin, D., Sharon, A., Saman, S., Hall, G.F., Moss, W.C., Cleveland, R.O., Tanzi, R.E., Stanton, P.K., AC, M.K., 2012. Chronic traumatic encephalopathy in blast-exposed military veterans and a blast neurotrauma mouse model. Sci. Transl. Med. 4 (134), 134ra60. https://doi.org/10.1126/scitranslmed.3003716.

Gu, Y., Zhang, J., Zhao, Y., Su, Y., Zhang, Y., 2016. Potassium aspartate attenuates brain injury induced by controlled cortical impact in rats through increasing adenosine triphosphate (ATP) Levels, Na+/K+-ATPase activity and reducing brain edema. Med. Sci. Monit. 22, 4894–4901. https://doi.org/10.12659/msm.898185.

Guerriero, R.M., Giza, C.C., Rotenberg, A., 2015. Glutamate and GABA imbalance following traumatic brain injury. Curr. Neurol. Neurosci. Rep. 15 (5), 27. https://doi.org/10.1007/s11910-015-0545-1.

Gulyaeva, N.V., 2017. Interplay between brain BDNF and glutamatergic systems: a brief state of the evidence and association with the pathogenesis of depression. Biochemistry (Mosc.) 82 (3), 301–307. https://doi.org/10.1134/S0006297917030087.

Gundersen, V., Fonnum, F., Ottersen, O.P., Storm-Mathisen, J., 2001. Redistribution of neuroactive amino acids in hippocampus and striatum during hypoglycemia: a quantitative immunogold study. J. Cereb. Blood Flow Metab. 21 (1), 41–51. https://doi.org/10.1097/00004647-200101000-00006.

Hadanny, A., Efrati, S., 2016. Treatment of persistent post-concussion syndrome due to mild traumatic brain injury: current status and future directions. Expert Rev. Neurother. 16 (8), 875–887. https://doi.org/10.1080/14737175.2016.1205487 (Epub 2016 Jul 4).

Hajiaghamemar, M., Kilbaugh, T., Arbogast, K.B., Master, C.L., Margulies, S.S., 2020. Using serum amino acids to predict traumatic brain injury: a systematic approach to utilize multiple biomarkers. Int. J. Mol. Sci. 21 (5), 1786. https://doi.org/10.3390/ijms21051786.

Hanada, T., 2020. Ionotropic glutamate receptors in epilepsy: a review focusing on AMPA and NMDA receptors. Biomolecules 10 (3), 464. https://doi.org/10.3390/biom10030464.

Harmon, K.G., Drezner, J.A., Gammons, M., Guskiewicz, K.M., Halstead, M., Herring, S.A., Kutcher, J.S., Pana, A., Putukian, M., Roberts, W.O., 2013. American Medical Society for Sports Medicine position statement: concussion in sport. Br. J. Sports Med. 47 (1), 15–26. https://doi.org/10.1136/bjsports-2012-091941.

Hasoon, J., 2017. Blast-associated traumatic brain injury in the military as a potential trigger for dementia and chronic traumatic encephalopathy. U.S. Army Med. Dep. J. 1–17, 102–105.

Hayes, R.L., Jenkins, L.W., Lyeth, B.G., 1992. Neurotransmitter-mediated mechanisms of traumatic brain injury: acetylcholine and excitatory amino acids. J. Neurotrauma 9 (Suppl. 1), S173–S187.

Heaney, C.F., Kinney, J.W., 2016. Role of GABA(B) receptors in learning and memory and neurological disorders. Neurosci. Biobehav. Rev. 63, 1–28. https://doi.org/10.1016/j.neubiorev.2016.01.007 (Epub 2016 Jan 24).

Heltemes, K.J., Dougherty, A.L., MacGregor, A.J., Galarneau, M.R., 2011. Inpatient hospitalizations of U.S. military personnel medically evacuated from Iraq and Afghanistan with combat-related traumatic brain injury. Mil. Med. 176 (2), 132–135. https://doi.org/10.7205/milmed-d-09-00238.

Herrera, C.G., Cadavieco, M.C., Jego, S., Ponomarenko, A., Korotkova, T., Adamantidis, A., 2016. Hypothalamic feedforward inhibition of thalamocortical network controls arousal and consciousness. Nat. Neurosci. 19 (2), 290–298. https://doi.org/10.1038/nn.4209 (Epub 2015 Dec 21).

Herrera-Marschitz, M., Goiny, M., You, Z.B., Meana, J.J., Pettersson, E., Rodriguez-Puertas, R., Xu, Z.Q., Terenius, L., Hökfelt, T., Ungerstedt, U., 1997. On the release of glutamate and aspartate in the basal ganglia of the rat: interactions with monoamines and neuropeptides. Neurosci. Biobehav. Rev. 21 (4), 489–495. https://doi.org/10.1016/s0149-7634(96)00033-4.

Herrera-Marschitz, M., Goiny, M., You, Z.B., Meana, J.J., Engidawork, E., Chen, Y., Rodriguez-Puertas, R., Broberger, C., Andersson, K., Terenius, L., Hökfelt, T., Ungerstedt, U., 1998. Release of endogenous excitatory amino acids in the neostriatum of the rat under physiological and pharmacologically-induced conditions. Amino Acids 14 (1–3), 197–203. https://doi.org/10.1007/BF01345262.

Hon, K.L., Leung, A.K.C., Torres, A.R., 2019. Concussion: a global perspective. Semin. Pediatr. Neurol. 30, 117–127. https://doi.org/10.1016/j.spen.2019.03.017 (Epub 2019 Mar 26).

Hook, V., Yoon, M., Mosier, C., Ito, G., Podvin, S., Head, B.P., Rissman, R., O'Donoghue, A.J., Hook, G., 2020. Cathepsin B in neurodegeneration of Alzheimer's disease, traumatic brain injury, and related brain disorders. Biochim. Biophys. Acta Proteins Proteomics 1868 (8), 140428. https://doi.org/10.1016/j.bbapap.2020.140428 (Epub 2020 Apr 17).

Hoshino, O., Zheng, M., Watanabe, K., 2018. Perceptual judgments via sensory-motor interaction assisted by cortical GABA. J. Comput. Neurosci. 44 (2), 233–251. https://doi.org/10.1007/s10827-018-0677-9 (Epub 2018 Jan 31).

Hosseini-Zare, M.S., Gu, F., Abdulla, A., Powell, S., Žiburkus, J., 2017. Effects of experimental traumatic brain injury and impaired glutamate transport on cortical spreading depression. Exp. Neurol. 295, 155–161. https://doi.org/10.1016/j.expneurol.2017.05.002 (Epub 2017 May 4).

Howell, D.R., Wilson, J.C., Kirkwood, M.W., Grubenhoff, J.A., 2019. Quality of life and symptom burden 1 month after concussion in children and adolescents. Clin. Pediatr. (Phila) 58 (1), 42–49. https://doi.org/10.1177/0009922818806308 (Epub 2018 Oct 12).

Hubertus, V., Marklund, N., Vajkoczy, P., 2019. Management of concussion in soccer. Acta Neurochir. 161 (3), 425–433. https://doi.org/10.1007/s00701-019-03807-6 (Epub 2019 Jan 28).

Ito, U., Tomita, H., Yamazaki, S., Takada, Y., Inaba, Y., 1986. Brain swelling and brain oedema in acute head injury. Acta Neurochir. 79 (2–4), 120–124. https://doi.org/10.1007/BF01407455.

Iverson, G.L., Gardner, A.J., Terry, D.P., Ponsford, J.L., Sills, A.K., Broshek, D.K., Solomon, G.S., 2017. Predictors of clinical recovery from concussion: a systematic review. Br. J. Sports Med. 51 (12), 941–948. https://doi.org/10.1136/bjsports-2017-097729.

Jackson, W.T., Starling, A.J., 2019. Concussion evaluation and management. Med. Clin. North Am. 103 (2), 251–261. https://doi.org/10.1016/j.mcna.2018.10.005 (Epub 2018 Dec 3).

Jagnoor, J., Cameron, I., 2015. Mild traumatic brain injury and motor vehicle crashes: limitations to our understanding. Injury 46 (10), 1871–1874. https://doi.org/10.1016/j.injury.2014.08.044 (Epub 2014 Sep 23).

Jenkins, P.O., Roussakis, A.A., De Simoni, S., Bourke, N., Fleminger, J., Cole, J., Piccini, P., Sharp, D., 2020. Distinct dopaminergic abnormalities in traumatic brain injury and Parkinson's disease. J. Neurol. Neurosurg. Psychiatry 91 (6), 631–637. https://doi.org/10.1136/jnnp-2019-321759 (Epub 2020 May 7).

Jha, R.M., Kochanek, P.M., Simard, J.M., 2019. Pathophysiology and treatment of cerebral edema in traumatic brain injury. Neuropharmacology 145 (Pt. B), 230–246. https://doi.org/10.1016/j.neuropharm.2018.08.004 (Epub 2018 Aug 4).

Jinguji, T.M., Krabak, B.J., Satchell, E.K., 2011. Epidemiology of youth sports concussion. Phys. Med. Rehabil. Clin. N. Am. 22 (4), 565–575. vii https://doi.org/10.1016/j.pmr.2011.08.001.

Johanson, C., Stopa, E., Baird, A., Sharma, H., 2011. Traumatic brain injury and recovery mechanisms: peptide modulation of periventricular neurogenic regions by the choroid plexus-CSF nexus. J. Neural Transm. (Vienna) 118 (1), 115–133. https://doi.org/10.1007/s00702-010-0498-0 (Epub 2010 Oct 10).

John, J., Kodama, T., Siegel, J.M., 2014. Caffeine promotes glutamate and histamine release in the posterior hypothalamus. Am. J. Physiol. Regul. Integr. Comp. Physiol. 307 (6), R704–R710.

Jones, K.M., Prah, P., Starkey, N., Theadom, A., Barker-Collo, S., Ameratunga, S., Feigin, V.L., BIONIC Study Group, 2019. Longitudinal patterns of behavior, cognition, and quality of life after mild traumatic brain injury in children: BIONIC study findings. Brain Inj. 33 (7), 884–893. https://doi.org/10.1080/02699052.2019.1606445 (Epub 2019 Apr 23).

Kanagal, S.G., Muir, G.D., 2008. The differential effects of cervical and thoracic dorsal funiculus lesions in rats. Behav. Brain Res. 187 (2), 379–386. https://doi.org/10.1016/j.bbr.2007.09.035 (Epub 2007 Oct 22).

Kawamata, T., Mori, T., Sato, S., Katayama, Y., 2007. Tissue hyperosmolality and brain edema in cerebral contusion. Neurosurg. Focus 22 (5), E5. https://doi.org/10.3171/foc.2007.22.5.6.

Kazl, C., Torres, A., 2019. Definition, classification, and epidemiology of concussion. Semin. Pediatr. Neurol. 30, 9–13. https://doi.org/10.1016/j.spen.2019.03.003 (Epub 2019 Mar 23).

Khoshnam, S.E., Winlow, W., Farzaneh, M., Farbood, Y., Moghaddam, H.F., 2017. Pathogenic mechanisms following ischemic stroke. Neurol. Sci. 38 (7), 1167–1186. https://doi.org/10.1007/s10072-017-2938-1 (Epub 2017 Apr 17).

Kierans, A.S., Kirov, I.I., Gonen, O., Haemer, G., Nisenbaum, E., Babb, J.S., Grossman, R.I., Lui, Y.W., 2014. Myoinositol and glutamate complex neurometabolite abnormality after mild traumatic brain injury. Neurology 82 (6), 521–528. https://doi.org/10.1212/WNL.0000000000000105 (Epub 2014 Jan 8).

Kim, J., Yang, J.H., Ryu, I.S., Sohn, S., Kim, S., Choe, E.S., 2019. Interactions of glutamatergic neurotransmission and Brain-derived neurotrophic factor in the regulation of behaviors after nicotine administration. Int. J. Mol. Sci. 20 (12), 2943. https://doi.org/10.3390/ijms20122943.

Kokiko-Cochran, O.N., Godbout, J.P., 2018. The inflammatory continuum of traumatic brain injury and Alzheimer's disease. Front. Immunol. 9, 672. https://doi.org/10.3389/fimmu.2018.00672 (eCollection 2018).

Krashia, P., Ledonne, A., Nobili, A., Cordella, A., Errico, F., Usiello, A., D'Amelio, M., Mercuri, N.B., Guatteo, E., Carunchio, I., 2016. Persistent elevation of D-Aspartate enhances NMDA receptor-mediated responses in mouse substantia nigra pars compacta dopamine neurons. Neuropharmacology 103, 69–78. https://doi.org/10.1016/j.neuropharm.2015.12.013 (Epub 2015 Dec 17).

Krzyżanowska, W., Pomierny, B., Bystrowska, B., Pomierny-Chamioło, L., Filip, M., Budziszewska, B., Pera, J., 2017. Ceftriaxone- and N-acetylcysteine-induced brain tolerance to ischemia: influence on glutamate levels in focal cerebral ischemia. PLoS One 12 (10), e0186243. https://doi.org/10.1371/journal.pone.0186243 (eCollection 2017).

Langer, L., Levy, C., Bayley, M., 2020. Increasing incidence of concussion: true epidemic or better recognition? J. Head Trauma Rehabil. 35 (1), E60–E66. https://doi.org/10.1097/HTR.0000000000000503.

Larkin, M.B., Graves, E.K.M., Boulter, J.H., Szuflita, N.S., Meyer, R.M., Porambo, M.E., Delaney, J.J., Bell, R.S., 2018. Two-year mortality and functional outcomes in combat-related penetrating brain injury: battlefield through rehabilitation. Neurosurg. Focus 45 (6), E4. https://doi.org/10.3171/2018.9.FOCUS18359.

Lasoń, W., 1997. The interaction between neuropeptides and excitatory amino acids in seizure phenomena. Pol. J. Pharmacol. 49 (1), 64–66.

Lau, A., Tymianski, M., 2010. Glutamate receptors, neurotoxicity and neurodegeneration. Pflugers Arch. 460 (2), 525–542. https://doi.org/10.1007/s00424-010-0809-1 (Epub 2010 Mar 14).

Leddy, J., Baker, J.G., Haider, M.N., Hinds, A., Willer, B., 2017. A physiological approach to prolonged recovery from sport-related concussion. J. Athl. Train. 52 (3), 299–308. https://doi.org/10.4085/1062-6050-51.11.08.

Leist, M., Nicotera, P., 1998. Apoptosis, excitotoxicity, and neuropathology. Exp. Cell Res. 239 (2), 183–201. https://doi.org/10.1006/excr.1997.4026.

Leonelli, M., Torrão, A.S., Britto, L.R., 2009. Unconventional neurotransmitters, neurodegeneration and neuroprotection. Braz. J. Med. Biol. Res. 42 (1), 68–75. https://doi.org/10.1590/s0100-879x2009000100011.

López-Picón, F., Snellman, A., Shatillo, O., Lehtiniemi, P., Grönroos, T.J., Marjamäki, P., Trigg, W., Jones, P.A., Solin, O., Pitkänen, A., Haaparanta-Solin, M., 2016. Ex vivo tracing of NMDA and GABA-A receptors in rat brain after traumatic brain injury using 18F-GE-179 and 18F-GE-194 autoradiography. J. Nucl. Med. 57 (9), 1442–1447. https://doi.org/10.2967/jnumed.115.167403 (Epub 2016 May 19).

Lovett-Barron, M., Kaifosh, P., Kheirbek, M.A., Danielson, N., Zaremba, J.D., Reardon, T.R., Turi, G.F., Hen, R., Zemelman, B.V., Losonczy, A., 2014. Dendritic inhibition in the hippocampus supports fear learning. Science 343 (6173), 857–863. https://doi.org/10.1126/science.1247485.

Lu, L.H., Reid, M.W., Cooper, D.B., Kennedy, J.E., 2019. Sleep problems contribute to post-concussive symptoms in service members with a history of mild traumatic brain injury without posttraumatic stress disorder or major depressive disorder. NeuroRehabilitation 44 (4), 511–521. https://doi.org/10.3233/NRE-192702.

Madhavarao, C.N., Moffett, J.R., Moore, R.A., Viola, R.E., Namboodiri, M.A., Jacobowitz, D.M., 2004. Immunohistochemical localization of aspartoacylase in the rat central nervous system. J. Comp. Neurol. 472, 318–329.

Maloney, K.J., Mainville, L., Jones, B.E., 1999. Differential c-Fos expression in cholinergic, monoaminergic, and GABAergic cell groups of the pontomesencephalic tegmentum after paradoxical sleep deprivation and recovery. J. Neurosci. 19 (8), 3057–3072. https://doi.org/10.1523/JNEUROSCI.19-08-03057.1999.

Maloney, K.J., Mainville, L., Jones, B.E., 2000. c-Fos expression in GABAergic, serotonergic, and other neurons of the pontomedullary reticular formation and raphe after paradoxical sleep deprivation and recovery. J. Neurosci. 20 (12), 4669–4679. https://doi.org/10.1523/JNEUROSCI.20-12-04669.2000.

Manser, C.E., Morris, T.H., Broom, D.M., 1995. An investigation into the effects of solid or grid cage flooring on the welfare of laboratory rats. Lab Anim. 29 (4), 353–363. https://doi.org/10.1258/002367795780740023.

Marshall, S., Bayley, M., McCullagh, S., Velikonja, D., Berrigan, L., Ouchterlony, D., Weegar, K., 2015. mTBI Expert Consensus Group.Updated clinical practice guidelines for concussion/mild traumatic brain injury and persistent symptoms. Brain Inj. 29 (6), 688–700. https://doi.org/10.3109/02699052.2015.1004755 (Epub 2015 Apr 14).

Martin, G., 2016. Traumatic brain injury: the first 15 milliseconds. Brain Inj. 30 (13–14), 1517–1524. https://doi.org/10.1080/02699052.2016.1192683 (Epub 2016 Sep 23).

Martin, L.J., Al-Abdulla, N.A., Brambrink, A.M., Kirsch, J.R., Sieber, F.E., Portera-Cailliau, C., 1998. Neurodegeneration in excitotoxicity, global cerebral ischemia, and target deprivation: a perspective on the contributions of apoptosis and necrosis. Brain Res. Bull. 46 (4), 281–309. https://doi.org/10.1016/s0361-9230(98)00024-0.

Marvin, J.S., Borghuis, B.G., Tian, L., Cichon, J., Harnett, M.T., Akerboom, J., Gordus, A., Renninger, S.L., Chen, T.W., Bargmann, C.I., Orger, M.B., Schreiter, E.R., Demb, J.B., Gan, W.B., Hires, S.A., Looger, L.L., 2013. An optimized fluorescent probe for visualizing glutamate neurotransmission. Nat. Methods 10 (2), 162–170. https://doi.org/10.1038/nmeth.2333 (Epub 2013 Jan 13).

Mattson, M.P., 1990. Excitatory amino acids, growth factors, and calcium: a teeter-totter model for neural plasticity and degeneration. Adv. Exp. Med. Biol. 268, 211–220. https://doi.org/10.1007/978-1-4684-5769-8_24.

Matute, C., 2011. Glutamate and ATP signalling in white matter pathology. J. Anat. 219 (1), 53–64. https://doi.org/10.1111/j.1469-7580.2010.01339.x (Epub 2011 Jan 20).

Mätzsch, T., Karlsson, B., 1986. Moped and motorcycle accidents—similarities and discrepancies. J. Trauma 26 (6), 538–543. https://doi.org/10.1097/00005373-198606000-00008.

McAllister, T., McCrea, M., 2017. Long-term cognitive and neuropsychiatric consequences of repetitive concussion and head-impact exposure. J. Athl. Train. 52 (3), 309–317. https://doi.org/10.4085/1062-6050-52.1.14.

McInnes, K., Friesen, C.L., MacKenzie, D.E., Westwood, D.A., Boe, S.G., 2017. Mild traumatic brain injury (mTBI) and chronic cognitive impairment: a scoping review. PLoS One 12 (4), e0174847. https://doi.org/10.1371/journal.pone.0174847 (eCollection 2017).

McKeon, A.B., Stocker, R.P.J., Germain, A., 2019. Traumatic brain injury and sleep disturbances in combat-exposed service members and veterans: where to go next? NeuroRehabilitation 45 (2), 163–185. https://doi.org/10.3233/NRE-192804.

Meana, J.J., Herrera-Marschitz, M., Brodin, E., Hökfelt, T., Ungerstedt, U., 1991. Simultaneous determination of cholecystokinin, dopamine, glutamate and aspartate in cortex and striatum of the rat using in vivo microdialysis. Amino Acids 1 (3), 365–373. https://doi.org/10.1007/BF00814005.

Menon, P.K., Muresanu, D.F., Sharma, A., Mössler, H., Sharma, H.S., 2012. Cerebrolysin, a mixture of neurotrophic factors induces marked neuroprotection in spinal cord injury following intoxication of engineered nanoparticles from metals. CNS Neurol. Disord. Drug Targets 11 (1), 40–49. https://doi.org/10.2174/187152712799960781.

Mercier, L.J., Fung, T.S., Harris, A.D., Dukelow, S.P., Debert, C.T., 2020. Improving symptom burden in adults with persistent post-concussive symptoms: a randomized aerobic exercise trial protocol. BMC Neurol. 20 (1), 46. https://doi.org/10.1186/s12883-020-1622-x.

Mez, J., Daneshvar, D.H., Kiernan, P.T., Abdolmohammadi, B., Alvarez, V.E., Huber, B.R., Alosco, M.L., Solomon, T.M., Nowinski, C.J., McHale, L., Cormier, K.A., Kubilus, C.A., Martin, B.M., Murphy, L., Baugh, C.M., Montenigro, P.H., Chaisson, C.E., Tripodis, Y., Kowall, N.W., Weuve, J., McClean, M.D., Cantu, R.C., Goldstein, L.E., Katz, D.I., Stern, R.A., Stein, T.D., McKee, A.C., 2017. Clinicopathological evaluation of chronic traumatic encephalopathy in players of American football. JAMA 318 (4), 360–370. https://doi.org/10.1001/jama.2017.8334.

Moffett, J.R., Ross, B., Arun, P., Madhavarao, C.N., Namboodiri, A.M., 2007. N-Acetylaspartate in the CNS: from neurodiagnostics to neurobiology. Prog. Neurobiol. 81, 89–131.

Mohanty, S., Dey, P.K., Sharma, H.S., Singh, S., Chansouria, J.P., Olsson, Y., 1989. Role of histamine in traumatic brain edema. An experimental study in the rat. J. Neurol. Sci. 90 (1), 87–97. https://doi.org/10.1016/0022-510x(89)90048-8.

Montenigro, P.H., Alosco, M.L., Martin, B.M., Daneshvar, D.H., Mez, J., Chaisson, C.E., Nowinski, C.J., Au, R., McKee, A.C., Cantu, R.C., McClean, M.D., Stern, R.A., Tripodis, Y., 2017. Cumulative head impact exposure predicts later-life depression, apathy, executive dysfunction, and cognitive impairment in former high school and college football players. J. Neurotrauma 34 (2), 328–340. https://doi.org/10.1089/neu.2016.4413 (Epub 2016 Jun 15).

Morland, C., Nordengen, K., Gundersen, V., 2012. Valproate causes reduction of the excitatory amino acid aspartate in nerve terminals. Neurosci. Lett. 527 (2), 100–104. https://doi.org/10.1016/j.neulet.2012.08.042 (Epub 2012 Aug 30).

Mu, W., Catenaccio, E., Lipton, M.L., 2017. Neuroimaging in blast-related mild traumatic brain injury. J. Head Trauma Rehabil. 32 (1), 55–69. https://doi.org/10.1097/HTR.0000000000000213.

Mullally, W.J., 2017. Concussion. Am. J. Med. 130 (8), 885–892. https://doi.org/10.1016/j.amjmed.2017.04.016 (Epub 2017 May 11).

Nathanson, J.A., Scavone, C., Scanlon, C., McKee, M., 1995. The cellular Na+ pump as a site of action for carbon monoxide and glutamate: a mechanism for long-term modulation of cellular activity. Neuron 14 (4), 781–794. https://doi.org/10.1016/0896-6273(95)90222-8.

Nicotera, P., Ankarcrona, M., Bonfoco, E., Orrenius, S., Lipton, S.A., 1997. Neuronal necrosis and apoptosis: two distinct events induced by exposure to glutamate or oxidative stress. Adv. Neurol. 72, 95–101.

Obrenovitch, T.P., Urenjak, J., 1997. Is high extracellular glutamate the key to excitotoxicity in traumatic brain injury? J. Neurotrauma 14 (10), 677–698. https://doi.org/10.1089/neu.1997.14.677.

Oja, S.S., Saransaari, P., 2009. Release of endogenous amino acids from the hippocampus and brain stem from developing and adult mice in ischemia. Neurochem. Res. 34 (9), 1668–1676. https://doi.org/10.1007/s11064-009-9961-4 (Epub 2009 Apr 3).

Olsson, Y., Sharma, H.S., Pettersson, A., Cervos-Navarro, J., 1992. Release of endogenous neurochemicals may increase vascular permeability, induce edema and influence cell changes in trauma to the spinal cord. Prog. Brain Res. 91, 197–203. https://doi.org/10.1016/s0079-6123(08)62335-3.

Olsson, Y., Sharma, H.S., Nyberg, F., Westman, J., 1995. The opioid receptor antagonist naloxone influences the pathophysiology of spinal cord injury. Prog. Brain Res. 104, 381–399. https://doi.org/10.1016/s0079-6123(08)61802-6.

Ommaya, A.K., 1995. Head injury mechanisms and the concept of preventive management: a review and critical synthesis. J. Neurotrauma 12 (4), 527–546. https://doi.org/10.1089/neu.1995.12.527.

Paine, T.A., Chang, S., Poyle, R., 2020. Contribution of GABA(A) receptor subunits to attention and social behavior. Behav. Brain Res. 378, 112261. https://doi.org/10.1016/j.bbr.2019.112261 (Epub 2019 Sep 24).

Patnaik, R., Mohanty, S., Sharma, H.S., 2000. Blockade of histamine H2 receptors attenuate blood-brain barrier permeability, cerebral blood flow disturbances, edema formation and cell reactions following hyperthermic brain injury in the rat. Acta Neurochir. Suppl. 76, 535–539. https://doi.org/10.1007/978-3-7091-6346-7_112.

Paukert, M., Huang, Y.H., Tanaka, K., Rothstein, J.D., Bergles, D.E., 2010. Zones of enhanced glutamate release from climbing fibers in the mammalian cerebellum. J. Neurosci. 30 (21), 7290–7299. https://doi.org/10.1523/JNEUROSCI.5118-09.2010.

Perry, D.C., Sturm, V.E., Peterson, M.J., Pieper, C.F., Bullock, T., Boeve, B.F., Miller, B.L., Guskiewicz, K.M., Berger, M.S., Kramer, J.H., Welsh-Bohmer, K.A., 2016. Association of traumatic brain injury with subsequent neurological and psychiatric disease: a meta-analysis. J. Neurosurg. 124 (2), 511–526. https://doi.org/10.3171/2015.2.JNS14503 (Epub 2015 Aug 28).

Pervez, M., Kitagawa, R.S., Chang, T.R., 2018. Definition of traumatic brain injury, neurosurgery, trauma orthopedics, neuroimaging, psychology, and psychiatry in mild traumatic brain injury. Neuroimaging Clin. N. Am. 28 (1), 1–13. https://doi.org/10.1016/j.nic.2017.09.010.

Petroff, O.A., 2002. GABA and glutamate in the human brain. Neuroscientist 8 (6), 562–573. https://doi.org/10.1177/1073858402238515.

Pilipenko, G., Sirko, A., Dzyak, L., Mizyakina, E., 2020. Results of brain injury primary surgical treatment in a complex care for patients with combat-related penetrating craniocerebral gunshot wound at a specialized medical facility. Georgian Med. News 301, 13–20.

Prehn, J.H., Welsch, M., Backhauss, C., Nuglisch, J., Ausmeier, F., Karkoutly, C., Krieglstein, J., 1993. Effects of serotonergic drugs in experimental brain ischemia: evidence for a protective role of serotonin in cerebral ischemia. Brain Res. 630 (1–2), 10–20. https://doi.org/10.1016/0006-8993(93)90636-2.

Provencher, M.T., Frank, R.M., Shubert, D.J., Sanchez, A., Murphy, C.P., Zafonte, R.D., 2019. Concussions in sports. Orthopedics 42 (1), 12–21. https://doi.org/10.3928/01477447-20181231-02.

Ramos-Cejudo, J., Wisniewski, T., Marmar, C., Zetterberg, H., Blennow, K., de Leon, M.J., Fossati, S., 2018. Traumatic brain injury and Alzheimer's disease: the cerebrovascular link. EBioMedicine 28, 21–30. https://doi.org/10.1016/j.ebiom.2018.01.021 (Epub 2018 Jan 31).

Randić, M., Kolaj, M., Kojić, L., Cerne, R., Cheng, G., Wang, R.A., 1995. Interaction of neuropeptides and excitatory amino acids in the rat superficial spinal dorsal horn. Prog. Brain Res. 104, 225–253. https://doi.org/10.1016/s0079-6123(08)61793-8.

Rebec, G.V., 2018. Corticostriatal network dysfunction in Huntington's disease: deficits in neural processing, glutamate transport, and ascorbate release. CNS Neurosci. Ther. 24 (4), 281–291. https://doi.org/10.1111/cns.12828 (Epub 2018 Feb 21).

Regasa, L.E., Agimi, Y., Stout, K.C., 2019. Traumatic brain injury following military deployment: evaluation of diagnosis and cause of injury. J. Head Trauma Rehabil. 34 (1), 21–29. https://doi.org/10.1097/HTR.0000000000000417.

Ruozi, B., Belletti, D., Sharma, H.S., Sharma, A., Muresanu, D.F., Mössler, H., Forni, F., Vandelli, M.A., Tosi, G., 2015. PLGA nanoparticles loaded cerebrolysin: studies on their preparation and investigation of the effect of storage and serum stability with reference to traumatic brain injury. Mol. Neurobiol. 52 (2), 899–912. https://doi.org/10.1007/s12035-015-9235-x.

Ryan, L.M., Warden, D.L., 2003. Post concussion syndrome. Int. Rev. Psychiatry 15 (4), 310–316. https://doi.org/10.1080/09540260310001606692.

Sager, T.N., Fink-Jensen, A., Hansen, A.J., 1997. Transient elevation of interstitial N-acetylaspartate in reversible global brain ischemia. J. Neurochem. 68, 675–682.

Scorza, K.A., Cole, W., 2019. Current concepts in concussion: initial evaluation and management. Am. Fam. Physician 99 (7), 426–434.

Seabury, S.A., Gaudette, É., Goldman, D.P., Markowitz, A.J., Brooks, J., McCrea, M.A., Okonkwo, D.O., Manley, G.T., TRACK-TBI Investigators, Adeoye, O., Badjatia, N., Boase, K., Bodien, Y., Bullock, M.R., Chesnut, R., Corrigan, J.D., Crawford, K., Diaz-Arrastia, R., Dikmen, S., Duhaime, A.C., Ellenbogen, R., Feeser, V.R., Ferguson, A., Foreman, B., Gardner, R., Giacino, J., Gonzalez, L., Gopinath, S., Gullapalli, R., Hemphill, J.C., Hotz, G., Jain, S., Korley, F., Kramer, J., Kreitzer, N., Levin, H., Lindsell, C., Machamer, J., Madden, C., Martin, A., McAllister, T., Merchant, R., Mukherjee, P., Nelson, L., Noel, F., Palacios, E., Perl, D., Puccio, A., Rabinowitz, M., Robertson, C., Rosand, J., Sander, A., Satris, G., Schnyer, D., Sherer, M., Stein, M., Taylor, S., Temkin, N., Toga, A., Valadka, A., Vassar, M., Vespa, P., Wang, K., Yue, J., Yuh, E., Zafonte, R., 2018. Assessment of follow-up care after emergency department presentation for mild traumatic brain injury and concussion: results from the TRACK-TBI study. JAMA Netw. Open 1 (1), e180210. https://doi.org/10.1001/jamanetworkopen.2018.0210.

Segev, S., Shorer, M., Rassovsky, Y., Pilowsky Peleg, T., Apter, A., Fennig, S., 2016. The contribution of posttraumatic stress disorder and mild traumatic brain injury to persistent post concussive symptoms following motor vehicle accidents. Neuropsychology 30 (7), 800–810. https://doi.org/10.1037/neu0000299 (Epub 2016 Aug 22).

Shah, U.H., González-Maeso, J., 2019. Serotonin and glutamate interactions in preclinical schizophrenia models. ACS Chem. Nerosci. 10 (7), 3068–3077. https://doi.org/10.1021/acschemneuro.9b00044 (Epub 2019 Mar 12).

Shannon, R.J., van der Heide, S., Carter, E.L., Jalloh, I., Menon, D.K., Hutchinson, P.J., 2016. Carpenter KL extracellular N-Acetylaspartate in human traumatic brain injury. J. Neurotrauma 33 (4), 319–329. https://doi.org/10.1089/neu.2015.3950 (Epub 2015 Aug 14).

Sharma, H.S., 1987. Effect of captopril (a converting enzyme inhibitor) on blood-brain barrier permeability and cerebral blood flow in normotensive rats. Neuropharmacology 26 (1), 85–92. https://doi.org/10.1016/0028-3908(87)90049-9.

Sharma, H.S., 2000. A bradykinin BK2 receptor antagonist HOE-140 attenuates blood-spinal cord barrier permeability following a focal trauma to the rat spinal cord. An experimental study using Evans blue, ^{131}I-sodium and lanthanum tracers. Acta Neurochir. Suppl. 76, 159–163. https://doi.org/10.1007/978-3-7091-6346-7_32.

Sharma, H.S., 2004a. Histamine influences the blood-spinal cord and brain barriers following injuries to the central nervous system. In: Sharma, H.S., Westman, J. (Eds.), The Blood-Spinal Cord and Brain Barriers in Health and Disease. Elsevier Academic Press, San Diego, pp. 159–190.

Sharma, H.S., 2004b. Influence of serotonin on the blood-brain and blood-spinal cord barriers. In: Sharma, H.S., Westman, J. (Eds.), The Blood-Spinal Cord and Brain Barriers in Health and Disease. Elsevier Academic Press, San Diego, pp. 117–158.

Sharma, H.S., 2004c. Pathophysiology of the blood-spinal cord barrier in traumatic injury. In: Sharma, H.S., Westman, J. (Eds.), The Blood-Spinal Cord and Brain Barriers in Health and Disease. Elsevier Academic Press, San Diego, pp. 437–518.

Sharma, H.S., 2006. Hyperthermia influences excitatory and inhibitory amino acid neurotransmitters in the central nervous system. An experimental study in the rat using behavioural, biochemical, pharmacological, and morphological approaches. J. Neural Transm. (Vienna) 113 (4), 497–519. https://doi.org/10.1007/s00702-005-0406-1.

Sharma, H.S., 2007a. Interaction between amino acid neurotransmitters and opioid receptors in hyperthermia-induced brain pathology. Prog. Brain Res. 162, 295–317. https://doi.org/10.1016/S0079-6123(06)62015-3.

Sharma, H.S., 2007b. Nanoneuroscience: emerging concepts on nanoneurotoxicity and nanoneuroprotection. Nanomedicine (Lond.) 2 (6), 753–758. https://doi.org/10.2217/17435889.2.6.753.

Sharma, H.S., 2007c. Neurotrophic factors in combination: a possible new therapeutic strategy to influence pathophysiology of spinal cord injury and repair mechanisms. Curr. Pharm. Des. 13 (18), 1841–1874. https://doi.org/10.2174/138161207780858410.

Sharma, H.S., 2011. Early microvascular reactions and blood-spinal cord barrier disruption are instrumental in pathophysiology of spinal cord injury and repair: novel therapeutic strategies including nanowired drug delivery to enhance neuroprotection. J. Neural Transm. (Vienna) 118 (1), 155–176. https://doi.org/10.1007/s00702-010-0514-4 (Epub 2010 Dec 16).

Sharma, H.S., Alm, P., 2002. Nitric oxide synthase inhibitors influence dynorphin A (1-17) immunoreactivity in the rat brain following hyperthermia. Amino Acids 23 (1–3), 247–259. https://doi.org/10.1007/s00726-001-0136-0.

Sharma, H.S., Cervós-Navarro, J., 1990. Brain oedema and cellular changes induced by acute heat stress in young rats. Acta Neurochir. Suppl. (Wien) 51, 383–386. https://doi.org/10.1007/978-3-7091-9115-6_129.

Sharma, H.S., Dey, P.K., 1986a. Influence of long-term immobilization stress on regional blood-brain barrier permeability, cerebral blood flow and 5-HT level in conscious normotensive young rats. J. Neurol. Sci. 72 (1), 61–76. https://doi.org/10.1016/0022-510x(86)90036-5.

Sharma, H.S., Dey, P.K., 1986b. Probable involvement of 5-hydroxytryptamine in increased permeability of blood-brain barrier under heat stress in young rats. Neuropharmacology 25 (2), 161–167. https://doi.org/10.1016/0028-3908(86)90037-7.

Sharma, H.S., Dey, P.K., 1987. Influence of long-term acute heat exposure on regional blood-brain barrier permeability, cerebral blood flow and 5-HT level in conscious normotensive young rats. Brain Res. 424 (1), 153–162. https://doi.org/10.1016/0006-8993(87)91205-4.

Sharma, H.S., Olsson, Y., 1990. Edema formation and cellular alterations following spinal cord injury in the rat and their modification with p-chlorophenylalanine. Acta Neuropathol. 79 (6), 604–610. https://doi.org/10.1007/BF00294237.

Sharma, A., Sharma, H.S., 2012a. Monoclonal antibodies as novel neurotherapeutic agents in CNS injury and repair. Int. Rev. Neurobiol. 102, 23–45. https://doi.org/10.1016/B978-0-12-386986-9.00002-8.

Sharma, H.S., Sharma, A., 2012b. Nanowired drug delivery for neuroprotection in central nervous system injuries: modulation by environmental temperature, intoxication of nanoparticles, and comorbidity factors. Wiley Interdiscip. Rev. Nanomed. Nanobiotechnol. 4 (2), 184–203. https://doi.org/10.1002/wnan.172. (Epub 2011 Dec 8). PMID: 22162425.

Sharma, H.S., Sharma, A., 2013. New perspectives of nanoneuroprotection, nanoneuropharmacology and nanoneurotoxicity: modulatory role of amino acid neurotransmitters, stress, trauma, and co-morbidity factors in nanomedicine. Amino Acids 45 (5), 1055–1071. https://doi.org/10.1007/s00726-013-1584-z.

Sharma, H.S., Sjöquist, P.O., 2002. A new antioxidant compound H-290/51 modulates glutamate and GABA immunoreactivity in the rat spinal cord following trauma. Amino Acids 23 (1–3), 261–272. https://doi.org/10.1007/s00726-001-0137-z.

Sharma, H.S., Westman, J., 1997. Prostaglandins modulate constitutive isoform of heat shock protein (72 kD) response following trauma to the rat spinal cord. Acta Neurochir. Suppl. 70, 134–137. https://doi.org/10.1007/978-3-7091-6837-0_41.

Sharma, H.S., Westman, J., Olsson, Y., Johansson, O., Dey, P.K., 1990a. Increased 5-hydroxytryptamine immunoreactivity in traumatized spinal cord. An experimental study in the rat. Acta Neuropathol. 80 (1), 12–17. https://doi.org/10.1007/BF00294216.

Sharma, H.S., Olsson, Y., Dey, P.K., 1990b. Early accumulation of serotonin in rat spinal cord subjected to traumatic injury. Relation to edema and blood flow changes. Neuroscience 36 (3), 725–730. https://doi.org/10.1016/0306-4522(90)90014-u.

Sharma, H.S., Cervós-Navarro, J., Dey, P.K., 1991a. Rearing at high ambient temperature during later phase of the brain development enhances functional plasticity of the CNS and induces tolerance to heat stress. An experimental study in the conscious normotensive young rats. Brain Dysfunction 4, 104–124.

Sharma, H.S., Cervós-Navarro, J., Dey, P.K., 1991b. Acute heat exposure causes cellular alteration in cerebral cortex of young rats. Neuroreport 2 (3), 155–158. https://doi.org/10.1097/00001756-199103000-00012.

Sharma, H.S., Westman, J., Nyberg, F., Cervós-Navarro, J., Dey, P.K., 1992a. Role of serotonin in heat adaptation: an experimental study in the conscious young rat. Endocr. Regul. 26 (3), 133–142.

Sharma, H.S., Zimmer, C., Westman, J., Cervós-Navarro, J., 1992b. Acute systemic heat stress increases glial fibrillary acidic protein immunoreactivity in brain: experimental observations in conscious normotensive young rats. Neuroscience 48 (4), 889–901. https://doi.org/10.1016/0306-4522(92)90277-9.

Sharma, H.S., Olsson, Y., Nyberg, F., Dey, P.K., 1993. Prostaglandins modulate alterations of microvascular permeability, blood flow, edema and serotonin levels following spinal cord injury: an experimental study in the rat. Neuroscience 57 (2), 443–449. https://doi.org/10.1016/0306-4522(93)90076-r.

Sharma, H.S., Olsson, Y., Cervós-Navarro, J., 1993a. Early perifocal cell changes and edema in traumatic injury of the spinal cord are reduced by indomethacin, an inhibitor of prostaglandin synthesis. Experimental study in the rat. Acta Neuropathol. 85 (2), 145–153. https://doi.org/10.1007/BF00227761.

Sharma, H.S., Olsson, Y., Cervós-Navarro, J., 1993b. p-Chlorophenylalanine, a serotonin synthesis inhibitor, reduces the response of glial fibrillary acidic protein induced by trauma to the spinal cord. An immunohistochemical investigation in the rat. Acta Neuropathol. 86 (5), 422–427. https://doi.org/10.1007/BF00228575.

Sharma, H.S., Olsson, Y., Persson, S., Nyberg, F., 1995a. Trauma-induced opening of the the blood-spinal cord barrier is reduced by indomethacin, an inhibitor of prostaglandin biosynthesis. Experimental observations in the rat using [131I]-sodium, Evans blue and lanthanum as tracers. Restor. Neurol. Neurosci. 7 (4), 207–215. https://doi.org/10.3233/RNN-1994-7403.

Sharma, H.S., Olsson, Y., Westman, J., 1995b. A serotonin synthesis inhibitor, p-chlorophenylalanine reduces the heat shock protein response following trauma to the spinal cord: an immunohistochemical and ultrastructural study in the rat. Neurosci. Res. 21 (3), 241–249. https://doi.org/10.1016/0168-0102(94)00855-a.

Sharma, H.S., Westman, J., Cervós-Navarro, J., Dey, P.K., Nyberg, F., 1997a. Opioid receptor antagonists attenuate heat stress-induced reduction in cerebral blood flow, increased blood-brain barrier permeability, vasogenic edema and cell changes in the rat. Ann. N. Y. Acad. Sci. 813, 559–571. https://doi.org/10.1111/j.1749-6632.1997.tb51747.x.

Sharma, H.S., Westman, J., Nyberg, F., 1997b. Topical application of 5-HT antibodies reduces edema and cell changes following trauma of the rat spinal cord. Acta Neurochir. Suppl. 70, 155–158. https://doi.org/10.1007/978-3-7091-6837-0_47.

Sharma, H.S., Westman, J., Nyberg, F., 1998a. Pathophysiology of brain edema and cell changes following hyperthermic brain injury. Prog. Brain Res. 115, 351–412. https://doi.org/10.1016/s0079-6123(08)62043-9.

Sharma, H.S., Alm, P., Westman, J., 1998b. Nitric oxide and carbon monoxide in the brain pathology of heat stress. Prog. Brain Res. 115, 297–333. https://doi.org/10.1016/s0079-6123(08)62041-5.

Sharma, H.S., Winkler, T., Stålberg, E., Mohanty, S., Westman, J., 2000a. p-Chlorophenylalanine, an inhibitor of serotonin synthesis reduces blood-brain barrier permeability, cerebral blood flow, edema formation and cell injury following trauma to the rat brain. Acta Neurochir. Suppl. 76, 91–95. https://doi.org/10.1007/978-3-7091-6346-7_19.

Sharma, H.S., Westman, J., Gordh, T., Nyberg, F., 2000b. Spinal cord injury induced c-fos expression is reduced by p-CPA, a serotonin synthesis inhibitor. An experimental study using immunohistochemistry in the rat. Acta Neurochir. Suppl. 76, 297–301. https://doi.org/10.1007/978-3-7091-6346-7_61.

Sharma, H.S., Westman, J., Nyberg, F., 2000c. Selective alteration of calcitonin gene related peptide in hyperthermic brain injury. An experimental study in the rat brain using immunohistochemistry. Acta Neurochir. Suppl. 76, 541–545. https://doi.org/10.1007/978-3-7091-6346-7_113.

Sharma, H.S., Patnaik, R., Patnaik, S., Mohanty, S., Sharma, A., Vannemreddy, P., 2007. Antibodies to serotonin attenuate closed head injury induced blood brain barrier disruption and brain pathology. Ann. N. Y. Acad. Sci. 1122, 295–312. https://doi.org/10.1196/annals.1403.022.

Sharma, H.S., Patnaik, R., Patnaik, S., Sharma, A., Mohanty, S., Vannemreddy, P., 2010. Antibodies to dynorphin a (1-17) attenuate closed head injury induced blood-brain barrier disruption, brain edema formation and brain pathology in the rat. Acta Neurochir. Suppl. 106, 301–306. https://doi.org/10.1007/978-3-211-98811-4_56.

Sharma, A., Muresanu, D.F., Mössler, H., Sharma, H.S., 2012. Superior neuroprotective effects of cerebrolysin in nanoparticle-induced exacerbation of hyperthermia-induced brain pathology. CNS Neurol. Disord. Drug Targets 11 (1), 7–25. https://doi.org/10.2174/187152712799960790.

Sharma, H.S., Muresanu, D.F., Lafuente, J.V., Nozari, A., Patnaik, R., Skaper, S.D., Sharma, A., 2016a. Pathophysiology of blood-brain barrier in brain injury in cold and hot environments: novel drug targets for neuroprotection. CNS Neurol. Disord. Drug Targets 15 (9), 1045–1071. https://doi.org/10.2174/1871527315666160902145145.

Sharma, A., Menon, P., Muresanu, D.F., Ozkizilcik, A., Tian, Z.R., Lafuente, J.V., Sharma, H.-S., 2016b. Nanowired drug delivery across the blood-brain barrier in central nervous system injury and repair. CNS Neurol. Disord. Drug Targets 15 (9), 1092–1117. https://doi.org/10.2174/1871527315666160819123059.

Sharma, H.S., Muresanu, D.F., Sharma, A., 2016c. Alzheimer's disease: cerebrolysin and nanotechnology as a therapeutic strategy. Neurodegener. Dis. Manag. 6 (6), 453–456. https://doi.org/10.2217/nmt-2016-0037 (Epub 2016 Nov 9).

Sharma, A., Menon, P.K., Patnaik, R., Muresanu, D.F., Lafuente, J.V., Tian, Z.R., Ozkizilcik, A., Castellani, R.J., Mössler, H., Sharma, H.S., 2017. Novel treatment strategies using TiO(2)-nanowired delivery of histaminergic drugs and antibodies to tau with cerebrolysin for superior neuroprotection in the pathophysiology of Alzheimer's disease. Int. Rev. Neurobiol. 137, 123–165. https://doi.org/10.1016/bs.irn.2017.09.002 (Epub 2017 Nov 6).

Sharma, B., Lawrence, D.W., Hutchison, M.G., 2018a. Branched chain amino acids (BCAAs) and traumatic brain injury: a systematic review. J. Head Trauma Rehabil. 33 (1), 33–45. https://doi.org/10.1097/HTR.0000000000000280.

Sharma, H.S., Muresanu, D.F., Lafuente, J.V., Patnaik, R., Tian, Z.R., Ozkizilcik, A., Castellani, R.J., Mössler, H., Sharma, A., 2018b. Co-administration of TiO2 nanowired mesenchymal stem cells with cerebrolysin potentiates neprilysin level and reduces brain pathology in Alzheimer's disease. Mol. Neurobiol. 55 (1), 300–311. https://doi.org/10.1007/s12035-017-0742-9.

Sharma, H.S., Muresanu, D.F., Nozari, A., Castellani, R.J., Dey, P.K., Wiklund, L., Sharma, A., 2019. Anesthetics influence concussive head injury induced blood-brain barrier breakdown, brain edema formation, cerebral blood flow, serotonin levels, brain pathology and functional outcome. Int. Rev. Neurobiol. 146, 45–81. https://doi.org/10.1016/bs.irn.2019.06.006 (Epub 2019 Jul 8).

Sharma, A., Muresanu, D.F., Castellani, R.J., Nozari, A., Lafuente, J.V., Sahib, S., Tian, Z.R., Buzoianu, A.D., Patnaik, R., Wiklund, L., Sharma, H.S., 2020a. Mild traumatic brain injury exacerbates Parkinson's disease induced hemeoxygenase-2 expression and brain pathology: neuroprotective effects of co-administration of TiO(2) nanowired mesenchymal stem cells and cerebrolysin. Prog. Brain Res. 258, 157–231. https://doi.org/10.1016/bs.pbr.2020.09.010 (Epub 2020 Nov 9).

Sharma, A., Muresanu, D.F., Sahib, S., Tian, Z.R., Castellani, R.J., Nozari, A., Lafuente, J.V., Buzoianu, A.D., Bryukhovetskiy, I., Manzhulo, I., Patnaik, R., Wiklund, L., Sharma, H.S., 2020b. Concussive head injury exacerbates neuropathology of sleep deprivation: superior neuroprotection by co-administration of TiO(2)-nanowired cerebrolysin, alpha-melanocyte-stimulating hormone, and mesenchymal stem cells. Prog. Brain Res. 258, 1–77. https://doi.org/10.1016/bs.pbr.2020.09.003 (Epub 2020 Nov 2).

Sharp, D.J., Jenkins, P.O., 2015. Concussion is confusing us all. Pract. Neurol. 15 (3), 172–186. https://doi.org/10.1136/practneurol-2015-001087.

Silverberg, N.D., Iaccarino, M.A., Panenka, W.J., Iverson, G.L., McCulloch, K.L., Dams-O'Connor, K., Reed, N., McCrea, M., American Congress of Rehabilitation Medicine Brain Injury Interdisciplinary Special Interest Group Mild TBI Task Force, 2020. Management of concussion and mild traumatic brain injury: a synthesis of practice guidelines. Arch. Phys. Med. Rehabil. 101 (2), 382–393. https://doi.org/10.1016/j.apmr.2019.10.179 (Epub 2019 Oct 23).

Somogyi, P., Takagi, H., 1982. A note on the use of picric acid-paraformaldehyde-glutaraldehyde fixative for correlated light and electron microscopic immunocytochemistry. Neuroscience 7 (7), 1779–1783. https://doi.org/10.1016/0306-4522(82)90035-5.

Steel, A., Mikkelsen, M., Edden, R.A.E., Robertson, C.E., 2020. Regional balance between glutamate+glutamine and GABA+ in the resting human brain. Neuroimage 220, 117112. https://doi.org/10.1016/j.neuroimage.2020.117112 (Epub 2020 Jun 30).

Su, I.C., Hung, C.F., Lin, C.N., Huang, S.K., Wang, S.J., 2019. Cycloheterophyllin inhibits the release of glutamate from nerve terminals of the rat hippocampus. Chem. Res. Toxicol. 32 (8), 1591–1598. https://doi.org/10.1021/acs.chemrestox.9b00121 (Epub 2019 Jul 16).

Taoka, M., Toda, T., Iwamura, Y., 1998. Representation of the midline trunk, bilateral arms, and shoulders in the monkey postcentral somatosensory cortex. Exp. Brain Res. 123 (3), 315–322. https://doi.org/10.1007/s002210050574.

Taylor, C.A., Bell, J.M., Breiding, M.J., Xu, L., 2017. Traumatic brain injury-related emergency department visits, hospitalizations, and deaths—United States, 2007 and 2013. MMWR Surveill. Summ. 66 (9), 1–16. https://doi.org/10.15585/mmwr.ss6609a1.

Temple, J.L., Struchen, M.A., Pappadis, M.R., 2016. Impact of pre-injury family functioning and resources on self-reported post-concussive symptoms and functional outcomes in persons with mild TBI. Brain Inj. 30 (13–14), 1672–1682. https://doi.org/10.3109/02699052.2015.1113561 (Epub 2016 Oct 14).

Tranberg, M., Stridh, M.H., Guy, Y., Jilderos, B., Wigström, H., Weber, S.G., Sandberg, M., 2004. NMDA-receptor mediated efflux of N-acetylaspartate: physiological and/or pathological importance? Neurochem. Int. 45 (8), 1195–1204. https://doi.org/10.1016/j.neuint.2004.06.005.

Troup, G.A., Thomas, M.D., Skilbeck, C.E., 2020. The factor structure of the quality of life inventory (QOLI) following traumatic brain injury. Neuropsychol. Rehabil. 30 (6), 1129–1149. https://doi.org/10.1080/09602011.2018.1564674 (Epub 2019 Jan 8).

Tsai, G., Coyle, J.T., 1995. N-acetylaspartate in neuropsychiatric disorders. Prog. Neurobiol. 46, 531–540.

van der Mast, R.C., Fekkes, D., 2000. Serotonin and amino acids: partners in delirium pathophysiology? Semin. Clin. Neuropsychiatry 5 (2), 125–131. doi: 10.153/SCNP00500125.

van Ierssel, J., Sveistrup, H., Marshall, S., 2018. Identifying the concepts contained within health-related quality of life outcome measures in concussion research using the international classification of functioning, disability, and health as a reference: a systematic review. Qual. Life Res. 27 (12), 3071–3086. https://doi.org/10.1007/s11136-018-1939-8 (Epub 2018 Jul 20).

VanItallie, T.B., 2019. Traumatic brain injury (TBI) in collision sports: possible mechanisms of transformation into chronic traumatic encephalopathy (CTE). Metabolism 100S, 153943. https://doi.org/10.1016/j.metabol.2019.07.007.

Vannemreddy, P., Ray, A.K., Patnaik, R., Patnaik, S., Mohanty, S., Sharma, H.S., 2006. Zinc protoporphyrin IX attenuates closed head injury-induced edema formation, blood-brain barrier disruption, and serotonin levels in the rat. Acta Neurochir. Suppl. 96, 151–156. https://doi.org/10.1007/3-211-30714-1_34.

Vazana, U., Veksler, R., Pell, G.S., Prager, O., Fassler, M., Chassidim, Y., Roth, Y., Shahar, H., Zangen, A., Raccah, R., Onesti, E., Ceccanti, M., Colonnese, C., Santoro, A., Salvati, M., D'Elia, A., Nucciarelli, V., Inghilleri, M., Friedman, A., 2016. Glutamate-mediated blood-brain barrier opening: implications for neuroprotection and drug delivery. J. Neurosci. 36 (29), 7727–7739. https://doi.org/10.1523/JNEUROSCI.0587-16.2016.

Vercelli, A., Biggi, S., Sclip, A., Repetto, I.E., Cimini, S., Falleroni, F., Tomasi, S., Monti, R., Tonna, N., Morelli, F., Grande, V., Stravalaci, M., Biasini, E., Marin, O., Bianco, F., di Marino, D., Borsello, T., 2015. Exploring the role of MKK7 in excitotoxicity and cerebral ischemia: a novel pharmacological strategy against brain injury. Cell Death Dis. 6 (8), e1854. https://doi.org/10.1038/cddis.2015.226.

Viano, D.C., Parenteau, C.S., 2015. Concussion, diffuse axonal injury, and AIS4+ head injury in motor vehicle crashes. Traffic Inj. Prev. 16 (8), 747–753. https://doi.org/10.1080/15389588.2015.1013188 (Epub 2015 Feb 9).

Voss, J.D., Connolly, J., Schwab, K.A., Scher, A.I., 2015. Update on the epidemiology of concussion/mild traumatic brain injury. Curr. Pain Headache Rep. 19 (7), 32. https://doi.org/10.1007/s11916-015-0506-z.

Wang, K., Cui, D., Gao, L., 2016. Traumatic brain injury: a review of characteristics, molecular basis and management. Front. Biosci. (Landmark Ed.) 21, 890–899. https://doi.org/10.2741/4426.

Webster, K.M., Sun, M., Crack, P., O'Brien, T.J., Shultz, S.R., Semple, B.D., 2017. Inflammation in epileptogenesis after traumatic brain injury. J. Neuroinflammation 14 (1), 10. https://doi.org/10.1186/s12974-016-0786-1.

Wenthold, R.J., 1981. Glutamate and aspartate as neurotransmitters for the auditory nerve. Adv. Biochem. Psychopharmacol. 27, 69–78.

Westman, J., Sharma, H.S., 1998. Heat shock protein response in the central nervous system following hyperthermia. Prog. Brain Res. 115, 207–239. https://doi.org/10.1016/s0079-6123(08)62038-5.

Wiese, S., Metzger, F., Holtmann, B., Sendtner, M., 1999. Mechanical and excitotoxic lesion of motoneurons: effects of neurotrophins and ciliary neurotrophic factor on survival and regeneration. Acta Neurochir. Suppl. 73, 31–39. https://doi.org/10.1007/978-3-7091-6391-7_5.

Wiklund, L., Toggenburger, G., Cuénod, M., 1982. Aspartate: possible neurotransmitter in cerebellar climbing fibers. Science 216 (4541), 78–80. https://doi.org/10.1126/science.6121375.

Willis, C., Lybrand, S., Bellamy, N., 2004. Excitatory amino acid inhibitors for traumatic brain injury. Cochrane Database Syst. Rev. 2003 (1), CD003986. https://doi.org/10.1002/14651858.CD003986.pub2.

Winkler, T., Sharma, H.S., Stålberg, E., Olsson, Y., Nyberg, F., 1994. Opioid receptors influence spinal cord electrical activity and edema formation following spinal cord injury: experimental observations using naloxone in the rat. Neurosci. Res. 21 (1), 91–101. https://doi.org/10.1016/0168-0102(94)90072-8.

Woitzik, J., Pinczolits, A., Hecht, N., Sandow, N., Scheel, M., Drenckhahn, C., Dreier, J.P., Vajkoczy, P., 2014. Excitotoxicity and metabolic changes in association with infarct progression. Stroke 45 (4), 1183–1185. https://doi.org/10.1161/STROKEAHA.113.004475 (Epub 2014 Mar 18).

Wu, Z., Wang, Z.H., Liu, X., Zhang, Z., Gu, X., Yu, S.P., Keene, C.D., Cheng, L., Ye, K., 2020. Traumatic brain injury triggers APP and tau cleavage by delta-secretase, mediating Alzheimer's disease pathology. Prog. Neurobiol. 185, 101730. https://doi.org/10.1016/j.pneurobio.2019.101730 (Epub 2019 Nov 25).

Yasen, A.L., Smith, J., Christie, A.D., 2018. Glutamate and GABA concentrations following mild traumatic brain injury: a pilot study. J. Neurophysiol. 120 (3), 1318–1322. https://doi.org/10.1152/jn.00896.2017 (Epub 2018 Jun 20).

Yasen, A.L., Lim, M.M., Weymann, K.B., Christie, A.D., 2020. Excitability, inhibition, and neurotransmitter levels in the motor cortex of symptomatic and asymptomatic individuals following mild traumatic brain injury. Front. Neurol. 11, 683. https://doi.org/10.3389/fneur.2020.00683 (eCollection 2020).

Ye, J.H., 2008. Regulation of excitation by glycine receptors. Results Probl. Cell Differ. 44, 123–143. https://doi.org/10.1007/400_2007_029.

You, Z.B., Herrera-Marschitz, M., Pettersson, E., Nylander, I., Goiny, M., Shou, H.Z., Kehr, J., Godukhin, O., Hökfelt, T., Terenius, L., Ungerstedt, U., 1996. Modulation of neurotransmitter release by cholecystokinin in the neostriatum and substantia nigra of the rat: regional and receptor specificity. Neuroscience 74 (3), 793–804. https://doi.org/10.1016/0306-4522(96)00149-2.

Yrondi, A., Brauge, D., LeMen, J., Arbus, C., Pariente, J., 2017. Depression and sports-related concussion: a systematic review. Presse Med. 46 (10), 890–902. https://doi.org/10.1016/j.lpm.2017.08.013 (Epub 2017 Sep 14).

Zetterberg, H., Blennow, K., 2018. Chronic traumatic encephalopathy: fluid biomarkers. Handb. Clin. Neurol. 158, 323–333. https://doi.org/10.1016/B978-0-444-63954-7.00030-6.

Zhang, H., Wang, Y., Lian, L., Zhang, C., He, Z., 2020. Glycine-histidine-lysine (GHK) alleviates astrocytes injury of intracerebral hemorrhage via the Akt/miR-146a-3p/AQP4 pathway. Front. Neurosci. 14, 576389. https://doi.org/10.3389/fnins.2020.576389 (eCollection 2020).

Zhong, Z., Wheeler, M.D., Li, X., Froh, M., Schemmer, P., Yin, M., Bunzendaul, H., Bradford, B., Lemasters, J.J., 2003. L-Glycine: a novel antiinflammatory, immunomodulatory, and cytoprotective agent. Curr. Opin. Clin. Nutr. Metab. Care 6 (2), 229–240. https://doi.org/10.1097/00075197-200303000-00013.

Zimmer, C., Sampaolo, S., Sharma, H.S., Cervós-Navarro, J., 1991. Altered glial fibrillary acidic protein immunoreactivity in rat brain following chronic hypoxia. Neuroscience 40 (2), 353–361. https://doi.org/10.1016/0306-4522(91)90125-8.

CHAPTER

7

Neuromodulation as a basic platform for neuroprotection and repair after spinal cord injury

Artur Biktimirov[a,*], Oleg Pak[b], Igor Bryukhovetskiy[a], Aruna Sharma[c], and Hari Shanker Sharma[c,*]

[a]*Department of Fundamental Medicine, School of Biomedicine, Far Eastern Federal University, Vladivostok, Russia*

[b]*Department of Neurosurgery, Medical Center, Far Eastern Federal University, Vladivostok, Russia*

[c]*International Experimental Central Nervous System Injury & Repair (IECNSIR), Department of Surgical Sciences, Anesthesiology & Intensive Care Medicine, Uppsala University Hospital, Uppsala University, Uppsala, Sweden*

**Corresponding authors: Tel.: +7-914-9651488 (Artur Biktimirov); Tel.: +46-18-243899 (Hari Shanker Sharma), e-mail address: biktimirov.ar@dvfu.ru; sharma@surgsci.uu.se*

Abstract

Spinal cord injury (SCI) is one of the most challenging medical issues. Spasticity is a major complication of SCI. A combination of spinal cord stimulation, new methods of neuroprotection and biomedical cellular products provides fundamentally new options for SCI treatment and rehabilitation. The paper attempts to critically analyze the effectiveness of using these procedures for patients with SCI, suggesting a protocol for a step-by-step personalized treatment of SCI, based on continuity of modern conservative and surgical methods. The study argues the possibility of using neuromodulation as a basis for rehabilitating patients with SCI.

Keywords

Neuromodulation, Spinal cord injury, Spasticity, Intrathecal baclofen therapy, Spinal cord stimulation

Progress in Brain Research, Volume 266, ISSN 0079-6123, https://doi.org/10.1016/bs.pbr.2021.06.012
Copyright © 2021 Elsevier B.V. All rights reserved.

Abbreviations

BBB blood-brain barrier
CNS central nervous system
CPG central pattern generator
FES functional electrostimulation
HIF hypoxia induced factor
ITB intrathecal baclofen therapy
MAS modified Ashworth scale
PS Penn scale
SC spinal cord
SCI spinal cord injury
SCS spinal cord stimulation
TS Tardieu scale

1 Introduction

Spinal cord injury (SCI) is one of the most challenging medical issues. The SCI incidence is 40–60 cases per a million people worldwide, and this number continues growing. Traffic accidents, falls from heights and other types of destructive mechanical impact are the main cause of SCI during the peacetime (Kumar et al., 2018). The average age of patients with SCI is 20–25 years old, and the cervical and dorsal areas of the spinal cord become damaged in 60% of cases (Sweis and Biller, 2017). SCI mortality rate is up to 64%, while the survivors become severely incapacitated and totally dependent on the external help.

Eighty percent of patients with SCI develop spasticity that significantly limits their rehabilitation options (Nagel et al., 2017), causes the development of contractures, pain syndrome, trophic disorders, poor quality of life and severe depression that frequently results in suicide. Modern approach to treating spasticity involves neuromuscular relaxants and physiotherapy, and in case of their ineffectiveness the protocol may be changed for the intrathecal baclofen therapy (IBT). Spinal cord stimulation (SCS)—another method of neuromodulation with a similar level of effectiveness—is frequently overlooked as a SC treatment option.

Meanwhile, SCS has significant advantages over ITB due to its being risk-free in terms of developing life-threatening complications. But its potential safety is not the only benefit of SCS. A considerable amount of clinical and experimental data (Yamazaki et al., 2020) indicate that combined application of new biomedical technologies could be very effective in the first few years, especially during the first year after the injury, and that requires development of a step-by-step treatment strategy, based on conservative and surgical methods that provide new opportunities and prospects for patients with SCI.

This study attempts to justify the usage of neuromodulation as a basis for rehabilitation of patients with SCI, implementing modern biomedical technologies.

2 Pathophysiology of SC injury and regeneration

There are several mechanisms of primary SC injuries (Sweis and Biller, 2017): mechanical force with constant SC compression, mechanical force with short-term SC compression, distraction, causing distortion in the axial plane, and SC severance due to a blast injury. A complex clinical picture and a type of damage in case of SCI precipitate several important pathophysiological mechanisms: necrosis, apoptosis, autophagy, swelling, cavitation and reactive gliosis.

Mechanical energy causes necrosis of the SC gray matter, damages axons, destroys intercellular connections and triggers secondary injury cascade both in close range, and in significant distance from the alteration area. Damage to the blood vessels is accompanied by the development of petechial hemorrhages that merge into extensive focal purpura. Lacunar ischemic strokes, resulting from angiospasm due to vasoconstrictors, released in the injured area, are followed by growing swelling and creation of disseminated necrotic foci beyond the area of initial damage. Accumulation of lactate, nitrogen oxide and other bioactive compounds result in microvascular palsy, blood sludging, microthrombosis and rapid growth of the secondary alteration area, edema and damage to the other SC segments.

Hypoxia, glutamate, Ca^2+ ions and various anti-inflammatory cytokines, produced by immunocytes and cells of resident microglia, trigger apoptosis, the first peak of which is detected in the area of the necrotic foci in 6–8 h after the SCI. The second peak of apoptosis starts in 48–72 h after the SCI, and the maximum amount of neurons and oligodendrocytes is destroyed in the injury-adjacent areas, followed by amyelination of axons and development of conduction disorders (Fig. 1).

SC reparation is connected with alterations of cellular composition and cytokine profile of the primary injury area. Hypoxia induces production of hypoxia induced factor (HIF) in CNS tissues, thus, stimulating expression of more than 80 genes that determine production of signaling molecules. Stromal cell-derived factor (SDF)-1α or chemokine (C-X-C motif) ligand 12 (CXCL12) is especially significant among those; it is a CXC chemokine that is encoded by CXCL12 gene in humans. SDF-1 connects with CXCR4 and CXCR7 receptors on the surface of neural stem cells and mononuclear cells of the red bone marrow, including the normal stem cells, attracting them into the area of damage. But stem cells migration and homing in the area of damage does not necessarily result in effective reparation of nervous tissue.

The possibilities of SC reparation are limited by the intracellular regeneration of neurons, partial recuperation and remyelination of nerve fibers, limited growth of axon length with a subsequent creation of functional connections with SC motoneurons that provides partial restoration of the lost functions. Despite the importance of endogenous neural stem cells for CNS regeneration and plasticity (Stenudd et al., 2015), the main element of SC reparation is proliferation of stromal cells, mostly fibroblasts, from damaged meninges. One of the key inducers of collagenous fibrosis is transforming growth factor (TGF)β—anti-inflammatory cytokine that coordinates

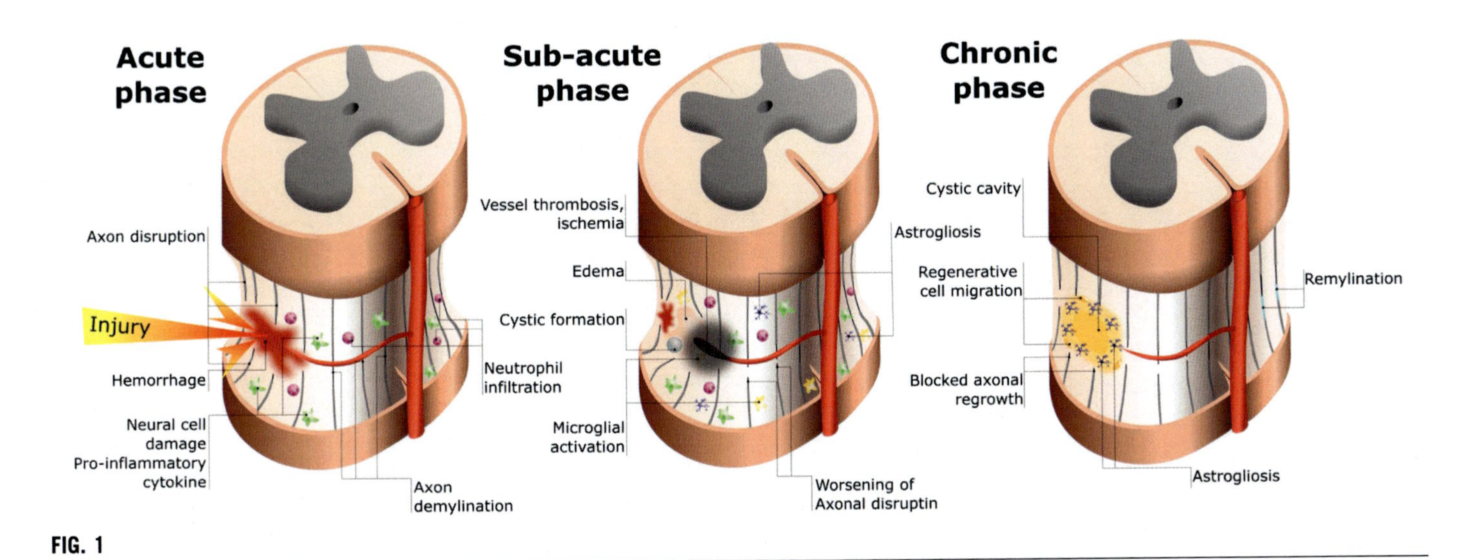

FIG. 1

Stages of development of spinal cord tissue damage as a result of primary spinal cord injury.

intercellular interaction among fibroblasts, astrocytes and immunocytes (Song et al., 2019), providing quick recovery of the tissue and creation of gliomesodermal scarring that becomes a literally impenetrable barrier for regenerating axons.

Therefore, the priority in early stages of SCI treatment is not only SC decompression, but the most active neuroprotection together with suppression of the cascade of suspended secondary CNS damage.

3 Strategy for treating SCI patients: Reanimation, neuroprotection and blood-brain barrier

The overwhelming majority of patients are in a critical state during the acute period of SCI. It goes without saying that effective neuroprotection with all available modern technologies is quite an issue to deal with while carrying out reanimation procedures. Nevertheless, it should be started as soon as the patient's condition allows. Lack of proper neuroprotection in first hours or even days after SCI results in spinal shock, ascending swelling, life-threatening arrhythmia, respiratory disorders and other severe cardiac and pulmonary complications (Yang et al., 2017) that frequently result in death of such patients.

Patients with cervical and upper-thoracic SC injuries run the highest risk of developing cardiopulmonary complications that could be accompanied by tetraplegia, paralysis of respiratory muscles and development of respiratory hypoxia. In its turn, hyperventilation of lungs due to artificial ventilation activates lipid peroxidation, leading to cell damage, dysfunction of cell membranes and subsequent destruction of neurons. Spinal and orthostatic hypotension, vegetative dysreflexia and bradycardia often occur in the patients with injuries of Th6 and higher (Hagen et al., 2012; Popa et al., 2010; Sweis and Biller, 2017) due to the inability of dorsomedial and paraventricular nuclei of hypothalamus to control sympathetic preganglionic neurons in SC. These incidences result in rapid reflectory hypertonus of vagus nerve, vasodilatation, decrease of venous return of blood back to the heart, hypotension, lower cardiac output and cardiac reserve, bradyarrhythmia and atrioventricular block.

Therefore, almost any factor, associated with or caused by SCI, like pain, shock, blood loss, artificial lung ventilation, enteroplegia, urinary tract infection, pressure ulcers and other reasons, aggravates the injury, resulting in damage being spread to the upper regions of the CNS.

That is why the adequate reanimation measures and intensive care in the first hours and days after the SCI predetermine the options for the subsequent rehabilitation of the patient, and they should be treated as an important component of the overall strategy for neuroprotection.

Meanwhile, the modern protocols for SCI treatment (Kwon et al., 2011) rarely include neuroprotection procedures to decrease the damage to nerve tissue in acute and subacute stages of the injury. Clinicians argue that the majority of the existing medication and procedures lack proven neuroprotective effect not only in relation to SCI, but also other types of SC injuries or vascular catastrophes.

The blood-brain barrier (BBB)—physiological barrier between blood circulation and central nervous system—is the main obstacle for neuroprotection in SCI. The methods of delivering neuroprotectors to CNS through the BBB are still undergoing the experimental phase. The possibility of targeted nanodelivery of cerebrolysin through the BBB has been proven (Muresanu et al., 2019; Niu et al., 2019). Also, there have been attempts at inhibiting epidural fibrosis via NF-kβ signaling pathway and nanodelivery of all-trans retinoic acids through the BBB (Zhang et al., 2020). Neuroprotective properties of HOE-140, a new B2 antagonist, after SCI were revealed (Sharma et al., 2019), the possibility of controlling permeability of microvessels, decreasing the swelling and activating nitrogen oxide, was proven. Neuroprotective effect of the targeted delivery of DL-3-n-butylphthalide by using titanium-oxide-based nanomaterials was demonstrated (Sahib et al., 2019).

In reality neuroprotection after SCI is a very pressing issue. Usually, the extent of support for patients in the acute SCI stage involves reanimation and a subsequent intensive care that is obviously insufficient. Lack of centralized standards forces medical specialists treat each case, based on empirical data and traditional neurosurgical protocols. The primary focus is on saving a patient's life and dealing with secondary complications, and only partly on preserving the damaged nerve tissue. Spasticity is one of the detrimental consequences of this clinical practice.

4 Spasticity

There is more than one approach to defining spasticity; its molecular and cellular mechanisms are relatively unknown. The expert panel, developing the Best Practices for Intrathecal Baclofen Therapy (Saulino et al., 2016a,b), defined this phenomenon as "disordered sensorimotor control resulting from an upper motor neuron (UMN) lesion, presenting as intermittent or sustained involuntary activation of muscles," and it is accompanied by a steady increase of muscle tonus and tendon reflexes. This definition is far from perfect since it does not show the contribution of reticulospinal, rubrospinal, cerebellospinal and other suprasegmental neurobiological mechanisms to spasticity development (Sangari and Perez, 2019). Spasticity is mainly considered to be caused by the loss of control over the segmental apparatus of the spinal cord that is frequently observed in patients with complete damage to this area of CNS, and it results in increased muscle tonus, impairment of tendon reflexes, synkinesis, pathological reflexes and functional abnormalities in the work of pelvic organs.

This simplified and even mechanical approach to the issue definitely makes clinical sense and is appropriate for the purposes of clinical neurology. However, there are some data (Ameer and Gillis, 2021; Badhiwala et al., 2020; Heald et al., 2017; Sangari et al., 2019) that a compete anatomical severance of the SC is a very rare and critical condition that becomes fatal in most cases. In experiments complete thoracic segmentectomy of rats resulted in increased mortality rates among animals with extending area of damage, reaching almost 100% in the cervical section of the SC (Yarygin et al., 2006). Taking that into account, it becomes obvious why the patriarch of modern neurobiology Santiago Ramón y Cajal decided against experiments with complete severance of SC due to the extremely high mortality rates and gravity of experimental animals' condition.

There are enough data (Heald et al., 2017) to believe that 60–70% of patients with complete damage of SC in reality have incomplete SC severance; such patients still retain control over the lower part of the body, while spasticity is more common among them. Therefore, clinically determined lack of control over the segmental apparatus of the spinal cord in the area below the injury does not necessarily mean that there is no centralized control over the neurons in that area. Hence, it is safe to say that spasticity is related to the imbalance between inhibitory and stimulatory supraspinal mechanisms, controlling intersegmental connections.

Some researchers (Dietz and Sinkjaer, 2012; Finnerup, 2017; Nielsen et al., 2007) discover many similarities in pain and spasticity pathogens since the same sensory systems can be involved in development of central pain. That is why in 2002 Sjolund introduced the term "sensory spasticity" (Sjolund, 2002). Experiments show that sensory pathways and trajectories of segmental reflex can overlap in the same interneurons (Sjolund, 2002). For instance, clinical signs of pain and spasticity can manifest themselves not right after SCI, but increase overtime and show reaction to a simple touch. Using medication for treating neuropathic pain (Pregabalin, Gabapentin) also decreases spasticity, while baclofen that is used for spasticity treatment does not have the same effect (Finnerup, 2017). Patients with musculo-articular pain have been described to demonstrate a high muscle tonus, and a higher level of neuropathic pain was attributed to the higher frequency of spasms (Andresen et al., 2016). In the light of the foregoing, it is obvious that, in terms of pathophysiology, spasticity is a more complex phenomenon than just a result of dissociation of central and motor neurons, as it was considered before.

Main complaints of patients with spasticity include uncontrollable increase of muscle tone and muscle spasms in the area below the injury that is accompanied by the pain and becomes both a physical, and an emotional burden. Spasticity is diagnosed with the classic and modified Ashworth scales (Ansari et al., 2006; Biering-Sørensen et al., 2006; Bohannon and Smith, 1987; Meseguer-Henarejos et al., 2018), frequency of spasms is determined with the Penn scale and its analogs (Penn et al., 1989; Snow et al., 1990). The Tardieu scale allows to give a quantitative characteristic of spasticity and distinguish it from contractures (Boyd and Graham, 1999; Gracies et al., 2000), but it is more time-consuming than using AS and MAS (Morris and Williams, 2018).

Isokinetic dynamometers allow to measure speed-dependent muscle resistance to passive movement that provides an opportunity to standardize the obtained results, based on stretch rates and range of motion (Biering-Sørensen et al., 2006). Apart from objectification of spasticity level, use of dynamometers allows to evaluate the real muscular strength since the severity of spasticity and muscular strength are inversely connected, i.e., if a spastic muscle is weaker, it should be taken into account in rehabilitation (Abdollahi et al., 2015).

It is not always possible to use electrophysiological, radiological and other objective methods for diagnosing spasticity in clinical practice. For instance, in myography such parameters as H- and M-reflex and F-wave do not correlate with the clinical indicators, discovered with traditional diagnostic scales. The method of motor evoked potentials (MEP) showed reliable correlation results when used to evaluate motor function of patients with SCI and their spasticity level (Sangari et al., 2019). Effectiveness of neurovisualization methods in spasticity diagnostics is doubtful, but these data could be potentially useful for estimation of a possible recovery level.

MRI with DTI and tractography could show the damage to SC matter both in acute and chronic stages of SCI (Alizadeh et al., 2018; Zhao et al., 2018). SC atrophy in the anterior-posterior width (APW) correlates with sensory deficit, and atrophy in the left-right width (LRW) correlates with motor deficit (Lundell et al., 2011). It was noted (Sangari et al., 2019) that patients with spasticity had the LRW parameters that correlated with MAS scores and scores of motor evoked potentials (MEP).

Therefore, presently there is no method, able to reflect all aspects of spasticity, moreover, clinicians should take into consideration all types of information from different sources when dealing with each case in particular. One possible solution to this issue is using artificial intelligence to analyze cause-and-effect connections and develop an appropriate protocol for spasticity treatment. Nowadays (Park et al., 2019) the results of spasticity severity assessment, performed by humans and artificial intelligence, have coincided in more than 80% of cases. If neural networks learn from the data, received from wearable devices (Kim et al., 2020), the accuracy of assessment and estimation reaches 95%, and the advantage of using modern advances becomes obvious.

Spasticity treatment involves a wide range of methods, such as physiotherapy, oral myorelaxants, botulin therapy and surgical correction with neuromodulation (ITB and SCS) and destruction (DREZ, neurotomy, etc.).

Daily passive stretching of muscles decreases muscular tonus, strengthens synergistic muscles and maintains articulator mobility and range of motion. If a necessary effect has not been achieved, functional electrostimulation is used up to 4h to decrease muscle tonus (Mills and Dossa, 2016; Sivaramakrishnan et al., 2018) in affected extremities. FES in the Parastep device allowed patients with lower paraplegia to partially maintain vertical posture and walk a distance up to 440m without external help (Graupe et al., 2008). However, application of functional electrostimulation, as well as robotic rehabilitation systems, is limited by patterns and standards of motion, and disregarding them might result in excessive stretching of tissues and injuries (Pizzolato et al., 2019), so usage of these methods should be thoroughly personalized for each patient.

Treatment always involves using oral muscle relaxants. In terms of the mechanisms of action, there are central and peripheral medical agents (Table 1, Fig. 2). Central agents affect mediators of segmental level and include GABA-B agonists (baclofen), GABA-A agonists—benzodiazepines (diazepam), α_2-adrenergic agonists (tizanidine), N-cholinoblockers (tolperisone). Peripheral agents affect neuromuscular transmission and include Dantrolene (Lapeyre et al., 2010).

Table 1 List of myorelaxant's.

	Mechanism of action	Side effects
Diazepam	GABA—A agonists	Sedation, hypotension, addiction
Baclofen	Central-acting muscle relaxant, GABA-B stimulator	Sedation, excessive weakness, dizziness, confusion, hallucinations, convulsions
Tizanidine	Agonist of central alpha-2-adrenergic receptors	Hypotension, dizziness, dry mouth, hepatotoxicity
Dantrolen	Selective inhibition of the release of Ca++ ions from the reticulum of the sarcoplasm of skeletal muscles	Diarrhea, nausea, vomiting, weakness, hepatotoxicity
Tolperizon	Central H-cholinolytic	Muscle weakness, headache, hypotension, nausea, vomiting, abdominal discomfort

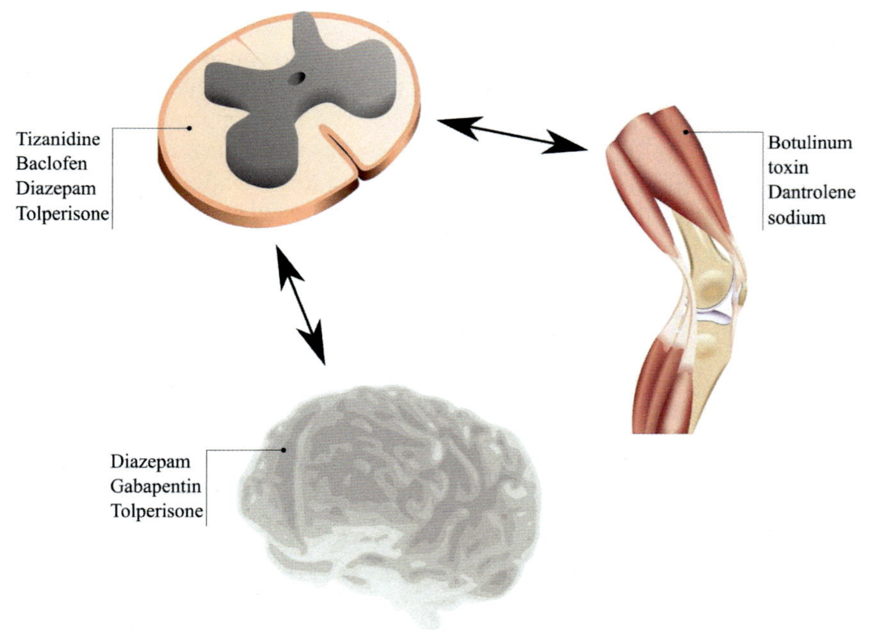

FIG. 2

Points of application of drugs with a muscle relaxant effect.

Focal spasticity of one or several muscle groups was treated with neuromuscular blockade with ethanol for the first time in 1960, but nowadays ethanol has been substituted for botulotoxin (Lapeyre et al., 2010; Lui et al., 2015). Its mechanism of action is determined by chemodenervation, i.e., direct peripheral effect on neuromuscular transmission due to blocking cholinergic transmission in neuromuscular junctions that relaxes muscles. Spasticity decreases in 1–2 weeks after the injection, and the effect remains for 3–6 months. Botulinotherapy is considered to be safe and effective for treating focal spasticity, according to the Guidelines developed by the American Academy of Neurology in 2016 (Picelli et al., 2019; Simpson et al., 2016), but treatment with botulinum neurotoxin requires regular injections and does not resolve the issues with generalized spasticity.

5 Surgical treatment of spasticity

Surgical treatment of spasticity after SCI involves intrathecal baclofen therapy, neurostimulation, as well as destructive methods (rhizotomy, neurotomy).

5.1 Intrathecal baclofen therapy (ITB)

Chronic intrathecal baclofen therapy is the most wide-spread method of surgical treatment for spasticity. Penn and Kroin were the first to announce the decreasing effect of bolus intrathecal baclofen injection on spasticity after SCI in 1984 (Penn and Kroin, 1984), and by 2010 about 44 pumps had been implanted in the USA to treat spasticity, and 25% of those were for patients with spinal injuries (Dvorak et al., 2010). ITB proved to be an effective method of treating all types of spasticity. ITB was approved by FDA and licensed in the EU, including for children aged 4–18, and today there is information about a successful baclofen treatment of an 11-month old child (Hagemann et al., 2020). However, despite the extensive history of this procedure, ITB can cause many complications, requiring subsequent surgeries (Table 2) in order to prevent more severe complications, including the fatal ones. In order to minimize those complications, ITB Therapy Best Practice Forum was created in 2004 in order to create guidelines, based on the international experience of using ITB (Albright et al., 2006).

These guidelines focus on choosing the level of implanting a catheter, depending on the symptoms of spasticity; also they describe the procedure of surgical implantation of the catheter and location of the pump pocket (with emphasis on subfascial implantation), insertion and fixation of the pump, prevention of liquorrhea and suppurative septic complications.

Later on, the importance of the catheter placement level was proven in experiments with animals (Bernards, 2006; Flack and Bernards, 2010) that showed that baclofen in its highest concentration could reach not further than 1 cm from the tip of the catheter, i.e., the distance should be about 2 cm for baclofen to be most effective.

In 2015 the German interdisciplinary research team (Dressler et al., 2015) published the results of their study where they used ITB for children, and in 2016 the expert panel of ITB Therapy Best Practices released new results in their four papers on main aspects of ITB: Patient selection (Saulino et al., 2016a,b), Screening test

Table 2 The number of complications associated with intrathecal baclofen therapy published between 2010 and 2020, in patients after spinal cord injury for the treatment of spastic syndrome.

	Number of patients (n pump)	Number of patients after spinal cord injury	Number of complications (n)	Number of complications per one implantation	Number of catheter-associated complications per one implantation
Stetkarova et al. (2010)	1352	36%	558	0.41	369 (0.27)
Dvorak et al. (2010)	167	33%			37 (0.22)
Draulans et al. (2013)	130	38%	102	0.78	78 (0.6)
Taira et al. (2013)	400 (406)	23%	148	0.36	34 (0.08)
Borrini et al. (2014)	158	42%	38	0.24	9 (0.06)
Stetkarova et al. (2015)	39 (54)	62%	24	0.44	14 (0.26)
Pittelkow et al. (2018)	88	45%	22	0.25	16 (0.18)
Pucks-Faes et al. (2018)	136	15%	32	0.28	23 (0.17)
Reis et al. (2019)	155 (251)		48	0.19	24 (0.1)
Gunnarsson et al. (2019)	113	23%	35	0.31	20 (0.18)
Kawano et al. (2018)	34	85%	5	0.15	3 (0.09)

(Boster et al., 2016a,b), Dosing and Long-term Management (Boster et al., 2016a,b) and Troubleshooting (Saulino et al., 2016a,b).

The expert panel showed that only 70% of doctors regularly use screening tests to select patients for ITB, and they argue that the test might not guarantee a long-term therapeutic effect; moreover, the test might demonstrate poorer motor functions of the patient due to muscle weakness Boster et al., 2016a,b. There are attempts to eliminate the drawbacks of bolus intrathecal baclofen injection by installing endolumbar catheter for a long-term continual infusion. But three-day and six-day testing showed that the frequency of implantation was 86% and 66%, respectively (Phillips et al., 2015; Pucks-Faes et al., 2019). Therefore, it could be concluded that permanent and/or prolonged testing does not indicate the level of therapy effectiveness in the long run, and in some cases could even discredit the therapy for the patient, since after experiencing the side effects of intrathecal treatment that mostly are temporary, a patient might reject the idea of ITB—the fact that is proven by a relatively small amount of actual implantations, performed after a prolonged test.

The main objective of ITB is decreasing negative consequences of spasticity and providing functional control over it, as well as withdrawal from taking oral muscle relaxants (Bostler et al., 2016). This objective might be reached in 12–18 months and requires multiple visits to the hospital for adjusting the dosage. According to the ITB Therapy Best Practices Guidelines, adjustment should be done once a week or two, but not more than once in 24 h (Bostler et al., 2016). However, the dosage might be increased even after 18 months at least in 20% of cases that is supposed to be considered as a manifestation of tolerance to ITB due to a decrease of the overall number of GABA-B receptors or their desensibilization after baclofen therapy (Heetla et al., 2009, 2010), and a possible solution to this issue is switching to bolus injections of the medication (Heetla et al., 2010, 2009; Krach et al., 2007).

When studying the ITB effectiveness in treating spasticity of patients with SCI, we encountered only one research on this topic (McIntyre et al., 2014), and its authors believe that there are no statistically significant differences in using ITB for spasticity treatment of patients with SCI and those with another cause of spasticity. ITB therapy of patients with SCI showed a tonus decrease of 2 points and more on the modified Ashworth scale, 1.5–2 points on the scale of spasm frequency and the reflex scale; baclofen dosage was 219–536 μg, therapy was adjusted for 6–12 months. McIntyre also notes that all studies show a significant improvement of the quality of life, easier care for patients in grave condition. Meanwhile, despite the high effectiveness of the therapy and long existence of this method, the results are still not conclusive enough (McIntyre et al., 2014).

Despite ITB advantages, the number of complications (Table 2), requiring a surgery, is still significant, namely, the incidence of complications per one implanted pump ranges from 0.15 to 0.78. Catheter issues remain the most frequent complication. Nevertheless, Pucks-Faes et al. (2019) and Kawano et al. (2018) in their research point out that new generation of catheters decreased the number of these incidences notably.

In the period of 2017–2019 the Manufacturer and User Facility Device Experience (MAUDE), as well as the U.S. Food and Drug Administration (FDA) databases registered more than two thousand cases of ITB complications, and only a quarter of them were device-related, 50% of them being catheter-related and 28% being classified as device or catheter malfunction (Abraham et al., 2020). Still, we have not discovered a comprehensive analysis, comparing the amount of complications with new and old generation catheters.

Therefore, the analysis of ITB-related research for the last 10 years shows that the medical community is at an impasse on this issue and starts losing its interest. Surely, there are still enough publications on the matter, but the majority of them are devoted to analyzing the results, obtained over a recent period for certain types of pathologies, and these results are relatively similar; there are almost no studies, comparing the effectiveness of modern catheters vs silicone ones, or different surgical methods. The only comparison of surgical techniques that could still be encountered in published materials is about ITB and selective rhizotomy in case of cerebral palsy. The same trend could be observed in research on pharmacological therapy of spasticity; almost all main types of medication and their schedules were developed in the second half of the last century and are well-studied, so there is no significant breakthrough to be expected in this area either. In this context special attention is drawn to publications on stimulation of spinal cord, especially via CPG, development of robotic systems with BMI technologies and artificial intelligence (Cutrone and Micera, 2019; Donati et al., 2016; Pizzolato et al., 2019), development of cellular methods of implantation, only on animal models at the moment (Csobonyeiova et al., 2019; Nagoshi and Okano, 2017; Teng, 2019).

5.2 Spinal cord stimulation (SCS)

The first application of spinal cord stimulation for spasticity treatment of patients with multiple sclerosis was described by the American scientists Cook and Weinstein in (1973). In 1979 another American, Richardson, was one of the pioneers in SCS for patients after SCI (Richardson et al., 1979), indicating a significant decrease of spasticity and describing neurotrophic effect of SCS, manifested in recovery of pressure ulcers and chronic ulcers, as well as some improvement of pelvic organs functioning in these patients (Table 3).

Almost a 50% decrease of spasm frequency after 6 months and a 60% decrease after 2 years of SCS was detected (Barolat et al., 1995) in the majority of patients with SCI complications who had been treated with that method. In turn, the possibilities of SCS application are not limited to spasticity alleviation. There has been reported (Waltz, 1997) improvement of motor functions in affected extremities, better neurotransmission and alleviation of pain in the majority of patients who received SCS after SCI.

A strategically important condition of SCS effectiveness is implanting an electrode below the injured area, because if an electrode is located above the injured area

Table 3 List of publications on spinal cord stimulation in patients after spinal cord injury for the treatment of spastic syndrome.

Author	Year of publication	Number of patients after spinal cord injury	Characteristics			Positive trial SCS	Observation (m)
			Frequency	Pulse width	Amplitude		
Richardson et al.	1978, 1979	5	33–75	100–200	0.2–2 B	3	2–18
M.M. Dimitrijevic et al.	1986	59	30–50	200	3–5 MA	30	120
Campos et al.	1987	8	30–50	200		8	
Barolat et al.	1995	48	2–130			48	24
Waltz J.M. et al.	1997	303	10–1500	200		303	300
Midsha et al.	1998	17				1	6
Pinter et al.	2000	8	50–100	210	2–7 B	8	14
Dekopov et al.	2015	13	100–130			13	12–108
Biktimirov et al.	2020	33	60–80	200–400		18	12

(Siegfried et al., 1981), it has no pronounced therapeutic effect. Some experiments (Chapman et al., 1983) on decerebrate cats proved the necessity of exact placement of a stimulating electrode above the back columns of the spinal cord and below the injury to decrease both tonic and phasic components of stretching reflex; without those it is impossible to achieve a stable effect over an extended period of time.

The American neurosurgeon Joseph M Waltz was among the first scholars to use a method of percutaneous stimulation with four cylindrical electrodes for spasticity treatment. The author emphasizes (Waltz, 1997) the importance of placing the electrode at the level, corresponding to clinical symptoms, and personalized adjustment of stimulation frequency. Placement of the electrode at the lumbar enlargement level (Campos et al., 1987) gives the patient an opportunity to control spasticity, personally increasing or decreasing the amplitude of stimulation.

There are several major principles of successful SCS (Pinter et al., 2000): epidural electrode must be placed right above the upper lumbar cord segment; stimulation frequency must be around 50–100 Hz, amplitude within 2–7 V and the stimulus width of 210 micross; the stimulus parameters must be optimized by clinically evaluating the effect of arbitrary combination of four electrodes; the amplitude of stimulation is to be adjusted in accordance with different body positions.

But not all studies showed high effectiveness of spinal cord stimulation in treating spasticity. For instance, in 1998 the Spinal Cord journal published a paper where its authors—neurosurgeons Midha and Schmitt—stated that their retrospective study had shown that, despite some obvious positive results, the overall effect of SCS on patients with SCI was short-term. The authors (Midha and Schmitt, 1998) believe that the effect of SCS decreases after 6 months since the stimulator implantation, and this is the main argument for choosing ITB. This fact explains that in the last 20 years there have been only a handful of papers (Biktimirov et al., 2020; Dekopov et al., 2015), proving the SCS effect on spasticity regulation.

Therefore, ITB and SCS are the most common methods of spasticity regulation for patients with SCI complications. Certainly, both of them are relatively effective and safe in terms of dealing with spasticity, but neither provides regeneration of nerve tissue in the area, damaged by SCI, even though it has been indicated (Siddiqui et al., 2020) that SCS could promote creation of new functional connections between neurons to some extent and more effective rehabilitation. The main objective of rehabilitating patients with SCI is to regenerate axons in the damaged area and create new functional connections with motoneurons of the damaged SC area below the injury (Gomes-Osman et al., 2016) that should be the focus for all approaches to motor rehabilitation of such patients.

6 Biomedical technologies in SCI treatment

Transplantation of stem cells for recovering SC functions is a dominant trend of the modern medicine that should not be discounted in clinical practice. More than a century passed since Alexander Maximow, professor of the Russian Imperial

Military Medical Academy, Petrograd University and University of Chicago, discovered stem cells in 1903, and today stem cells transplantation is a routine medical procedure, giving new hope to patients with SCI (Konstantinov, 2000).

The cornerstone of developing biomedical products for SCI treatment is using paracrine potential of stem and progenitor cells. Neural and mesenchymal stem cells of humans produce exosomes—nanodimentional microvesicles with multiple factor of intercellular communication that suppress apoptosis, block proinflammatory activation of neuroglia and autophagy. Experimental intrathecal administration of exosomes of neural stem cells (Hawryluk et al., 2012; Ritfeld et al., 2015; Rong et al., 2019) significantly reduces the damaged area of SC tissue, decreases the amount of pro-apoptotic Bax protein, caspase-3 and proinflammatory cytokines TNF-α, IL-1β, and IL-6 in the cerebrospinal fluid, inhibits swelling, cavitation and reactive gliosis. Neuroprotective effect of exosomes is based on stem cells producing miR-216a-5p, miRNA-29b (Yu et al., 2019), MiR-126, miR-133b and some other key micro RNA, suppressing expression of proinflammatory response genes in neurons (Sugaya and Vaidya, 2018) and activation of the complement system (Zhao et al., 2019).

Exosomes of neural stem cells (Vogel et al., 2018; Zhong et al., 2020) contain vascular endothelial growth factor (VEGF), insulin-like growth factor-1 (Ma et al., 2019), nerve growth factor (NGF), brain-derived neurotrophic factor (BDNF) and some other crucial proteins for functional recuperation. Development of biomedical products, based on secreted microvesicles, is possible with using both autologous neural stem and progenitor cells of the patient, mesenchymal stem cells (Liu et al., 2020), and allogenic, multiplied and induced stem cells (Silvestro et al., 2020).

Theoretically speaking, exosomes of stem cells, cultured in hypoxia conditions (Luo et al., 2019; Zhu et al., 2020) or in a culture medium with glucocorticosteroids and other immunosuppressors, could contain an immense amount of anti-inflammatory factors, and combining them could produce a unique biomedical product for SCI treatment. This bioproduct could be administered intrathecally together with SCS, and this would promote remyelination of nerve guides, angiogenesis and neuroprotection. Additionally, overall neurotrophic potential of exosomal fractions of stem cells could be complemented by autologous hematopoietic stem cells. Implantation of such exosome-cellular product with hydrogel matrix into the area of SC severance allows to suppress local gliosis (Li et al., 2020), create loose connective tissue in the damaged area and avoid retraction of dissociated fragments of SC with a subsequent destruction of anatomic connections in the spinal canal.

In 2015 our research team (Bryukhovetskiy and Bryukhovetskiy, 2015) proved the possibility of using hydrogel matrixes with incorporated stem cell culture to create prosthetic tissue of a better quality between the caudal and cranial regions of the spinal cord and not only avoid their retraction, but also create favorable conditions for creation of a glial border. Our study, as well as some others (Gao et al., 2019; Llorens-Bobadilla et al., 2020), show the ability of neural and mesenchymal stem cells to differentiate into neurons and gliocytes with some external conditions being maintained which is possible due to hydrogel matrix that becomes a promising object to be considered in cellular treatment of SCI.

Transplantation of biopolymer cellular systems promotes axon growth and creation of axon bridges—multicellular structures, comprised of Schwann cells, olfactory ensheathing cells, neural stem and progenitor cells, and mesenchymal stem cells (Assinck et al., 2017). These bridges allow axons to travel through the damaged area and stimulate remyelination. Alternative mechanism of the damaged tissue regeneration involves creation of neural relays between transplanted cells and host cells.

Undoubtedly, such complicated mechanisms of SC regeneration impose high standards on biopolymer cell-based composites, used for nerve tissue engineering. In the last 10 years biopolymer cell-based materials and technologies, used for SC restoration, have really advanced, but the most fundamental issues remain unresolved. For instance, it was shown that growth of regenerating nerve conductors through a prosthetic connective tissue, that is locally created via implantation of biopolymer cell-based products, depends on induction of proliferation, maintenance and chemoattraction of axons. The first objective is reached by high concentration of steopontin, insulin-like growth factor 1 and ciliary-derived neurotrophic factor in the damaged area, the second one is by the presence of fibroblast growth factor-2 and epidermal growth factor, and chemoattraction is provided by the dynamic increase of glial-derived neurotrophic factor concentration in the damaged area, but all this only works in theory.

Even hypothetically speaking, in order to achieve these objectives newly formed connective tissue should have a complex three-dimensional structure where these factors must be appropriately allocated at the moment of matrix implantation, and then they must be released from some sort of containers in response to various stimuli, for instance, electric stimulation of different intensity. A more likely alternative option involves SC stimulation based on different therapeutic modes, providing pre-programmed stimulation, substrate and mechanical maintenance, as well as chemoattraction of regenerating axons.

However, despite some promising results, the majority of studies are still far from being ready for introduction into clinical practice for a number of objective reasons. Firstly, the most effective methods of biotherapy are still not determined, including such aspects as the type of stem cells, dosage, methods of injection and transplantation periods. Secondly, the situation is aggravated by the fact that the overwhelming majority of pre-clinical trials on animals are performed in the acute period of SCI (Assinck et al., 2017), while the most clinical trials of biomedical cell-based products and technologies are conducted with patients, having SCI in its chronic stage (Cheng et al., 2014; El-Kheir et al., 2014; Gomes-Osman et al., 2016; Yamazaki et al., 2020), even though there are singular cases of using stem cells in the acute stage of SCI (Xiao et al., 2018). It could be explained by the fact that in the acute SCI stage there is usually an issue of saving a patient's life, and it is impossible to obtain a necessary amount of autologous cells. Also, it would require a long period of time for the cells to proliferate, that is why the strategies of using biomedical cell-based methods for SCI patients require certain adjustment.

7 Prospects of integrating biomedical and bionic technologies

Determination of strategy for spasticity treatment relies on answering two major questions—when to consider spasticity as a severe one and when to offer a surgery. Spasticity that was defined by a clinician as a non-severe one could still significantly impact on everyday life of a patient, causing pain, depressive condition, affecting sleep, interactions with other people. Expert panel of ITB Therapy Best Practices (Saulino et al., 2016a,b) suggested the following answer to the abovementioned questions: any patient, experiencing spasticity that interferes with their comfort, active or passive functions, everyday life, mobility, should be considered as a potential candidate for treatment, including surgery.

Without any doubt, spasticity negatively affects many factors, but it could have positive influence on intestinal and bladder reflexes, help to maintain posture and mobility. Spasticity can stimulate blood circulation and prevent thrombosis of deep veins, as well as retain bone mass (Halpern et al., 2013). Therefore, spasticity should be treated not only as a pathological symptom to be eliminated, but as an adaptive mechanism of the body to existing neurological deficiency, and its positive effects should be preserved in each certain case. We agree with those authors who believe that the first approach to spasticity treatment should involve physiotherapy with stretching and strengthening exercises, orthopedic splints and electric stimulation (Halpern et al., 2013).

When physiotherapy does not provide the necessary result, pharmacological treatment should be added. Pharmacotherapy definitely has some advantages in prescribing and using, it does not require special equipment. Nevertheless, pharmacotherapy involves a wide range of side effects that impede the recovery process. Oral administration of baclofen was proven to cause weakening of muscles and muscle fatigue that could adversely affect the recovery of muscular activity and development of motor skills (Thomas et al., 2010; Willerslev-Olsen et al., 2011). Muscle relaxants produce multiple side effects in terms of CNS and other organs and systems (Lapeyre et al., 2010); up to 40% of patients are not able to handle these side effects (Ward, 2008). Moreover, one disadvantage of pharmacotherapy that could not be ignored is that patients are required to infallibly follow the prescriptions of their doctor, but they frequently fail to do so (Halpern et al., 2013).

We believe that SCS remains underrated in terms of spasticity treatment. Indeed, in the early days of using this method, specialists encountered many complications due to imperfections of stimulation systems and lack of experience in the area of technology. Apart from decreasing spasticity, spinal cord stimulation produces notable neuroplastic effect (Duffell and Donaldson, 2020), improves rehabilitation results for patients with dysfunction of lower extremities after SCI (Gill et al., 2018; Harkema et al., 2011), For instance, discovery of the central pattern generator (CPG) for locomotion (Dimitrijevic et al., 1998) brought about the whole new branch of science, devoted to studying this phenomenon. A series of electric stimuli of

25–60 Hz and amplitude of 5–9 V proved to be effective in induction of rhythmic locomotor activity and stimulation of walking-like movements, Therefore, this discovery brings us to a conclusion that human SC is able to generate motion even without brain control (Dimitrijevic et al., 1998). Later it was shown that electrode placement on the lumbar enlargement generates tonic and rhythmic patterns of motor activity in patients with complete loss of motor function and allows to improve the quality of walking for patients with incomplete damage of the spinal cord (Lavrov et al., 2008; Wenger et al., 2016). It was in 2018 when using SCS for a patient with lower paraplegia allowed him to move vertically and walk independently (Gill et al., 2018; Wagner et al., 2018). That is why we believe that using SCS for patients with spasticity is more advantageous than ITB: firstly, patients become independent from a hospital where they receive refills; secondly, there are no side effects, associated with using baclofen; thirdly, they have additional motor rehabilitation options due to activation of CPG (Fig. 3). Needless to say, patients with implanted neurostimulator also run the risks of complications, but they are mostly associated with electrode migration and could be resolved by a simple stimulator reprogramming, while only about 10% of complications require a surgery (Eldabe et al., 2016).

Many researchers proved that spinal cord stimulation is less sensitive compared to ITB in cases of treating severe spasticity (Barolat et al., 1995; Biktimirov et al., 2020; Dimitrijevic et al., 1986a,b; Pinter et al., 2000). Richardson and Barolat demonstrated the increase of stimulation effect overtime and weakening of spasms, the fact that has not been either proved or disproved at this point. Our work with patients showed that the time of stimulation is a crucial factor—overtime some of the patients with implanted stimulators decrease the amplitude and frequency of stimulation during the day. This allows us to conclude that a longer stimulation, accompanied by rehabilitation with CPG stimulation, could produce better results even for patients

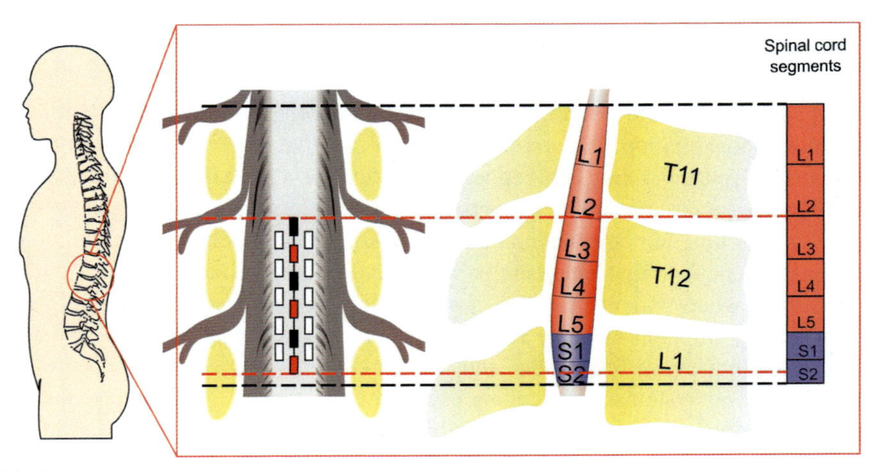

FIG. 3

CPG location.

with negative response in the trial stimulation period. However, this assumption requires further studies and proof in clinical trials.

Undoubtedly, neither SCS, nor ITB are able to completely restore lost functions, since they do not recover the missing nerve tissue. This issue could be resolved with advancing cell-based technologies. Stem cells transplantation is very promising in terms of regenerative medicine that aims to discover treatment of spinal cord injuries. Clinical trials showed that stem cells as an instrument of therapy improve motor functions and neurological condition (Csobonyeiova et al., 2019; Nagoshi and Okano, 2017; Silvestro et al., 2020; Yamazaki et al., 2020; Yang, 2019). A combination of neurotransplantation with SCS will promote a better integration of donor cells with patient's neurons and accelerate regeneration, as it was demonstrated in experiments with animals (Siddiqui et al., 2020).

It is the need for creating a new neuromodulation platform for integration of innovative biomedical technologies that is the fundamental basis for clinical analysis of these methods effect on patients with SCI. In 2016 we published a short paper on the issue (Biktimirov et al., 2016), while in 2020 the results of a research on a step-by-step approach to selecting patients for surgery was published (Biktimirov et al., 2020). The first stage involved trial spinal cord stimulation that was used to decide whether to implant the system for chronic SC stimulation or to use intrathecal baclofen therapy. The patients with a positive response to the trial stimulation retained the effect of decreasing spasticity over a long period without significant differences from the patients who received ITB.

8 Conclusion

The conducted analysis allows us to suggest the following scheme of step-by-step spasticity treatment (Fig. 4) that is fundamentally different from others in introduction of spinal cord stimulation as a first stage of selecting patients for surgical treatment of spasticity before sending them to ITB; also, cell-based methods should be given more consideration. We would like to indicate that patients with positive response to anti-spasticity drugs still experience side effects from their administration, that is why we suggest providing patients with information about avoiding these side effects by switching to neuromodulation and offering them to choose the method of treatment. We believe that medicine is being revolutionized due to modern global digitalization when patients become more trusting toward invasive devices, and people consider implants not only as a final recourse and absolute necessity, but also as a tool for improving life quality and facilitating everyday activities. There are neuromodulation systems with feedback; prosthetics of new generation with personalized models of neuromuscular dysfunctions are being developed (Pizzolato et al., 2019). A French research team published (Benabid et al., 2019) the results describing the usage of invasive brain machine interface (BMI) method to control exoskeleton of a tetraplegic patient with a cervical spinal cord injury.

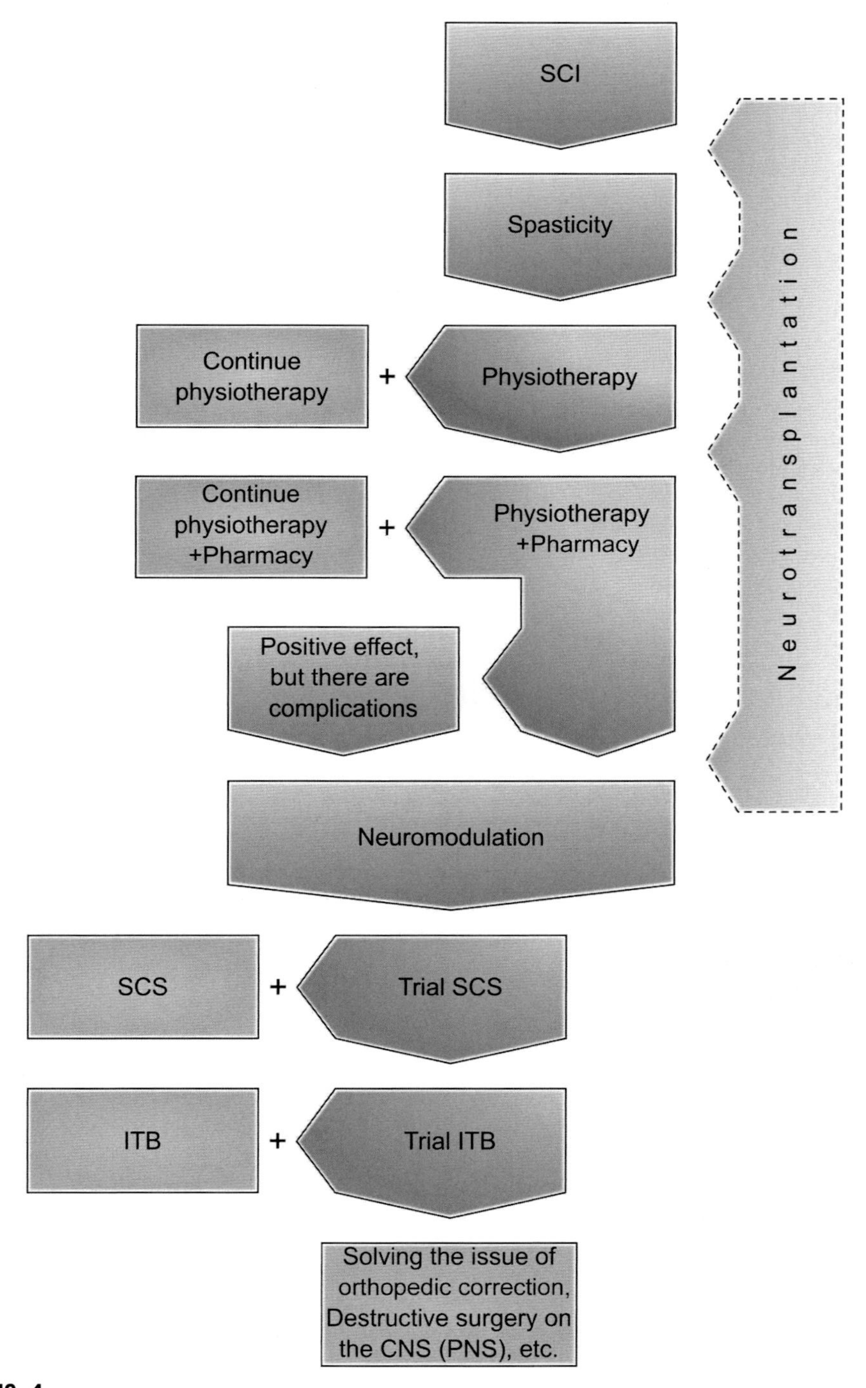

FIG. 4

Scheme of step-by-step selection of therapy for spastic syndrome in a patient with spinal cord injury.

Acknowledgments

Funding

This work was supported by Far Eastern Federal University Grant "Neurotechnologies, virtual and augmented reality technologies," an agreement between the Fund for Support of Projects of the National Technological Initiative (NTI) and Far Eastern Federal University dated 05/08/2019 No. 2/1251/2019, identifier of the agreement on the provision of subsidies for state support of NTI Centers No. 0000000007518P240002.

Authors' contributions

AB and OP wrote the manuscript, proposed the study idea, designed the study, offered support with the experiments, organized the scientific team, performed an analysis of literature and own scientific reports. IB provided scientific guidance, validation, formal analysis, data curation, visualization, supervision. AS and HS discussed, analyzed and interpreted the results of the study, and also worked on the manuscript. All authors read and approved the manuscript and agree to be accountable for all aspects of the research in ensuring that the accuracy or integrity of any part of the work are appropriately investigated and resolved.

Patient consent for publication

Not applicable.

Competing interests

The authors declare that they have no competing interests.

References

Abdollahi, I., Taghizadeh, A., Shakeri, H., Eivazi, M., Jaberzadeh, S.J., 2015. The relationship between isokinetic muscle strength and spasticity in the lower limbs of stroke patients. J. Bodyw. Mov. Ther. 2, 284–290. https://doi.org/10.1016/j.jbmt.2014.07.002.

Abraham, M., Gold, J., Dweck, J., Ward, M., Gendreau, J., Panse, N., Holani, K., Gupta, P., Mammis, A., 2020. Classifying device-related complications associated with intrathecal baclofen pumps: a MAUDE STUDY. World Neurosurg. 24, 1878–8750. https://doi.org/10.1016/j.wneu.2020.04.070.

Albright, A.L., Turner, M., Pattisapu, J.V., 2006. Best-practice surgical techniques for intrathecal baclofen therapy. J. Neurosurg. 104 (4), 233–239. https://doi.org/10.3171/ped.2006.104.4.233.

Alizadeh, M., Fisher, J., Saksena, S., Sultan, Y., Conklin, C.J., Middleton, D.M., Finsterbusch, J., Krisa, L., Flanders, A.E., Faro, S.H., Mulcahey, M.J., Mohamed, F.B., 2018. Reduced field of view diffusion tensor imaging and fiber tractography of the pediatric cervical and thoracic spinal cord injury. J. Neurotrauma 35 (3), 452–460. https://doi.org/10.1089/neu.2017.5174.

Ameer, M.A., Gillis, C.C., 2021. Central Cord Syndrome. In: StatPearls [Internet]. Treasure Island (FL): StatPearls Publishing. Bookshelf ID, NBK441932.

Andresen, S.R., Biering-Sorensen, F., Hagen, E.M., Nielsen, J.F., Bach, F.W., Finnerup, N.B., 2016. Pain, spasticity and quality of life in individuals with traumatic spinal cord injury in Denmark. Spinal Cord 54, 973–979. https://doi.org/10.1038/sc.2016.46.

Ansari, N.N., Naghdi, S., Moammeri, H., Jalaie, S., 2006. Ashworth scales are unreliable for the assessment of muscle spasticity. Physiother. Theory Pract. 22, 119–125. https://doi.org/10.1080/09593980600724188.

Assinck, P., Duncan, G.J., Hilton, B.L., Plemel, J.R., Tetzlaff, W., 2017. Cell transplantation therapy for spinal cord injury. Nat. Neurosci. 20, 637–647. https://doi.org/10.1038/nn.4541.

Badhiwala, J.H., Wilson, J.R., Fehlings, M.G., 2020. The case for revisiting central cord syndrome. Spinal Cord 58 (1), 125–127. https://doi.org/10.1038/s41393-019-0354-5.

Barolat, G., Singh-Sahni, K., Staas, W.E., Shatin, D., Ketcik, B., Allen, K., 1995. Epidural spinal cord stimulation in the management of spasms in spinal cord injury: a prospective study. Stereotact. Funct. Neurosurg. 64, 153–164. https://doi.org/10.1159/000098744.

Benabid, A.L., Costecalde, T., Eliseyev, A., Charvet, G., Verney, A., Karakas, S., Foerster, M., Lambert, A., Morinière, B., Abroug, N., Schaeffer, M.-C., Moly, A., Sauter-Starace, F., Ratel, Moro, C., Torres-Martinez, N., Langar, L., Oddoux, M., Polosan, M., Pezzani, S., Auboiroux, V., Aksenova, T., Mestais, C., Chabardes, S., 2019. An exoskeleton controlled by an epidural wireless brain-machine interface in a tetraplegic patient: a proof-of-concept demonstration. Lancet Neurol. 18 (12), 1112–1122. https://doi.org/10.1016/S1474-4422(19)30321-7.

Bernards, C.M., 2006. Cerebrospinal fluid and spinal cord distribution of baclofen and bupivacaine during slow intrathecal infusion in pigs. Anesthesiology 105 (1), 169–178. https://doi.org/10.1097/00000542-200607000-00027.

Biering-Sørensen, F., Nielsen, J.B., Klinge, K., 2006. Spasticity-assessment: a review. Spinal Cord 44 (12), 708–722. https://doi.org/10.1038/sj.sc.3101928.

Biktimirov, A., Pak, O., Kalinsky, P., Totorkulov, R., 2016. Spinal cord stimulation and chronic intrathecal baclofen therapy in the treatment of drug-resistant forms of spasticity in patients after spinal cord injury, XXIInd congress of the European Society for Stereotactic and Functional Neurosurgery. Madrid, Spain, September 28-October 1, 2016: abstract. Stereotact. Funct. Neurosurg. 94 (1), 1–132. https://doi.org/10.1159/000448961.

Biktimirov, A., Bryukhovetskiy, I., Sharma, A., Sharma, H.S., 2020. Spinal cord stimulation and intrathecal baclofen therapy for patients with severe spasticity after spinal cord injury. Prog. Brain Res. 258, 79–99. https://doi.org/10.1016/bs.pbr.2020.09.007.

Bohannon, R.W., Smith, M.B., 1987. Interrater reliability of a modified ashworth scale of muscle spasticity. Phys. Ther. 67, 206–207. https://doi.org/10.1093/ptj/67.2.206.

Borrini, L., Bensmail, D., Thiebaut, J.-B., Hugeron, C., Rech, C., Jourdan, C., 2014. Occurrence of adverse events in long-term intrathecal baclofen infusion: a 1-year follow-up study of 158 adults. Arch. Phys. Med. Rehabil. 95 (6), 1032–1038. https://doi.org/10.1016/j.apmr.2013.12.019.

Boster, A.L., Bennett, S.E., Bilsky, G.S., Gudesblatt, M., Koelbel, S.F., McManus, M., Saulino, M., 2016a. Best practices for intrathecal baclofen therapy: screening test. Neuromodulation 19 (6), 616–622. https://doi.org/10.1111/ner.12437.

Boster, A.L., Adair, R.L., Gooch, J.L., Nelson, M.E., Toomer, A., Urquidez, J., Saulino, M., 2016b. Best practices for intrathecal baclofen therapy: dosing and long–term management. Neuromodulation 19 (6), 623–631. https://doi.org/10.1111/ner.12388.

Boyd, R.N., Graham, H.K., 1999. Objective measurement of clinical findings in the use of botulinum toxin type a for the management of children with cerebral palsy. Eur. J. Neurol. 6, 23–35. https://doi.org/10.1111/j.1468-1331.1999.tb00031.x.

Bryukhovetskiy, A.S., Bryukhovetskiy, I.S., 2015. Effectiveness of repeated transplantations of hematopoietic stem cells in spinal cord injury. World J. Transplant. 5 (3), 110–128. https://doi.org/10.5500/wjt.v5.i3.110.

Campos, R.J., Dimitrijevic, M.R., Sharkey, P.C., Sherwood, A.M., 1987. Epidural spinal cord stimulation in spastic spinal cord injury patients. Appl. Neurophysiol. 50 (1–6), 453–454. https://doi.org/10.1159/000100763.

Chapman, C.E., Ruegg, D.G., Wiesendanger, M., 1983. Effects of dorsal cord stimulation on stretch reflexes. Brain Res. 258 (2), 211–215. https://doi.org/10.1016/0006-8993(83)91144-7.

Cheng, H., Liu, X., Hua, R., Dai, G., Wang, X., Gao, J., An, Y., 2014. Clinical observation of umbilical cord mesenchymal stem cell transplantation in treatment for sequelae of thoracolumbar spinal cord injury. J. Transl. Med. 12, 253. https://doi.org/10.1186/s12967-014-0253-7.

Cook, W., Weinstein, S.P., 1973. Chronic dorsal column stimulation in multiple sclerosis. N. Y. State J. Med. 73, 2868–2872.

Csobonyeiova, M., Polak, S., Zamborsky, R., Danisovic, L., 2019. Recent progress in the regeneration of spinal cord injuries by induced pluripotent stem cells. Int. J. Mol. Sci. 20 (15), 3838. https://doi.org/10.3390/ijms20153838.

Cutrone, A., Micera, S., 2019. Implantable neural interfaces and wearable tactile systems for bidirectional neuroprosthetics systems. Adv. Health Mater. 8 (24). https://doi.org/10.1002/adhm.201801345, e1801345.

Dekopov, A.V., Shabalov, V.A., Tomsky, A.A., Hit, M.V., Salova, E.M., 2015. Chronic spinal cord stimulation in the treatment of cerebral and spinal spasticity. Stereotact. Funct. Neurosurg. 93, 133–139. https://doi.org/10.1159/000368905.

Dietz, V., Sinkjaer, T., 2012. Spasticity. Handb. Clin. Neurol. 109, 197–211. https://doi.org/10.1016/B978-0-444-52137-8.00012-7.

Dimitrijevic, D.D., Dimitrijevic, M.R., Illis, L.S., Nakajima, K., Sharkey, P.C., Sherwood, A.M., 1986a. Spinal cord stimulation for the control of spasticity in patients with chronic spinal cord injury: I. Clinical observations. Cent. Nerv. Syst. Trauma 3 (2), 129–144. https://doi.org/10.1089/cns.1986.3.129.

Dimitrijevic, M.R., Illis, L.S., Nakajima, K., Sharkey, P.C., Sherwood, A.M., 1986b. Spinal cord stimulation for the control of spasticity in patients with chronic spinal cord injury: II. Neurophysiologic observations. Cent. Nerv. Syst. Trauma 3 (2), 145–152. https://doi.org/10.1089/cns.1986.3.145.

Dimitrijevic, M.R., Gerasimenko, Y., Pinter, M.M., 1998. Evidence for a spinal central pattern generator in humans. Ann. N. Y. Acad. Sci. 860, 360–376. https://doi.org/10.1111/j.1749-6632.1998.tb09062.x.

Donati, A.R., Shokur, S., Morya, E., Campos, D., Moioli, R.C., Gitti, C.M., Augusto, P.B., Tripodi, S., Pires, C.G., Pereira, G.A., Brasil, F.L., Gallo, S., Lin, A.A., Takigami, A.K., Aratanha, M.A., Joshi, S., Bleuler, H., Cheng, G., Rudolph, A., Nicolelis, M.A., 2016. Long-term training with a brain-machine Interface-based gait protocol induces partial neurological recovery in paraplegic patients. Sci. Rep. 6, 30383. https://doi.org/10.1038/srep30383.

Draulans, N., Vermeersch, K., Degraeuwe, B., Meurrens, T., Peers, K., Nuttin, B., Kiekens, C., 2013. Intrathecal baclofen in multiple sclerosis and spinal cord injury: complications and long-term dosage evolution. Clin. Rehabil. 27 (12), 1137–1143. https://doi.org/10.1177/0269215513488607.

Dressler, D., Berweck, S., Chatzikalfas, A., Ebke, M., Frank, B., Hesse, S., Huber, M., Krauss, J.K., Mücke, K.-H., Nolte, A., Oelmann, H.-D., Schönle, P.W., Schmutzler, M., Pickenbrock, H., Van der Ven, C., Veelken, N., Vogel, M., Vogt, T., Saberi, F.A., 2015. Intrathecal baclofen therapy in Germany: proceedings of the IAB-interdisciplinary working group for movement disorders consensus meeting. J. Neural Transm. 122 (11), 1573–1579. https://doi.org/10.1007/s00702-015-1425-1.

Duffell, L.D., Donaldson, N.N., 2020. A comparison of FES and SCS for Neuroplastic recovery after SCI: historical perspectives and future directions. Front. Neurol. 11, 607. https://doi.org/10.3389/fneur.2020.00607.

Dvorak, E.M., McGuire, J.R., Elizabeth, M., Nelson, S., 2010. Incidence and identification of intrathecal baclofen catheter malfunction. PM R 2 (8), 751–756. https://doi.org/10.1016/j.pmrj.2010.01.016.

Eldabe, S., Buchser, E., Duarte, R.V., 2016. Complications of spinal cord stimulation and peripheral nerve stimulation techniques: a review of the literature. Pain Med. 17 (2), 325–336. https://doi.org/10.1093/pm/pnv025.

El-Kheir, W.A., Gabr, H., Awad, M.R., Ghannam, O., Barakat, Y., Farghali, H., Maadawi, Z., Ewes, I., Sabaawy, H.E., 2014. Autologous bone marrow-derived cell therapy combined with physical therapy induces functional improvement in chronic spinal cord injury patients. Cell Transplant. 23 (6), 729–745. https://doi.org/10.3727/096368913X664540.

Finnerup, N.B., 2017. Neuropathic pain and spasticity: intricate consequences of spinal cord injury. Spinal Cord 55 (12), 1046–1050. https://doi.org/10.1038/sc.2017.70.

Flack, S.H., Bernards, C.M., 2010. Cerebrospinal fluid and spinal cord distribution of hyperbaric bupivacaine and baclofen during slow intrathecal infusion in pigs. Anesthesiology 112 (1), 165–173. https://doi.org/10.1097/ALN.0b013e3181c38da5.

Gao, S.H., Guo, X., Zhao, S., Jin, Y., Zhou, F., Yuan, P., et al., 2019. Differentiation of human adipose-derived stem cells into neuron/motoneuron-like cells for cell replacement therapy of spinal cord injury. Cell Death and Disease 108, 597. https://doi.org/10.1038/s41419-019-1772-1.

Gill, M.L., Grahn, P.J., Calvert, J.S., Linde, M.B., Lavrov, I.A., Strommen, J.A., Beck, L.A., Sayenko, D.G., Van Straaten, M.G., Drubach, D.I., Veith, D.D., Thoreson, A.R., Lopez, C., Gerasimenko, Y.P., Edgerton, V.R., Lee, K.H., Zhao, K.D., 2018. Neuromodulation of lumbosacral spinal networks enables independent stepping after complete paraplegia. Nat. Med. 24 (11), 1677–1682. https://doi.org/10.1038/s41591-018-0175-7.

Gomes-Osman, J., Cortes, M., Guest, J., Pascual-Leone, A., 2016. A systematic review of experimental strategies aimed at improving motor function after acute and chronic spinal cord injury. J. Neurotrauma 33 (5), 425–438. https://doi.org/10.1089/neu.2014.3812.

Gracies, J.-M., Marosszeky, J., Renton, R., Sandaman, J., Gandevia, S., Burke, D., 2000. Short term effects of dynamic lycra splints on upper limb in hemiplegia patients. Arch. Phys. Med. Rehabil. 81, 1547–1555. https://doi.org/10.1053/apmr.2000.16346.

Graupe, D., Cerrel-Bazo, H., Kern, H., Carraro, U., 2008. Walking performance, medical outcomes and patient training in FES of innervated muscles for ambulation by thoracic-level complete paraplegics. Neurol. Res. 30 (2), 123–130. https://doi.org/10.1179/174313208X281136.

Gunnarsson, S., Alehagen, S., Lemming, D., Ertzgaard, P., Berntsson, S.G., Samuelsson, K., 2019. Experiences from intrathecal baclofen treatment based on medical records and patient-and proxy-reported outcome: a multicentre study. Disabil. Rehabil. 41 (9), 1037–1043. https://doi.org/10.1080/09638288.2017.1419291.

Hagemann, C., Schmitt, I., Lischetzki, G., Kunkel, P., 2020. Intrathecal baclofen therapy for treatment of spasticity in infants and small children under 6 years of age. Childs Nerv. Syst. 36 (4), 767–773. https://doi.org/10.1007/s00381-019-04341-7.

Hagen, E.M., Rekand, T., Grønning, M., Færestrand, S., 2012. Cardiovascular complications of spinal cord injury. Tidsskr. Nor. Laegeforen. 132 (9), 1115–1120. https://doi.org/10.4045/tidsskr.11.0551.

Halpern, R., Gillard, P., Graham, G.D., Varon, S.F., Zorowitz, R.D., 2013. Adherence associated with oral medications in the treatment of spasticity. PM R 5 (9), 747–756. https://doi.org/10.1016/j.pmrj.2013.04.022.

Harkema, S., Gerasimenko, Y., Hodes, J., Burdick, J., Angeli, C., Chen, Y., Ferreira, C., Willhite, A., Rejc, E., Grossman, R.G., Edgerton, V.R., 2011. Effect of epidural stimulation of the lumbosacral spinal cord on voluntary movement, standing, and assisted stepping after motor complete paraplegia: a case study. Lancet 377 (9781), 1938–1947. https://doi.org/10.1016/S0140-6736(11)60547-3.

Hawryluk, G.W., Mothe, A., Wang, J., Wang, S., Tator, C., Fehlings, M.G., 2012. An in vivo characterization of trophic factor production following neural precursor cell or bone marrow stromal cell transplantation for spinal cord injury. Stem Cells Dev. 21 (12), 2222–2238. https://doi.org/10.1089/scd.2011.0596.

Heald, E.H., Hart, R., Kilgore, K., Peckham, P.H., 2017. Characterization of volitional electromyographic signals in the lower extremity after motor complete SCI. Neurorehabil. Neural Repair 31 (6), 583–591. https://doi.org/10.1177/1545968317704904.

Heetla, H.W., Staal, M.J., Kliphuis, C., Laar, T., 2009. The incidence and management of tolerance in intrathecal baclofen therapy. Spinal Cord 47 (10), 751–756. https://doi.org/10.1038/sc.2009.34.

Heetla, H.W., Staal, M.J., Laar, T., 2010. Tolerance to continuous intrathecal baclofen infusion can be reversed by pulsatile bolus infusion. Spinal Cord 48 (6), 483–486. https://doi.org/10.1038/sc.2009.156.

Kawano, O., Masuda, M., Takao, T., Sakai, H., Morishita, Y., Hayashi, T., Ueta, T., Maeda, T., 2018. The dosage and administration of long-term intrathecal baclofen therapy for severe spasticity of spinal origin. Spinal Cord 56 (10), 996–999. https://doi.org/10.1038/s41393-018-0153-4.

Kim, J.Y., Park, G., Lee, S.A., Nam, Y., 2020. Analysis of machine learning-based assessment for elbow spasticity using inertial sensors. Sensors 20 (6), 1622. https://doi.org/10.3390/s20061622.

Konstantinov, E., 2000. In search of Alexander a. Maximow: the man behind the unitarian theory of hematopoiesis. Perspect. Biol. Med. 43 (2), 269–276. https://doi.org/10.1353/pbm.2000.0006.

Krach, L.E., Kriel, R.L., Nugent, A.C., 2007. Complex dosing schedules for continuous intrathecal baclofen infusion. Pediatr. Neurol. 37 (5), 354–359. https://doi.org/10.1016/j.pediatrneurol.2007.06.020.

Kumar, R., Lim, J., Mekary, R.A., Rattani, A., Dewan, M.C., Sharif, S.Y., Osorio-Fonseca, E., Park, K.B., 2018. Traumatic spinal injury: global epidemiology and worldwide. World Neurosurg. 113, e345–e363. https://doi.org/10.1016/j.wneu.2018.02.033.

Kwon, B.K., Okon, E., Hillyer, J., Mann, C., Baptiste, D., Weaver, L.C., Fehlings, M.G., Tetzlaff, W., 2011. A systematic review of non-invasive pharmacologic neuroprotective treatments for acute spinal cord injury. J. Neurotrauma 28 (8), 1545–1588. https://doi.org/10.1089/neu.2009.1149.

Lapeyre, E., Kuks, J.K., Meijler, W.J., 2010. Spasticity: revisiting the role and the individual value of several pharmacological treatments. NeuroRehabilitation 27 (2), 193–200. https://doi.org/10.3233/nre-2010-0596.

Lavrov, I., Grégoire, C., Dy, J.C., van den Brand, R., Fong, A.J., Gerasimenko, Y., Zhong, H., Roy, R.R., Edgerton, V.R., 2008. Facilitation of stepping with epidural stimulation in spinal rats: role of sensory input. J. Neurosci. 28 (31), 7774–7780. https://doi.org/10.1523/JNEUROSCI.1069-08.2008.

Li, L., Zhang, Y., Mu, J., et al., 2020. Transplantation of human mesenchymal stem-cell-derived exosomes immobilized in an adhesive hydrogel for effective treatment of spinal cord injury. Nano Lett. 20 (6), 4298–4305. https://doi.org/10.1021/acs.nanolett.0c00929.

Liu, W., Rong, Y., Wang, J., Zhou, Z.H., Ge, X., Ji, C.H., et al., 2020. Exosome-shuttled miR-216a-5p from hypoxic preconditioned mesenchymal stem cells repair traumatic spinal cord injury by shifting microglial M1/M2 polarization. Journal of Neuroinflammation 17 (1), 47. https://doi.org/10.1186/s12974-020-1726-7.

Llorens-Bobadilla, E., Chell, J.M., Merre, P.L., Wu, Y., Zamboni, M., Bergenstråhle, J., Stenudd, M., Sopova, E., Lundeberg, J., Shupliakov, O., Carlén, M., Frisén, J., 2020. A latent lineage potential in resident neural stem cells enables spinal cord repair. Science 370 (6512), 36. https://doi.org/10.1126/science.abb8795.

Lui, J., Sarai, M., Mills, P.B., 2015. Chemodenervation for treatment of limb spasticity following spinal cord injury: a systematic review. Spinal Cord 53, 252–264. https://doi.org/10.1038/sc.2014.241.

Lundell, H., Barthelemy, D., Skimminge, A., Dyrby, T.B., Biering-Sørensen, F., Nielsen, J.B., 2011. Independent spinal cord atrophy measures correlate to motor and sensory deficits in individuals with spinal cord injury. Spinal Cord 49 (1), 70–75. https://doi.org/10.1038/sc.2010.87.

Luo, Z., Wu, F., Xue, E., Huang, L., Yan, P., Pan, X., 2019. Hypoxia preconditioning promotes bone marrow mesenchymal stem cells survival by inducing HIF-1α in injured neuronal cells derived exosomes culture system. Cell Death and Disease 102, 134. https://doi.org/10.1038/s41419-019-1410-y.

Ma, K., Xu, H., Zhang, J., Zhao, F., Liang, H., Sun, H., et al., 2019. Insulin-like growth factor-1 enhances neuroprotective effects of neural stem cell exosomes after spinal cord injury via an miR-219a-2-3p/YY1 mechanism. Aging (Albany NY) 1124, 12278–12294. https://doi.org/10.18632/aging.102568.

McIntyre, A., Mays, R., Mehta, S., Janzen, S., Townson, A., Hsieh, J., Wolfe, D., Teasell, R., 2014. Examining the effectiveness of intrathecal baclofen on spasticity in individuals with chronic spinal cord injury: a systematic review. J. Spinal Cord Med. 37 (1), 11–18. https://doi.org/10.1179/2045772313Y.0000000102.

Meseguer-Henarejos, A.B., Sánchez-Meca, J., López-Pina, J.A., Carles-Hernández, R., 2018. Inter- and intra-rater reliability of the modified Ashworth scale: a systematic review and meta-analysis. Eur. J. Phys. Rehabil. Med. 54 (4), 576–590. https://doi.org/10.23736/S1973-9087.17.04796-7.

Midha, M., Schmitt, J.K., 1998. Epidural spinal cord stimulation for the control of spasticity in spinal cord injury patients lacks long-term efficacy and is not cost-effective. Spinal Cord 36 (3), 190–192. https://doi.org/10.1038/sj.sc.3100532.

Mills, P.B., Dossa, F., 2016. Transcutaneous electrical nerve stimulation for management of limb spasticity: a systematic review. Am. J. Phys. Med. Rehabil. 95 (4), 309–318. https://doi.org/10.1097/PHM.0000000000000437.

Morris, S.L., Williams, G., 2018. A historical review of the evolution of the Tardieu scale. Brain Inj. 32 (5), 665–669. https://doi.org/10.1080/02699052.2018.1432890.

Muresanu, D.F., Sharma, A., Patnaik, R., Menon, P.K., Mössler, H., Sharma, H.S., 2019. Exacerbation of blood-brain barrier breakdown, edema formation, nitric oxide synthase upregulation and brain pathology after heat stroke in diabetic and hypertensive rats. Potential neuroprotection with cerebrolysin treatment. Int. Rev. Neurobiol. 146, 83–102. https://doi.org/10.1016/bs.irn.2019.06.007.

Nagel, S.J., Wilson, S., Johnson, M.D., Machado, A., Frizon, L., Chardon, M.K., Reddy, C.G., Gillies, G.T., Howard, M.A., 2017. Spinal cord stimulation for spasticity: historical approaches, current status, and future directions. Neuromodulation 20 (4), 307–321. https://doi.org/10.1111/ner.12591.

Nagoshi, N., Okano, H., 2017. Applications of induced pluripotent stem cell technologies in spinal cord injury. J. Neurochem. 141 (6), 848–860. https://doi.org/10.1111/jnc.13986.

Nielsen, J.B., Crone, C., Hultborn, H., 2007. The spinal pathophysiology of spasticity—from a basic science point of view. Acta Physiol. 189, 171–180. https://doi.org/10.1111/j.1748-1716.2006.01652.x.

Niu, F., Sharma, A., Feng, L., Ozkizilcik, A., Muresanu, D.F., Lafuente, J.F., Tian, Z.R., Nozari, A., Sharma, H.S., 2019. Nanowired delivery of DL-3-n-butylphthalide induces superior neuroprotection in concussive head injury. Prog. Brain Res. 245, 89–118. https://doi.org/10.1016/bs.pbr.2019.03.008.

Park, J.H., Kim, Y., Lee, K.J., Yoon, Y.S., Kang, S.H., Kim, H., Park, H., 2019. Artificial neural network learns clinical assessment of spasticity in modified Ashworth scale. Arch. Phys. Med. Rehabil. 100 (10), 1907–1915. https://doi.org/10.1016/j.apmr.2019.03.016.

Penn, R.D., Kroin, J.S., 1984. Intrathecal baclofen alleviates spinal cord spasticity. Lancet 1, 1078. https://doi.org/10.1016/s0140-6736(84)91487-9.

Penn, R.D., Savoy, S.M., Corcos, D., Latash, M., Gottlieb, G., Parke, B., Kroin, J.S., 1989. Intrathecal baclofen for severe spinal spasticity. N. Engl. J. Med. 320 (23), 1517–1521. https://doi.org/10.1056/NEJM198906083202303.

Phillips, M.M., Miljkovic, N., Ramos-Lamboy, M., Moossy, J.J., Horton, J., Buhari, A.M., Munin, M.C., 2015. Clinical experience with continuous intrathecal baclofen trials prior to pump implantation. PM R 7 (10), 1052–1058. https://doi.org/10.1016/j.pmrj.2015.03.020.

Picelli, A., Santamato, A., Chemello, E., Cinone, N., Cisari, C., Gandolfi, M., Ranieri, M., Smania, N., Baricich, A., 2019. Adjuvant treatments associated with botulinum toxin injection for managing spasticity: an overview of the literature. Ann. Phys. Rehabil. Med. 62 (4), 291–296. https://doi.org/10.1016/j.rehab.2018.08.004.

Pinter, M.M., Gerstenbrand, F., Dimitrijevic, M.R., 2000. Epidural electrical stimulation of posterior structures of the human lumbosacral cord: 3. Control of spasticity. Spinal Cord 38, 524–531. https://doi.org/10.1038/sj.sc.3101040.

Pittelkow, T.P., Bendel, M.A., Lueders, D.R., Beck, L.A., Pingree, M.J., Hoelzer, B.C., 2018. Quantifying the change of spasticity after intrathecal baclofen administration: a descriptive retrospective analysis. Clin. Neurol. Neurosurg. 171, 163–167. https://doi.org/10.1016/j.clineuro.2018.06.010.

Pizzolato, C., Saxby, D.J., Palipana, D., Diamond, L.E., Barrett, R.S., Teng, Y.T., Lloyd, D.G., 2019. Neuromusculoskeletal modeling-based prostheses for recovery after spinal cord injury. Front. Neurorobot. 13, 97. https://doi.org/10.3389/fnbot.2019.00097.

Popa, C., Popa, F., Grigorean, V.T., Onose, G., Sandu, A.M., Popescu, M., Burnei, G., Strambu, V., Sinescu, C.J., 2010. Vascular dysfunctions following spinal cord injury. J. Med. Life 3 (3), 275–285. https://pubmed.ncbi.nlm.nih.gov/20945818/.

Pucks-Faes, E., Hitzenberger, G., Matzak, H., Fava, E., Verrienti, G., Laimer, I., et al., 2018. Eleven years' experience with Intrathecal Baclofen - Complications, risk factors. Brain and Behavior 8 (5), e00965. https://doi.org/10.1002/brb3.965.

Pucks-Faes, E., Matzak, H., Hitzenberger, G., Genelin, E., Halbmayer, L.-M., Fava, E., Fritz, J., Saltuari, L., 2019. Intrathecal baclofen trial before device implantation: 12-year experience with continuous administration. Arch. Phys. Med. Rehabil. 100 (5), 837–843. https://doi.org/10.1016/j.apmr.2018.09.124.

Reis, P.V., Vieira, C.R., Midões, A.C., Rebelo, V., Barbosa, P., Gomes, A., 2019. Intrathecal baclofen infusion pumps in the treatment of spasticity: a retrospective cohort study in a portuguese centre. Acta Medica Port. 32 (12), 754–759. https://doi.org/10.20344/amp.10482.

Richardson, R.R., Cerullo, L.J., McLone, D.J., Gutierrez, F.A., Lewis, V., 1979. Percutaneous epidural neurostimulation in modulation of paraplegic spasticity. Six case reports. Acta Neurochir. 49 (3–4), 235–243. https://doi.org/10.1007/BF01808963.

Ritfeld, G.J., Patel, A., Chou, A., Novosat, T.L., Castillo, D.J., Roos, R.A., Oudega, O., 2015. The role of brain-derived neurotrophic factor in bone marrow stromal cell-mediated spinal cord repair. Cell Transplant. 24 (11), 2209–2220. https://doi.org/10.3727/096368915X686201.

Rong, Y., Liu, W., Wang, J., et al., 2019. Neural stem cell-derived small extracellular vesicles attenuate apoptosis and neuroinflammation after traumatic spinal cord injury by activating autophagy. Cell Death Dis. 10 (5), 340. https://doi.org/10.1038/s41419-019-1571-8.

Sahib, S., Niu, F., Sharma, A., Feng, L., Tian, R.T., Muresanu, D.F., Nozari, A., Sharma, H.S., 2019. Potentiation of spinal cord conduction and neuroprotection following nano-delivery of DL-3-n-butylphthalide in titanium implanted nanomaterial in a focal spinal cord injury induced functional outcome, blood-spinal cord barrier breakdown and edema formation. Int. Rev. Neurobiol. 146, 153–188. https://doi.org/10.1016/bs.irn.2019.06.009.

Sangari, S., Lundell, H., Kirshblum, S., Perez, M.A., 2019. Residual descending motor pathways influence spasticity after spinal cord injury. Ann. Neurol. 86 (1), 28–41. https://doi.org/10.1002/ana.25505.

Sangari, S., Perez, M.A., 2019. Imbalanced corticospinal and reticulospinal contributions to spasticity in humans with spinal cord injury. The Journal of Neuroscience 39 (40), 7872–7881. https://doi.org/10.1523/JNEUROSCI.1106-19.2019.

Saulino, M., Anderson, D.J., Doble, J., Farid, R., Gul, F., Konrad, P., Boster, A.L., 2016a. Best practices for intrathecal baclofen therapy: troubleshooting. Neuromodulation 19 (6), 632–641. https://doi.org/10.1111/ner.12467.

Saulino, M., Ivanhoe, C.B., McGuire, J.R., Ridley, B., Shilt, J.S., Boster, A.L., 2016b. Best practices for intrathecal baclofen therapy: patient selection. Neuromodulation 19 (6), 607–615. https://doi.org/10.1111/ner.12447.

Sharma, H.S., Feng, L., Muresanu, D.F., Castellani, R.J., Sharma, A., 2019. Neuroprotective effects of a potent bradykinin B2 receptor antagonist HOE-140 on microvascular permeability, blood flow disturbances, edema formation, cell injury and nitric oxide synthase upregulation following trauma to the spinal cord. Int. Rev. Neurobiol. 146, 103–152. https://doi.org/10.1016/bs.irn.2019.06.008.

Siddiqui, A.M., Islam, R., Cuellar, C.A., Silvernail, J.L., Knudsen, B., Curley, D.E., Strickland, T., Manske, E., Suwan, P.T., Latypov, T., Akhmetov, N., Zhang, S., Summer, P., Nesbitt, J.J., Chen, B.K., Grahn, P.J., Madigan, N.N., Yaszemski, M.J., Windebank, A.J., Lavrov, I., 2020. Newly regenerated axons through a cell-containing biomaterial scaffold promote reorganization of spinal circuitry and restoration of motor functions with epidural electrical stimulation. bioRxiv. preprint https://doi.org/10.1101/2020.09.09.288100.

Siegfried, J., Lazorthes, Y., Broggi, G., 1981. Electrical spinal cord stimulation for spastic movement disorders. Appl. Neurophysiol. 44 (1–3), 77–92. https://doi.org/10.1159/000102187.

Silvestro, S., Bramanti, P., Trubiani, O., Mazzon, E., 2020. Stem cells therapy for spinal cord injury: an overview of clinical trials. Int. J. Mol. Sci. 21 (2), 659. https://doi.org/10.3390/ijms21020659.

Simpson, D.M., Hallett, M., Ashman, E.J., Comella, C.L., Green, M.W., Gronseth, G.S., Armstrong, M.J., Gloss, D., Potrebic, S., Jankovic, J., Karp, B.P., Naumann, M., So, Y.T., Yablon, S.A., 2016. Practice guideline update summary: botulinum neurotoxin for the treatment of blepharospasm, cervical dystonia, adult spasticity, and headache: report of the guideline development subcommittee of the American Academy of Neurology. Neurology 86 (19), 1818–1826. https://doi.org/10.1212/WNL.0000000000002560.

Sivaramakrishnan, A., Solomon, J.M., Manikandan, N., 2018. Comparison of transcutaneous electrical nerve stimulation (TENS) and functional electrical stimulation (FES) for spasticity in spinal cord injury—a pilot randomized cross-over trial. J. Spinal Cord Med. 41 (4), 397–406. https://doi.org/10.1080/10790268.2017.1390930.

Sjolund, B.H., 2002. Pain and rehabilitation after spinal cord injury: the case of sensory spasticity? Brain Res. Brain Res. Rev. 40 (1–3), 250–256. https://doi.org/10.1016/s0165-0173(02)00207-2.

Snow, B.J., Tsui, J.C., Bhatt, M.H., Varelas, M., Hashimoto, S.A., Calne, D.B., 1990. Treatment of spasticity with botulinum toxin: a double-blind study. Ann. Neurol. 28, 512–515. https://doi.org/10.1002/ana.410280407.

Song, G., Yang, R., Zhang, Q., Chen, L., Huang, D., Zeng, J., Yang, C., Zhang, T., 2019. TGF-beta secretion by M2 macrophages induces glial scar formation by activating astrocytes in vitro. J. Mol. Neurosci. 69 (2), 324–332. https://doi.org/10.1007/s12031-019-01361-5.

Stenudd, M., Sabelström, H., Frisén, J., 2015. Role of endogenous neural stem cells in spinal cord injury and repair. JAMA Neurol. 72 (2), 235–237. https://doi.org/10.1001/jamaneurol.2014.2927.

Stetkarova, I., Yablon, S.A., Kofler, M., Stokic, D.S., 2010. Procedure- and device-related complications of intrathecal baclofen administration for management of adult muscle hypertonia: a review. Neurorehabil. Neural Repair 24 (7), 609–619. https://doi.org/10.1177/1545968310363585.

Stetkarova, I., Brabec, K., Vasko, P., Mencl, L., 2015. Intrathecal baclofen in spinal spasticity: frequency and severity of withdrawal syndrome. Pain Physician 18 (4), 633–641. https://pubmed.ncbi.nlm.nih.gov/26218954/.

Sugaya, K., Vaidya, M., 2018. Chapter 5: Stem Cell Therapies for Neurodegenerative Diseases. Springer International Publishing AG, part of Springer Nature 61 - 84 K. L. Mettinger et al. (eds.), Exosomes, Stem Cells and MicroRNA, Advances in Experimental Medicine and Biology 1056, https://doi.org/10.1007/978-3-319-74470-4_5.

Sweis, R., Biller, J., 2017. Systemic complications of spinal cord injury. Curr. Neurol. Neurosci. Rep. 17 (2). https://doi.org/10.1007/s11910-017-0715-4.

Taira, T., Ueta, T., Katayama, Y., Kimizuka, M., Nemoto, A., Mizusawa, H., Liu, M., Koito, M., Hiro, Y., Tanabe, H., 2013. Rate of complications among the recipients of intrathecal baclofen pump in Japan: a multicenter study. Neuromodulation 16 (3), 266–272. https://doi.org/10.1111/ner.12010.

Teng, Y.D., 2019. Functional multipotency of stem cells and recovery neurobiology of injured spinal cords. Cell Transplantation 284, 451–459. https://doi.org/10.1177/0963689719850088.

Thomas, C.K., Häger-Ross, C.K., Klein, C.S., 2010. Effects of baclofen on motor units paralyzed by chronic cervical spinal cord injury. Brain 133 (1), 117–125. https://doi.org/10.1093/brain/awp285.

Vogel, A., Upadhya, R., Shetty, A.K., 2018. Neural stem cell derived extracellular vesicles: Attributes and prospects for treating neurodegenerative disorders. EBioMedicine 38, 273–282. https://doi.org/10.1016/j.ebiom.2018.11.026.

Wagner, F.B., Mignardot, J.-B., Le Goff-Mignardot, C.G., Demesmaeker, R., Komi, S., Capogrosso, M., Rowald, A., Seáñez, I., Caban, M., Pirondini, E., Vat, M., McCracken, L.-A., Heimgartner, R., Fodor, I., Watrin, A., Seguin, P., Paoles, E., Van Den Keybus, K., Eberle, G., Schurch, B., Pralong, E., Becce, F., Prior, J., Buse, N., Buschman, R., Neufeld, E., Kuster, N., Carda, S., von Zitzewitz, J., Delattre, V., Denison, T., Lambert, H., Minassian, K., Bloch, J., Courtine, G., 2018. Targeted neurotechnology restores walking in humans with spinal cord injury. Nature 563 (7729), 65–71. https://doi.org/10.1038/s41586-018-0649-2.

Waltz, J.W., 1997. Spinal cord stimulation: a quarter century of development and investigation. A review of its development and effectiveness in 1,336 cases. Stereotact. Funct. Neurosurg. 69 (1–4 Pt. 2), 288–299. https://doi.org/10.1159/000099890.

Ward, A.B., 2008. Spasticity treatment with botulinum toxins. J. Neural Transm. 115, 607–616. https://doi.org/10.1007/s00702-007-0833-2.

Wenger, N., Moraud, E.M., Gandar, J., Musienko, P., Capogrosso, M., Baud, L., Le Goff, C.G., Barraud, Q., Pavlova, N., Dominici, N., Minev, I.R., Asboth, L., Hirsch, A., Duis, S., Kreider, J., Mortera, A., Haverbeck, O., Kraus, S., Schmitz, F., DiGiovanna, J., van den Brand, R., Bloch, J., Detemple, P., Lacour, S.P., Bézard, E., Micera, S., Courtine, G., 2016. Spatiotemporal neuromodulation therapies engaging muscle synergies improve motor control after spinal cord injury. Nat. Med. 22, 138–145. https://doi.org/10.1038/nm.4025.

Willerslev-Olsen, M., Lundbye-Jensen, J., Petersen, T.H., Nielsen, J.B., 2011. The effect of baclofen and diazepam on motor skill acquisition in healthy subjects. Exp. Brain Res. 213 (4), 465–474. https://doi.org/10.1007/s00221-011-2798-5.

Xiao, Z., Tang, F., Zhao, Y., Han, G., Yin, N., Li, X., Chen, B., Han, S., Jiang, X., Yun, C., Zhao, C., Cheng, S., Zhang, S., Dai, J., 2018. Significant improvement of acute complete spinal cord injury patients diagnosed by a combined criteria implanted with NeuroRegen scaffolds and mesenchymal stem cells. Cell Transplant. 27 (6), 907–915. https://doi.org/10.1177/0963689718766279.

Yamazaki, K., Kawabori, M., Seki, T., Houkin, K., 2020. Clinical trials of stem cell treatment for spinal cord injury. Int. J. Mol. Sci. 21 (11), 3994. https://doi.org/10.3390/ijms21113994.

Yang, X.X., Huang, Z.Q., Li, Z.H., Ren, D.F., Tang, J.G., 2017. Risk factors and the surgery affection of respiratory complication and its mortality after acute traumatic cervical spinal cord injury. Medicine (Baltimore) 96 (36), e7887. https://doi.org/10.1097/MD.0000000000007887.

Yarygin, V.N., Banin, V.V., Yarygin, K.N., 2006. Regeneration of the rat spinal cord after thoracic segmentectomy: restoration of the anatomical integrity of the spinal cord. Neurosci. Behav. Physiol. 36 (5), 483–490.

Yu, T., Zhao, C., Hou, Sh., Zhou, W., Wang, B., Chen, Y., 2019. Exosomes secreted from miRNA-29b-modified mesenchymal stem cells repaired spinal cord injury in rats. Brazilian Journal of Medical and Biological Research 52 (12), e8735. https://doi.org/10.1590/1414-431X20198735.

Zhang, C., Kong, X., Ning, G., Liang, Z., Qu, T., Chen, F., Cao, D., Wang, T.W., Sharma, H.S., Feng, S., 2020. Corrigendum to "all-trans retinoic acid prevents epidural fibrosis through NF-kB signaling pathway in post-laminectomy rats" [neuropharmacology 79 (2014) 275–281]. Neuropharmacology 172, 107927. https://doi.org/10.1016/j.neuropharm.2019.107927.

Zhao, C., Rao, J.S., Pei, X.J., Lei, J.F., Wang, Z.J., Zhao, W., Wei, R.H., Yang, Z.Y., Li, X.G., 2018. Diffusion tensor imaging of spinal cord parenchyma lesion in rat with chronic spinal cord injury. Magn. Reson. Imaging 47, 25–32. https://doi.org/10.1016/j.mri.2017.11.009.

Zhao, C.H., Zhou, X., Qiu, J., Xin, D., Li, T., Chu, X., et al., 2019. Exosomes derived from bone marrow mesenchymal stem cells inhibit complement activation in rats with spinal cord injury. Drug Design, Development and Therapy 13, 3693–3704. https://doi.org/10.2147/DDDT.S209636.

Zhong, D., Cao, Y., Li, Ch-J., Li, M., Rong, Z-J., Jiang, L., et al., 2020. Neural stem cell-derived exosomes facilitate spinal cord functional recovery after injury by promoting angiogenesis. Experimental Biology and Medicine (Maywood) 2451, 54–65. https://doi.org/10.1177/1535370219895491.

Zhu, S., Chen, M., Deng, L., Zhang, J., Ni, W., Wang, X., et al., 2020. The repair and autophagy mechanisms of hypoxia-regulated bFGF-modified primary embryonic neural stem cells in spinal cord injury. Stem Cells Translational Medicine 9 (5), 603–619. https://doi.org/10.1002/sctm.19-0282.

CHAPTER

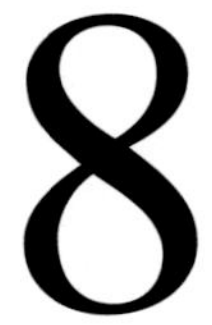

Superior antioxidant and anti-ischemic neuroprotective effects of cerebrolysin in heat stroke following intoxication of engineered metal Ag and Cu nanoparticles: A comparative biochemical and physiological study with other stroke therapies

Hari Shanker Sharma[a,*], Dafin F. Muresanu[c,d], Asya Ozkizilcik[e], Seaab Sahib[f], Z. Ryan Tian[f], José Vicente Lafuente[b], Rudy J. Castellani[g], Ala Nozari[h], Lianyuan Feng[i], Anca D. Buzoianu[j], Preeti K. Menon[k], Ranjana Patnaik[l], Lars Wiklund[a], and Aruna Sharma[a,*]

[a]*International Experimental Central Nervous System Injury & Repair (IECNSIR), Department of Surgical Sciences, Anesthesiology & Intensive Care Medicine, Uppsala University Hospital, Uppsala University, Uppsala, Sweden*

[b]*LaNCE, Department of Neuroscience, University of the Basque Country (UPV/EHU), Leioa, Bizkaia, Spain*

[c]*Department of Clinical Neurosciences, University of Medicine & Pharmacy, Cluj-Napoca, Romania*

[d]*"RoNeuro" Institute for Neurological Research and Diagnostic, Cluj-Napoca, Romania*

[e]*Department of Biomedical Engineering, University of Arkansas, Fayetteville, AR, United States*

[f]*Department of Chemistry & Biochemistry, University of Arkansas, Fayetteville, AR, United States*

[g]*Department of Pathology, University of Maryland, Baltimore, MD, United States*

[h]*Anesthesiology & Intensive Care, Massachusetts General Hospital, Boston, MA, United States*

[i]*Department of Neurology, Bethune International Peace Hospital, Shijiazhuang, Hebei Province, China*

[j]*Department of Clinical Pharmacology and Toxicology, "Iuliu Hatieganu" University of Medicine*

Progress in Brain Research, Volume 266, ISSN 0079-6123, https://doi.org/10.1016/bs.pbr.2021.06.014
Copyright © 2021 Elsevier B.V. All rights reserved.

and Pharmacy, Cluj-Napoca, Romania
[k]*Department of Biochemistry and Biophysics, Stockholm University, Stockholm, Sweden*
[l]*Department of Biomaterials, School of Biomedical Engineering, Indian Institute of Technology, Banaras Hindu University, Varanasi, India*
**Corresponding authors: Tel.: +46-70-2011-801; Fax: +46-18-24-38-99 (Hari Shanker Sharma); Tel.: +46-70-21-95-963; Fax: +46-18-24-38-99 (Aruna Sharma),*
e-mail address: sharma@surgsci.uu.se; harishanker_sharma55@icloud.com; hssharma@aol.com; aruna.sharma@surgsci.uu.se

Abstract

Military personnel are often exposed to high environmental heat associated with industrial or ambient abundance of nanoparticles (NPs) affecting brain function. We have shown that engineered metal NPs Ag and Cu exacerbate hyperthermia induced brain pathology. Thus, exploration of novel drug therapy is needed for effective neuroprotection in heat stroke intoxicated with NPs. In this investigation neuroprotective effects of cerebrolysin, a balanced composition of several neurotrophic factors and active peptides fragments exhibiting powerful antioxidant and anti-ischemic effects was examined in heat stroke after NPs intoxication. In addition, its efficacy is compared to currently used drugs in post-stroke therapies in clinics. Thus, levertiracetam, pregabalin, topiramat and valproate were compared in standard doses with cerebrolysin in heat stroke intoxicated with Cu or Ag NPs (50–60 nm, 50 mg/kg, i.p./day for 7 days). Rats were subjected to 4 h heat stress (HS) in a biological oxygen demand incubator at 38 °C (Relative Humidity 45–47%; Wind velocity 22.4–25.6 cm/s) that resulted in profound increase in oxidants Luminol, Lucigenin, Malondialdehyde and Myeloperoxidase, and a marked decrease in antioxidant Glutathione. At this time severe reductions in the cerebral blood flow (CBF) was seen together with increased blood-brain barrier (BBB) breakdown and brain edema formation. These pathophysiological responses were exacerbated in NPs treated heat-stressed animals. Pretreatment with cerebrolysin (2.5 mL/kg, i.v.) once daily for 3 days significantly attenuated the oxidative stress, BBB breakdown and brain edema and improved CBF in the heat stressed group. The other drugs were least effective on brain pathology following heat stroke. However, in NPs treated heat stressed animals 5 mL/kg conventional cerebrolysin and 2.5 mL/kg nanowired cerebrolysin is needed to attenuate oxidative stress, BBB breakdown, brain edema and to improve CBF. Interestingly, the other drugs even in higher doses used are unable to alter brain pathologies in NPs and heat stress. These observations are the first to demonstrate that cerebrolysin is the most superior antioxidant and anti-ischemic drug in NPs exposed heat stroke, not reported earlier.

Keywords

Cerebrolysin, Levetiracetam, Pregabalin, Topiramat, Valproate, Heat stroke, Cerebral blood flow, Brain edema, Heat exposure, Brain pathology, Neuroproetction, Nanoparticles from metals, Ag, Cu toxicity

1 Introduction

Heat stroke is a devastating disease inducing mortality in more than 50% of the victims (Al Mahri and Bouchama, 2018; Epstein and Yanovich, 2019; Higareda-Basilio et al., 2019; Xia et al., 2021). The clinical manifestation of heat stroke includes hyperthermia (<40 °C), various organ dysfunction and brain damage (Hemmelgarn and Gannon, 2013; Kenny et al., 2018; Knapik and Epstein, 2019; Lawton et al., 2019; Leon and Helwig, 1985; Morris and Patel, 2021; Shimada et al., 2020). The possible causes of death are likely due to sever brain ischemia, blood-brain barrier (BBB) damage, brain edema, and neuronal injuries associated with multi-organ dysfunction in heat stroke (Lu et al., 2004; Sharma, 2005a,b,c,d, 2006; Sharma et al., 1992a,b; Szold et al., 2002; Thulesius, 2006; Wang et al., 2012).

There are many similarities between ischemic stroke and heat stroke (Chang et al., 2007; Chen et al., 2013; Hsuan et al., 2016; Tseng et al., 2019; Wang et al., 2019; Zhou et al., 2017). Thus, it appears that similar mechanisms are operating in ischemic stroke and heat stroke. This is apparent from the findings that nanoparticles exposure from the environment or other sources exacerbates the brain pathology in ischemic stroke and heat stroke (Liu et al., 2016; Miller et al., 2017; Olaru et al., 2019; Sharma and Sharma, 2007; Xin et al., 2015). These observations suggest that nanoparticles exacerbates ischemic stroke and heat stroke induced pathology.

Ischemic stroke is associated with oxidative stress, epilepsy and central pain (Cheng et al., 2017; Cichoń et al., 2018; Delpont et al., 2018; Klit et al., 2009; Nabavi et al., 2015; Plecash et al., 2019; Tanaka and Ihara, 2017). To manage these post-stroke syndromes several antiepileptic agents are prescribed in clinics. However, these drugs offering neuroprotection following nanoparticles exposure in ischemic stroke is still not well known. In ischemic post stroke several drugs are prescribed in clinical situations to control symptoms like epilepsy, pain and related disorders. Often the drugs used are levetiracetam, pregabalin, topiramat and valproate (Alberti et al., 2020; Belcastro et al., 2011; Faggi et al., 2018; Hansen et al., 2018; Jungehulsing et al., 2013; Kim et al., 2011; Lee et al., 2014; Onouchi et al., 2014; Powers et al., 2017; Rahman and Uddin, 2014; Romoli et al., 2019; Scotton et al., 2019; Tao et al., 2020; Yoon et al., 2011; Zhao et al., 2020).

Heat stroke is also associated with oxidative stress, epilepsy and central pain (Abraira et al., 2020; Bruchim et al., 2019; Garcia-Larrea and Hagiwara, 2019; Hsu et al., 2014; Lee et al., 2017; Liu et al., 2020). Our previous studies show that nanoparticles exacerbate heat stroke induced brain pathology are significantly attenuated by cerebrolysin—a balanced composition of several neurotrophic factors and active peptide fragments (Muresanu et al., 2010, 2019; Sharma et al., 2010a,b, 2011, 2012a,b). In one study, we have compared cerebrolysin with levetiracetam, pregabalin, topiramat and valproate in two powerful doses in heat stress intoxicated with engineered metal nanoparticles (NPs) from Ag or Cu (50–60 nm in size, 50 mg/kg, i.p. once for 1 week) administration in a rat model (Sharma et al., 2011). Our findings show that cerebrolysin appears to be the most potent in reducing brain pathology in NPs intoxicated heat stressed rats as compared to even high doses of the levetiracetam, pregabalin, topiramat and valproate (Sharma et al., 2011). It appears that cerebrolysin could exerts its neuroprotective effects in heat stroke by reducing oxidative stress and attenuating cerebral ischemia in clinical and preclinical studies (Amiri-Nikpour et al., 2014; Formichi et al., 2013; Muresanu et al., 2010; Rockenstein et al., 2005; Schwab et al., 1997).

In this investigation, we explored effects of cerebrolysin in comparison with levetiracetam, pregabalin, topiramat and valproate on heat stroke induced oxidative stress and cerebral blood flow in heat stress model. Furthermore, we also examined the effects of Ag or Cu NPs intoxication on the above parameters under identical conditions in heat stress. In addition, nanowired cerebrolysin was delivered in this model of heat stress to explore the possibility of superior neuroprotective effects on brain pathology.

2 Materials and methods

2.1 Animals

Experiments were carried out on inbred Charles Foster Male Rats (age 9–12 weeks, body weight 150–200 g) housed at controlled ambient temperature (21 ± 1 °C) with 12 h light and 12 h dark schedule. The rat feed and tap water were supplied ad libitum.

All experiments were conducted according to the Handling and Care of Experimental animals according to the National Institute of Health (NIH) Guidelines and approved by the Local Institutional Ethics Committee.

2.2 Exposure to heat stress

All experiments were commenced at 8:00 AM to avoid circadian variation induced altered physiological conditions among individual animals. These animals were handled daily for 1 week before the heat stress experiment to place them in the Biological Oxygen Demand incubator (BOD, Adair Dutt & Co., New Delhi, India) at 21 ± 1 °C with relative humidity (45–47%) and wind velocity 20–26 cm/s to make them accustomed in placing in the chamber (Sharma, 2005b, 2007a,b,c; Sharma and Dey, 1984, 1986, 1987; Sharma et al., 1991a,b).

At the time of heat exposure the BOD was maintained at 38 °C with relative humidity and wind velocity kept constant. The animals were placed in the BOD incubator for 4 h (Sharma and Dey, 1986, 1987). The control group of rats either placed at room temperature (21 ± 1 °C) or at the BOD incubator (21 ± 1 °C) under identical conditions (Sharma, 2005b, 2007a,b,c; Sharma et al., 1991a,b).

2.3 Nanoparticles exposure

Commercial engineered metal nanoparticles from Ag, and Cu (50–60nm) were obtained from IoLiTec Ionic Liquids Technologies (Denzlingen, Germany). The NPs were suspended in 0.05% Tween 80 in 0.7% NaCl solution (Sharma, 1987; Sharma and Dey, 1986) and administered intraperitoneally in a dose of 50mg/kg each once daily for 7 days (Sharma et al., 2009). In a group of 4–6 animals, the NPs were suspended in 0.9% physiological saline and administered into the animals (50mg/kg, per day for 7 days, i.p.) under identical conditions.

The control groups were administered either 0.05% Tween 80 in 0.7% NaCl or 0.9% physiological saline at room temperature (Sharma et al., 2011).

On the 8th day, control or NPs treated rats were exposed to HS for 4h at 38 °C in the BOD incubator as described earlier (Sharma et al., 2011).

2.4 Drug treatments

The following drugs were used to evaluate their neuroprotective efficacy in control or HS animals as a pretreatment (Sharma et al., 2011). All drug doses were used based on the recommendations for treatment in stroke in clinical cases. These drugs were administered in NPs treated group from 5th day and continued for 3 days once daily. On the 8th day these drug treated with NPs treatment or without NPs were subjected to HS for 4h in BOD incubator maintained at 38 °C (Sharma et al., 2011).

(a) Cerebrolysin (Ever NeuroPharma, Austria): Cerebrolysin ampoule (2.5 or 5mL/kg, i.v.) was administered once daily for 3 days before subjection of animals with NPs treated group or saline to HS. Normal intact controls received physiological saline instead of cerebrolysin under identical condition.

(b) Levetiracetam (Cat. Nr. 2839, Tocris, Madrid, Spain): Levetiracetam ((2*S*)-(2-Oxopyrrolidin-1-yl)butyramide) is dissolved in sterile 0.9% NaCl or Dimethyl sulfoxide (DMSO) and administered intravenously in a dose of 44mg or 88mg/kg once for 3 days before 4h HS experiments in control or NPs intoxicated group. Intact control group treated identically with DMSO or 0.9% saline.

(c) Pregabalin ([S]-3-(Aminomethyl)-5-methylhexanic acid): Pregabalin (CAS nr. 148,553–50-8, Merck, Madrid, Spain) dissolved in 0.9% and administered intravenously (after adjusting pH 7.4) in a dose of either 200mg or 400mg/kg one daily for 3 days in control and NPs intoxicated group daily for 3 days before HS experiments. Intact control animals received physiological saline instead of pregabalin under identical condition.

(d) Topiramate: Topiramate (2,3:4,5-*Bis*-*O*-(1-methylethylidene)-β-D-fructopyranose sulfamate, Tocris, Cat No. 3620, Madrid Spain) was dissolved in Dimethyl sulfoxide (DMSO) and injected intraperitoneally in a dose of (40mg or

80 mg/kg) once daily for 3 days in control or NPs administered rats before 4 h HS exposure at 38 °C. The intact control were administered DMSO vehicle under identical conditions.

(e) Valproate: Valproate sodium (Sodium 2-propylpentanoate; Cat Nr. 2815, Tocris, Madrid, Spain) was dissolved in 0.9% NaCl or dimethyl sulfoxide (DMSO) and administered intravenously in a dose of 400 or 800 mg/kg in control or NPs treated animals once daily for 3 days before HS exposure at 38 °C. Intact control rats were identical treated with saline or DMSO instead of valproate sodium.

In general, the drug treatments commenced from the day 5th of NPs administration and continued until 7th day of treatment. On the 8th day control or NPs treated rats were subjected to 4 h HS in BOD incubator at 38 °C. The intact control group with or without drug or vehicle treatment kept at room temperature (21 ± 1 °C) or in the BOD incubator (21 °C) for 4 h (Sharma et al., 2011).

2.5 Nanowired delivery of cerebrolysin (NWCBL)

In separate group of rats TiO2-labeled cerebrolysin (NWCBL 2.5 or 5 mL/kg, i.v.) was examined in heat stroke induced brain pathology as described earlier (Sharma et al., 2012a,b).

2.6 Parameters measured

The following parameters are measured in control, NPs treated HS rats with or without drug treatments and compared the data obtained from the intact controls (see Sharma et al., 2011).

2.7 Stress symptoms

The stress symptoms in normal, heat stroke and NPs treated rats were assessed using core body temperature, behavioral salivation, prostration and occurrence of gastric hemorrhagic spots in the stomach at postmortem examination as described earlier (Sharma and Dey, 1986).

For this purpose, thermistor probes (Yellow Springfield, USA) were inserted into the rectum that was connected to a telethermometer (Aplab Electronics, USA) before and after the end of the experiment. The occurrence of behavioral salivation was observed during heat exposure by measuring the spread of saliva over the snout of animals (Sharma and Dey, 1986, 1987).

When rats were exposed to heat stroke sometimes after they develop prostration as seen by lying on the belly on the floor of the cage and do not move even after gentle pushing. The status of the prostration was manually recorded.

After the end of the experiment, animals were anesthetized and duodenal ulcers or any hemorrhagic spots within the stomach was examined at postmortem using a magnifying lens and the number of spots were counted manually (Sharma and Dey, 1986).

2.8 Physiological variables

In control, experimental or drug treated animals physiological variables were measured using mean arterial blood pressure (MABP), arterial pH and arterial blood gases PaO2 and PaCO2 as described earlier (Sharma and Dey, 1986). In brief, for MABP measurement a sterile polyethylene cannula PE10 was implanted into the left common carotid artery retrogradely toward heart and secure there about 1 week before the experiment (Sharma and Dey, 1986). Another cannula (PE10) was implanted into the right jugular vein for administration of drugs aseptically about 1 week before the experiments.

At the time of MABP measurement, the arterial cannula is connected to a Strain Gauge Pressure transducer (Statham P23, USA) connected to a chart recorder (Electromed, UK). Immediately before arterial cannula connected to be pressure transducer about 1 mL of arterial blood was withdrawn for later measurement of the arterial pH and blood gases using a Radiometer Apparatus (Copenhagen, Denmark). Recording of heart rate (beats per minute) and respiration (cycle per minute) was also done suing separate leads connected with the chart recorder as described earlier (Dey et al., 1980).

2.9 Blood-brain barrier breakdown

The BBB permeability was examined using Evans blue albumin (EBA) and radioiodine ([131]-I-Na) tracers (Sharma and Dey, 1986). For this purpose, EBA 2% solution in physiological saline (pH 7.4) 3 mL/kg was administered through jugular vein cannula 5–10 min before end of the experiment. Radioiodine (100 μCi/kg) was administered into the right femoral vein cannula implanted aseptically 1 week before the experiment. The tracers are allowed to circulate at least 5–10 min. These tracers bind to serum albumin in vivo and thus, their leakage into the brain represent tracer-protein complex (Sharma et al., 1990).

At the end of the experiment, the animals were anesthetized deeply with Equithesin (3 mL/kg, i.p.) and the chest was rapidly cut and the heart was exposed. A butterfly cannula (31 G) was inserted into the left cardiac ventricle and the right auricle was cut. The intravascular tracer was washed out using cold physiological saline (4 °C, about 50 mL) at 90 mm Torr (Sharma and Dey, 1987). Immediately before perfusion about 1 mL of whole blood was sample was withdrawn from the left ventricle cannula for later determination of the whole blood radioactivity and EBA content. After perfusion, the brain were removed and examined for the extravasation of the blue dye on the dorsal and ventral side of the brain. After a mid sagittal section the dye penetration into the cerebral ventricles was also examined. After that the desired brain regions were dissected weighed immediately and the Radioactivity counted in a 3-in Gamma counter (Packard, USA).

After counting the radioactivity the brain samples were homogenized in a mixture of acetone and 0.5% sedum sulfate (pH 7.0) to extract the Evans blue dye from the samples. The samples were then centrifuged at 900 × *g* and the supernatant was taken out. The dye entered into the samples was measured calorimetrically in a

Spectrophotometer at 620 nm. The dye entered into the samples was calculated using a standard curve of Evans blue as described earlier (Sharma and Dey, 1986).

The radioactivity extravasated into the brain is expressed as percentage of whole blood radioactivity. The EBA entered into the brain is expressed as mg % (Sharma et al., 1990).

2.10 Cerebral blood flow

The cerebral blood flow (CBF) is measured in the brain of control, experimental and drug treated arts using radiolabelled ($^{[125]}$-Iodine) carbonized microspheres (o.d. $15 \pm 0.6\,\mu m$) using peripheral reference blood flow techniques as described earlier (Sharma and Dey, 1987, 1988). In brief, about one million microspheres were suspended in 0.05% Tween 80 and 0.7% NaCl solution at room temperature and administered as a bolus (within 45 s) retrogradely toward the heart using the left carotid artery cannula. Peripheral serial arterial blood samples at the rate of 0.8 mL/min from right femoral artery implanted about 1 week before aseptically were collected starting from 30 s before microsphere administration at every 30 s that was continued up to 90 s after completion of microsphere administration (Sharma, 1987).

After 90 s of microsphere administration, the animals were decapitated and the brains were removed and placed on a cold saline wetted filter paper. The large superficial blood vessels and blood clots if any were discarded. The desired brain regions were quickly dissected out and weighed immediately and the radioactivity counted in a 3-in Gamma Counter (Packard, USA). The serial blood samples obtained from the femoral artery were also counted for the radioactivity.

The CBF in mL/g/min is calculated from the radioactivity of bran samples CPM per gram multiplied by reference blood flow (0.8 mL/g/min) divided by the whole blood radioactivity per gram obtained from the serial samples (Sharma, 1987; Sharma and Dey, 1987). The CBF values are expressed as brain regions mL/g/min (Sharma et al., 1990).

2.11 Brain edema and volume swelling (%*f*)

Brain edema was determined from the brain water content of the samples from control, experimental or drug treated groups. The volume swelling is calculated from the differences of brain water content between the control and experimental groups as described earlier (Sharma and Cervós-Navarro, 1990; Sharma and Olsson, 1990). In brief, after the end of the experiment, animals ware decapitated and the brains were removed and kept on a filter paper wetted by cold physiological saline. The large superficial vessels and blood clots if any were removed immediately. The desired brain regions were dissected quickly and placed on preweighed filter papers (Whatman Nr. 1, USA). The brain samples were weighed immediately in an analytical balance (Mettler, USA, sensitivity 0.1 mg). After that the samples were placed in a Laboratory oven maintained at 90 °C for 72 h in order to remove water from the samples. When the dry weight of the samples became constant in at least 3 determinations, the brain water content was calculated from the differences between dry and wet weight of the samples (Sharma and Olsson, 1990). For volume swelling (%*f*) calculation, differences between control and experimental brain water content were calculated according to the formula of Elliott and Jasper (1949) described earlier (Sharma and Cervós-Navarro, 1990). In general about 1% increase in brain water results in about 4% increase in volume swelling (Sharma and Cervós-Navarro, 1990; Sharma et al., 1991a,b).

2.12 Oxidative stress parameters

The oxidative stress parameters in the control, experimental and drug treated animals were evaluated by measuring Luminol, Lucigenin, Myeloperoxidase, Malondialdehyde and Glutathione content in the brain using standard biochemical methods as described earlier (Sharma et al., 2018).

In brief, myeloperoxidase assay was carried out using commercial protocol [see 33]. For this purpose, tissue samples (200–300mg) were homogenized in 10 volumes of ice-cold potassium phosphate buffer (50mM K_2HPO_4, pH 6.0) containing hexadecyl-trimethyl-ammonium bromide (HETAB; 0.5%, w/v). The homogenized samples were then centrifuged at 41,400 × *g* for 10min and the pellets were suspended in 50mM PB containing HETAB; 0.5%. After that the aliquots (0.3mL) were added to 2.3mL of reaction mixture containing 50mM PB, o-dianisidine, and 20mM H_2O_2 solution (see Sharma et al., 2018). The samples were measured in a spectrophotometer at 460nm for 3min. The level of myeloperoxidase is measures as units where one unit of enzyme activity corresponds to the amount that caused changes in absorbance measured at 460nm and expressed the activity as U/g tissue.

For malondialdehyde and glutathione measurements, the brain tissues were homogenized in ice-cold 150mM KCl and levels were measured for products of lipid peroxidation according to commercial protocol (Dahle et al., 1962). Malondialdehyde levels were expressed as $nM\,g^{-1}$ tissue whereas the glutathione was determined using spectrophotometric method employing Ellman's reagent (Paglia and Valentine, 1967) and values are expressed as $\mu mol\,g^{-1}$ tissue.

Luminol and Lucigenin are reactive oxygen species that can be measures using chemiluminescent (CL) probes as described earlier (Sharma et al., 2018). For this purpose, brain tissues were washed with saline and Luminescence of the samples was recorded at room temperature using a luminometer (Bad Wildbad, Germany) in the presence of enhancers. Tissue specimens are placed in PBS-HEPES buffer (0.5mol/L phosphate buffered saline containing 20mmol/L HEPES, pH 7.2) (Guzik and Channon, 2005; Katsuragi et al., 2000) and signals were measured after addition of the enhancer (lucigenin or luminol) to a final concentration of 0.2mmol/L. After the measurements, the tissues were dried on filter papers and weighed. All chemiluminometric counts were obtained at 1-min intervals for 5min (Sharma et al., 2018). The results are expressed as relative light units (rlu) for 5min per mg of tissue (see Sharma et al., 2018).

2.13 Brain pathology

Morphological changes following heat stroke in the brain were examined using Light and Transmission electron Microscopy (TEM) using standard procedures (Sharma and Cervós-Navarro, 1990; Sharma and Olsson, 1990; Sharma et al., 1991a,b).

2.13.1 Perfusion and fixation

In brief, after the end of the experiments, animal ware deeply anesthetized with Equithesin (3 mL/kg, i.p.) and chest was rapidly opened and heart was exposed and right auricle cut. A butterfly cannula (21 G) was inserted into the left cardiac ventricle and intravascular blood was washed out with about 100 cold 0.1 M phosphate buffer saline (PBS, pH 7.0 at 4 °C). This was followed by perfusion of cold 4% paraformaldehyde (200 mL, 4 °C) for brain fixation in situ. After perfusion, the animals were wrapped into an aluminum foil and kept at a refrigerator for overnight (Sharma, 2000; Sharma and Alm, 2002; Sharma et al., 2000).

2.13.2 Light microscopy

On the next day the brains were dissected out and coronal sections passing through hippocampus, caudate nucleus, and cerebellum were cut and processes for paraffin embedding using automated tissue processer (Tissue-Tek, Sakura, FineTek, Tokyo, Japan). For light microscopy, about 3-μm thick sections were cut and processed for standard histopathological processing using Nissl or Hematoxylin & Eosin (H&E) staining. The slides were analyzed and images captured on Carl-Zeiss (Germany) Microscope attached with a digital camera. The images obtained at original magnification of ×20 or ×40 were stored at Apple Macintosh PowerBook computer Mac Os El Capitan version 10.11.6 and processed using commercial software Photoshop Extended 12.4 × 64 using identical color balance and brightness control (Sharma and Alm, 2002; Sharma and Sjöquist, 2002).

2.13.3 Transmission electron microscopy

For transmission electron microscopy (TEM) small tissue pieces from the desired areas of the brain were post fixed in Osmium Tetraoxide (OsO4) and embedded in Plastic (Epon 812) (Sharma and Olsson, 1990). About 1 μm semithin sections ware cut and stained with Toluidine blue for gross examination of cellular location under a Leica (Jena, Germany) Bench microscope. After identification of the area for examination at TEM, the tissue blocks were trimmed and ultrathin sections were cut on an Ultramicrotome (LKB, Sweden) about 50 nm thick using diamond knife. The serial ultrathin sections were placed on either a copper mesh grid or one whole copper grid and counterstained with Uranyl acetate and Lead citrate as described earlier (Sharma and Olsson, 1990). Some sections ware unstained for comparison. These ultrathin sections were examined under a Phillips 400 TEM attached with a digital camera (Gatan System, USA) and photographed at ×4k, ×8k or ×12k original magnification. The images captured were stored in an Apple Macintosh PowerBook computer

(Os 12.11.6) and processed using commercial software Photoshop extended (PS 12.04 × 64) using identical filters and brightness control from control and experimental specimens (Sharma et al., 1991a,b).

2.14 Statistical analyses of the data obtained

All data were analyzed for Statistical Significance using Commercial Software StatView 5 (Abacus Concepts, Inc., USA) using Apple Macintosh Computer for quantitative data using ANOVA followed by Dunnett's test for multiple comparisons from one control group. The semiquantittative data were analyzed using non-parametric test Chi-Square test. A P-value less than 0.05 is considered significant.

3 Results

3.1 Effect of heat stroke on stress symptoms

Heat stress induced profound symptoms of hyperthermia and behavioral salivation and prostration. Thus, subjection of animals to heat stress resulted in profound hyperthermia, degree of salivation and behavioral prostration (Table 1 and Fig. 1). These stress symptoms were further aggravated by NPs intoxication from metals, i.e., Cu and Ag (Table 1 and Fig. 1). The stress symptoms were slightly aggravated by exposure to Cu and Ag NPs in heat stress. At postmortem, gastric ulcerations and hemorrhagic spots examined in the stomach showed profound increase in 4h HS (Table 1). These stress symptoms of ulceration in the stomach appears to be further aggravated by intoxication of Cu and Ag NPs (Table 1). These observations suggest that NPs aggravate stress symptoms in heat stress.

3.2 Physiological variables

Heat stress affects physiological variables such as MABP, Arterial pH and blood gases PaO2 and PaCO2 significantly from the control group. Thus, there was a significant reduction in the MABP after 4h heat stress from the control group (Table 2). These changes in MABP were slightly modified by intoxication with Cu or Ag NPs (Table 2). Saline control group also showed some minor changes in control MABP (Table 2). The arterial pH showed slight alterations after 4h heat stress and is also affected by Cu or Ag NPs intoxication following 4h heat stress and also in saline treated control groups (Table 2). In general, the arterial PaCO2 showed significant increase from control values after 4h heat stress as compared to the control group. These changes in PaCO2 were also affected a little bit by intoxication with Cu or Ag NPs in heat stress and saline treated groups. The PaO2 showed significant increase in heat stress as compared to the control saline treated group (Table 2). The values of PaO2 was mildly altered following Cu or Ag NPs intoxication in both control and heat stressed group (Table 2).

Table 1 Effect of Drugs on Stress Symptoms, Blood-Brain Barrier Permeability, Cerebral Blood Flow, Brain Edema, Volume Swelling and Brain Pathology in Heat Stress with Ag or Cu Nanoparticles Treatment

	Stress Symptoms			BBB		Brain Edema	Volume Swelling	Brain Pathology (No. of Cases)		
Type of Experiment	Salivation	Prostration	Ulceration	[131]-I %	CBF (mL/g/min)	Brain water (%)	%f	Neuronal Damage	Myelin Vesiculation	Capillary Damage
(A) Control										
Saline nil	Nil	Nil	Nil	0.34±0.06	1.25±0.06	74.48±0.21	Nil	2±4	1±2	1±2
Ag NPs	Nil	Nil	1±2	0.56±0.08*	1.18±0.04*	74.67±0.10*	+0.5	8±3*	3±1*	3±1*
Cu NPs	Nil	Nil	2±2	0.48±0.03*	1.16±0.06*	74.69±0.09*	+0.5	10±6*	4±3*	4±1*
(B). 4 h heats stress 38°C										
Heats stress 4 h	++++	+++	35±8*	2.56±0.10*	0.76±0.08*	78.96±0.13*	+18	239±21*	54±8*	42±6*
HS+Ag NPs	++++	++++	46±9*	3.12±0.14*	0.72±0.06*	79.80±0.21*	+21	319±18*	64±7*	53±9*
HS+Cu NPs	+++	++++	43±9*	2.96±0.12*	0.70±0.05*	79.46±0.15*	+20	314±12*	60±10*	52±13*
HS+CBL 2.5 mL	+++	+++	12±4*#	1.43±0.08*#	0.98±0.05*#	75.89±0.14*#	+5.5	39±12*#	14±6*#	12±5*#
HS+CBL 5 mL	+++	++	8±4*#	0.98±0.08*#	1.03±0.06*#	75.05±0.23*#	+3.5	26±8*#	12±5*#	8±4*#
HS+NWCBL 2.5 mL	+++	++	7±3*#	0.87±0.12*#	1.18±0.06*#	75.01±0.14*#	+3	18±6*#	10±6*#	8±5*#
HS+LVRTM 44 mg	++++	++++	56±12*	1.89±0.21*	0.70±0.12*	78.67±0.22*	+24	280±27*	65±13*	54±11*
HS+LVRTM 88 mg	+++	+++	43±14*	1.25±0.21*	0.78±0.24*	78.17±0.23*	+17	212±23*	54±11*	49±9*
HS+PGBL 200 mg	+++	+++	40±8*	1.34±0.08*	0.82±0.08*	78.03±0.13*	+16	204±28*	50±12*	48±14*
HS+PGBL 400 ng	+++	+++	34±8*	1.08±0.12*	0.85±0.12*	77.89±0.13*	+14	188±18*	45±8*	43±10*
HS+TPRM 40 mg	++++	++++	54±23*	1.28±0.09*	0.72±0.08*	78.45±012*	+16	238±23*	56±18*	54±13*
HS+TPRM 80 mg	++++	+++	38±8*	1.19±0.12*	0.78±0.07*	78.05±0.13*	+16	210±18*	45±13*	40±12*
HS+VLPRT 400 mg	+++	+++	38±8*	1.23±0.21*	0.70±0.08*	78.34±0.13*	+20	278±12*	56±13*	49±14*
HS+VLPRT 800 mg	++++	++++	48±12*	1.34±0.08*	0.76±0.09*	77.89±0.12	+14	205±13*	48±12*	56±14*
(C). Ag NPs+4 h heat stress										
HS+NWCBL 2.5 mL	+++	++	6±4*	0.67±0.09*	1.08±0.06*	75.14±0.32*	+2	28±8*	9±6*	8±4*
HS+LVRTM 88 mg	+++	+++	36±7*	1.34±0.12*	0.71±0.07*	78.23±0-34*	+16	201±23*	45±8*	37±5*
HS+PGBL 400 ng	+++	+++	33±8*	1.12±0.09*	0.70±0.05*	78.14±0.21*	+16	180±14*	40±6*	38±8*
HS+TPRM 80 mg	++++	+++	38±9*	1.34±0.21*	0.68±0.08*	78.23±0.24*	+16	148±12*	38±7*	42±6*
HS+VLPRT 800 mg	+++	+++	36±5*	1.21±0.18*	0.72±0.12*	77.89±0.23*	+14	138±14*	32±6*	36±4*

In this group the drugs were given on the fifth day for 3 days and all parameters were examined 24 h after the last injection of the drug.
Values are mean±SD of 6–8 rats at each point. CBL, cerebrolysin; NWCBL, nanowired cerebrolysin.
**P < 0.05 from saline control; #P < 0.05 from experimental control, ANOVA followed by Dunnett's test for multiple group comparison from one control group. For details see text. Salivtion = +++ moderate, ++++ = Severe, Prostration ++ = mild; +++ = moderate; ++++ = Severe (for details see text).*

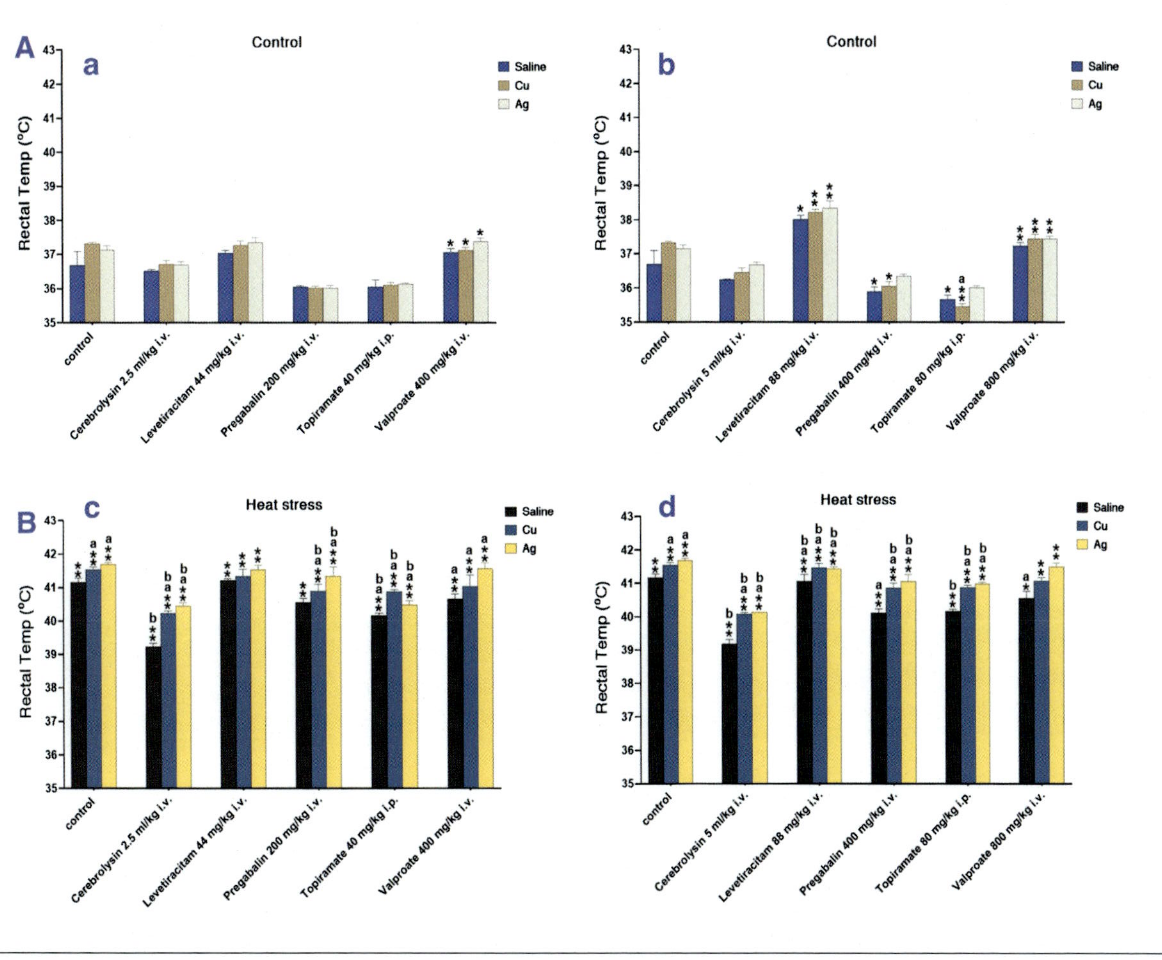

FIG. 1

Rectal temperature changes in control (A) and heat stressed rats (B) after treatment with drugs at low dose (a, c) and high dose (b, d) intoxicated with engineered metal nanoparticles Cu or Ag (50–60 nm, 50 mg/kg, i.p. once daily for 7 days). Animals were heat exposed on the 8th day in a BOD (Biological Oxygen Demand) incubator at 38 °C for 4 h. For details see text. Values are Mean ± SD from 6 to 8 rats at each point. $^{*}P<0.05$; $^{**}P<0.01$ significantly different from saline control group, Over the bars symbol $^{a}P<0.05$, significantly different from saline control from same group; Over the bars symbol $^{b}P<0.05$ significantly different from respective control group, ANOVA followed by Dunnett's test from one control group.

Table 2 Physiological Variables in Heat Stress with Ag or Cu Nanoparticles and Modification with Drugs

Type of Experiment	Physiological Variables			
	MABP Torr	Arterial pH	PaCO2 Torr	PaO2 Torr
(A) Control				
Saline	118±6	7.38±0.08	34.57±0.13	79.67±0.14
Ag NPs	120±8	7.36±0.07	34.63±0.08	79.89±0.12
Cu NPs	124±7*	7.34±0.05	34.73±0.14	79.87±0.12
(B) 4 h heats stress 38°C				
Heats stress 4 h	75±6*	7.34±0.06	35.67±0.14*	80.24±0.16*
HS+Ag NPs	78±8*	7.32±0.13	35.62±0.11*	80.18±0.14*
HS+Cu NPs	76±7*	7.30±0.02*	35.54±0.12*	80-36±0.12*
HS+CBL 2.5 mL				
HS+CBL 5 mL	84±6*	7.34±0.12	34.95±0.07#	79.93±0.06#
HS+NWCBL 2.5 mL	88±6*	7.34±0.13	34.87±0.12#	80.05±0.14
HS+LVRTM 44 mg	70±8*	7.32±0.14	35.76±0.12*	80.43±0.17*
HS+LVRTM 88 mg	74±6*	7.33±0.12	35.65±0.14*	80.28±0.14*
HS+PGBL 200 mg	78±8*	7.31±0.08*	35.54±0.10*	80.34±0.13*
HS+PGBL 400 mg	75±6*	7.30±0.21	35.14±0.21*	80.38±0.11*
HS+TPRM 40 mg	75±8*	7.31±0.12	35.16±0.08*	80.30±0.12*
HS+TPRM 80 mg	77±6*	7.30±0.07*	35.23±0.10*	80.34±0.13*
HS+VLPRT 400 mg	78±9*	7.32±0.08	35.24±0.08*	79.98±0.17
HS+VLPRT 800 mg	80±5*	7.34±0.12	35.20±0.14	79.87±0.16
(C) Ag NPs+4 h heat stress				
HS+NWCBL 2.5 mL	82±6*	7.35±0.05	34.98±0.08	79.89±0.06
HS+LVRTM 88 mg	74±5*	7.31±0.12	35.23±0.18*	80.56±0.15
HS+PGBL 400 mg	70±7*	7.30±0.18	35.28±0.16	80.67±0.17
HS+TPRM 80 mg	75±8*	7.32±0.19	35.81±0.12*	80.32±0.17
HS+VLPRT 800 mg	76±6*	7.32±0.34	35.56±0.15	80.17±0.15

In this group the drugs were given on the fifth day for 3 days and all parameters were examined 24 h after the last injection of the drug.
Values are mean±SD of 6–8 rats at each point. CBL, cerebrolysin; NWCBL, *nanowired cerebrolysin.*
*P < 0.05 *from saline control;* #P < 0.05 *from experimental control, ANOVA followed by Dunnett's test for multiple group comparison from one control group.*

3.3 Blood-brain barrier breakdown

The BBB breakdown to EBA and radioiodine (Fig. 2 and Table 1) showed significant increase from the control value after 4 h heat stress as compared to the saline treated group. In general the radioiodine leakage as slightly higher following 4 h heat stress group as compared to the EBA leakage under identical conditions. Treatment with Ag or Cu NPs further increased the leakage of EBA and radioiodine in the brain after heat stress (Table 1 and Fig. 2). NPs treatment alone in control group significantly increased the leakage of EBA and radioiodine extravasation in the brain (Table 1 and Fig. 2).

The leakage of EBA was seen on the cingulate, parietal and occipital cortices in moderately deep stain while frontal, temporal and piriform cortices show mild blue staining. Looking at mid sagittal sections of the heat stressed brain exhibited penetration of dye in the ventricles indicating disruption of the blood-CSF-barrier (BCSFB) as well. The subcortical areas that exhibited blue staining included, hippocampus, caudate nucleus, massa intermedia, thalamus and hypothalamus. Interestingly, cerebellar vermis and cerebellar cortices also took mild to moderate blue staining. The brain stem region, colliculi and reticular activating system were also stained mild blue. This suggests that 4 h heat stress induces global penetration of EBA dye in the brain.

NPs intoxication has further enhanced the EBA penetration in similar areas in moderately deep blue as compared to the without NPs rats subjected to heat stress (results not shown).

Leakage of radioactivity was slightly further extended into the same brain regions showing blue staining. However, some areas do not show blue staining was also exhibited radioactivity leakage. This indicates that radioiodine permeability in 4 h heat stress is more widespread than the EBA in the brain. NPs administration has further exacerbated the BBB leakage of radioiodine in the brain (Fig. 3).

3.4 Brain edema

Heat stress induces brain edema formation. Thus, measurement of brain water content in 4 h heat stress significantly enhanced brain water from the control group (Table 1). NPs intoxication further aggravated this brain water content in heat stressed rats indicating the brain edema formation was enhanced in heat stress after NPs Cu or Ag intoxication (Table 1).

Volume swelling calculated from the difference between control and heat stress brain revealed about 18% increase in brain swelling after 4 h heat stress. This brain swelling was further exacerbated by Ag NPs to 21% and following Cu NPs to 20% as compared to the control group. Interestingly, in normal animals Ag or Cu NPs also enhanced brain swelling by 0.5% each from the control group. This suggests that NPs intoxication aggravates brain edema formation and volume swelling (Table 1).

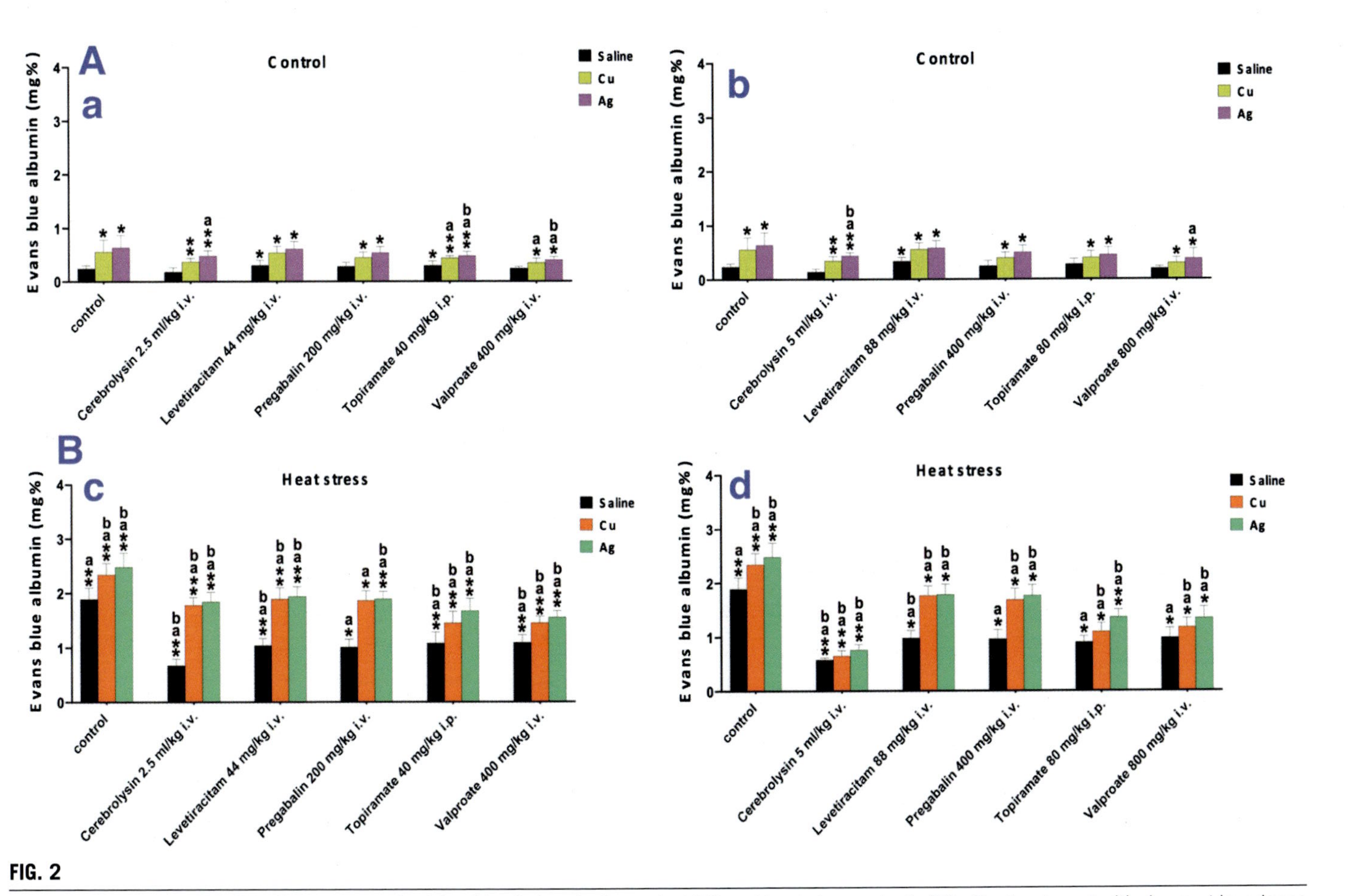

FIG. 2

Evans blue albumin (EBA) leakage across the blood-brain barrier in control (A) and heat stressed rats (B) after treatment with drugs at low doses (a, c) and high doses (b, d) intoxicated with engineered metal nanoparticles Cu or Ag (50–60 nm, 50 mg/kg, i.p. once daily for 7 days). Animals were heat exposed on the 8th day in a BOD (Biological Oxygen Demand) incubator at 38 °C for 4 h. For details see text. Values are Mean ± SD from 6 to 8 rats at each point. *$P<0.05$; **$P<0.01$ significantly different from saline control group, Over the bars symbol $^{a}P<0.05$, significantly different from saline control from same group; Over the bars symbol $^{b}P<0.05$ significantly different from respective control group, ANOVA followed by Dunnett's test from one control group.

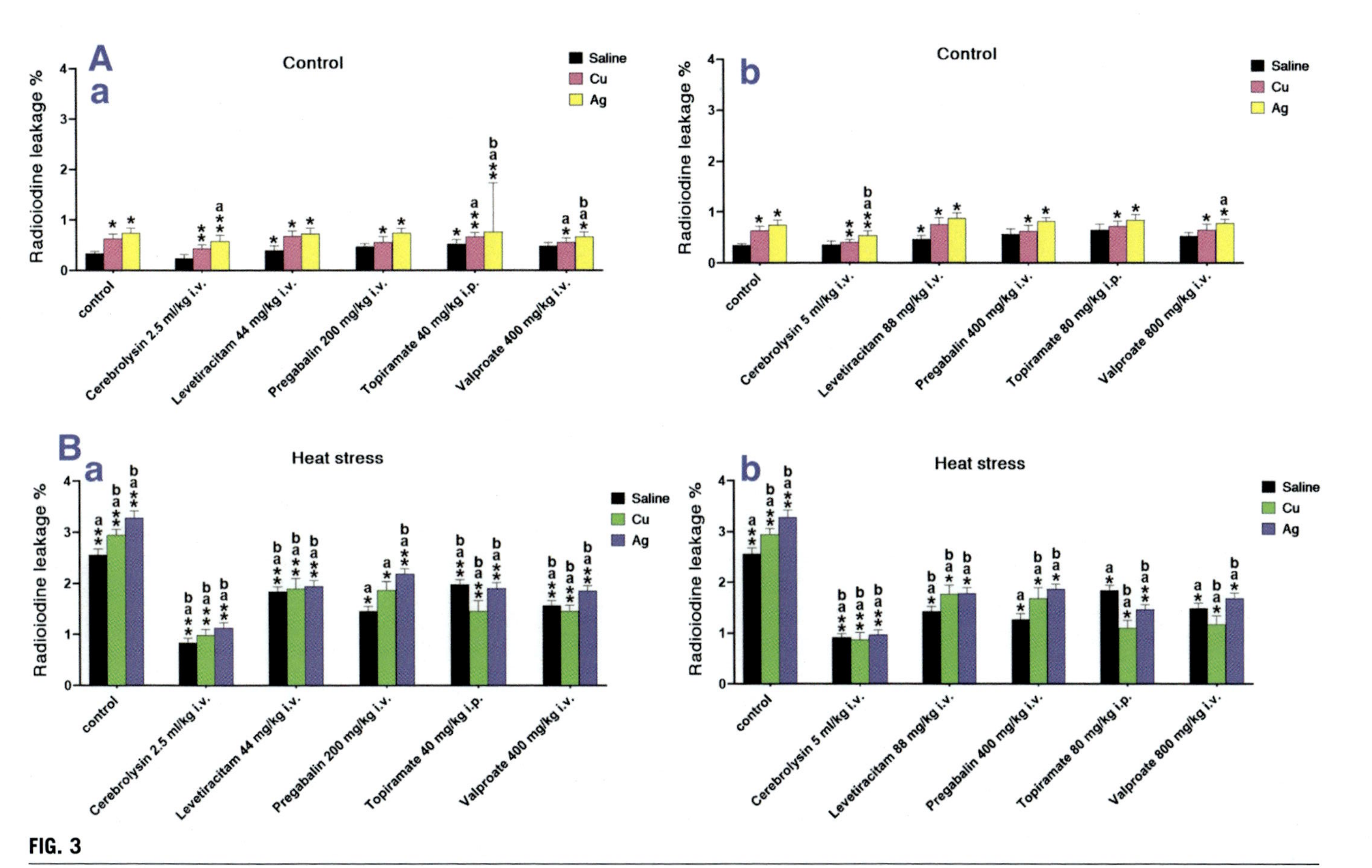

FIG. 3

Radioiodine ([131]-Iodine) leakage across the blood-brain barrier in control (A) and heat stressed rats (B) after treatment with drugs at low doses (a, c) and high doses (b, d) intoxicated with engineered metal nanoparticles Cu or Ag (50–60 nm, 50 mg/kg, i.p. once daily for 7 days). Animals were heat exposed on the 8th day in a BOD (Biological Oxygen Demand) incubator at 38 °C for 4 h. For details, see text. Values are Mean ± SD from 6 to 8 rats at each point. $^{*}P<0.05$; $^{**}P<0.01$ significantly different from saline control group, Over the bars symbol $^{a}P<0.05$, significantly different from saline control from same group; Over the bars symbol $^{b}P<0.05$ significantly different from respective control group, ANOVA followed by Dunnett's test from one control group.

3.5 Cerebral blood flow

Heat stress significantly reduces CBF (Table 1). Thus, subjection of rats to 4h heat stress resulted in decline of CBF to 0.76 ± 0.08 mL/g/min from the control value of 1.25 mL/g/min ($P < 0.05$). This decrease was further exacerbated in heat stress following Cu or Ag NPs intoxication (Table 1). Interestingly NPs intoxication in saline treated control groups also showed significant but mild decrease in normal animals (Table 1). This suggests that CBF is significantly reduced by heat stress and NPs treatment further aggravated this reduction in heat stress (Fig. 4).

3.6 Oxidative stress parameters

We measured oxidative parameters in heat stress and their modification with NPs intoxication. The oxidative stress parameters measured include Glutathione, Malondialdehyde, Lucigenin, Luminol and Myeloperoxidse (Figs. 5–9). Glutathione is needed for antioxidant defense whereas all other oxidative stress parameters including malondialdehyde, myeloperoxidase, luminol and lucigenin activate reactive oxygen species and induce oxidative stress induced cellular damage (Frijhoff et al., 2015; Jacobs et al., 2020; Šćepanović et al., 2018).

Our observations show that following 4h heat stress the antioxidative stress marker glutathione significantly reduced from the control level (Figs. 5–9). Whereas, the other oxidants such as myeloperoxidase, malondialdehyde, luminol and lucigenin are significantly enhanced after heat stress as compared to the control group (Figs. 5–9).

Interestingly, NPs intoxication further exacerbates these increases in myeloperoxidase, malondialdehyde, luminol and lucigenin after heat stress. This was the most pronounced by Ag NPs intoxication in heat stress (Figs. 5–9). On the other hand, NPs treatment in heat stress significantly exacerbated the decrease of glutathione level that was the most pronounced by Ag NPs treatment as compared to the Cu NPs intoxication (Figs. 5–9). These observations suggest that NPs intoxication in heat stress exacerbates oxidative stress and reduces antioxidative defense mechanisms.

3.7 Brain pathology

Heat stress induced marked brain pathology seen at light microscopy and transmission electron microscopy (TEM) (Sharma and Hoopes, 2003; Sharma et al., 1991a,b, 1992a,b, 1998a,b). Our results further show significant brain pathology in the cerebral cortex, hippocampus, cerebellum, thalamus, hypothalamus, brain stem and spinal cord by light and electron microscopy (Figs. 10–13; results not shown).

Interestingly, NPs intoxication further exacerbated the brain pathology in heat stress both at light microscopy (Fig. 14) and at TEM (Fig. 15). Ag NPs appears to have exacerbated brain damage in heat stress as compared to the Cu NPs. Thus, widespread degeneration of synaptic cleft, axons, dendrites and neuronal cells in the cerebral cortex and hippocampus is seen after Ad or Cu NPs intoxication (Fig. 14).

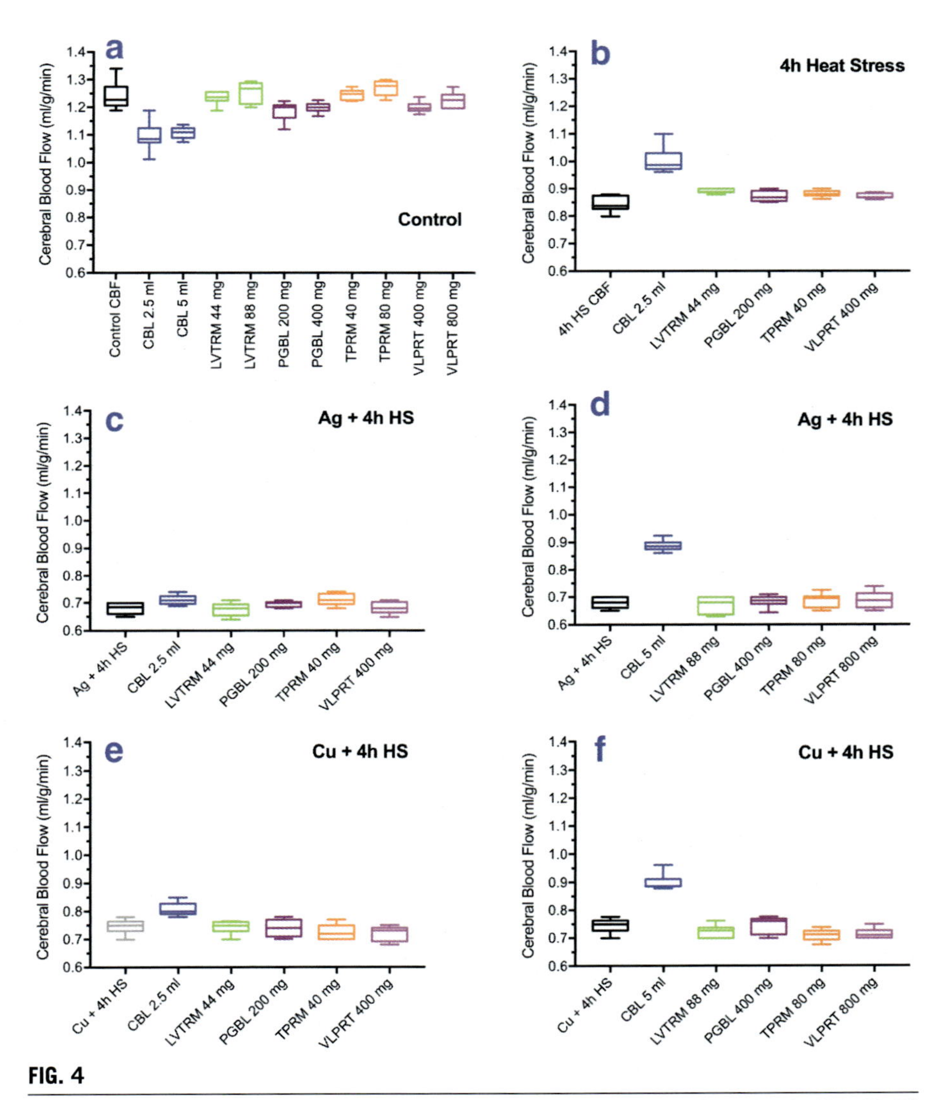

FIG. 4

Changes in cerebral blood flow (CBF) in control (A) heat stress (B) and following Ag NPs (C, D) or Cu NPs (E, F) with cerebrolysin (CBL), levetiracetam (LVTRM), pregabalin (PGBL), topiramate (TPRM) and valproate (VLPRT) in low doses (C, E) and in high doses (D, F) treatment. CBL treatment in heat stress has superior effects on CBF in heat stress. For details see text.

Ag TEM also these cellular damages, vacuolation, axonal degeneration and neuronal degeneration were present (Fig. 15). These degenerating changes appear more prominent in Ag NPs intoxication in heat stress as compared to the Cu NPs treatment (Figs. 10 and 13).

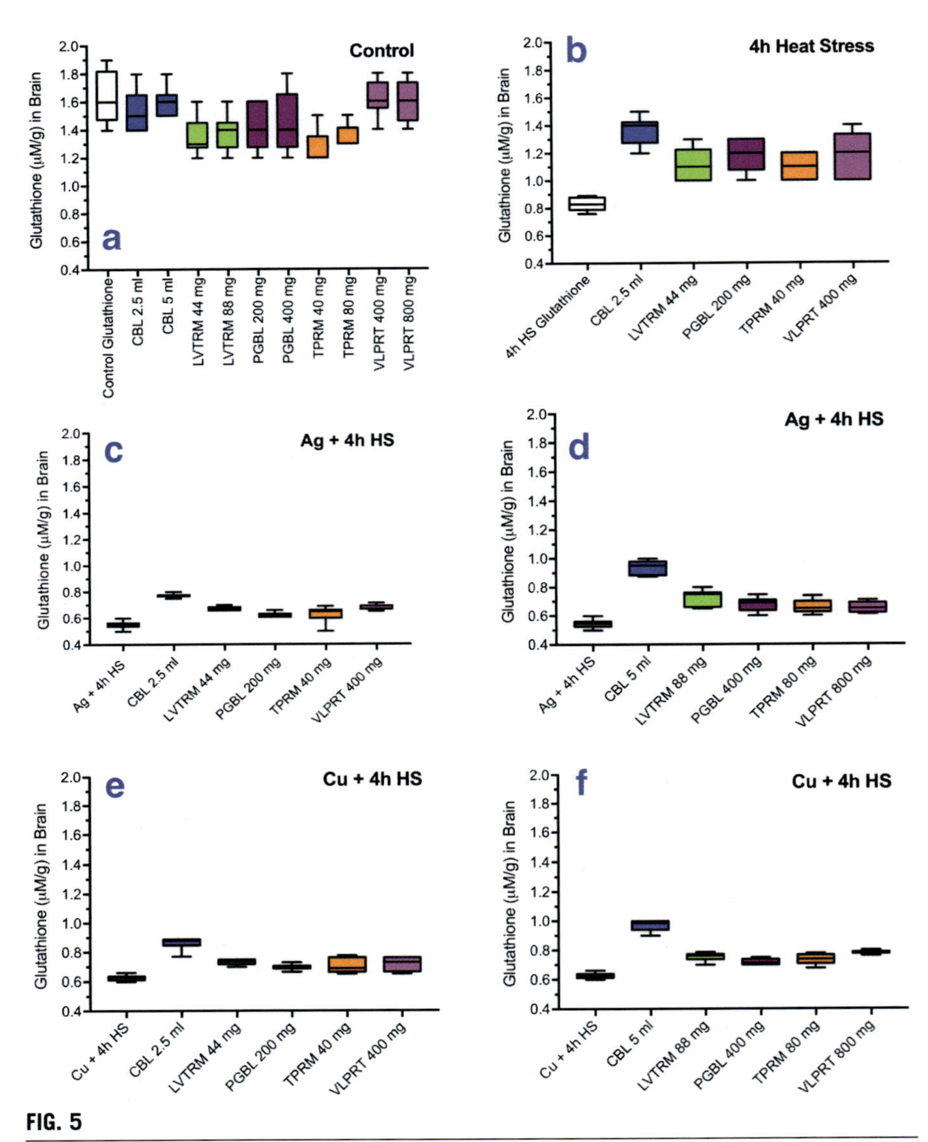

FIG. 5

Changes Glutathione levels in control (A) heat stress (B) and following Ag NPs (C, D) or Cu NPs (E, F) with cerebrolysin (CBL), levetiracetam (LVTRM), pregabalin (PGBL), topiramate (TPRM) and valproate (VLPRT) in low doses (C, E) and in high doses (D, F) treatment. CBL treatment in heat stress has superior effects on Glutathione in maintaining high levels in heat stress. For details see text.

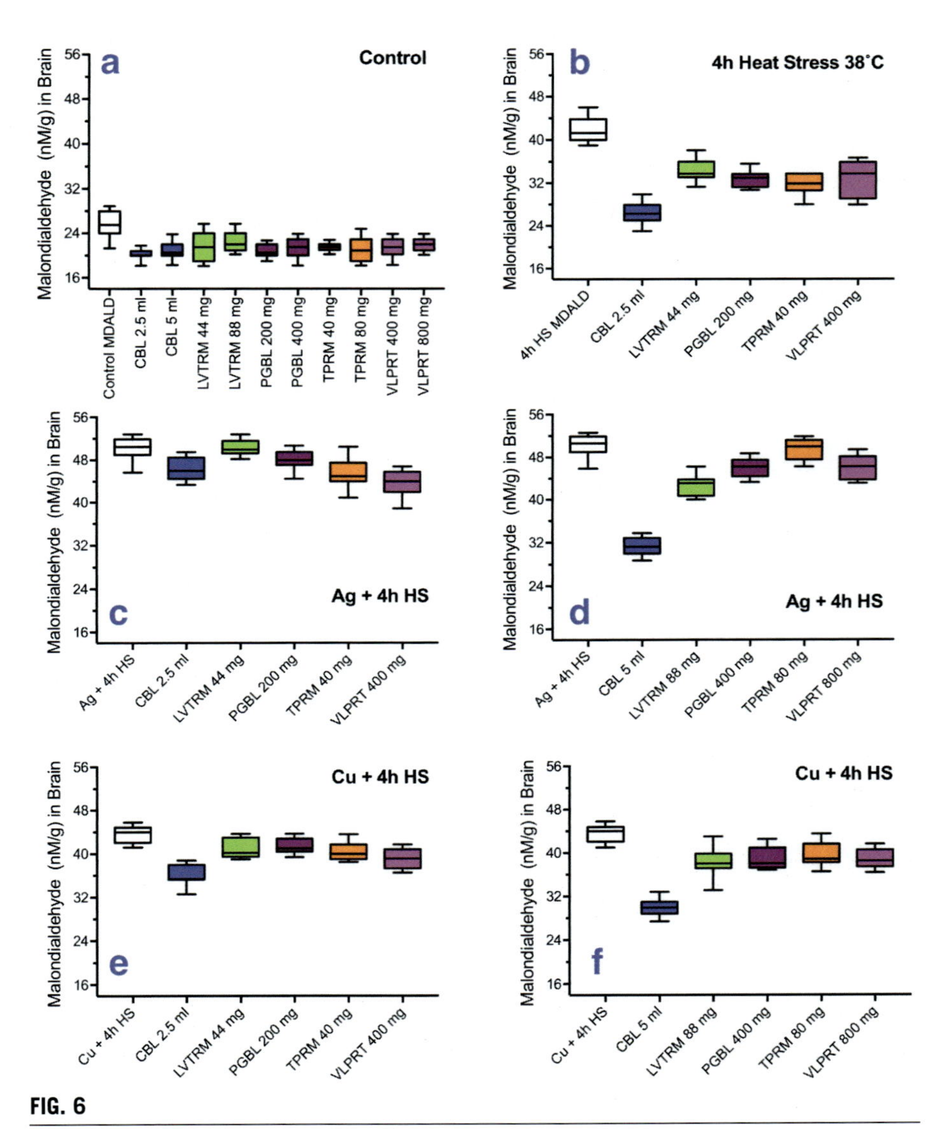

FIG. 6

Changes Malondialdehyde levels in control (A) heat stress (B) and following Ag NPs (C, D) or Cu NPs (E, F) with cerebrolysin (CBL), levetiracetam (LVTRM), pregabalin (PGBL), topiramate (TPRM) and valproate (VLPRT) in low doses (C, E) and in high doses (D, F) treatment. CBL treatment in heat stress has superior effects in lowering Malondialdehyde levels in heat stress. For details see text.

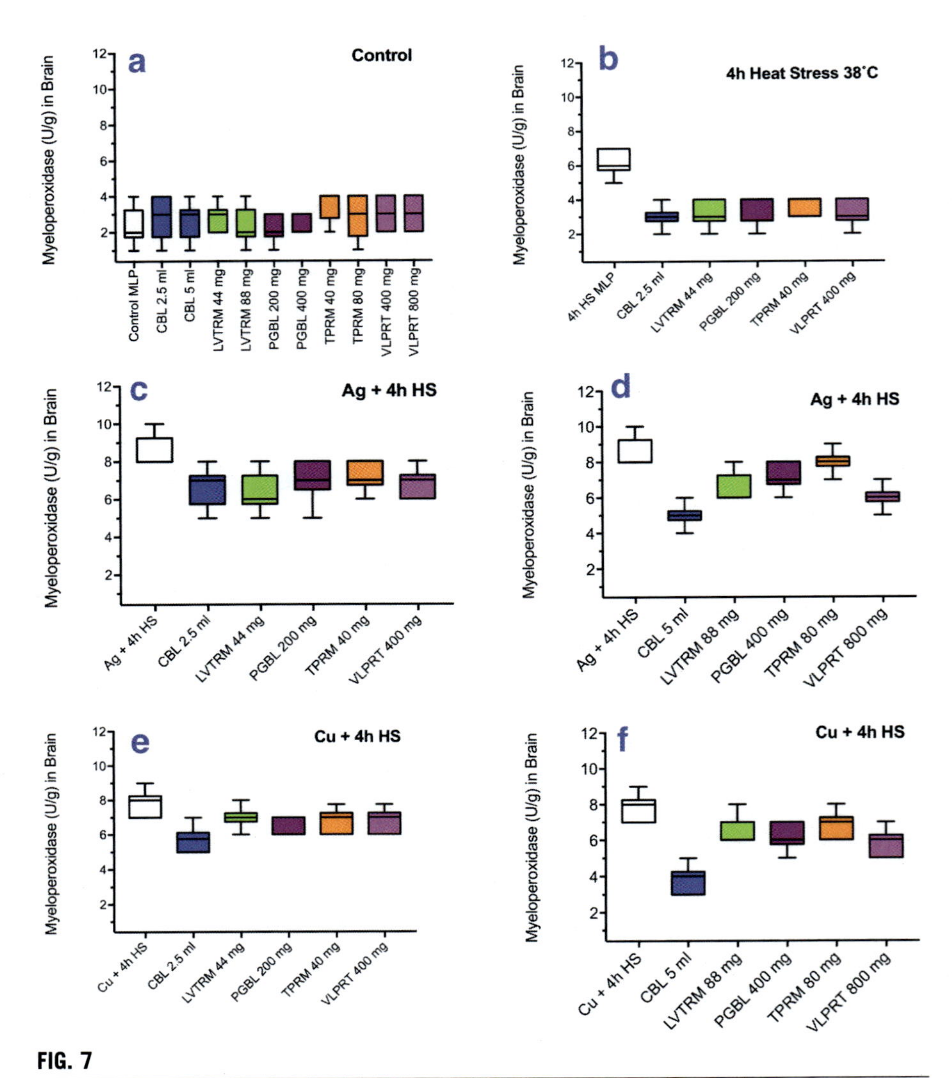

FIG. 7

Changes Myeloperoxidase levels in control (A) heat stress (B) and following Ag NPs (C, D) or Cu NPs (E, F) with cerebrolysin (CBL), levetiracetam (LVTRM), pregabalin (PGBL), topiramate (TPRM) and valproate (VLPRT) in low doses (C, E) and in high doses (D, F) treatment. CBL treatment in heat stress has superior effects in lowering Myeloperoxidase levels in heat stress. For details see text.

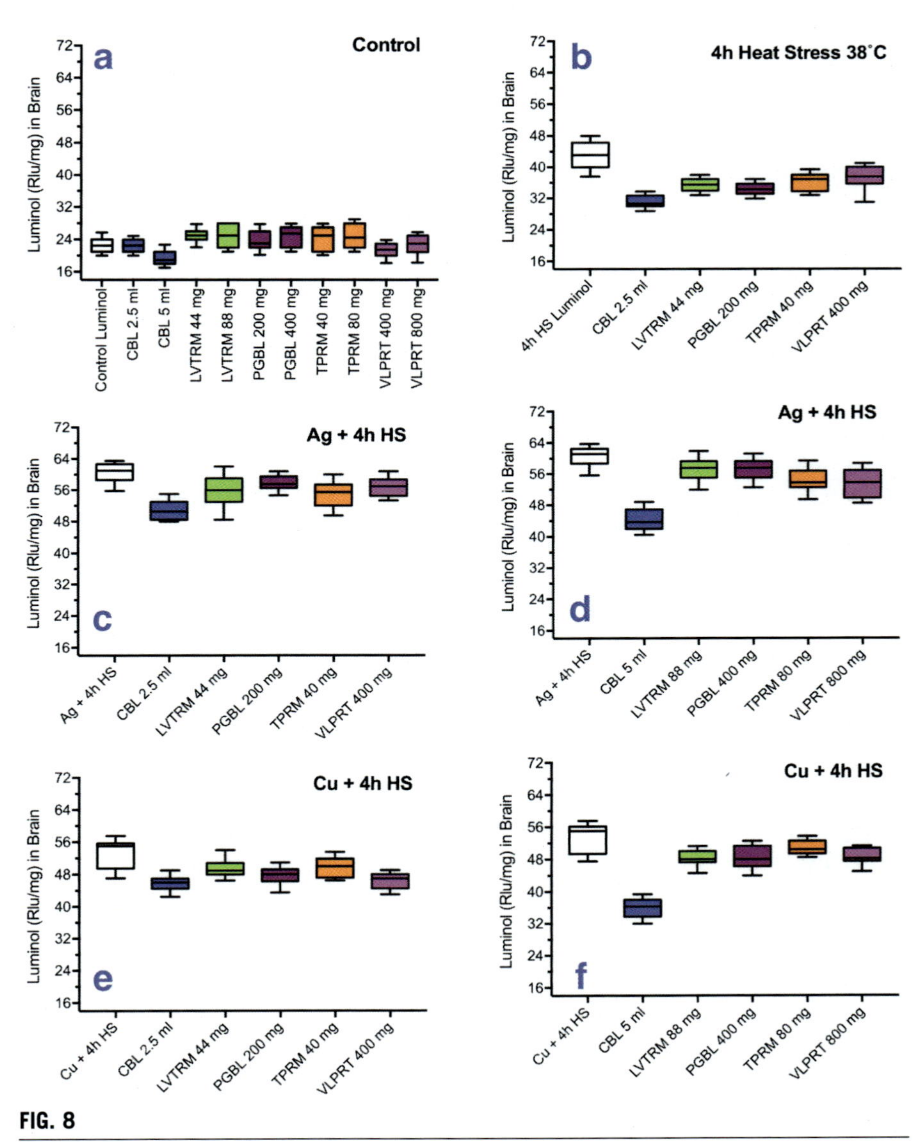

FIG. 8

Changes Luminol levels in control (A) heat stress (B) and following Ag NPs (C, D) or Cu NPs (E, F) with cerebrolysin (CBL), levetiracetam (LVTRM), pregabalin (PGBL), topiramate (TPRM) and valproate (VLPRT) in low doses (C, E) and in high doses (D, F) treatment. CBL treatment in heat stress has superior effects in lowering Luminol levels in heat stress. For details see text.

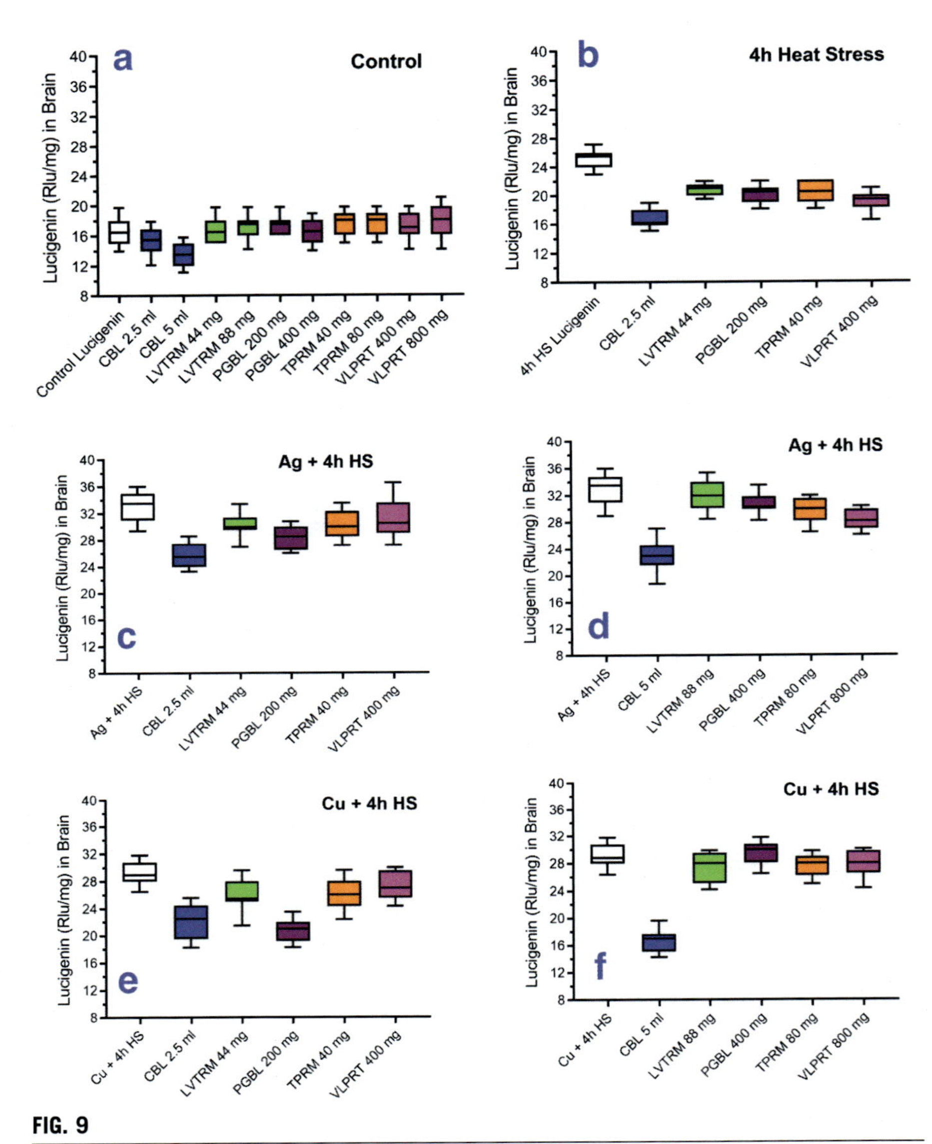

FIG. 9

Changes Lucigenin levels in control (A) heat stress (B) and following Ag NPs (C, D) or Cu NPs (E, F) with cerebrolysin (CBL), levetiracetam (LVTRM), pregabalin (PGBL), topiramate (TPRM) and valproate (VLPRT) in low doses (C, E) and in high doses (D, F) treatment. CBL treatment in heat stress has superior effects in lowering Lucigenin levels in heat stress. For details see text.

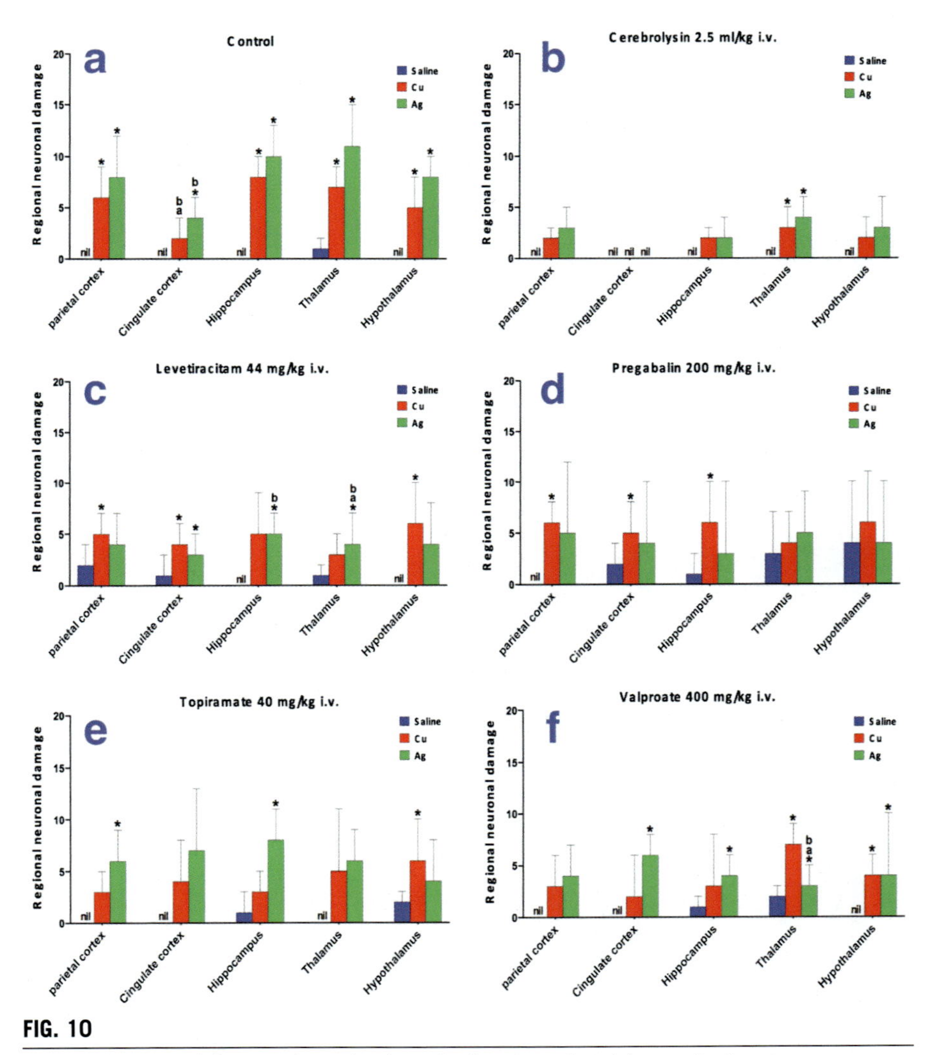

FIG. 10

Regional neuronal damage in parietal cerebral cortex, cingulate cerebral cortex, hippocampus, thalamus and hypothalamus in control (A) following low dose treatment with cerebrolysin (B), Levetiracetam (C), Pregabalin (D), Topiramate (E) and Valproate (F) following 4h heat stress in saline treated and Ag or Cu NPs intoxicated rats. Values are Mean ± SD of 6–8 rats at each point. $*P<0.05$ from saline treated group. Over the bars symbol $^{a}P<0.05$, significantly different from saline control from same group; Over the bars symbol $^{b}P<0.05$ significantly different from respective control group, ANOVA followed by Dunnett's test from one control group.

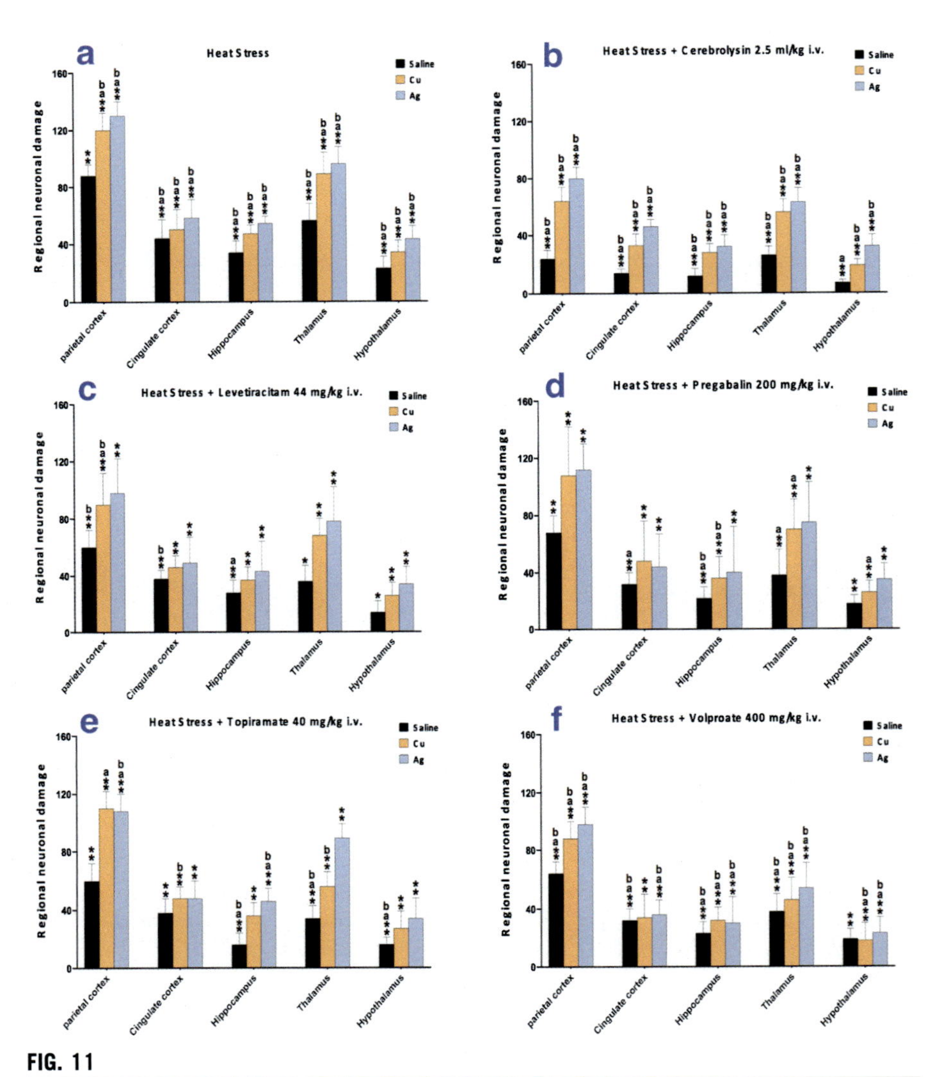

FIG. 11

Regional neuronal damage in parietal cerebral cortex, cingulate cerebral cortex, hippocampus, thalamus and hypothalamus in 4 h heat stress at 38 °C control (A) following low dose treatment with cerebrolysin (B), Levetiracetam (C), Pregabalin (D), Topiramate (E) and Valproate (F) following 4 h heat stress in saline treated and Ag or Cu NPs intoxicated rats. Values are Mean ± SD of 6 to 8 rats at each point. $*P < 0.05$ from saline treated group. Over the bars symbol $^{a}P < 0.05$, significantly different from saline control from same group; Over the bars symbol $^{b}P < 0.05$ significantly different from respective control group, ANOVA followed by Dunnett's test from one control group.

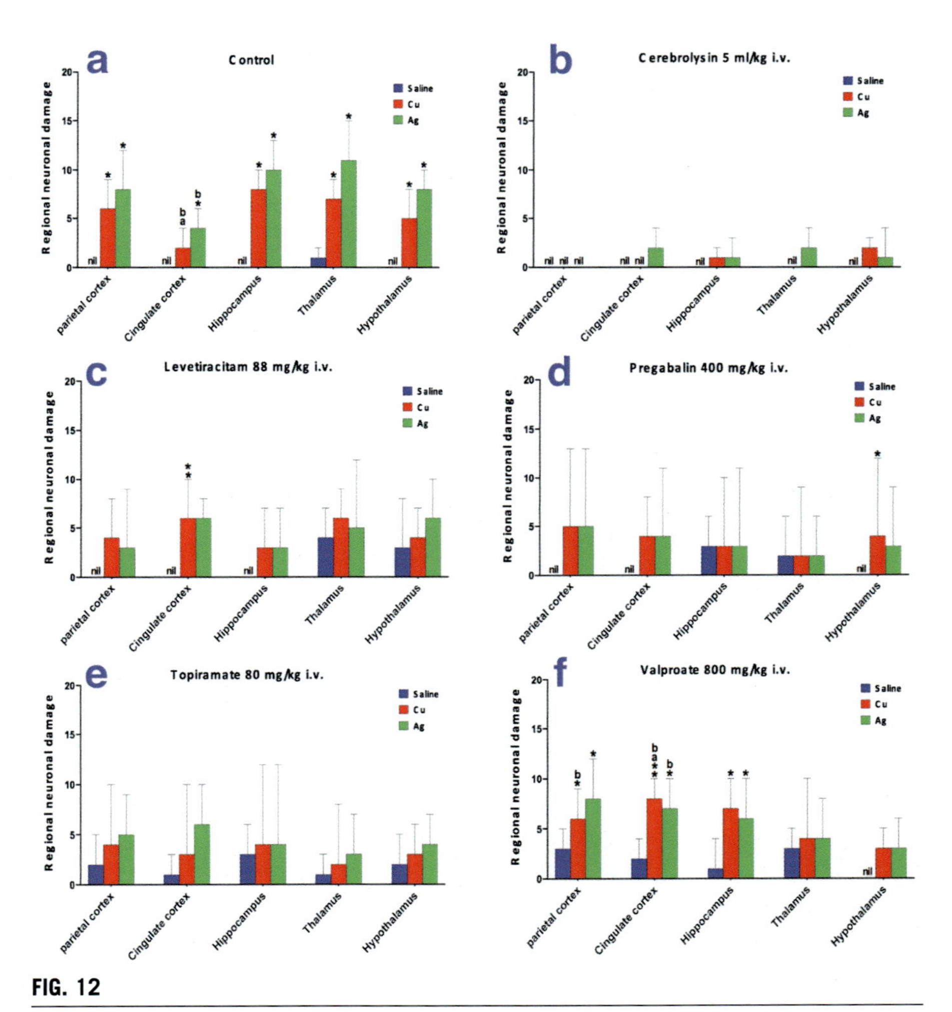

FIG. 12

Regional neuronal damage in parietal cerebral cortex, cingulate cerebral cortex, hippocampus, thalamus and hypothalamus in control (A) following high dose treatment with cerebrolysin (B), Levetiracetam (C), Pregabalin (D), Topiramate (E) and Valproate (F) following 4h heat stress in saline treated and Ag or Cu NPs intoxicated rats. Values are Mean ± SD of 6–8 rats at each point. $*P<0.05$ from saline treated group. Over the bars symbol $^{a}P<0.05$, significantly different from saline control from same group; Over the bars symbol $^{b}P<0.05$ significantly different from respective control group, ANOVA followed by Dunnett's test from one control group.

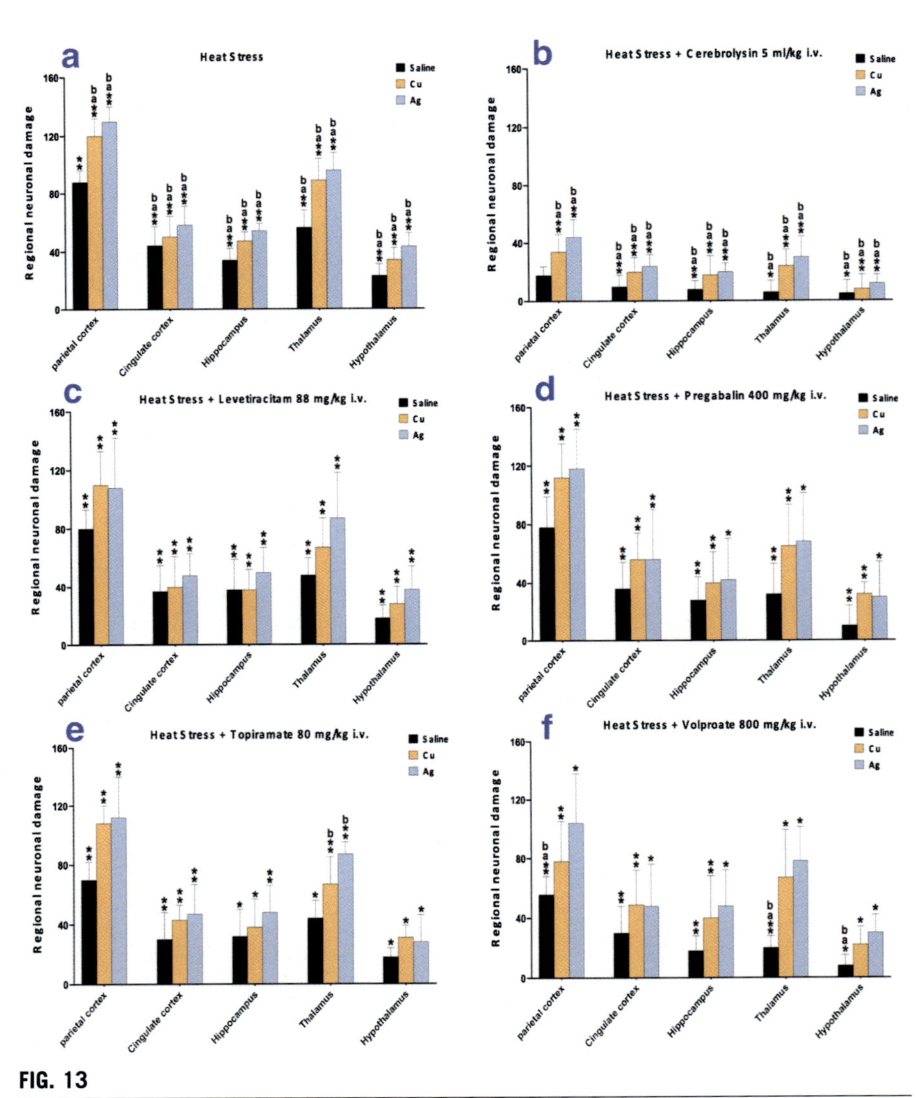

FIG. 13

Regional neuronal damage in parietal cerebral cortex, cingulate cerebral cortex, hippocampus, thalamus and hypothalamus in 4h heat stress at 38°C control (A) following high dose treatment with cerebrolysin (B), Levetiracetam (C), Pregabalin (D), Topiramate (E) and Valproate (F) following 4h heat stress in saline treated and Ag or Cu NPs intoxicated rats. Values are Mean ± SD of 6–8 rats at each point. $*P < 0.05$ from saline treated group. Over the bars symbol $^{a}P < 0.05$, significantly different from saline control from same group; Over the bars symbol $^{b}P < 0.05$ significantly different from respective control group, ANOVA followed by Dunnett's test from one control group.

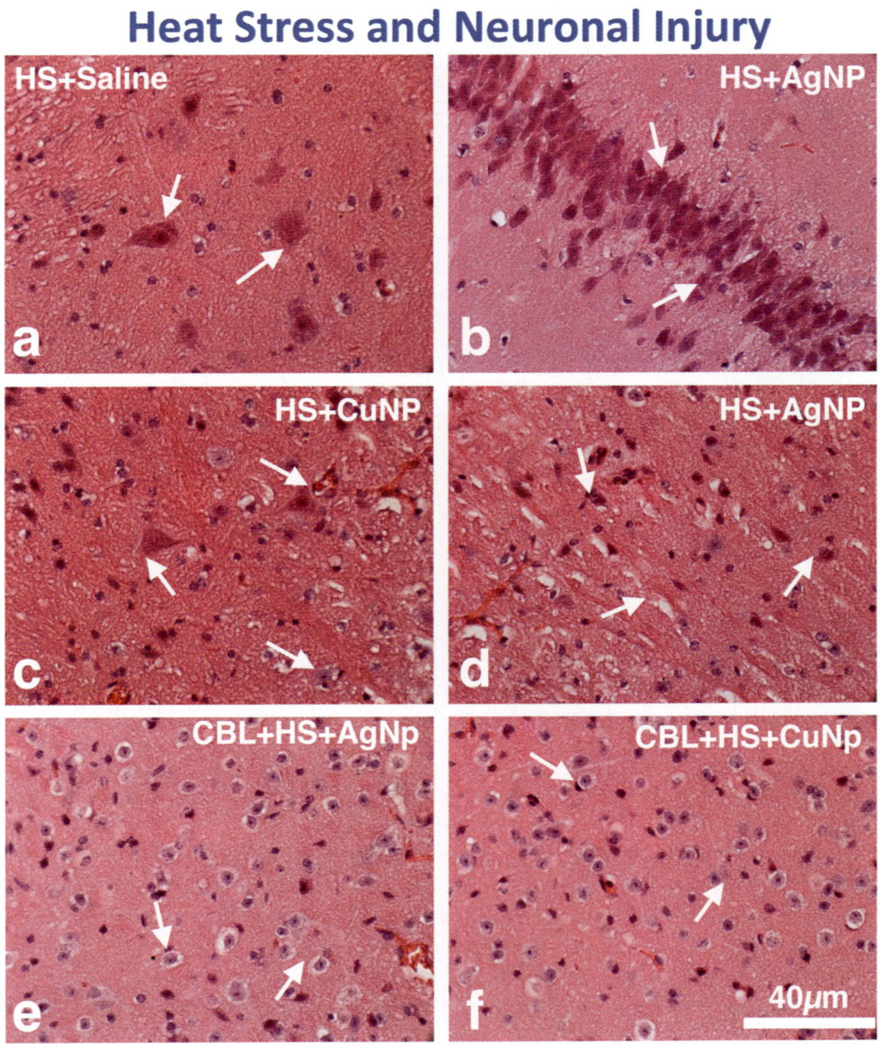

Heat Stress and Neuronal Injury

HS+Saline a

HS+AgNP b

HS+CuNP c

HS+AgNP d

CBL+HS+AgNp e

CBL+HS+CuNp f

40μm

FIG. 14

A representative example of light micrograph of cerebral cortex (A), hippocampus (B) in heat stress intoxicated with Ag (B, D) or Cu (C) NPs showing profound neuronal injury (arrows) with dark appearance of neurons, sponginess and edematous expansion of neuropil. Ag or Cu NPs exacerbated heat stress induced neuronal pathologies (B–D). Treatment with cerebrolysin (CBL) reduces neuronal injury in Ag (E) or Cu (F) NPs intoxication induced exacerbation of brain pathology in heat stress. Only few dark neurons with perineuronal edema (arrows) are seen in CBL treated group with compact neuropil in the cerebral cortex (E, F). Bar = 40 μm. Paraffin sections 3-μm, H&E stain.

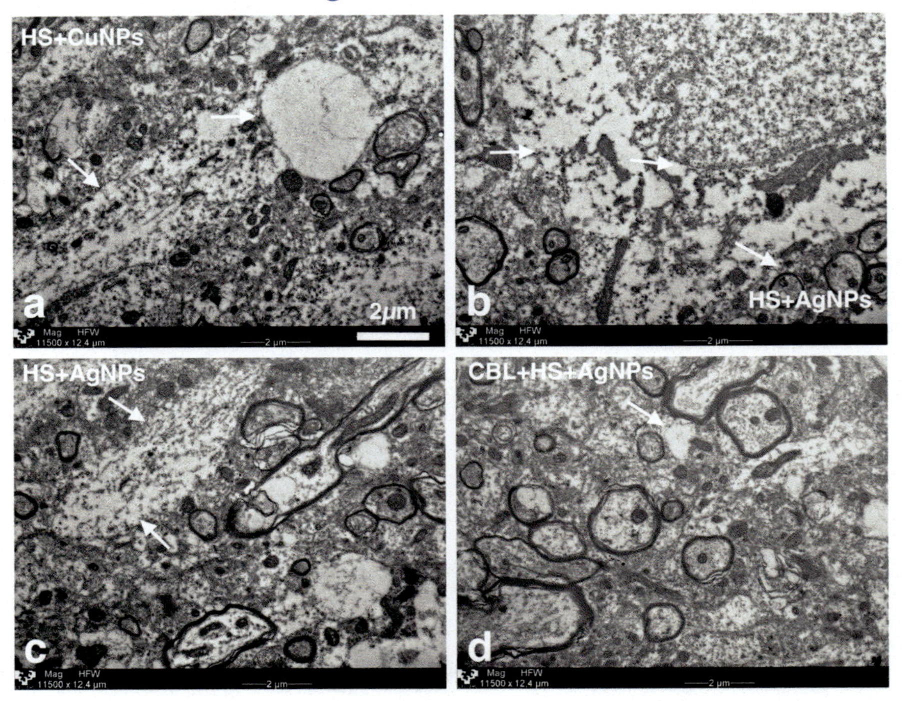

FIG. 15

A representative example of transmission electron micrograph (TEM) from cerebral cortex of 4h heat stressed at 38°C rat intoxicated with Cu (A) or Ag (B, C) NPs and their modification with cerebrolysin (CBL) treatment (D). Axonal neurodegeneration (A, C) are exacerbated by NPs in heat stress (arrows). Disintegration of cellular nucleus (B) is clearly seen (arrow) in Ag NPs treated heat stressed rat from cerebral cortex. Membrane vacuolation, synaptic damages and axonal or dendritic swelling or degeneration is clearly evident (A–C). Treatment with cerebrolysin (CBL) attenuated heat stress induced brain pathology exacerbated by Ag NPs in the cerebral cortex (D). Deformation of axonal morphology in the cerebral cortex is much less severe and cellular degeneration (arrow) is markedly reduced (D). Bar=2μm.

3.8 Effect of drug treatments on nanoparticles intoxication in heat stroke

We have compared cerebrolysin treatment in heat stress with 4 clinically used drugs in post-stroke treatment. These drugs are levetiracetam (Santana-Gómez et al., 2018), pregabalin (Lee et al., 2021), topiramate (Landucci et al., 2018) and valproate (Faggi et al., 2018). These individual drugs are basically anti-epileptic drugs but all has significant neuroprotective effects in stroke. In general we administered two different doses of drugs that are currently being used in clinical practice in heat stress alone or with Ag or Cu NPs. The results are shown below.

3.8.1 Effect of levetiracetam

Low or high doses of levetiracetam mildly affected body temperature changes and stress symptoms after 4 h heat stress (Table 2 and Fig. 2). The physiological variables were also altered mildly after heat stress alone or intoxication with Ag or Cu NPs (Table 2 and Fig. 2). The BBB breakdown to EBA and radioiodine were slightly reduced (Fig. 2 and Table 1). However, the CBF, brain edema and cell changes were not much different from the heat stress alone or intoxicated with Ag or Cu NPs (Fig. 4 and Tables 1 and 2). The oxidative parameters also did not affect much following heat stress alone and associated with Cu or AG NPs intoxication (Figs. 5–9). Brain pathology in heat stress was also only slightly affected (Figs. 10–13). This suggests that levetiracetam has only limited protective effects on heat stress induced pathology exacerbated with Ag or Cu NPs treatment.

3.8.2 Effect of pregabalin

Effects of low and high doses of pregabalin significantly reduced BBB leakage to EBA and radioiodine in 4 h heat stress and also reduced the extravasation of these tracers in Ag or Cu NPs intoxication following heat stress as compared to the untreated heat stressed group (Figs. 2 and 3 and Table 1). However, the other parameters of tress symptoms, physiological variable, CBF, brain edema and cell changes were only slightly affected (Tables 1 and 2 and Fig. 4). Changes in oxidative stress parameters in heat stress alone or treated with Ag or Cu NPs has only sight beneficial effects in pregabalin treated heat exposed rats (Figs. 5–9). Neuronal pathology in heat stress is also only mildly affected (Figs. 10–13). These observations suggest that pregabalin has only limited beneficial effects in heat stress either alone or following Ag or Cu NPs intoxication in heat stress.

3.8.3 Effect of topiramate

Treatment with topiramate in low or high doses significantly reduced the BBB breakdown to EBA and radioiodine tracers after 4 h heat stress (Figs. 2 and 3). This reduction was also evident in Ag or Cu NPs intoxicated heat stress group (Table 1). However, the other parameters such as CBF, brain edema formation and cell changes were only slightly modified after heat stress alone or with Ag or Cu NPs administration (Fig. 4 and Table 1). The stress symptoms and physiological variables were also showed only mild changes following pregabalin treatment in low or high doses (Tables 1 and 2). The oxidative stress parameters following heat stress either alone or together with Ag or Cu NPs intoxication were only slightly altered by pregabalin as compared to the untreated stressed groups (Figs. 5–9). Neuronal pathologies in heat stress was only moderately affected (Figs. 10–13). These observations support the idea that pregabalin also has very limited neuroprotective ability in heat stress associated with Ag or Cu NPs administration.

3.8.4 Effect of valproate

Treatment with valproate significantly reduced BBB permeability to EBA and radioiodine tracers (Figs. 2 and 3 and Table 1) and edema formation in 4 h heat stress group either alone or intoxicated with Ag or Cu NPs (Table 1). The cell changes were also markedly reduced significantly but its effects on CBF does not appear promising in heat stress alone or intoxicated with Ag or Cu NPs (Fig. 4 and Table 1). The stress symptoms and physiological variables were not greatly affected by valproate treatment in either low or high doses in heat stress associated with Ag or Cu NPs (Tables 1 and 2). The oxidative stress parameters were slightly attenuated after heat stress and treatment with Ag or Cu NPs by valproate treatment (Figs. 5–9). Neuronal pathologies were moderately affected by valproate in heat stress (Figs. 10–13). In general, these observations support that valproate has also limited ability to induce brain protection in heat stress either alone or with NPs treatment.

3.8.5 Effect of cerebrolysin

Treatment with cerebrolysin with two doses significantly attenuated stress symptoms and physiological variables in heat stress associated with Ag or Cu NPs intoxication (Tables 1 and 2). The BBB breakdown to EBA and radioiodine was significantly and most efficiently reduced s compared to all other drug treatment after 4h heat stress (Figs. 2 and 3). Also exacerbation of BBB breakdown by NPs in heat stress was also attenuated depending on the doses used (Figs. 2 and 3 and Tables 1 and 2). The CBF levels declined in heat stress and intoxication with NPs were significantly attenuated by cerebrolysin treatment that was dose dependent (Fig. 4; Table 1). The brain edema and volume swelling is also significantly reduced in cerebrolysin treatment depending on the doses used (Table 1). Significant reduction in oxidative stress parameters is seen in heat stress alone or following NPs treatments (Figs. 5–9).

Interestingly, the cell changes were also markedly reduced after heat stress by cerebrolysin treatment (Figs. 10–13 and Table 1). High doses of cerebrolysin were also able to attenuate cell injury exacerbated by Ag or Cu NPs in heat stress (Figs. 10–13; Table 1). An example of cell changes in the cerebral cortex and hippocampus was show profound upregulation of cell damage in heat stress by intoxication with Ag or Cu NPs (Fig. 14). Treatment with high doses of cerebrolysin attenuated these exacerbated cell injury by NPs in heat stress (Fig. 14). Also ultrastructural changes in neuropil by high doses of cerebrolysin are markedly reduced following heat stress associated with NPs treatment (Fig. 15).

These are possible by cerebrolysin treatment to induce effective neuroprotection possible by the drugs ability to reduce oxidative parameters in a dose related manner (Figs. 5–9). Thus, the oxidants malondialdehyde, myeloperoxidase, luminol and lucigenin were significantly reduced in 4h heat stress with NPs by cerebrolysin treatment depending on the dose of the drug used (Figs. 5–9). These observations clearly support the idea that cerebrolysin in quite effective in reducing heat stress induced brain pathology and its exacerbation by NPs intoxication.

3.8.6 Effects of nanowired cerebrolysin

Based on our previous experiences that nanowired cerebrolysin induces superior neuroprotection in several neurodegenerative diseases (Sharma et al., 2017, 2018, 2019a,b,c,d, 2020a,b), we have used TiO2 nanowired cerebrolysin in 4h heat stress associated with Ag or Cu NPs intoxication. Nanowired cerebrolysin treatment is quite effective in low doses to induce superior neuroprotection (Sharma et al., 2016a,b). Thus, we have used low doses of nanowired cerebrolysin (2.5 or 5 mL doses) in 4 h heat stress associated with Ag or Cu NPs treated group. Our results clearly show that nanowired cerebrolysin in low doses have significant effect on neuroprotection on BBB breakdown, brain edema formation and cerebral blood restoration in heat stress either alone or following Ag or Cu NPs intoxication (Table 1). Also, the stress symptoms and physiological variables were also protected by nanowired cerebrolysin treatment in heat stress associated with Ag or Cu NPs (Table 1). The cell changes at light microscopy and at electron microscopy in 4 h heat stress together with Ag or Cu NPs also significantly reduced by nanowired cerebrolysin treatment (Table 1). The oxidative stress parameters after 4 h heat stress in Ag or Cu NPs intoxication were further significantly attenuated by nanowired cerebrolysin (results not shown, H.S. Sharma, unpublished observations).

These observations further conform that nanowired delivery of cerebrolysin in heat stress either alone or with Ag or Cu NPs significantly attenuated brain pathology on various parameters. This also suggests that cerebrolysin appears to be most potent in reducing brain pathology in heat stress as compared to the currently used other antiepileptic drugs in post-stroke treatment for neuroprotection.

4 Discussion

The present investigation for the first time shows that cerebrolysin either given in high doses or delivered through nanowired technology is able to reduce oxidative stress parameters following heat stress. This is evident from the findings that cerebrolysin treatment enhanced glutathione levels in the brain and reduced the level of malondialdehyde, myeloperoxidase, luminol and lucigenin. This suggests that cerebrolysin is a powerful antioxidant compound in heat stress not reported earlier.

A reduction in oxidative stress is also associated with a reduction in brain ischemia. This is evident with the findings that heat stress significantly reduced CBF by almost 50% after heat stress and cerebrolysin treatment almost restored this global ischemic episode. This suggests that cerebrolysin improves CBF significantly following heat stressed brain ischemia.

When ischemia and oxidative stress are reduced by cerebrolysin this treatment also repaired the BBB breakdown in heat stress (Sharma, 2003). This is evident from the observation that in cerebrolysin treated heat stressed group the breakdown of the BBB to EBA and radioiodine is significantly reduced. This suggests that cerebrolysin is able to maintain BBB permeability to large molecules tracer such as radioiodine and EBA.

EBA and radioiodine are large molecule protein tracers (Kuroiwa et al., 1985; Sharma, 1987, 2005a,b,c,d; Wolman et al., 1981). These tracers bind to serum albumin when introduced into the circulation (Brightman and Reese, 1969; Clasen et al., 1970; Olsson et al., 1971; Sharma, 1987, 2005a,b,c,d; Sun et al., 2021). These tracers when administered into the systemic circulation, they bind with different fractions of serum proteins (Sharma, 1987, 2005a,b,c,d; Sharma and Dey, 1986, 1987;

Sharma et al., 1990). However, these tracers bind largely to serum albumin (Olsson et al., 1971; Wolman et al., 1981). Thus, about 68% albumin is bound to Evans blue and about 52% albumin binds with radioiodine (Sharma et al., 1990). This suggests that the extravasation of EBA and radioiodine in the brain represents the leakage of tracer protein complex (Sharma and Dey, 1986, 1987). Due to differences in albumin binding by EBA and radioiodine the molecular diameter of iodine-proteins complex is smaller than Evan blue-albumin complex (Sharma, 1987). This appears to be one of the main regions that iodine leakage was higher in all regions in control and experimental groups as compared to the EBA tracer (Sharma, 1987).

When the BBB breakdown occurs for large molecular tracers such as EBA-protein complex (MW about 68 kDa) or radioiodine-protein complex (MW about 52 kDa) (Sharma and Dey, 1986, 1987; Sharma et al., 1990) in to the brain results in vasogenic brain edema formation (Hatashita and Hoff, 1990; Katzman et al., 1977; Klatzo, 1967, 1987; Maeda et al., 2021; Nagashima et al., 1990; Rapoport, 1978; Sharma, 2007a,b,c; Sharma et al., 1998). Brain edema leads to volume swelling in close cranial compartment and leads to compression of vital centers within the brain. Massive volume swelling inside the brain could be fatal (Maxeiner and Behnke, 2008; Nehring et al., 2021; Pasantes-Morales and Tuz, 2006). This is one of the reasons that heat stress that results in massive brain edema and volume swelling, death in more than 50% of victims occur (Leyk et al., 2019; Lundesgaard Eidahl et al., 2019; Robles and Cuevas-Solórzano, 2018; von Kummer and Dzialowski, 2017).

In present investigations heat stress resulted in massive volume swelling within the cranium to about 18%. This results in tissue softening and expansion of the brain volume in heat stress (Sharma et al., 1992a,b). There are several mechanisms by which BBB breakdown and leakage of proteins inside brain extra cellular or intracellular microenvironment leads to volume swelling (Sharma et al., 1998). There alterations in osmotic pressure between vascular and cerebral compartment due to shift of proteins from blood to brain compartment. This leads to transfer of water from the vascular compartment to the cerebral compartment due to change in osmotic pressure (Hatashita and Hoff, 1990; Klatzo, 1967; Nagashima et al., 1990; Rapoport, 1978). In addition, breakdown of the BBB opens the barrier for ion and water transport into the brain from vascular compartment (Sharma and Sharma, 2019; Sharma et al., 1998c, 2019a,b). Our results further show that cerebrolysin therapy significantly attenuated brain edema and volume swelling in heat stress. Thus, treatment with cerebrolysin resulted in significant reduction of brain edema and volume swelling by more than 16% in heat stress (Table 1). This shows that cerebrolysin has potent anti-edema effects in heat stress. This anti-edema effect of cerebrolysin is also seen following traumatic brain or spinal cord injury, sleep deprivation, Parkinson's disease and Alzheimer's disease earlier (Ozkizilcik et al., 2019; Sahib et al., 2020; Sharma et al., 2019c,d, 2020a,b).

Reduction in BBB permeability and prevention of oxidative stress by cerebrolysin in heat stress is crucial in reducing the brain edema formation in heat stress associated with brain pathology. A reduction in BBB breakdown in heat stress could occur by cerebrolysin due to enhancing endothelial cell membrane structure in heat stress. This is evident from the findings that several neurotrophic factors present in cerebrolysin are capable of membrane stabilization in heat stress resulting in BBB restoration (Sharma and Johanson, 2007; Sharma et al., 2010a,b,c). Neurotrophins support membrane stabilization and cell membrane repair thus strengthening the BBB function (Gómez-Palacio-Schjetnan and Escobar, 2013; Menon et al., 2012; Sharma, 2005a,b,c,d, 2007a,b). These observations are in line with the idea that cerebrolysin has the capability to strengthen cell membrane and reduces BBB permeability in heat stress and other neurological diseases.

Brain edema, oxidative stress and breakdown of the BBB to plasma proteins induces various immunological reactions and allow several cytokines, neurochemicals and harmful agents to enter into the brain together with edema fluid (Sharma et al., 1993, 1995, 2016a,b). We have previously shown that heat stress induces profound increase in serotonin in plasma and brain (Sharma and Dey, 1986, 1987). Enhanced serotonin level is well known to induced BBB breakdown, reduction in CBF, brain edema formation and oxidative stress induced brain pathology (Lindqvist et al., 2017; Kao and Lin, 1985; Sharma et al., 1990, 1994, 2019a,b,c,d, 2020a,b). Thus, it appears that BBB breakdown is one of the main consequences responsible for brain pathology in heat stress or neurological disease (Sharma, 2004, 2009). The consequences of BBB breakdown are spread of edema fluid and induce cellular injury (Sharma, 2004). The cell injury was normally seen in the areas stained with EBA in brains of heat stress (Sharma and Dey, 1986, 1987). This suggests that breakdown of the BBB is crucial for brain pathology (Sharma et al., 1991a,b, 1992a,b).

In present observations we found that intoxication with engineered NPs from Ag or Cu further exacerbated the brain pathology in heat stress. This observation is well in line with our previous observations of exacerbation of brain pathology in heat stress (Sharma and Sharma, 2007; Sharma et al., 2009, 2011, 2012a,b, 2013). This suggests that NPs exacerbate heat stress induced brain pathology. It is quite likely that NPs itself induce oxidative stress leading to increased BBB breakdown and brain pathology. Increased BBB breakdown to EBA and radioiodine in saline treated normal animals also exhibited in NPs intoxicated rats as compared to naïve rats in this investigation as seen earlier (Sharma et al., 2011). It is likely that additional heat exposure in NPs treated group further enhances the oxidative stress and aggravates brain pathology induced by metal NPs (Sharma and Sharma, 2007; Sharma et al., 2009). In present investigation we observed higher BBB breakdown, increased brain edema and volume swelling together greater reduction in CBF in NPS treated heat stressed rats. This supports our hypothesis that NPs exacerbate brain pathology in heat stress (Sharma et al., 2011, 2012a,b).

Interestingly, NPs induced exacerbation of brain pathology after heat stress was also significantly reduced by cerebrolysin. However, in case of NPs intoxication in heat stress either higher doses of cerebrolysin is need or nanowired delivery of cerebrolysin even in low doses achieved similar neuroprotection. Thus, NWCBL when administered in heat stress group intoxicated with metal NPs significant reduction in BBB breakdown, CBF changes, edema formation and cell changes are seen. This suggests that NWCBL has significant superior neuroprotective effects in NPs induced heat stress.

It appears that nanodelivery of cerebrolysin induces superior neuroprotection in heat stress intoxicated with metal NPs as compared to the conventional cerebrolysin treatment. The possible mechanisms of enhanced neuroprotection by NWCBL are still unclear. However, available evidences support the idea that nanodelivery of cerebrolysin could penetrate the cerebral compartments by making a tiny hole without damaging the cell and spread quickly into the brain parenchyma. When nanowired loaded with cerebrolysin enter into brain fluid microenvironment they release their drug content slowly and effectively over a long period of time (Sharma et al., 2016a,b). Furthermore, nanowired drugs are protected for raid catabolism in the brain microenvironment because of their binding with titanium nanowires. All these factors either alone or in combination enhance superior neuroprotective effects of NWCBL in heat stress.

We also found that NWCBL significantly enhances glutathione level in heat stress intoxicated with metal NPs and reduced the accumulation of malondialdehyde, myeloperoxidase, luminol and lucigenin (unpublished observations). These effects of NWCBL is also associated with restoration of CBF, decrease in BBB breakdown, brain edema formation and brain pathology. This suggests that NWCBL has superior ability to induce neuroprotection on BBB breakdown, brain ischemia,brain edema formation and brain pathology.

On the other hand, the other drugs levetiracetam, pregabalin, topiramate and valproate when given either in low or high doses induces small beneficial effects on some parameters where as other parameters are only lightly affected. These observations suggest that the drugs have only marginal and small neuroprotective ability in heat stress associated with NPs treatment. However, this is still unclear whether these drugs will be delivered through nanowired technique they could do further enhanced neuroprotection in heat stress intoxicated with Ag or Cu NPs. This is a feature correctly being investigated in our laboratory.

As compared to the Ag and Cu NPs we found that Ag NPs induce much more adverse effects in heat stress on various parameters as compared to the Cu NPs. This is possibly due to the inherent properties of Ag and its zeta potentials that induce greater damage in biological tissues as compared to the Cu NPs in identical doses. However, it remains to be seen whether other NPs such as Al, Mn, and Si in identical doses could also have afferent effects in heat stress induced exacerbation of brain pathology. Another way of comparing different NPs in heat stress is to use equimolar doses, a feature that requires additional investigations in heat stress.

In summary, our observations are the first to show that cerebrolysin has superior beneficial effects to thwart oxidative stress in heat stress and restores CBF as compared to other ischemic stroke drugs such as levetiracetam, pregabalin, topiramate and valproate not reported earlier. Reduction in oxidative stress and ischemia in heat stress associated with NPs intoxication appears to play crucial roles in attenuating brain edema formation and brain pathology by reducing the BBB breakdown. NPs intoxicated heat stress. Taken together our observations strongly indicate superior neuroprotective effects of cerebrolysin in heat stroke induced brain pathology together with Ag or Cu Nps intoxication.

Acknowledgments

This investigation is supported by grants from the Air Force Office of Scientific Research (EOARD, London, UK), and Air Force Material Command, USAF, under grant number FA8655-05-1-3065; Grants from the CNCSIS "UEFISCSU, project number PNII" IDEI 787/2007, Romania; EverNeuroPharma, Austria, Alzheimer's Association (IIRG-09-132087), the National Institutes of Health (R01 AG028679) and the Dr. Robert M. Kohrman Memorial Fund (R.J.C.); Swedish Medical Research Council (Nr 2710-HSS), Göran Gustafsson Foundation, Stockholm, Sweden (HSS), Astra Zeneca, Mölndal, Sweden (HSS/AS), Alexander von Humboldt Foundation, Bonn, Germany (HSS), The University Grants Commission, New Delhi, India (HSS/AS), Ministry of Science & Technology, Govt. of India (HSS/AS), Indian Medical Research Council, New Delhi, India (HSS/AS) and India-EU Co-operation Program (R.P./A.S./H.S.S.) and IT-901/16 (J.V.L.), Government of Basque Country and PPG 17/51 (J.V.L.), J.V.L. thanks to the support of the University of the Basque Country (UPV/EHU) PPG 17/51 and 14/08, the Basque Government (IT-901/16 and CS-2203) Basque Country, Spain; and Foundation for Nanoneuroscience and Nanoneuroprotection (FSNN), Romania. Technical assistance of Kärstin Flink, Ingmarie Olsson, Uppsala University and Franzisca Drum, Katja Deparade, Free University Berlin, Germany are highly appreciated. Technical and human support provided by Dr. Ricardo Andrade from SGIker (UPV/EHU) is gratefully acknowledged. Dr. Seaab Sahib is supported by Research Fellowship at University of Arkansas, Fayetteville, AR, USA, by Department of Community Health, Middle Technical University, Wassit, Iraq, and The Higher Committee for Education Development in Iraq, Baghdad, Iraq. We thank Suraj Sharma, Blekinge Inst. Technology, Karlskrona, Sweden and Dr. Saja Alshafeay, University of Arkansas, Fayetteville, AR, USA, affiliated with University of Baghdad, Baghdad IQ, for computer and graphic support. The U.S. Government is authorized to reproduce and distribute reprints for Government purpose notwithstanding any copyright notation thereon. The views and conclusions contained herein are those of the authors and should not be interpreted as necessarily representing the official policies or endorsements, either expressed or implied, of the Air Force Office of Scientific Research or the U.S. Government.

Conflict of interests

The authors have no conflict of interests with any funding agency or entity reported here.

References

Abraira, L., Santamarina, E., Cazorla, S., Bustamante, A., Quintana, M., Toledo, M., Fonseca, E., Grau-López, L., Jiménez, M., Ciurans, J., Luis Becerra, J., Millán, M., Hernández-Pérez, M., Cardona, P., Terceño, M., Zaragoza, J., Cánovas, D., Gasull, T., Ustrell, X., Rubiera, M., Castellanos, M., Montaner, J., Álvarez-Sabín, J., 2020. Blood biomarkers predictive of epilepsy after an acute stroke event. Epilepsia 61 (10), 2244–2253. https://doi.org/10.1111/epi.16648. Epub 2020 Aug 28.

Al Mahri, S., Bouchama, A., 2018. Heatstroke. Handb. Clin. Neurol. 157, 531–545. https://doi.org/10.1016/B978-0-444-64074-1.00032-X.

Alberti, P., Canta, A., Chiorazzi, A., Fumagalli, G., Meregalli, C., Monza, L., Pozzi, E., Ballarini, E., Rodriguez-Menendez, V., Oggioni, N., Sancini, G., Marmiroli, P., Cavaletti, G., 2020. Topiramate prevents oxaliplatin-related axonal hyperexcitability and oxaliplatin induced peripheral neurotoxicity. Neuropharmacology 164, 107905. https://doi.org/10.1016/j.neuropharm.2019.107905. (Epub 2019 Dec 4).

Amiri-Nikpour, M.R., Nazarbaghi, S., Ahmadi-Salmasi, B., Mokari, T., Tahamtan, U., Rezaei, Y., 2014. Cerebrolysin effects on neurological outcomes and cerebral blood flow in acute ischemic stroke. Neuropsychiatr. Dis. Treat. 10, 2299–2306. https://doi.org/10.2147/NDT.S75304. eCollection 2014.

Belcastro, V., Pierguidi, L., Tambasco, N., 2011. Levetiracetam in brain ischemia: clinical implications in neuroprotection and prevention of post-stroke epilepsy. Brain Dev. 33 (4), 289–293. https://doi.org/10.1016/j.braindev.2010.06.008. (Epub 2010 Jul 13).

Brightman, M.W., Reese, T.S., 1969. Junctions between intimately apposed cell membranes in the vertebrate brain. J. Cell Biol. 40, 648–677. https://doi.org/10.1083/jcb.40.3.648.

Bruchim, Y., Aroch, I., Nivy, R., Baruch, S., Abbas, A., Frank, I., Fishelson, Y., Codner, C., Horowitz, M., 2019. Impacts of previous heatstroke history on physiological parameters eHSP72 and biomarkers of oxidative stress in military working dogs. Cell Stress Chaperones 24 (5), 937–946. https://doi.org/10.1007/s12192-019-01020-z. (Epub 2019 Aug 11).

Chang, C.K., Chang, C.P., Liu, S.Y., Lin, M.T., 2007. Oxidative stress and ischemic injuries in heat stroke. Prog. Brain Res. 162, 525–546. https://doi.org/10.1016/S0079-6123(06)62025-6.

Chen, S.H., Lin, M.T., Chang, C.P., 2013. Ischemic and oxidative damage to the hypothalamus may be responsible for heat stroke. Curr. Neuropharmacol. 11 (2), 129–140. https://doi.org/10.2174/1570159X11311020001.

Cheng, Y.C., Sheen, J.M., Hu, W.L., Hung, Y.C., 2017. Polyphenols and oxidative stress in atherosclerosis-related ischemic heart disease and stroke. Oxid. Med. Cell. Longev. 2017, 8526438. https://doi.org/10.1155/2017/8526438. (Epub 2017 Nov 26).

Cichoń, N., Rzeźnicka, P., Bijak, M., Miller, E., Miller, S., Saluk, J., 2018. Extremely low frequency electromagnetic field reduces oxidative stress during the rehabilitation of post-acute stroke patients. Adv. Clin. Exp. Med. 27 (9), 1285–1293. https://doi.org/10.17219/acem/73699.

Clasen, R.A., Pandolfi, S., Hass, G.M., 1970. Vital staining, serum albumin and the blood-brain barrier. J. Neuropathol. Exp. Neurol. 29, 266–284.

Dahle, L.K., Hill, E.G., Holman, R.T., 1962. The thiobarbituric acid reaction and the autoxidations of polyunsaturated fatty acid methylesters. Arch. Biochem. Biophys. 98, 53–261.

Delpont, B., Blanc, C., Osseby, G.V., Hervieu-Bègue, M., Giroud, M., Béjot, Y., 2018. Pain after stroke: a review. Rev. Neurol. (Paris) 174 (10), 671–674. https://doi.org/10.1016/j.neurol.2017.11.011. (Epub 2018 Jul 24).

Dey, P.K., Sharma, H.S., Rao, K.S., 1980. Effect of indomethacin (a prostaglandin synthetase inhibitor) on the permeability of blood-brain and blood-CSF barriers in rat. Indian J. Physiol. Pharmacol. 24 (1), 25–36.

Elliott, K.A., Jasper, H., 1949. Measurement of experimentally induced brain swelling and shrinkage. Am. J. Physiol. 157 (1), 122–129. https://doi.org/10.1152/ajplegacy.1949.157.1.122.

Epstein, Y., Yanovich, R., 2019. Heatstroke. N. Engl. J. Med. 380 (25), 2449–2459. https://doi.org/10.1056/NEJMra1810762.

Faggi, L., Pignataro, G., Parrella, E., Porrini, V., Vinciguerra, A., Cepparulo, P., Cuomo, O., Lanzillotta, A., Mota, M., Benarese, M., Tonin, P., Annunziato, L., Spano, P., Pizzi, M., 2018. Synergistic association of valproate and resveratrol reduces brain injury in ischemic stroke. Int. J. Mol. Sci. 19 (1), 172. https://doi.org/10.3390/ijms19010172.

Formichi, P., Radi, E., Battisti, C., Di Maio, G., Dotti, M.T., Muresanu, D., Federico, A., 2013. Effects of cerebrolysin administration on oxidative stress-induced apoptosis in lymphocytes from CADASIL patients. Neurol. Sci. 34 (4), 553–556. https://doi.org/10.1007/s10072-012-1174-y. (Epub 2012 Aug 10).

Frijhoff, J., Winyard, P.G., Zarkovic, N., Davies, S.S., Stocker, R., Cheng, D., Knight, A.R., Taylor, E.L., Oettrich, J., Ruskovska, T., Gasparovic, A.C., Cuadrado, A., Weber, D., Poulsen, H.E., Grune, T., Schmidt, H.H., Ghezzi, P.J., Winyard, P.G., Zarkovic, N., Davies, S.S., Stocker, R., Cheng, D., Knight, A.R., Taylor, E.L., Oettrich, J., Ruskovska, T., Gasparovic, A.C., Cuadrado, A., Weber, D., Poulsen, H.E., Grune, T., Schmidt, H.H., Ghezzi, P., 2015. Clinical relevance of biomarkers of oxidative stress. Antioxid. Redox Signal. 23 (14), 1144–1170. https://doi.org/10.1089/ars.2015.6317. Epub 2015 Oct 26.

Garcia-Larrea, L., Hagiwara, K., 2019. Electrophysiology in diagnosis and management of neuropathic pain. Rev. Neurol. (Paris) 175 (1–2), 26–37. https://doi.org/10.1016/j.neurol.2018.09.015. (Epub 2018 Oct 26).

Gómez-Palacio-Schjetnan, A., Escobar, M.L., 2013. Neurotrophins and synaptic plasticity. Curr. Top. Behav. Neurosci. 15, 117–136. https://doi.org/10.1007/7854_2012_231.

Guzik, T.J., Channon, K.M., 2005. Measurement of vascular reactive oxygen species production by chemiluminescence. Methods Mol. Med. 108, 73–89.

Hansen, C.C., Ljung, H., Brodtkorb, E., Reimers, A., 2018. Mechanisms underlying aggressive behavior induced by antiepileptic drugs: focus on Topiramate, Levetiracetam, and Perampanel. Behav. Neurol. 2018, 2064027. https://doi.org/10.1155/2018/2064027 (eCollection 2018).

Hatashita, S., Hoff, J.T., 1990. Brain edema and cerebrovascular permeability during cerebral ischemia in rats. Stroke 21 (4), 582–588. https://doi.org/10.1161/01.str.21.4.582.

Hemmelgarn, C., Gannon, K., 2013. Heatstroke: clinical signs, diagnosis, treatment, and prognosis. Compend. Contin. Educ. Vet. 35 (7), E3.

Higareda-Basilio, A.E., Trujillo-Narvaez, F.A., Jaramillo-Ramirez, H.J., 2019. Mortality and functional disability in heat stroke. Salud Publica Mex. 61 (2), 99–100. https://doi.org/10.21149/9849.

Hsu, C.C., Chen, L.F., Lin, M.T., Tian, Y.F., 2014. Honokiol protected against heatstroke-induced oxidative stress and inflammation in diabetic rats. Int. J. Endocrinol. 2014, 134575. https://doi.org/10.1155/2014/134575. Hsu CC, Chen LF, Lin MT, Tian YF. Epub 2014 Feb 17.

Hsuan, Y.C., Lin, C.H., Chang, C.P., Lin, M.T., 2016. Mesenchymal stem cell-based treatments for stroke, neural trauma, and heat stroke. Brain Behav. 6 (10), e00526. eCollection 2016 Oct. https://doi.org/10.1002/brb3.526.

Jacobs, P.J., Oosthuizen, M.K., Mitchell, C., Blount, J.D., Bennett, N.C., 2020. Heat and dehydration induced oxidative damage and antioxidant defenses following incubator heat stress and a simulated heat wave in wild caught four-striped field mice Rhabdomys dilectus. PLoS One 15 (11), e0242279. eCollection 2020. https://doi.org/10.1371/journal.pone.0242279.

Jungehulsing, G.J., Israel, H., Safar, N., Taskin, B., Nolte, C.H., Brunecker, P., Wernecke, K.D., Villringer, A., 2013. Levetiracetam in patients with central neuropathic post-stroke pain- -a randomized, double-blind, placebo-controlled trial. Eur. J. Neurol. 20 (2), 331–337. https://doi.org/10.1111/j.1468-1331.2012.03857.x. (Epub 2012 Aug 27).

Kao, T.Y., Lin, M.T., 1996. Brain serotonin depletion attenuates heatstroke-induced cerebral ischemia and cell death in rats. J. Appl. Physiol. (1985) 80 (2), 680–684. https://doi.org/10.1152/jappl.1996.80.2.680.

Katsuragi, H., Takahashi, K., Suzuki, H., Maeda, M., 2000. Chemiluminescent measurement of peroxidase activity and its application using a lucigenin CT-complex. Luminescence 15 (1), 1–7.

Katzman, R., Clasen, R., Klatzo, I., Meyer, J.S., Pappius, H.M., Waltz, A.G., 1977. Report of joint Committee for Stroke Resources. IV. Brain edema in stroke. Stroke 8 (4), 512–540. https://doi.org/10.1161/01.str.8.4.512.

Kenny, G.P., Wilson, T.E., Flouris, A.D., Fujii, N., 2018. Heat exhaustion. Handb. Clin. Neurol. 157, 505–529. https://doi.org/10.1016/B978-0-444-64074-1.00031-8.

Kim, J.S., Bashford, G., Murphy, K.T., Martin, A., Dror, V., Cheung, R., 2011. Safety and efficacy of pregabalin in patients with central post-stroke pain. Pain 152 (5), 1018–1023. https://doi.org/10.1016/j.pain.2010.12.023. (Epub 2011 Feb 12).

Klatzo, I., 1967. Presidential address. Neuropathological aspects of brain edema. J. Neuropathol. Exp. Neurol. 26 (1), 1–14. https://doi.org/10.1097/00005072-196701000-00001.

Klatzo, I., 1987. Pathophysiological aspects of brain edema. Acta Neuropathol. 72 (3), 236–239. https://doi.org/10.1007/BF00691095.

Klit, H., Finnerup, N.B., Jensen, T.S., 2009. Central post-stroke pain: clinical characteristics, pathophysiology, and management. Lancet Neurol. 8 (9), 857–868. https://doi.org/10.1016/S1474-4422(09)70176-0.

Knapik, J.J., Epstein, Y., 2019. Exertional heat Stroke: pathophysiology, epidemiology, diagnosis, treatment, and prevention. J. Spec. Oper. Med. 19 (2), 108–116.

von Kummer, R., Dzialowski, I., 2017. Imaging of cerebral ischemic edema and neuronal death. Neuroradiology 59 (6), 545–553. https://doi.org/10.1007/s00234-017-1847-6. (Epub 2017 May 24).

Kuroiwa, T., Cahn, R., Juhler, M., Goping, G., Campbell, G., Klatzo, I., 1985. Role of extracellular proteins in the dynamics of vasogenic brain edema. Acta Neuropathol. 66 (1), 3–11. https://doi.org/10.1007/BF00698288.

Landucci, E., Filippi, L., Gerace, E., Catarzi, S., Guerrini, R., Pellegrini-Giampietro, D.E., 2018. Neuroprotective effects of topiramate and memantine in combination with hypothermia in hypoxic-ischemic brain injury in vitro and in vivo. Neurosci. Lett. 668, 103–107. https://doi.org/10.1016/j.neulet.2018.01.023. (Epub 2018 Jan 12).

Lawton, E.M., Pearce, H., Gabb, G.M., 2019. Review article: environmental heatstroke and long-term clinical neurological outcomes: a literature review of case reports and case series 2000-2016. Emerg. Med. Australas. 31 (2), 163–173. https://doi.org/10.1111/1742-6723.12990. Epub 2018 May 31.

Lee, J.T., Chou, C.H., Cho, N.Y., Sung, Y.F., Yang, F.C., Chen, C.Y., Lai, Y.H., Chiang, C.I., Chu, C.M., Lin, J.C., Hsu, Y.D., Hong, J.S., Peng, G.S., Chuang, D.M., 2014. Post-insult valproate treatment potentially improved functional recovery in patients with acute middle cerebral artery infarction. Am. J. Transl. Res. 6 (6), 820–830 (eCollection 2014).

Lee, W.G., Huh, S.Y., Lee, J.H., Yoo, B.G., Kim, M.K., 2017. Status epilepticus as an unusual manifestation of heat stroke. J. Epilepsy Res. 7 (2), 121–125. https://doi.org/10.14581/jer.17020. eCollection 2017 Dec.

Lee, J., Kang, C.G., Park, C.R., Hong, I.K., Kim, D.Y., 2021. The neuroprotective effects of pregabalin after cerebral ischemia by occlusion of the middle cerebral artery in rats. Exp. Ther. Med. 21 (2), 165. https://doi.org/10.3892/etm.2020.9596. (Epub 2020 Dec 21).

Leon, L.R., Helwig, B.G., 2010. Heat stroke: role of the systemic inflammatory response. J. Appl. Physiol. (1985) 109 (6), 1980–1988. https://doi.org/10.1152/japplphysiol.00301.2010. Epub 2010 Jun 3.

Leyk, D., Hoitz, J., Becker, C., Glitz, K.J., Nestler, K., Piekarski, C., 2019. Health risks and interventions in exertional heat stress. Dtsch. Arztebl. Int. 116 (31 – 32), 537–544. https://doi.org/10.3238/arztebl.2019.0537.

Lindqvist, D., Dhabhar, F.S., James, S.J., Hough, C.M., Jain, F.A., Bersani, F.S., Reus, V.I., Verhoeven, J.E., Epel, E.S., Mahan, L., Rosser, R., Wolkowitz, O.M., Mellon, S.H., 2017. Oxidative stress, inflammation and treatment response in major depression. Psychoneuroendocrinology 76, 197–205. https://doi.org/10.1016/j.psyneuen.2016.11.031. Epub 2016.

Liu, Q., Babadjouni, R., Radwanski, R., Cheng, H., Patel, A., Hodis, D.M., He, S., Baumbacher, P., Russin, J.J., Morgan, T.E., Sioutas, C., Finch, C.E., Mack, W.J., 2016. Stroke damage is exacerbated by nano-size particulate matter in a mouse model. PLoS One 11 (4), e0153376. eCollection 2016. https://doi.org/10.1371/journal.pone.0153376.

Liu, Z., Chen, J., Hu, L., Li, M., Liang, M., Chen, J., Lin, H., Zeng, Z., Yin, W., Dong, Z., Weng, J., Yao, W., Yi, G., 2020. Expression profiles of genes associated with inflammatory responses and oxidative stress in lung after heat stroke. Biosci. Rep. 40 (6). https://doi.org/10.1042/BSR20192048, BSR20192048.

Lu, T.S., Chen, H.W., Huang, M.H., Wang, S.J., Yang, R.C., 2004. Heat shock treatment protects osmotic stress-induced dysfunction of the blood-brain barrier through preservation of tight junction proteins. Cell Stress Chaperones 9 (4), 369–377. https://doi.org/10.1379/csc-45r1.1.

Lundesgaard Eidahl, J.M., Opdal, S.H., Rognum, T.O., Stray-Pedersen, A., 2019. Postmortem evaluation of brain edema: an attempt with measurements of water content and brain-weight-to-inner-skull-circumference ratio. J. Forensic Leg. Med. 64, 1–6. https://doi.org/10.1016/j.jflm.2019.03.003. (Epub 2019 Mar 5).

Maeda, K.J., McClung, D.M., Showmaker, K.C., Warrington, J.P., Ryan, M.J., Garrett, M.R., Sasser, J.M., 2021. Endothelial cell disruption drives increased blood-brain barrier permeability and cerebral edema in the dahl SS/jr rat model of superimposed preeclampsia. Am. J. Physiol. Heart Circ. Physiol. 320 (2), H535–H548. https://doi.org/10.1152/ajpheart.00383.2020. (Epub 2020 Dec 4).

Maxeiner, H., Behnke, M., 2008. Intracranial volume, brain volume, reserve volume and morphological signs of increased intracranial pressure—a post-mortem analysis. Leg. Med. (Tokyo) 10 (6), 293–300. https://doi.org/10.1016/j.legalmed.2008.04.001. (Epub 2008 Jul 26).

Menon, P.K., Muresanu, D.F., Sharma, A., Mössler, H., Sharma, H.S., 2012. Cerebrolysin, a mixture of neurotrophic factors induces marked neuroprotection in spinal cord injury following intoxication of engineered nanoparticles from metals. CNS Neurol. Disord. Drug Targets 11 (1), 40–49. https://doi.org/10.2174/187152712799960781.

Miller, M.R., Raftis, J.B., Langrish, J.P., McLean, S.G., Samutrtai, P., Connell, S.P., Wilson, S., Vesey, A.T., Fokkens, P.H.B., Boere, A.J.F., Krystek, P., Campbell, C.J., Hadoke, P.W.F., Donaldson, K., Cassee, F.R., Newby, D.E., Duffin, R., Mills, N.L., 2017. Inhaled nanoparticles accumulate at sites of vascular disease. ACS Nano 11 (5), 4542–4552. https://doi.org/10.1021/acsnano.6b08551. Epub 2017 Apr 26.

Morris, A., Patel, G., 2021. Heat stroke. In: StatPearls. StatPearls Publishing, Treasure Island, FL. Jun 26 [Internet] 2021 Jan.

Muresanu, D.F., Zimmermann-Meinzingen, S., Sharma, H.S., 2010. Chronic hypertension aggravates heat stress-induced brain damage: possible neuroprotection by cerebrolysin. Acta Neurochir. Suppl. 106, 327–333. https://doi.org/10.1007/978-3-211-98811-4_61.

Muresanu, D.F., Sharma, A., Patnaik, R., Menon, P.K., Mössler, H., Sharma, H.S., 2019. Exacerbation of blood-brain barrier breakdown, edema formation, nitric oxide synthase upregulation and brain pathology after heat stroke in diabetic and hypertensive rats. Potential neuroprotection with cerebrolysin treatment. Int. Rev. Neurobiol. 146, 83–102. https://doi.org/10.1016/bs.irn.2019.06.007. (Epub 2019 Jul 18).

Nabavi, S.F., Dean, O.M., Turner, A., Sureda, A., Daglia, M., Nabavi, S.M., 2015. Oxidative stress and post-stroke depression: possible therapeutic role of polyphenols? Curr. Med. Chem. 22 (3), 343–351.

Nagashima, T., Horwitz, B., Rapoport, S.I., 1990. A mathematical model for vasogenic brain edema. Adv. Neurol. 52, 317–326.

Nehring, S.M., Tadi, P., Tenny, S., 2021. Cerebral edema. In: StatPearls. StatPearls Publishing, Treasure Island (FL). Jan 31 2021 [Internet] Jan.

Olaru, D.G., Olaru, A., Kassem, G.H., Popescu-Drigă, M.V., Pinoşanu, L.R., Dumitraşcu, D.I., Popescu, E.L., Hermann, D.M., Popa-Wagner, A., 2019. Toxicity and health impact of nanoparticles. Basic biology and clinical perspective. Rom. J. Morphol. Embryol. 60 (3), 787–792.

Olsson, Y., Crowell, R.M., Klatzo, I., 1971. The blood-brain barrier to protein tracers in focal cerebral ischemia and infarction caused by occlusion of the middle cerebral artery. Acta Neuropathol. 18 (2), 89–102. https://doi.org/10.1007/BF00687597.

Onouchi, K., Koga, H., Yokoyama, K., Yoshiyama, T., 2014. An open-label, long-term study examining the safety and tolerability of pregabalin in Japanese patients with central neuropathic pain. J. Pain Res. 7, 439–447. https://doi.org/10.2147/JPR.S63028. eCollection 2014.

Ozkizilcik, A., Sharma, A., Lafuente, J.V., Muresanu, D.F., Castellani, R.J., Nozari, A., Tian, Z.R., Mössler, H., Sharma, H.S., 2019. Nanodelivery of cerebrolysin reduces pathophysiology of Parkinson's disease. Prog. Brain Res. 245, 201–246. https://doi.org/10.1016/bs.pbr.2019.03.014. (Epub 2019 Apr 2).

Paglia, D.E., Valentine, W.N., 1967. Studies on the quantitative and qualitative characterization of erythrocyte glutathione peroxidase. J. Lab. Clin. Med. 70 (1), 158–169.

Pasantes-Morales, H., Tuz, K., 2006. Volume changes in neurons: hyperexcitability and neuronal death. Contrib. Nephrol. 152, 221–240. https://doi.org/10.1159/000096326.

Plecash, A.R., Chebini, A., Ip, A., Lai, J.J., Mattar, A.A., Randhawa, J., Field, T.S., 2019. Updates in the treatment of post-stroke pain. Curr. Neurol. Neurosci. Rep. 19 (11), 86. https://doi.org/10.1007/s11910-019-1003-2.

Powers, S.W., Coffey, C.S., Chamberlin, L.A., Ecklund, D.J., Klingner, E.A., Yankey, J.W., Korbee, L.L., Porter, L.L., Hershey, A.D., CHAMP Investigators, 2017. Trial of amitriptyline, topiramate, and placebo for pediatric migraine. N. Engl. J. Med. 376 (2), 115–124. https://doi.org/10.1056/NEJMoa1610384. Epub 2016 Oct 27.

Rahman, M.S., Uddin, M.T., 2014. Comparative efficacy of pregabalin and therapeutic ultrasound versus therapeutic ultrasound alone on patients with post stroke shoulder pain. Mymensingh Med. J. 23 (3), 456–460.

Rapoport, S.I., 1978. A mathematical model for vasogenic brain edema. J. Theor. Biol. 74, 439–467.

Robles, L.A., Cuevas-Solórzano, A., 2018. Massive brain swelling and death after cranioplasty: a systematic review. World Neurosurg. 111, 99–108. https://doi.org/10.1016/j.wneu.2017.12.061. (Epub 2017 Dec 18).

Rockenstein, E., Adame, A., Mante, M., Larrea, G., Crews, L., Windisch, M., Moessler, H., Masliah, E., 2005. Amelioration of the cerebrovascular amyloidosis in a transgenic model of Alzheimer's disease with the neurotrophic compound cerebrolysin. J. Neural Transm. (Vienna) 112 (2), 269–282. https://doi.org/10.1007/s00702-004-0181-4. (Epub 2004 Jul 7).

Romoli, M., Mazzocchetti, P., D'Alonzo, R., Siliquini, S., Rinaldi, V.E., Verrotti, A., Calabresi, P., Costa, C., 2019. Valproic acid and epilepsy: from molecular mechanisms to clinical evidences. Curr. Neuropharmacol. 17 (10), 926–946. https://doi.org/10.2174/1570159X17666181227165722.

Šćepanović, V., Tasić, G., Repac, N., Nikolić, I., Janićijević, A., Todorović, D., Stojanović, M., Šćepanović, R., Mitrović, D., Šćepanović, T., Borozan, S., Šćepanović, L., 2018. The role of oxidative stress as a risk factor for rupture of posterior inferior cerebellar artery aneurysms. Mol. Biol. Rep. 45 (6), 2157–2165. https://doi.org/10.1007/s11033-018-4374-6. (Epub 2018 Sep 20).

Sahib, S., Sharma, A., Menon, P.K., Muresanu, D.F., Castellani, R.J., Nozari, A., Lafuente, J.V., Bryukhovetskiy, I., Tian, Z.R., Patnaik, R., Buzoianu, A.D., Wiklund, L., Sharma, H.S., 2020. Cerebrolysin enhances spinal cord conduction and reduces blood-spinal cord barrier breakdown, edema formation, immediate early gene expression and cord pathology after injury. Prog. Brain Res. 258, 397–438. https://doi.org/10.1016/bs.pbr.2020.09.012. (Epub 2020 Nov 12).

Santana-Gómez, C.E., Valle-Dorado, M.G., Domínguez-Valentín, A.E., Hernández-Moreno, A., Orozco-Suárez, S., Rocha, L., 2018. Neuroprotective effects of levetiracetam, both alone and combined with propylparaben, in the long-term consequences induced by lithium-pilocarpine status epilepticus. Neurochem. Int. 120, 224–232. https://doi.org/10.1016/j.neuint.2018.09.004. (Epub 2018 Sep 10).

Schwab, M., Bauer, R., Zwiener, U., 1997. Physiological effects and brain protection by hypothermia and cerebrolysin after moderate forebrain ischemia in rats. Exp. Toxicol. Pathol. 49 (1–2), 105–116. https://doi.org/10.1016/S0940-2993(97)80078-4.

Scotton, W.J., Botfield, H.F., Westgate, C.S., Mitchell, J.L., Yiangou, A., Uldall, M.S., Jensen, R.H., Sinclair, A.J., 2019. Topiramate is more effective than acetazolamide at lowering intracranial pressure. Cephalalgia 39 (2), 209–218. https://doi.org/10.1177/0333102418776455. Epub 2018 Jun 13.

Sharma, H.S., 1987. Effect of captopril (a converting enzyme inhibitor) on blood-brain barrier permeability and cerebral blood flow in normotensive rats. Neuropharmacology 26 (1), 85–92. https://doi.org/10.1016/0028-3908(87)90049-9.

Sharma, H.S., 2000. A bradykinin BK2 receptor antagonist HOE-140 attenuates blood-spinal cord barrier permeability following a focal trauma to the rat spinal cord. An experimental study using Evans blue, ^{131}I-sodium and lanthanum tracers. Acta Neurochir. Suppl. 76, 159–163. https://doi.org/10.1007/978-3-7091-6346-7_32.

Sharma, H.S., 2003. Neurotrophic factors attenuate microvascular permeability disturbances and axonal injury following trauma to the rat spinal cord. Acta Neurochir. Suppl. 86, 383–388. https://doi.org/10.1007/978-3-7091-0651-8_81.

Sharma, H.S., 2004. Influence of serotonin on the blood-brain and blood-spinal cord barriers. In: Sharma, H.S., Westman, J. (Eds.), The Blood-Spinal Cord and Brain Barriers in Health and Disease. Elsevier Academic Press, San Diego, pp. 117–158.

Sharma, H.S., 2005a. Heat-related deaths are largely due to brain damage. Indian J. Med. Res. 121 (5), 621–623.

Sharma, H.S., 2005b. Methods to produce brain hyperthermia. Curr. Protoc. Toxicol. Chapter 11. https://doi.org/10.1002/0471140856.tx1114s23, Unit11.14.

Sharma, H.S., 2005c. Neuroprotective effects of neurotrophins and melanocortins in spinal cord injury: an experimental study in the rat using pharmacological and morphological approaches. Ann. N. Y. Acad. Sci. 1053, 407–421. https://doi.org/10.1111/j.1749-6632.2005.tb00050.x.

Sharma, H.S., 2005d. Pathophysiology of blood-spinal cord barrier in traumatic injury and repair. Curr. Pharm. Des. 11 (11), 1353–1389. https://doi.org/10.2174/1381612053507837.

Sharma, H.S., 2006. Hyperthermia induced brain oedema: current status and future perspectives. Indian J. Med. Res. 123 (5), 629–652.

Sharma, H.S., 2007a. A select combination of neurotrophins enhances neuroprotection and functional recovery following spinal cord injury. Ann. N. Y. Acad. Sci. 1122, 95–111. https://doi.org/10.1196/annals.1403.007.

Sharma, H.S., 2007b. Methods to produce hyperthermia-induced brain dysfunction. Prog. Brain Res. 162, 173–199. https://doi.org/10.1016/S0079-6123(06)62010-4.

Sharma, H.S., 2007c. Neurotrophic factors in combination: a possible new therapeutic strategy to influence pathophysiology of spinal cord injury and repair mechanisms. Curr. Pharm. Des. 13 (18), 1841–1874. https://doi.org/10.2174/138161207780858410.

Sharma, H.S., 2009. Blood–central nervous system barriers: the gateway to neurodegeneration, neuroprotection and neuroregeneration. In: Lajtha, A., Banik, N., Ray, S.K. (Eds.), Handbook of Neurochemistry and Molecular Neurobiology: Brain and Spinal Cord Trauma. Springer Verlag, Berlin, Heidelberg, New York, pp. 363–457.

Sharma, H.S., Alm, P., 2002. Nitric oxide synthase inhibitors influence dynorphin A (1-17) immunoreactivity in the rat brain following hyperthermia. Amino Acids 23 (1–3), 247–259. https://doi.org/10.1007/s00726-001-0136-0.

Sharma, H.S., Cervós-Navarro, J., 1990. Brain oedema and cellular changes induced by acute heat stress in young rats. Acta Neurochir. Suppl. (Wien) 51, 383–386. https://doi.org/10.1007/978-3-7091-9115-6_129.

Sharma, H.S., Dey, P.K., 1984. Role of 5-HT on increased permeability of blood-brain barrier under heat stress. Indian J. Physiol. Pharmacol. 28 (4), 259–267.

Sharma, H.S., Dey, P.K., 1986. Probable involvement of 5-hydroxytryptamine in increased permeability of blood-brain barrier under heat stress in young rats. Neuropharmacology 25 (2), 161–167. https://doi.org/10.1016/0028-3908(86)90037-7.

Sharma, H.S., Dey, P.K., 1987. Influence of long-term acute heat exposure on regional blood-brain barrier permeability, cerebral blood flow and 5-HT level in conscious normotensive young rats. Brain Res. 424 (1), 153–162. https://doi.org/10.1016/0006-8993(87)91205-4.

Sharma, H.S., Dey, P.K., 1988. EEG changes following increased blood-brain barrier permeability under long-term immobilization stress in young rats. Neurosci. Res. 5 (3), 224–239. https://doi.org/10.1016/0168-0102(88)90051-x.

Sharma, H.S., Hoopes, P.J., 2003. Hyperthermia induced pathophysiology of the central nervous system. Int. J. Hyperthermia 19 (3), 325–354. https://doi.org/10.1080/0265673021000054621.

Sharma, H.S., Johanson, C.E., 2007. Intracerebroventricularly administered neurotrophins attenuate blood cerebrospinal fluid barrier breakdown and brain pathology following whole-body hyperthermia: an experimental study in the rat using biochemical and morphological approaches. Ann. N. Y. Acad. Sci. 1122, 112–129. https://doi.org/10.1196/annals.1403.008.

Sharma, H.S., Olsson, Y., 1990. Edema formation and cellular alterations following spinal cord injury in the rat and their modification with p-chlorophenylalanine. Acta Neuropathol. 79 (6), 604–610. https://doi.org/10.1007/BF00294237.

Sharma, H.S., Sharma, A., 2007. Nanoparticles aggravate heat stress induced cognitive deficits, blood-brain barrier disruption, edema formation and brain pathology. Prog. Brain Res. 162, 245–273. https://doi.org/10.1016/S0079-6123(06)62013-X.

Sharma, H.S., Sharma, A., 2019. New Therapeutic Strategies for Brain Edema and Cell Injury. International Review of Neurobiology, vol. 146 Academic Press, New York, pp. 1–319, https://doi.org/10.1016/S0074-7742(19)30061-3.

Sharma, H.S., Sjöquist, P.O., 2002. A new antioxidant compound H-290/51 modulates glutamate and GABA immunoreactivity in the rat spinal cord following trauma. Amino Acids 23 (1–3), 261–272. https://doi.org/10.1007/s00726-001-0137-z.

Sharma, H.S., Olsson, Y., Dey, P.K., 1990. Changes in blood-brain barrier and cerebral blood flow following elevation of circulating serotonin level in anesthetized rats. Brain Res. 517 (1–2), 215–223. https://doi.org/10.1016/0006-8993(90)91029-g.

Sharma, H.S., Cervós-Navarro, J., Dey, P.K., 1991a. Acute heat exposure causes cellular alteration in cerebral cortex of young rats. Neuroreport 2 (3), 155–158. https://doi.org/10.1097/00001756-199103000-00012.

Sharma, H.S., Cervós-Navarro, J., Dey, P.K., 1991b. Rearing at high ambient temperature during later phase of the brain development enhances functional plasticity of the CNS and induces tolerance to heat stress. An experimental study in the conscious normotensive young rats. Brain Dysfunct. 4, 104–124.

Sharma, H.S., Kretzschmar, R., Cervós-Navarro, J., Ermisch, A., Rühle, H.J., Dey, P.K., 1992a. Age-related pathophysiology of the blood-brain barrier in heat stress. Prog. Brain Res. 91, 189–196. https://doi.org/10.1016/s0079-6123(08)62334-1.

Sharma, H.S., Westman, J., Nyberg, F., Cervós-Navarro, J., Dey, P.K., 1992b. Role of serotonin in heat adaptation: an experimental study in the conscious young rat. Endocr. Regul. 26 (3), 133–142.

Sharma, H.S., Olsson, Y., Cervós-Navarro, J., 1993. Early perifocal cell changes and edema in traumatic injury of the spinal cord are reduced by indomethacin, an inhibitor of prostaglandin synthesis. Experimental study in the rat. Acta Neuropathol. 85 (2), 145–153. https://doi.org/10.1007/BF00227761.

Sharma, H.S., Westman, J., Nyberg, F., Cervos-Navarro, J., Dey, P.K., 1994. Role of serotonin and prostaglandins in brain edema induced by heat stress. An experimental study in the young rat. Acta Neurochir. Suppl. (Wien) 60, 65–70. https://doi.org/10.1007/978-3-7091-9334-1_17.

Sharma, H.S., Olsson, Y., Persson, S., Nyberg, F., 1995. Trauma-induced opening of the the blood-spinal cord barrier is reduced by indomethacin, an inhibitor of prostaglandin biosynthesis. Experimental observations in the rat using [131I]-sodium, Evans blue and lanthanum as tracers. Restor. Neurol. Neurosci. 7 (4), 207–215. https://doi.org/10.3233/RNN-1994-7403.

Sharma, H.S., Westman, J., Nyberg, F., 1998a. Pathophysiology of brain edema and cell changes following hyperthermic brain injury. Prog. Brain Res. 115, 351–412. https://doi.org/10.1016/s0079-6123(08)62043-9.

Sharma, H.S., Alm, P., Westman, J., 1998b. Nitric oxide and carbon monoxide in the brain pathology of heat stress. Prog. Brain Res. 115, 297–333. https://doi.org/10.1016/s0079-6123(08)62041-5.

Sharma, H.S., Nyberg, F., Westman, J., Alm, P., Gordh, T., Lindholm, D., 1998c. Brain derived neurotrophic factor and insulin like growth factor-1 attenuate upregulation of nitric oxide synthase and cell injury following trauma to the spinal cord. An immunohistochemical study in the rat. Amino Acids 14 (1–3), 121–129. https://doi.org/10.1007/BF01345252.

Sharma, H.S., Westman, J., Nyberg, F., 2000. Selective alteration of calcitonin gene related peptide in hyperthermic brain injury. An experimental study in the rat brain using immunohistochemistry. Acta Neurochir. Suppl. 76, 541–545. https://doi.org/10.1007/978-3-7091-6346-7_113.

Sharma, H.S., Ali, S.F., Tian, Z.R., Hussain, S.M., Schlager, J.J., Sjöquist, P.O., Sharma, A., Muresanu, D.F., 2009. Chronic treatment with nanoparticles exacerbate hyperthermia induced blood-brain barrier breakdown, cognitive dysfunction and brain pathology in the rat. Neuroprotective effects of nanowired-antioxidant compound H-290/51. J. Nanosci. Nanotechnol. 9 (8), 5073–5090. https://doi.org/10.1166/jnn.2009.gr10.

Sharma, H.S., Muresanu, D., Sharma, A., Zimmermann-Meinzingen, S., 2010a. Cerebrolysin treatment attenuates heat shock protein overexpression in the brain following heat stress: an experimental study using immunohistochemistry at light and electron microscopy in the rat. Ann. N. Y. Acad. Sci. 1199, 138–148. https://doi.org/10.1111/j.1749-6632.2009.05330.x.

Sharma, H.S., Zimmermann-Meinzingen, S., Johanson, C.E., 2010b. Cerebrolysin reduces blood-cerebrospinal fluid barrier permeability change, brain pathology, and functional deficits following traumatic brain injury in the rat. Ann. N. Y. Acad. Sci. 1199, 125–137. https://doi.org/10.1111/j.1749-6632.2009.05329.x.

Sharma, H.S., Zimmermann-Meinzingen, S., Sharma, A., Johanson, C.E., 2010c. Cerebrolysin attenuates blood-brain barrier and brain pathology following whole body hyperthermia in the rat. Acta Neurochir. Suppl. 106, 321–325. https://doi.org/10.1007/978-3-211-98811-4_60.

Sharma, H.S., Muresanu, D.F., Patnaik, R., Stan, A.D., Vacaras, V., Perju-Dumbrav, L., Alexandru, B., Buzoianu, A., Opincariu, I., Menon, P.K., Sharma, A., 2011. Superior neuroprotective effects of cerebrolysin in heat stroke following chronic intoxication of cu or ag engineered nanoparticles. A comparative study with other neuroprotective agents using biochemical and morphological approaches in the rat. J. Nanosci. Nanotechnol. 11 (9), 7549–7569. https://doi.org/10.1166/jnn.2011.5114.

Sharma, A., Muresanu, D.F., Mössler, H., Sharma, H.S., 2012a. Superior neuroprotective effects of cerebrolysin in nanoparticle-induced exacerbation of hyperthermia-induced brain pathology. CNS Neurol. Disord. Drug Targets 11 (1), 7–25. https://doi.org/10.2174/187152712799960790.

Sharma, H.S., Sharma, A., Mössler, H., Muresanu, D.F., 2012b. Neuroprotective effects of cerebrolysin, a combination of different active fragments of neurotrophic factors and peptides on the whole body hyperthermia-induced neurotoxicity: modulatory roles of co-morbidity factors and nanoparticle intoxication. Int. Rev. Neurobiol. 102, 249–276. https://doi.org/10.1016/B978-0-12-386986-9.00010-7.

Sharma, H.S., Muresanu, D.F., Patnaik, R., Sharma, A., 2013. Exacerbation of brain pathology after partial restraint in hypertensive rats following SiO2 nanoparticles exposure at high ambient temperature. Mol. Neurobiol. 48 (2), 368–379. https://doi.org/10.1007/s12035-013-8502-y. (Epub 2013 Jul 6).

Sharma, A., Menon, P., Muresanu, D.F., Ozkizilcik, A., Tian, Z.R., Lafuente, J.V., Sharma, H.S., 2016a. Nanowired drug delivery across the blood-brain barrier in central nervous system injury and repair. CNS Neurol. Disord. Drug Targets 15 (9), 1092–1117. https://doi.org/10.2174/1871527315666160819123059.

Sharma, H.S., Muresanu, D.F., Lafuente, J.V., Nozari, A., Patnaik, R., Skaper, S.D., Sharma, A., 2016b. Pathophysiology of blood-brain barrier in brain injury in cold and hot environments: novel drug targets for neuroprotection. CNS Neurol. Disord. Drug Targets 15 (9), 1045–1071. https://doi.org/10.2174/1871527315666160902145145.

Sharma, H.S., Patnaik, R., Muresanu, D.F., Lafuente, J.V., Ozkizilcik, A., Tian, Z.R., Nozari, A., Sharma, A., 2017. Histaminergic receptors modulate spinal cord injury-induced neuronal nitric oxide synthase upregulation and cord pathology: new roles of nanowired drug delivery for neuroprotection. Int. Rev. Neurobiol. 137, 65–98. https://doi.org/10.1016/bs.irn.2017.09.001. Epub 2017 Nov 3.

Sharma, A., Muresanu, D.F., Lafuente, J.V., Sjöquist, P.O., Patnaik, R., Ryan Tian, Z., Ozkizilcik, A., Sharma, H.S., 2018. Cold environment exacerbates brain pathology and oxidative stress following traumatic brain injuries: potential therapeutic effects of nanowired antioxidant compound H-290/51. Mol. Neurobiol. 55 (1), 276–285. https://doi.org/10.1007/s12035-017-0740-y.

Sharma, A., Castellani, R.J., Smith, M.A., Muresanu, D.F., Dey, P.K., Sharma, H.S., 2019a. 5-Hydroxytryptophan: a precursor of serotonin influences regional blood-brain barrier breakdown, cerebral blood flow, brain edema formation, and neuropathology. Int. Rev. Neurobiol. 146, 1–44. https://doi.org/10.1016/bs.irn.2019.06.005. (Epub 2019 Jul 18).

Sharma, H.S., Muresanu, D.F., Castellani, R.J., Nozari, A., Lafuente, J.V., Tian, Z.R., Ozkizilcik, A., Manzhulo, I., Mössler, H., Sharma, A., 2019b. Nanowired delivery of cerebrolysin with neprilysin and p-Tau antibodies induces superior neuroprotection in Alzheimer's disease. Prog. Brain Res. 245, 145–200. https://doi.org/10.1016/bs.pbr.2019.03.009. (Epub 2019 Apr 2).

Sharma, H.S., Muresanu, D.F., Nozari, A., Castellani, R.J., Dey, P.K., Wiklund, L., Sharma, A., 2019c. Anesthetics influence concussive head injury induced blood-brain barrier breakdown, brain edema formation, cerebral blood flow, serotonin levels, brain pathology and functional outcome. Int. Rev. Neurobiol. 146, 45–81. https://doi.org/10.1016/bs.irn.2019.06.006. (Epub 2019 Jul 8).

Sharma, A., Muresanu, D.F., Ozkizilcik, A., Tian, Z.R., Lafuente, J.V., Manzhulo, I., Mössler, H., Sharma, H.S., 2019d. Sleep deprivation exacerbates concussive head injury induced brain pathology: neuroprotective effects of nanowired delivery of cerebrolysin with alpha-melanocyte-stimulating hormone. Prog. Brain Res. 245, 1–55. https://doi.org/10.1016/bs.pbr.2019.03.002. (Epub 2019 Apr 2).

Sharma, A., Muresanu, D.F., Castellani, R.J., Nozari, A., Lafuente, J.V., Sahib, S., Tian, Z.R., Buzoianu, A.D., Patnaik, R., Wiklund, L., Sharma, H.S., 2020a. Mild traumatic brain injury exacerbates Parkinson's disease induced hemeoxygenase-2 expression and brain pathology: neuroprotective effects of co-administration of TiO(2) nanowired mesenchymal stem cells and cerebrolysin. Prog. Brain Res. 258, 157–231. https://doi.org/10.1016/bs.pbr.2020.09.010. (Epub 2020 Nov 9).

Sharma, A., Muresanu, D.F., Sahib, S., Tian, Z.R., Castellani, R.J., Nozari, A., Lafuente, J.V., Buzoianu, A.D., Bryukhovetskiy, I., Manzhulo, I., Patnaik, R., Wiklund, L., Sharma, H.S., 2020b. Concussive head injury exacerbates neuropathology of sleep deprivation: superior neuroprotection by co-administration of TiO(2)-nanowired cerebrolysin, alpha-melanocyte-stimulating hormone, and mesenchymal stem cells. Prog. Brain Res. 258, 1–77. https://doi.org/10.1016/bs.pbr.2020.09.003. (Epub 2020 Nov 2).

Shimada, T., Miyamoto, N., Shimada, Y., Watanabe, M., Shimura, H., Ueno, Y., Yamashiro, K., Hattori, N., Urabe, T., 2020. Analysis of clinical symptoms and brain

MRI of heat Stroke: 2 case reports and a literature review. J. Stroke Cerebrovasc. Dis. 29 (2), 104511. https://doi.org/10.1016/j.jstrokecerebrovasdis.2019.104511. (Epub 2019 Nov 26).

Sun, H., Hu, H., Liu, C., Sun, N., Duan, C., 2021. Methods used for the measurement of blood-brain barrier integrity. Metab. Brain Dis. 36, 723–735. https://doi.org/10.1007/s11011-021-00694-8. Online ahead of print.

Szold, O., Reider-Groswasser, I.I., Ben Abraham, R., Aviram, G., Segev, Y., Biderman, P., Sorkine, P., 2002. Gray-white matter discrimination—a possible marker for brain damage in heat stroke? Eur. J. Radiol. 43 (1), 1–5. https://doi.org/10.1016/s0720-048x(01)00467-3.

Tanaka, T., Ihara, M., 2017. Post-stroke epilepsy. Neurochem. Int. 107, 219–228. https://doi.org/10.1016/j.neuint.2017.02.002. (Epub 2017 Feb 12).

Tao, S., Sun, J., Hao, F., Tang, W., Li, X., Guo, D., Liu, X., 2020. Effects of sodium valproate combined with lamotrigine on quality of life and serum inflammatory factors in patients with Poststroke secondary epilepsy. J. Stroke Cerebrovasc. Dis. 29 (5), 104644. https://doi.org/10.1016/j.jstrokecerebrovasdis.2020.104644. (Epub 2020 Feb 18).

Thulesius, O., 2006. Thermal reactions of blood vessels in vascular stroke and heatstroke. Med. Princ. Pract. 15 (4), 316–321. https://doi.org/10.1159/000092999.

Tseng, M.F., Chou, C.L., Chung, C.H., Chien, W.C., Chen, Y.K., Yang, H.C., Chu, P., 2019. Association between heat stroke and ischemic heart disease: a national longitudinal cohort study in Taiwan. Eur. J. Intern. Med. 59, 97–103. https://doi.org/10.1016/j.ejim.2018.09.019. (Epub 2018 Oct 5).

Wang, Q., Ishikawa, T., Michiue, T., Zhu, B.L., Guan, D.W., Maeda, H., 2012. Evaluation of human brain damage in fatalities due to extreme environmental temperature by quantification of basic fibroblast growth factor (bFGF), glial fibrillary acidic protein (GFAP), S100beta and single-stranded DNA (ssDNA) immunoreactivities. Forensic Sci. Int. 219 (1–3), 259–264. https://doi.org/10.1016/j.forsciint.2012.01.015. (Epub 2012 Jan 28).

Wang, J.C., Chien, W.C., Chu, P., Chung, C.H., Lin, C.Y., Tsai, S.H., 2019. The association between heat stroke and subsequent cardiovascular diseases. PLoS One 14 (2), e0211386. eCollection 2019. https://doi.org/10.1371/journal.pone.0211386.

Wolman, M., Klatzo, I., Chui, E., Wilmes, F., Nishimoto, K., Fujiwara, K., Spatz, M., 1981. Evaluation of the dye-protein tracers in pathophysiology of the blood-brain barrier. Acta Neuropathol. 54 (1), 55–61. https://doi.org/10.1007/BF00691332.

Xia, D.M., Wang, X.R., Zhou, P.Y., Ou, T.L., Su, L., Xu, S.G., 2021. Research progress of heat stroke during 1989-2019: a bibliometric analysis. Mil. Med. Res. 8 (1), 5. https://doi.org/10.1186/s40779-021-00300-z.

Xin, L., Wang, J., Wu, Y., Guo, S., Tong, J., 2015. Increased oxidative stress and activated heat shock proteins in human cell lines by silver nanoparticles. Hum. Exp. Toxicol. 34 (3), 315–323. https://doi.org/10.1177/0960327114538988. Epub 2014 Jun 30.

Yoon, J.S., Lee, J.H., Son, T.G., Mughal, M.R., Greig, N.H., Mattson, M.P., 2011. Pregabalin suppresses calcium-mediated proteolysis and improves stroke outcome. Neurobiol. Dis. 41 (3), 624–629. https://doi.org/10.1016/j.nbd.2010.11.011. (Epub 2010 Nov 24).

Zhao, T., Ding, Y., Feng, X., Zhou, C., Lin, W., 2020. Effects of atorvastatin and aspirin on post-stroke epilepsy and usage of levetiracetam. Medicine (Baltimore) 99 (50), e23577. https://doi.org/10.1097/MD.0000000000023577.

Zhou, L., Chen, K., Chen, X., Jing, Y., Ma, Z., Bi, J., Kinney, P.L., 2017. Heat and mortality for ischemic and hemorrhagic stroke in 12 cities of Jiangsu Province, China. Sci. Total Environ. 601–602, 271–277. https://doi.org/10.1016/j.scitotenv.2017.05.169. (Epub 2017 May 27).

CHAPTER

A clinical study of high-dose urokinase for the treatment of the patients with hypertension induced ventricular hemorrhage

Chao He[a,†], Gang Yang[a,*], Ming Zhao[a], Qilin Wu[a], Lin Wang[b], and Hari Shanker Sharma[c,*]

[a]*Department of Neurosurgery, Zhuji People' Hospital of Zhejiang Province, Zhuji Affiliated Hospital of Shaoxing University, Zhuji, Zhejiang, China*

[b]*Department of Neurosurgery, Second Affiliated Hospital of Zhejiang University School of Medicine, Hangzhou, Zhejiang, China*

[c]*International Experimental Central Nervous System Injury & Repair (IECNSIR), Department of Surgical Sciences, Anesthesiology & Intensive Care Medicine, Uppsala University Hospital, Uppsala University, Uppsala, Sweden*

**Corresponding authors: Tel.: +86-13646829750; Fax: +86-0571-87068001 (Gang Yang); Tel.: + 46-70-20-11-801; Fax: + 46-18-24-38-99 (Hari Shanker Sharma), e-mail address: tekeyang@sina.com; sharma@surgsci.uu.se*

Hypertension ventricular hemorrhage is characterized by sudden onset and serious symptoms. The combined treatment of external ventricular drainage together with urokinase therapy is the main therapeutic approach for hypertension ventricular hemorrhage at present. The clinical effects of combined external ventricular drainage with urokinase treatment at different doses were comprehensively evaluated to determine the appropriate urokinase dose, the incidence of complications, and patient prognosis.

Abstract

Objective: This study discusses the therapeutic effect of high-dose urokinase treatment for hypertension ventricular hemorrhage. *Methods*: A total of 60 patients with hypertension ventricular hemorrhage were randomly assigned to two groups: treatment group ($n=30$)

[†]Chao He and Gang Yang have equally contributed to this work as co-first authors.

Progress in Brain Research, Volume 266, ISSN 0079-6123, https://doi.org/10.1016/bs.pbr.2021.06.017
Copyright © 2021 Elsevier B.V. All rights reserved.

and control group ($n = 30$). Both groups received bilateral external ventricular drain. The treatment group was injected with 50,000 IU urokinase to the lateral ventricle every day; the total injection volume per day was 100,000 IU. The control group was injected with 20,000 IU urokinase to the lateral ventricle every day with a total injection volume per day of 40,000 IU. Lumbar puncture was performed in both groups after the later ventricular drain was removed to release cerebrospinal fluid (CSF). Head Computed tomography(CT) examination was performed regularly to observe changes in the ventricular hematoma as well as the occurrence of complications such as intracranial infection and hydrocephalus. Patient prognosis 6 weeks after surgery was compared between the two groups. *Results*: In the treatment group, the intraventricular hemorrhage clearance time and the number of instances of urokinase treatment were significantly less than those of the control group ($P < 0.05$). The total urokinase dosage of the treatment group was significantly higher than that of the control group ($P < 0.05$). With respect to post-surgery complications, in the treatment group, there were three cases of hydrocephalus and one case of intracranial infection. In the control group, there were four cases of hydrocephalus and three cases of intracranial infection. Intraventricular re-hemorrhage was not observed in either group. Intracranial infection was relieved after strengthened anti-infective therapy and continuous drainage. There was a statistically significant difference in the occurrence of complications between the treatment group and the control group ($P < 0.05$). The rate of good prognosis in the treatment group was higher than that of the control group ($P < 0.05$), and the inefficiency rate was lower ($P < 0.05$). *Conclusions*: High-dose urokinase treatment produces a significant therapeutic effect in hypertension ventricular hemorrhage. This treatment can quickly eliminate intraventricular hemorrhage, shorten the ventricular drain tube indwelling time, decrease the occurrence of intracranial infection, and increase the likelihood of a good prognosis.

Keywords

Intraventricular hemorrhage, High-dose urokinase, External ventricular drain, Cerebrospinal fluid drain

1 Data and methods

1.1 General data

A total of 60 patients with primary intraventricular hemorrhage who were diagnosed by head CT examination between September 1st, 2017, and July 1st, 2020, were evaluated. The Graeb scores for intraventricular hemorrhage ranged between 5 and 12. Patients were randomly assigned to the treatment group or the control group based on the order of hospitalization. The treatment group comprised 30 patients, of which 17 were male and 13 were female; these patients ranged in age from 45 to 75 years (mean 51.3 ± 11.2). Seven patients in the treatment group had Glasgow Coma Scale (GCS) scores ranging between 3 and 5, 16 patients had scores between 6 and 8, and seven patients had scores between 9 and 13. The volume of intraventricular hemorrhage in the treatment group ranged from 20 to 44 mL, with an average of 35.3 mL. The control group comprised 30 cases, including 18 males and 12 females, ranging in age from 47 to 75 years (mean 52.2 ± 12.1). In the control group, 6 patients had CGS scores between 3 and 5, 18 patients had scores between 6 and 8, and 6 patients had scores between 9 and 13. The volume of intraventricular hemorrhage ranged from 22 to 47 mL, with an average of 37.1 mL. There were no statistically significant differences between the two groups in terms of general information and the severity of symptoms.

1.2 Inclusion criteria

Patients who met the following criteria were eligible for participation: (1) hospitalized within 6 h of the attack; (2) cerebral hemorrhage symptoms such as headache, nausea and vomiting, disturbance of consciousness, and limb dyskinesia; (3) Graeb score ranging between 5 and 12 at hospitalization.

1.3 Exclusion criteria

Patients who met the following criteria were excluded: (1) age > 75 years old; (2) bilateral mydriasis and fixed, very unstable vital signs or brainstem disorder; (3) complications with history of serious heart, lung, liver, or kidney diseases; (4) died within a short period of surgery; (5) computerized tomography angiography (CTA), digital subtraction angiography (DSA), and magnetic resonance angiography (MRA) diagnosis of intraventricular hemorrhage caused by cerebral aneurysm, moyamoya disease, cerebral arteriovenous malformation, or brain tumor during the perioperative period.

1.4 Therapy

All patients were administrated a bilateral external ventricular drain through the puncture at the former cornu performed during emergency treatment. Head CT examination was performed again 24 h after surgery to confirm the position of the ventricular drain tube. Cerebral vascular angiography data, such as CTA, DSA and MRA, was obtained. The control group received 20,000 IU urokinase/side, 40,000 IU per day, bilateral alternative intraventricular treatment every day. The treatment group received 50,000 IU urokinase/side, 100,000 IU per day, bilateral alternative intraventricular treatment every day. Urokinase was diluted by normal saline to 5 mL and the external ventricular drain tube was closed for 2 h after urokinase treatment. The number of urokinase treatments was calculated from opening of the ventricular drain tube at one side to closing at completion of drainage. Head CT examination and conventional examination of CSF were performed at 1d, 3d, 5d, 7d, 10d, and 14d after surgery. Urokinase was ceased after the intraventricular hemorrhage was eliminated. The drain tube was removed after meeting the extubation indicators. Several lumbar punctures were performed after removing the tube to release bloody CSF. Relevant measures were taken if there was intraventricular re-hemorrhage, delayed hydrocephalus, or intracranial infection.

1.5 Extubation indicators

(1) Head CT examination showing that the ventricular system has become smooth; in particular, that the hematocele in the third and fourth ventricular systems is clean, without evident dilated ventricles; (2) CSF is faint yellow or clean; (3) no evident discomfort or increased intracranial pressure within 24h of drainage tube closure.

1.6 Assessment of therapeutic effect

(1) Post-surgery intraventricular hematoma elimination time, number of instances of urokinase treatment, and total urokinase use were compared. (2) Prognosis of patients 6 weeks following surgery was compared between the groups according to the Karnofsky scores: good prognosis (70–100), effectiveness (40–60), and invalidity ($\leq$30). (3) Occurrence of major complications was compared between the groups, including re-hemorrhage, delayed hydrocephalus, and intracranial infection.

1.7 Statistical method

Data were processed using SPSS 19.0 statistical software. Observed data were expressed as means ± standard deviations (x- ± s) and frequency data were expressed as percentages (%). Group differences in measurement data and frequency data were tested using *t*-tests and χ^2tests, respectively. $P<0.05$ indicated a statistically significant difference.

2 Results

2.1 Post-surgery intraventricular hemorrhage clearance times, number of urokinase treatments, and total urokinase dosages of the two groups are shown in Table 1. The intraventricular hemorrhage clearance time and the number of urokinase treatments in the treatment group were significantly less than those of the control group ($P<0.05$). The total urokinase dosage of the treatment group was significantly higher than that of the control group ($P<0.05$).

Table 1 Comparison of post-surgery intraventricular hemorrhage clearance time, number of urokinase treatments, and total urokinase dosage between the two groups.

	Cases	Intraventricular hemorrhage clearance time (d)	Time of urokinase treatment	Total urokinase dosage (10,000 IU)
Treatment group	30	4.20±1.12	7.2	36.0
Control group	30	10.23±2.52	14.3	28.6

Table 2 Comparison of complications rate between the two groups.

	Hydrocephalus (cases)	Intracranial infection (cases)	Intracranial re-hemorrhage (cases)	Total (cases)
Treatment group	3(10.0%)	1(3.3%)	0	4 (13.3%)
Control group	4(13.3%)	3(10.0%)	0	7 (23.3%)

Table 3 Comparison of Karnofsky scores between the two groups.

	100–70	60–40	30–0
Treatment group	6 cases (53.3%)	11 cases(36.7%)	3 cases(10.0%)
Control group	10cases c (33.3%)	12 cases(40.0%)	8 cases(26.7%)

2.2 Incidence rates of delayed hydrocephalus, intracranial infection, and intracranial re-hemorrhagein the two groups were compared (Table 2). In the treatment group, there were three cases of hydrocephalus and one case of intracranial infection, but no cases of intracranial re-hemorrhage. In the control group, there were four cases of hydrocephalus and three cases of intracranial infection, but no cases of intracranial re-hemorrhage. Intracranial infection was relieved after strengthened anti-infective therapy and continuous drainage. There was a statistically significant difference between the treatment group and the control group in terms of occurrence of complications ($P < 0.05$).

2.3 Prognosis of the two groups 6 weeks after surgery was assessed by Karnofsky scores: good prognosis (70–100), effectiveness (40–60), and invalidity (≤ 30). The results are shown in Table 3. The proportion of patients in the treatment group with a good prognosis was significantly higher than that of the control group, while the proportion of patients with invalidity was significantly lower ($P < 0.05$).

3 Discussion

About 40% of patients with hypertension ventricular hemorrhage experience hypertensive cerebral hemorrhage. External ventricular drainage is the main surgical therapy for intraventricular hemorrhage at present. External ventricular drainage can quickly eliminate acute hydrocephalus caused by intraventricular hemorrhage. The combined therapy of external ventricular drainage together with urokinase treatment can address the series of pathological changes caused by hematoma oppression of important brain tissues surrounding the ventricle. Fast elimination of the effects of intraventricular hematoma on deep brain structures is key to treatment of intraventricular hemorrhage (Atsumi et al., 2011).

The therapeutic effect of external ventricular drainage combined with urokinase treatment for intraventricular hemorrhage has been controversial. Some studies have reported that this combined therapy can lower the death rate among patients with intraventricular hemorrhage, but it cannot improve neurological function (King et al., 2012; Wang et al., 2017). Other studies argue that external ventricular drainage together with urokinase treatment does not lower the death rate or improve neurological function (Liu et al., 2017). These previous studies generally involve small doses of urokinase, which might influence the therapeutic effect of this drug for intraventricular hemorrhage.

In this study, large doses of urokinase (50,000 IU/side) were injected into the lateral ventricle and this assisted in the elimination of intraventricular hemorrhage. The intraventricular hemorrhage clearance time of the treatment group was 4.20 ± 1.12 days, which is about 6 days shorter than that of the control group (10.23 ± 2.52 days). As such, there was a significant difference in intraventricular hemorrhage clearance time between the two groups ($P < 0.05$). Although the total urokinase dosage of the treatment group was increased relative to the control group, the treatment group had decreased urokinase treatment instances relative to the control group. The number of urokinase treatment instances in the treatment group was 7.2, on average, which was significantly different from that of the control group (14.3 instances) ($P < 0.05$). Moreover, the high-dose urokinase treatment reduced the time of open ventricular system operation, the drainage tube indwelling time, and the incidence of intracranial infection, which are the major complications of external ventricular drainage. The treatment group had one case of intracranial infection and the control group has three cases of intracranial infection; this difference was statistically significant ($P < 0.05$).

The Karnofsky scores of the two groups 6 weeks after surgery were compared. The proportion of patients in the treatment group with a good was higher than the control group, while the proportion of patients with invalidity was lower. There were statistically significant differences between the two groups ($P < 0.05$).

Intracranial re-hemorrhage and neurotoxicity are major side effects of urokinase treatment. Torres et al. (2008) reported that urokinase treatment of moderate and severe intraventricular hemorrhage can quickly eliminate intraventricular hematoma and prevent intracranial re-hemorrhage. Many studies have demonstrated that coagulation function remains basically the same before and after urokinase treatment (Brott et al., 2007). In this study, high-dose urokinase treatment was applied, but no patients suffered intracranial re-hemorrhageor re-hemorrhage of other systems. Based on this analysis, the re-hemorrhage caused by urokinase treatmentin previous studies may be related to other diseases associated with a high risk of hemorrhage, such as cerebral aneurysm, moyamoya disease, and cerebral arteriovenous malformation, among others. Clinically, intraventricular hemorrhage caused by these diseases requires early etiological treatment. Attention should be paid to urokinase treatment before pathogenesis elimination. It should also be noted that a comparative study of urokinase and alteplase in an intraventricular hemorrhage model of rats reported that urokinase can significantly improve functional recovery without causing neurotoxicity (Gaberel et al., 2014; Lin and Chen, 2019).

For patients with hypertension ventricular hemorrhage, ventricular drainage combined with high-dosage urokinase treatment can quickly eliminate intraventricular hemorrhage, shorten the drainage tube indwelling time, and decrease the duration of the open ventricular system, thus enabling a decreased incidence of post-surgery intracranial infection and an increased likelihood of good prognosis.

Funding

This work was supported by Projects of Zhejiang Medical and Health Science and Technology (No. 2015ZA209,No. 2019ZD058), Shaoxing Science and Technology Bureau (No. 2018C30146) and Zhejiang Science and Technology (No. GF18H09008).

References

Atsumi, H., Matsumae, M., Hirayama, A., et al., 2011. Newly developed electromagnet tracked flexible neuroendoscope. Neurol. Med. Chir. (Tokyo) 51 (8), 611.

Brott, T., Broderick, J., Kothari, R., et al., 2007. Early hemorrhage growth in patients with intracerebral hemorrhage. Stroke 28 (1), 1–5.

Gaberel, T., Montagne, A., Lesept, F., et al., 2014. Urokinase versus Alteplase for intraventricular hemorrhage fibrinolysis. Neuropharmacology 85, 158–165.

King, N.K., Lai, J.L., Tan, L.B., et al., 2012. A randomized placebo-controlled pilot study of patients with spontaneous intraventricular haemorrhage treated with intraventricular thrombolysis. J. Clin. Neurosci. 19 (7), 961–964.

Lin, C.-Q., Chen, L.-K., 2019. Effect of hemorrhagic cerebrospinal fluid drainage on cognitive function after intracranial aneurysm clipping. Brain Sci. Adv. 5 (1), 65–72. https://doi.org/10.26599/bsa.2019.9050006.

Liu, Y.F., Qiu, H.C., Zhao, G.F., et al., 2017. Retrospective analysis of the discharge outcome of externalventricular drainage and conservative treatment in patients with severe intraventricular hemorrhage. Zhonghua Yi Xue Za Zhi 97 (29), 2257–2260.

Torres, A., Plans, G., Martino, J., et al., 2008. Fibrinolytic therapy in spontaneous intraventricular haemorrhage: efficacy and safety of the treatment. Br. J. Neurosurg. 22, 269.

Wang, D., Liu, J., Norton, C., et al., 2017. Local fibrinolytic therapy for intraventricular hemorrhage: a meta-analysis of randomized controlled trials. World Neurosurg. 107, 1016–1024.

CHAPTER

Topical application of CNTF, GDNF and BDNF in combination attenuates blood-spinal cord barrier permeability, edema formation, hemeoxygenase-2 upregulation, and cord pathology

10

Aruna Sharma[a,*], Lianyuan Feng[b], Dafin F. Muresanu[c,d], Hongyun Huang[e], Preeti K. Menon[f], Seaab Sahib[g], Z. Ryan Tian[g], José Vicente Lafuente[h], Anca D. Buzoianu[i], Rudy J. Castellani[j], Ala Nozari[k], Lars Wiklund[a], and Hari Shanker Sharma[a,*]

[a]*International Experimental Central Nervous System Injury & Repair (IECNSIR), Department of Surgical Sciences, Anesthesiology & Intensive Care Medicine, Uppsala University Hospital, Uppsala University, Uppsala, Sweden*

[b]*Department of Neurology, Bethune International Peace Hospital, Shijiazhuang, Hebei Province, China*

[c]*Department of Clinical Neurosciences, University of Medicine & Pharmacy, Cluj-Napoca, Romania*

[d]*"RoNeuro" Institute for Neurological Research and Diagnostic, Cluj-Napoca, Romania*

[e]*Institute of Neurorestoratology, Third Medical Center of General Hospital of PLA, Beijing, China*

[f]*Department of Biochemistry and Biophysics, Stockholm University, Stockholm, Sweden*

[g]*Department of Chemistry & Biochemistry, University of Arkansas, Fayetteville, AR, United States*

[h]*LaNCE, Department of Neuroscience, University of the Basque Country (UPV/EHU), Leioa, Bizkaia, Spain*

[i]*Department of Clinical Pharmacology and Toxicology, "Iuliu Hatieganu" University of Medicine and Pharmacy, Cluj-Napoca, Romania*

[j]*Department of Pathology, University of Maryland, Baltimore, MD, United States*

[k]*Anesthesiology & Intensive Care, Massachusetts General Hospital, Boston, MA, United States*

**Corresponding authors: Tel.: +46-70-2195963 (AS)/+46-18-243899 (HSS); Fax: +46-18-243899 (AS)/+46-18-243899 (HSS), e-mail address: aruna.sharma@surgsci.uu.se; sharma@surgsci.uu.se; harishanker_sharma55@icloud.com; hssharma@aol.com*

Progress in Brain Research, Volume 266, ISSN 0079-6123, https://doi.org/10.1016/bs.pbr.2021.06.013

Copyright © 2021 Elsevier B.V. All rights reserved.

Abstract

Spinal cord injury (SCI) is one of the leading causes of disability in Military personnel for which no suitable therapeutic strategies are available till today. Thus, exploration of novel therapeutic measures is highly needed to enhance the quality of life of SCI victims. Previously, topical application of BDNF and GDNF in combination over the injured spinal cord after 90min induced marked neuroprotection. In present investigation, we added CNTF in combination with BDNF and/or GDNF treatment to examine weather the triple combination applied over the traumatic cord after 90 or 120min could thwart cord pathology. Since neurotrophins attenuate nitric oxide (NO) production in SCI, the role of carbon monoxide (CO) production that is similar to NO in inducing cell injury was explored using immunohistochemistry of the constitutive isoform of enzyme hemeoxygenase-2 (HO-2). SCI inflicted over the right dorsal horn of the T10–11 segments by making an incision of 2mm deep and 5mm long upregulated the HO-2 immunostaining in the T9 and T12 segments after 5h injury. These perifocal segments are associated with breakdown of the blood-spinal cord barrier (BSCB), edema development and cell injuries. Topical application of CNTF with BDNF and GDNF in combination (10ng each) after 90 and 120min over the injured spinal cord significantly attenuated the BSCB breakdown, edema formation, cell injury and overexpression of HO-2. These observations are the first to show that CNTF with BDNF and GDNF induced superior neuroprotection in SCI probably by downregulation of CO production, not reported earlier.

Keywords

Hemeoxygenase 2, Spinal cord injury, BDNF, GDNF, CNTF, Blood-spinal cord barrier, Spinal cord edema

1 Introduction

The cellular and molecular mechanisms of spinal cord injury (SCI) induced pathophysiology leading to sensory motor disturbances and cord dysfunction is still not well known (Afshary et al., 2020; Enck et al., 2006; Grabher et al., 2015; Grulova et al., 2013; Hearn and Cross, 2020; Noamani et al., 2020; Papa et al., 2020; Sadowsky et al., 2013; see Sharma, 2011, 2005b). A focal trauma to the spinal cord results in production of several neurodestructive factors, e.g., nitric oxide (NO), carbon monoxide (CO), oxygen radicals, lipid peroxidation and other injury inducing agents leading to cell and tissue destruction and long term functional disability (Christie et al., 2008; Conti et al., 2007; de Rivero Vaccari, 2019; Gao and Li, 2017; Hall et al., 2016; Li et al., 2019; Liu et al., 2020; Marsala et al., 2007; Panahian and Maines, 2001; Sámano and Nistri, 2019; Sharma, 2006; Sharma and Westman, 2003; Sharma et al., 1998, 2000a,b,c, 2017, 2019; Zheng et al., 2019).

There are reasons to believe that several tissue injury factors that are released after trauma act in coherence to produce cell damage (Sharma, 2011, 2005b). Thus, a possibility exists that blocking of one factor following trauma will thwart the injury cascades and slowing the progression and persistence of cell injuries by reducing the consequences of cord pathology effectively (see Sharma, 2011, 2007b, 2006). Previous observations from our laboratory showed that SCI induces marked production of NO and CO and blockade of these radicals production with pharmacological agents resulted in pronounced neuroprotection (Sharma and Westman, 2003; Sharma et al., 2000a,b,c). This suggests that several neurodestructive factors in SCI are working in coherence to induce cellular damage.

This idea is further supported by our experiments in which topical application of brain derived neurotrophic factor (BDNF) either given alone or in combination with other neurotrophins, e.g., glial derived neurotrophic factor (GDNF) or insulin like growth factor1 (IGF-1) topically reduces cell injury after SCI in a rat model (Sharma, 2011, 2007b, 2006; Sharma and Westman, 2003; Sharma et al., 2000a,b,c). Interestingly topical application of BDNF or IGF-1 given separately are able to thwart upregulation of neuronal nitric oxide synthase (NOS) or hemeoxugenase-2 (HO-2), the enzymes responsible for NO or CO production, respectively within the cord and induced profound neuroprotection (Sharma et al., 2000a,b,c). This suggests that both NO and CO production contribute to spinal cord cell and tissue damage in SCI. Our observations further show that a select combination of BDNF and GDNF when administered topically over the injured cord up to 90min after SCI resulted in good neuroprotection (Sharma, 2010, 2007a,b, 2006, 2005a,b, 2003; Sharma et al., 2000a,b,c). Thus, novel strategies and new combination of neurotrophins are needed to extend this post-trauma application period of drugs to achieve neuroprotection that is needed for trauma victims in clinical situations. In present investigation we examined triple combinations of neurotrophins, i.e., BDNF, GDNF and ciliary neurotrophic factor (CNTF) given 90 or 120min after SCI on CO production and cell injury in our rat model. In addition, breakdown of the blood-spinal cord barrier (BSCB) and edema formation were also examined in these neurotrophins treated injured animals.

2 Materials and methods

2.1 Animals

Experiments were carried out on Male Sprague Dawley Rats (250–300g body weight) housed at controlled room temperature 21 ± 1 °C with 12h light and dark schedule. The rat feed and tap water were supplied ad libitum.

All experiments were carried out according to National Institute of health (NIH) Guidelines and Care of animals and approved by local Institutional Ethics Committee.

2.2 Spinal cord injury

Under Equithesin anesthesia (3mL/kg, i.p.) one segment laminectomy was done over the T10–11 segments. Spinal cord injury (SCI) was inflicted using a longitudinal lesion (2mm deep and 5mm long) over the right dorsal horn of the T10–11 segment using a sterile scalpel blade (Vannemreddy et al., 2006; Winkler et al., 2000). The deepest portion of the lesion was limed to the Rexed laminae VIII (Sharma and Olsson, 1990; Sharma et al., 1990a,b, 1993, 1996). The wound was covered with cotton soaked in saline and animals were allowed to survive 5 h after injury. Normal animals were used as controls.

2.3 Neurotrophins treatment

BDNF, CNTF and GDNF (Sigma Chemical Co., St. Louis, MO, USA) were dissolved in saline in a concentration of 1 μg/mL each. At the time of experiment, 10 ng of each solution (10 μL) were applied over the injured spinal cord using a Hamilton 50 μL Syringe over 10 s at 90 or 120 min after SCI (Sharma and Westman, 2003; Sharma, 2010, 2007a,b, 2006, 2005a,b, 2003; Sharma et al., 2000a,b,c). All three neurotrophins were given immediately after each other in the order of BDNF (10 μL) followed by GDNF (10 μL) and CNTF (10 μL). Reversing the order of treatment was not examined in this study. Control rats received saline in identical manner after SCI (see Sharma et al., 1998).

2.4 Morphological investigations of the spinal cord

For spinal cord morphological investigations, after the end of the experiment, animals were perfused with 4% buffered paraformaldehyde through cardiac puncture preceded with a saline rinse. After perfusion, the spinal cord was dissected out and tissue pieces from the T9, T10–11 and T12 were divided and embedded in paraffin. About 3-μm thick sections were cut and stained with Hematoxylin or Eosin or Nissl for investigation neuronal injuries (Sharma et al., 1998; Vannemreddy et al., 2006).

2.5 Hemeoxygenase 2 immunostaining of the spinal cord

For immunostaining of hemeoxygenase-2 (HO-2) the spinal cord tissues were fixed on a Vibratome (Oxford Instruments, UK) and free-floating sections (40 μm thick) were cut and incubated with HO-2 antibodies (monoclonal 1:500) for overnight at 4 °C under continuous agitation (Alm et al., 2000; Gordh et al., 2000; Sharma et al., 2003, 2000a,b,c). The immunostaining was developed on the next day using peroxidase-antiperoxidase method. Some paraffin sections were also immunostained without HO-2 antibodies on Vibratome sections separately (Sharma et al., 2000c).

2.6 Cell injury in the spinal cord

The histological data for nerve cell injury and immunocytochemical sections for HO-2 immunoreactions were analyzed under an Inverted Zeiss microscope and photographed using a digital camera. All imaged were processed using Photoshop 7 version on a Mac OS X.10.6 system using identical color filters for control and experimental groups (see Sharma, 2011, 2007b).

2.7 Blood-spinal cord barrier permeability

The blood-spinal cord barrier (BSCB) was examined using Evans blue and radioiodine tracers as described eelier (Sharma et al., 2000a,b,c). In brief, both tracers were administered into the right femoral vein under anesthesia and after 5 min the intravascular tracer was washed out using 0.9% saline at 90 mmHg arterial pressure. The spinal cord was dissected out and radioactivity measured in a Gamma counter. Before saline perfusion, 1 mL of whole blood was taken out after cardiac puncture from the left vernicle for measuring blood radioiodine concentration at the time of killing (Sharma et al., 1998; Vannemreddy et al., 2006). After that, the spinal cord tissue was homogenized in a mixture of sodium sulfate and acetone to extract the Evans blue due (Alm et al., 2000; Gordh et al., 2000; Sharma et al., 2003). The extracted dye after centrifugation of the samples was measured in a spectrophotometer at 620 nm and the dye concentration was calculated from the standard curve of Evans blue as described earlier.

2.8 Spinal cord edema

The spinal cord edema formation was determined from the changes in the water content of the cord. For this purpose, spinal cord tissue from T9 to T12 were dissected out, weighed immediately and placed in an incubator at 90 °C for 72 h to obtain dry weight (Alm et al., 2000; Sharma, 2011, 2010, 2007a,b, 2006, 2005a). The cord water content was calculated from the difference in wet and dry weights of the cord samples (Sharma, 2003; Sharma and Westman, 2003; Sharma et al., 2003).

2.9 Spinal cord blood flow

The spinal cord blood flow (SCBF) was measured using [125]-I labeled carbonized microspheres ($15 \pm 0.6\,\mu m$, o.d.) administered into the heart through left common carotid artery near the left ventricle within 45 s (about 90k microspheres) and reference blood flow was withdrawn from the right femoral arterly at the rate of 0.8 mL/g/min as described earlier (Sharma et al., 1990a,b). The SCBF was expressed as mL/g/min (Sharma et al., 1990a,b).

2.10 Statistical analyses of the data

ANOVA followed by Dunnett's test for multiple group comparison with one control was used to analyze statistical significance of the data obtained. For semiquantittative data on morphological analyses, non-parametric Chi-Square test was used. A p value <0.05 was considered significant.

3 Results

3.1 Effect of neurotrophins on HO-2 expression in the cord after trauma

SCI resulted in profound upregulation of HO-2 expression in the cord that was most marked in the ipsilateral side in all the spinal cord segments T9, T10–11 or T12 examined (Table 1). The magnitude and intensity of HO-2 immunostaining was most

Table 1 Blood-spinal cord barrier, edema formation, volume swelling, spinal cord blood flow, neuronal injuries and hemeoxynesae immunohistochemistry in 5 h spinal cord injury (SCI).

Type of experiment	Spinal cord segments		
	T9	T10–11	T12
A. BSCB permeability			
Control	0.20 ± 0.02	0.18 ± 0.03	0.19 ± 0.04
SCI EBA, mg %	1.34 ± 0.12**	1.45 ± 0.21**	1.23 ± 0.18**
SCI [131]-I, %	1.58 ± 0.14**#	2.12 ± 0.18**#	1.78 ± 0.16**#
B. Spinal cord edema			
Control	65.76 ± 0.34	65.59 ± 0.23	65.34 ± 0.28
SCI water content, %	68.34 ± 0.28**	69.18 ± 0.17**	67.98 ± 0.23**
SCI, % *f*	+10%	+14%	+9%
C. Spinal cord blood flow			
Control	1.25 ± 0.05	1.23 ± 0.08	1.21 ± 0.06
SCI, mL/g/min	0.72 ± 0.05**	0.70 ± 0.06**	0.73 ± 0.08**
D. Neuronal injury			
Control	3 ± 2	4 ± 3	2 ± 2
SCI nr.	42 ± 6§§	49 ± 4§§	38 ± 7§§
E. HO-2 immunostaining			
Control	4 ± 3	3 ± 2	2 ± 2
SCI nr.	34 ± 4§§	41 ± 6§§	32 ± 3§§

*SCI was inflicted on the T10–T11 segments on the right dorsal horn by making a longitudinal incision 2 mm deep and 5 mm long. The deepest part of the lesion is within the laminae VIII of the right dorsal horn. For details, see text. Values are Mean ± SD of 5–6 rats at each point. **P < 0.01 from control; #P < 0.05 from EBA, ANOVA followed by Dunnett's test for multiple group comparison from on control. Semiquantitative data analyzed using non-parametric Chi-Square test §§P < 0.01. % f = Volume swelling calculated from differences between control and experimental brain water content according to Formulae of Elliott and Jasper (1949). About 1% changes in the water content is equal to approximately 4% volume swelling. HO-2 = Hemeoxygenesae-2.*

severe in the injured cord segment followed by the segment below the lesion (T12) and the rostral to the injury site (T9) (Fig. 1). This upregulation of HO-2 expression was thwarted by a combination of BDNF and GDNF or GDNF or CNTF slightly but significantly when applied over the cord 90 min after trauma (Fig. 2). However, these combinations failed to reduce HO-2 expression when given after 120 min injury (Results not shown). On the other hand, a combination of BDNF, GDNF and CNTF if given 90 or 120 min after trauma, the reduction in HO-2 expression was markedly attenuated (Fig. 1). This effect of the triple combination of neurotrophins was most marked when they were administered 90 min after trauma but a significant and pronounced decrease in HO-2 expression was still seen following their application over the cord at 120 min after the SCI.

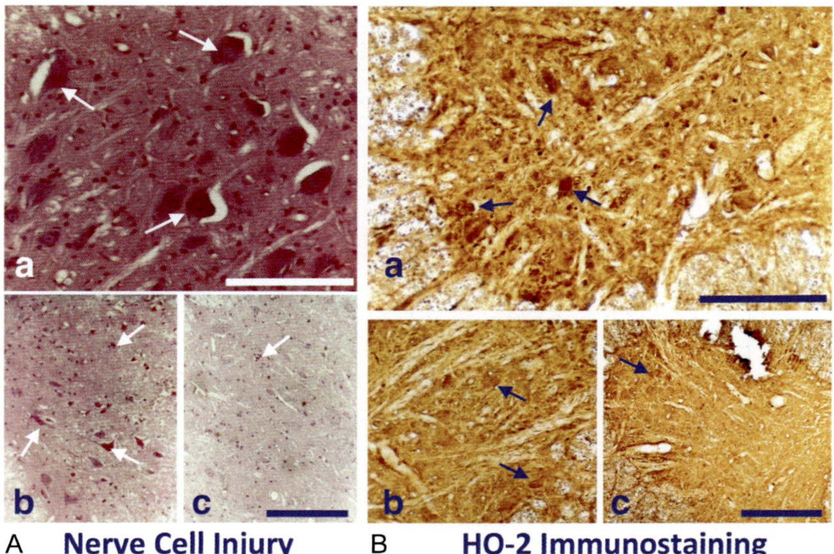

FIG. 1

Morphological evaluation of spinal cord neuronal injuries in the left ventral horn of the T9 segment (A) and hemeoxygenase-2 (HO-2) immunohistochemistry in the spinal cord left ventral horn of the T9 segment (B) after 5 h spinal cord injury (SCI). SCI was inflicted on the T10–11 segments by making longitudinal incision over the right dorsal horn (2 mm deep and 5 mm long). Neuronal damage and perineuronal edema, expansion of the neuropil and distorted neurons are present in the left T9 ventral horn (A: a, b arrows). HO-2 upregulation is also seen in the left T9 ventral horn in the areas showing neuronal damages and regions that are stained with Evans blue albumin (EBA, B: a arrows). Topical administration of ciliary neurotrophic factor (CNTF) with brain derived neurotrophic factor (BDNF) and glial cell line-derived neurotrophic factor (GDNF) in combination significantly attenuating HO-2 immunoreactive cells in the spinal cord within the T9 segment of the ventral horn (B: c arrows) after injury and neuronal injuries (A: c). Topical application of other combinations of neurotrophins (CNTF + BDNF) was also reduced the spinal cord HO-2 immunoreactive cells in the T9 segment (B: b arrows) as compared to the untreated injured spinal cord (B: a). SCI = spinal cord injury; CNT = ciliary neurotrophic factor; BD = brain derived neurotrophic factor; GD = glial cell line-derived neurotrophic factor. For details, see text. The HO-2 immunostaining was done Vibratome sections (40 μm) and neuronal injuries were examined on Paraffin sections (3-μm thick) using standard protocol. Bar = (A) a = 40 μm; b, c = 60 μm; (B) a = 40 μm; (B) a, c = 30 μm.

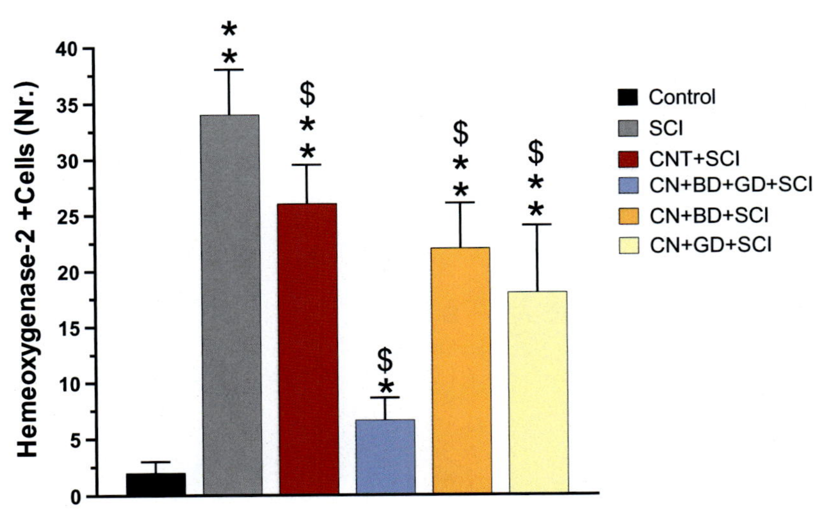

FIG. 2

Semiquantitative evaluation of hemeoxygenase-2 (HO-2) immunohistochemistry in the T9 segment of spinal cord after 5 h spinal cord injury (SCI). SCI was inflicted on the T10–11 segments by making longitudinal incision over the right dorsal horn (2 mm deep and 5 mm long). Topical administration of ciliary neurotrophic factor (CNTF) with brain derived neurotrophic factor (BDNF) and glial cell line-derived neurotrophic factor (GDNF) in combination has the most pronounced effects in attenuating HO-2 immunoreactive cells in the spinal cord within the T9 segment of the spinal cord after injury. Topical application of other combinations of neurotrophins were also reduced the spinal cord HO-2 immunoreactive cells in the T9 segment significantly. Values are Mean ± SD of 5–6 rats at each point. **$P<0.01$ from control; \$$P<0.05$ from SCI 5 h, ANOVA followed by Dunnett's test for multiple group comparison from one control. SCI = spinal cord injury; CNT = ciliary neurotrophic factor; BD = brain derived neurotrophic factor; GD = glial cell line-derived neurotrophic factor. For details, see text.

3.2 Effect of neurotrophins on neuronal injury in the cord after trauma

A close parallelism was seen between neuronal injuries and HO-2 expression in the cord in untreated animals after SCI (Table 1). Thus, neurons that could express HO-2 upregulation or in its vicinity were markedly damaged as evident form their distorted shape, dark cytoplasm and karyoplasm with signs of chromatolysis. In many of these neurons the cell nucleus is also distorted containing and eccentric nucleolus. Perineuronal edema and sponginess of the adjacent neuropil are clearly visible (Fig. 1).

Treatment with a combination of BDNF with GDNF or GDNF with CNTF after 90 min of SCI resulted in only mild reduction in neuronal damages. However, a triple combination of these neurotrophins, i.e., BDNF, CNTF and GDNF resulted in marked reduction in neuronal injury even when applied after 120 min of SCI. This

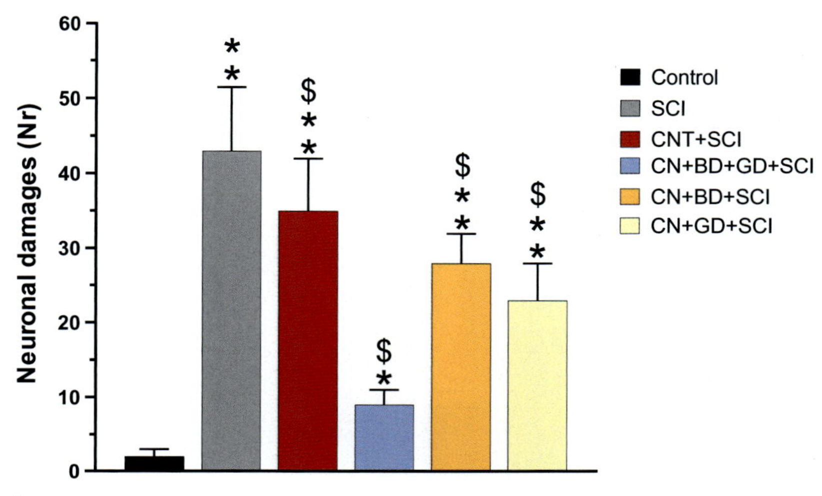

FIG. 3

Semiquantitative evaluation of neuronal distortion in the T9 segment of spinal cord after 5 h spinal cord injury (SCI). SCI was inflicted on the T10–11 segments by making longitudinal incision over the right dorsal horn (2 mm deep and 5 mm long). Topical administration of ciliary neurotrophic factor (CNTF) with brain derived neurotrophic factor (BDNF) and glial cell line-derived neurotrophic factor (GDNF) in combination has the most pronounced effects in attenuating neuronal injuries in the spinal cord within the T9 segment of the spinal cord after injury. Topical application of other combinations of neurotrophins were also reduced the spinal cord neuronal injuries in the T9 segment significantly. Values are Mean ± SD of 5–6 rats at each point. ** $P < 0.01$ from control; \$ $P < 0.05$ from SCI 5 h, ANOVA followed by Dunnett's test for multiple group comparison from one control. SCI = spinal cord injury; CNT = ciliary neurotrophic factor; BD = brain derived neurotrophic factor; GD = glial cell line-derived neurotrophic factor. For details, see text.

triple combination of neurotrophins almost thwarted neuronal injuries if applied 90 min after trauma (Figs. 1 and 3). The most marked neuronal protection with triple combination of neurotrophins was evident in the T9 and T12 segments as compared to T10–11 segment. These neuroprotective effects of triple combination of neurotrophins were most pronounced in the contralateral side of the cord as compared to the ipsilateral cord.

3.3 Effect of neurotrophins on blood-spinal cord barrier permeability after trauma

SCI resulted in profound leakage of Evans blue albumin and radioiodine across the T9, T10–11 and the T12 segments. The most marked increase in BSCB permeability were seen in the injured T10–11 segments followed by T12 and T9 segments (Table 1). A combination of BDNF and GDNF or GDNF and CNTF reduced this

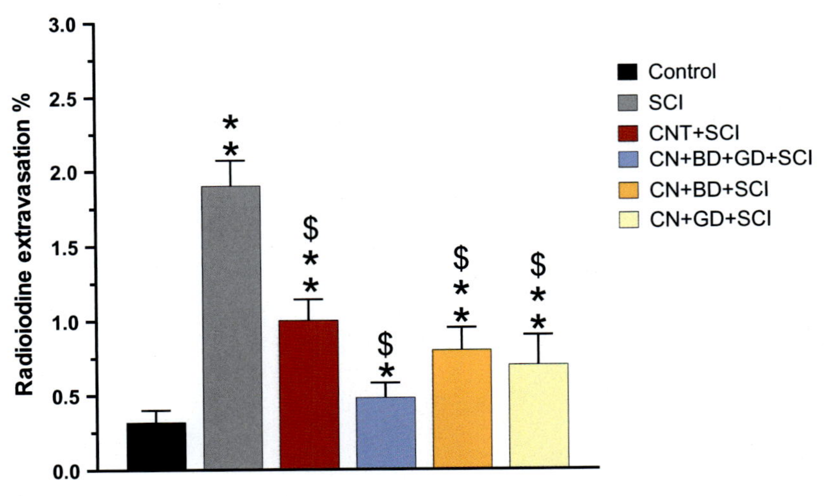

FIG. 4

Extravasation of [131]-Iodine in the T9 spinal cord segment after 5 h spinal cord injury (SCI)-SCI was inflicted on the T10–11 segments by making longitudinal incision over the right dorsal horn (2 mm deep and 5 mm long). Topical administration of ciliary neurotrophic factor (CNTF) with brain derived neurotrophic factor (BDNF) and glial cell line-derived neurotrophic factor (GDNF) in combination has the most pronounced effects in attenuating extravasation of radioiodine within the T9 segment of the spinal cord after injury. Topical application of other combinations of neurotrophins were also reduced the blood-spinal cord barrier (BSCB) permeability to radioiodine in the T9 segment significantly. Values are Mean ± SD of 5–6 rats at each point. $^{**}P<0.01$ from control; $\$P<0.05$ from SCI 5 h, ANOVA followed by Dunnett's test for multiple group comparison from one control. SCI = spinal cord injury; CNT = ciliary neurotrophic factor; BD = brain derived neurotrophic factor; GD = glial cell line-derived neurotrophic factor. For details, see text.

leakage slightly but significantly only when applied 90 min after SCI (Fig. 4). However, these combinations of neurotrophins did not reduce BSCB leakage to radioiodine or Evans's blue when applied 120 min after trauma. On the other hand, the triple combination of neurotrophins, e.g., BDNF, GDNF and CNTF resulted in not only a pronounced reduction in the BSCB breakdown when applied after 90 min of trauma but also powerfully to thwart BSCB disturbances after their application at 120 min after injury.

3.4 Effect of neurotrophins on spinal cord edema after trauma

SCI resulted in profound edema formation in the injured (T10–11) segment that was most pronounced in the injured T10–11 segments followed by T12 and T9 segments. A combination of BDNF and GDNF or GDNF and CNTF 90 min after SCI reduced the edema formation mildly but significantly (Table 1 and Fig. 5). However, if these

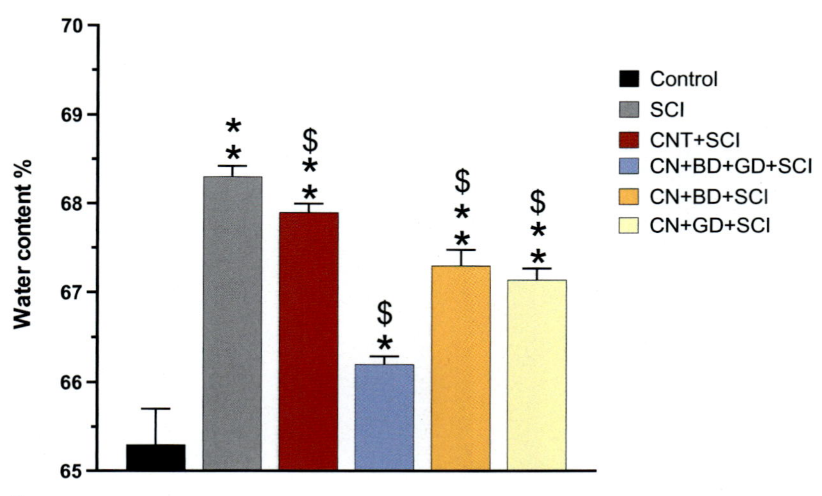

FIG. 5

Measurement of T9 segment of spinal cord water content after 5 h spinal cord injury (SCI). SCI was inflicted on the T10–11 segments by making longitudinal incision over the right dorsal horn (2 mm deep and 5 mm long). Topical administration of ciliary neurotrophic factor (CNTF) with brain derived neurotrophic factor (BDNF) and glial cell line-derived neurotrophic factor (GDNF) in combination has the most pronounced effects in attenuating edematous swelling and the spinal cord water content within the T9 segment of the spinal cord after injury. Topical application of other combinations of neurotrophins were also reduced the spinal cord water content in the T9 segment significantly. Values are Mean ± SD of 5–6 rats at each point. ***P*<0.01 from control; $*P*<0.05 from SCI 5 h, ANOVA followed by Dunnett's test for multiple group comparison from one control. SCI = spinal cord injury; CNT = ciliary neurotrophic factor; BD = brain derived neurotrophic factor; GD = glial cell line-derived neurotrophic factor. For details, see text.

combination of neurotrophins were given 120 min after injury no beneficial effects of these growth factors could be noted. On the other hand, when triple combination of neurotrophins were given either 90 or 120 min after SCI profound reduction of spinal cord edema formation was seen in these group of animals. This antiedematous effect of triple neurotrophins was most marked after 90 min SCI although; remarkable increase in the SCBF was also seen after their application at 120 min after trauma.

3.5 Effect of neurotrophins on spinal cord blood flow after trauma

SCI resulted in significant reductions in the SCBF in the injured (T10–11) segment that was most pronounced followed by T12 and T9 segments (Fig. 6 and Table 1). A combination of BDNF and GDNF or GDNF and CNTF 90 min after SCI reduced the edema formation mildly but significantly. However, if these combinations of neurotrophins were given 120 min after injury no beneficial effects of these growth

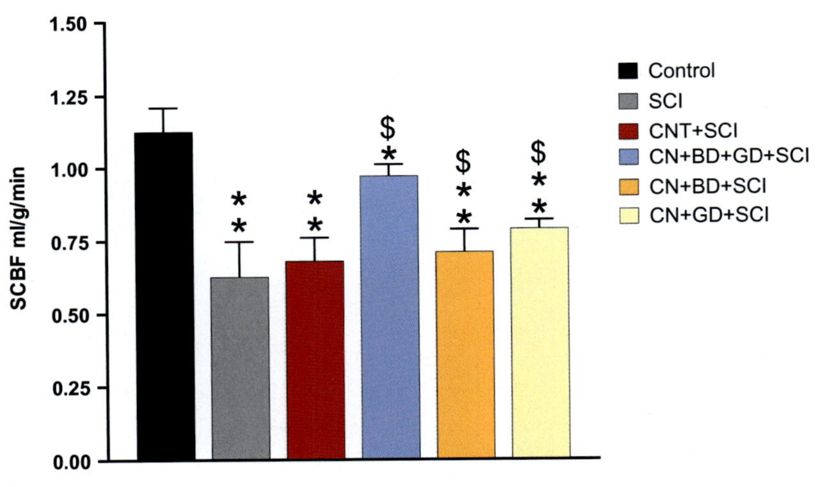

FIG. 6

Measurement of T9 segment of spinal cord blood flow using radiolabeled carbonized microspheres ($15 \pm 0.6\,\mu m$ o.d.) after 5h spinal cord injury (SCI). SCI was inflicted on the T10–11 segments by making longitudinal incision over the right dorsal horn (2mm deep and 5mm long). Topical administration of ciliary neurotrophic factor (CNTF) with brain derived neurotrophic factor (BDNF) and glial cell line-derived neurotrophic factor (GDNF) in combination has the most pronounced effects on improving the spinal cord blood flow (SCBF) within the T9 segment of the spinal cord after injury. Topical application of other combinations of neurotrophins were also improved the SCBF in the T9 segment significantly. Values are Mean ± SD of 5–6 rats at each point. $**P<0.01$ from control; $\$P<0.05$ from SCI 5h, ANOVA followed by Dunnett's test for multiple group comparison from one control. SCI = spinal cord injury; CNT = ciliary neurotrophic factor; BD = brain derived neurotrophic factor; GD = glial cell line-derived neurotrophic factor. For details, see text.

factors could be noted. On the other hand, when triple combination of neurotrophins were given either 90 or 120min after SCI profound reduction in edema formation was seen in these group of animals (Fig. 6). This antiedematous effect of triple neurotrophins was most marked after 90min SCI although; remarkable reduction in edema development was also seen after their application at 120min after trauma.

4 Discussion

The salient findings of this investigation show that a triple combination of neurotrophins, e.g., BDNF, GDNF and CNTF (but not any other two combinations) if applied over the traumatized cord up to 90 or 120min after SCI resulted in remarkable neuroprotection. This suggests that adding CNTF in association with BDNF and GDNF results in enhanced neuroprotection of the cells and tissues in the spinal cord even applied at the late stage of SCI, not reported earlier (Chen et al., 2019, 2015;

Cao et al., 2010; Hodgetts et al., 2018; Hu et al., 2020; Xiong et al., 2017). Previous studies form our laboratory showed that BDNF if given alone induces marked neuroprotection when administered within 30 min after injury (Sharma, 2003). Adding GDNF with BDNF is able to protect the spinal cord cells and tissue damage when applied over the cord 60–90 min after injury (Sharma, 2010, 2007a,b, 2006, 2005a). However, the present investigation for the first time showed that adding CNTF could further induce remarkable neuroprotection if this triple combination of neurotrophins is applied 90–120 min after trauma. This suggests that suitable mixture of neurotrophins is needed to induce neuroprotection at later stages of traumatic injuries to the cord.

It appears that this combination of neurotrophins is able to thwart tissue damage for at least up to 5 h after the initial insult to the cord. This observation is supported by the fact that cerebrolysin—a balanced composition of several neurotrophic factors and active peptide fragments induced profound neuroprotection in SCI (Haninec et al., 2003; Menon et al., 2012; Sahib et al., 2020; Sharma et al., 2010, 2014). Thus, it remains to be seen whether prolongation of the survival period from 5 to 12 h and beyond could also exhibit neuroprotection caused by these neurotrophins. Alternatively, whether repeated treatment of neurotrophins is needed following prolongation of survival periods after injury. This is a subject that is currently being examined in our laboratory.

In spite of these shortcomings of the present investigation, the functions significance of our findings is immense. This is evident form the facts that normally in clinical situations patients with SCI form the injury site requires at least 90–120 min to reach hospitals depending on the location and region of the traumatic insult. By the time victims reaches neurotrauma centers for specialized treatments, damage to the cord already worsen leading to profound disability at the time of admission to the hospital (Sharma, 2011, 2010, 2007a,b, 2006). Based on our observations it is possible that victims may be given a cocktail of neurotrophins at the site of injury to arrest rapid pathological changes in the cord while transferred to the hospitals for better treatment facilities. This is especially true for military personnel who get injured during combat operations and require specialized care hospitals or trauma centers for effective treatments (Ayaz et al., 2014; Fritz et al., 2015; Furlan et al., 2020; Hanigan and Sloffer, 2004; Kallakuri et al., 2015; Leslie, 1946; McDonald et al., 2018; Schoenfeld et al., 2011; Wolman, 1967; Wyndaele, 2011). In this way one could delay the post trauma damage of the cords for some time before therapeutic intervention by specialists could take place.

The possible molecular or cellular mechanisms of neuroprotection caused by addition of CNTF with BDNF and GDNF in SCI at the late phase of injury is unclear from this investigation. However, it appears that CNTF addition is needed at the late phase of SCI to achieve neuroprotection when BDNF or GDNF effects are dwindling (Cao et al., 2010; Han et al., 2011; Nagahara and Tuszynski, 2011; Panahian and Maines, 2001). This suggests that CNTF signaling pathways in the spinal cord are activated at a later phase and exogenous supplement of CNTF could enhance neuroprotective strategies together with BDNF and GDNF in the cord (Cao et al., 2010;

Nagahara and Tuszynski, 2011; Sharma, 2011, 2010, 2007a,b; Sharma et al., 1990a,b). However, our results also showed that a combination of CNTF with BDNF or GDNF alone did not result sufficient neuroprotection in the cord if applied after 90–120 min (Askvig and Watt, 2015; Azadi et al., 2007; Ip and Yancopoulos, 1992; Jia et al., 2018; Stankoff et al., 2002). This suggests that a combination of these three neurotrophins are essential in reducing cell and tissue injury in SCI after 90–120 min of insult.

This neuroprotection caused by triple actions of neurotrophins is closely related with breakdown of the BSCB permeability. This suggests that neurotrophins in combination are able to strengthen the endothelial cell membrane function leading to restoration the endothelial cell membrane integrity (Sharma, 2011, 2007a,b, 2006, 2005a,b, 2003). Obviously, restoration of the BSCB results in reduction of vasogenic edema formation and cell injury (Sharma, 2007b, 2005b). The possible mechanisms of restoration of BSCB function by neurotrophins is far from clear. However, it appears that SCI induced oxidative stress and generation of free radicals, e.g., NO or CO could contribute to the endothelial cell damage (Sharma et al., 2003, 2000a,b,c, 1998). Blockade of NO production by BDNF and IGF-1 in SCI and resulting neuroprotection is in line with this idea (Sharma, 2011, 2010, 2007a,b, 2006, 2005a; Sharma et al., 2003, 2000a,b,c, 1998). In present investigation these neurotrophins are also able to thwart CO production as evident with downregulation of HO-2 expression in the cord after that showed a close parallelism between neuronal injury, BSCB leakage and edema formation. This suggest that HO-2 expression representing formation of CO could be related with endothelial cells injury, edema formation and tissue damage (see Panahian and Maines, 2001; Sharma et al., 2000a,b,c). Upregulation of free radicals could easily contribute to cell membrane damages including endothelial, neuronal or glial cells (Alm et al., 2000; Gordh et al., 2000; Sharma et al., 2003; Vannemreddy et al., 2006; Winkler et al., 2000). It appears that these neurotrophins could also strengthen the cell membranes of neuronal, glial and endothelium simultaneously. Addition of CNTF at the late stage could act on glial cells especially on astrocytes to improve their functions (Alm et al., 2000; Cao et al., 2010; Panahian and Maines, 2001). Since astrocytes are regulators of endothelia and neuronal function, strengthening of the astrocytic function by CNTF at this late stage could be another added value for these neurotrophins resulting in neuroprotection (Sharma, 2007b, 2005b). To further support this idea, specific receptor blockers of neurotrophins are needed in this model to test this hypothesis for details.

In conclusion, our results are the first to show that CNTF in combination of BDNF and GDNF is able thwart SCI induced cells and tissue injury even applied after 120 min of primary insult. This injury caused by SCI is closely related with HO-2 expression indicates that production of CO is an important injury factor in SCI.

Acknowledgments

This investigation is supported by grants from the Air Force Office of Scientific Research (EOARD, London, UK), and Air Force Material Command, USAF, under grant number

FA8655-05-1-3065; Ministry of Science & Technology, People Republic of China, the Alzheimer's Association (IIRG-09-132087), the National Institutes of Health (R01 AG028679) and the Dr. Robert M. Kohrman Memorial Fund (RJC); Swedish Medical Research Council (Nr 2710-HSS), Göran Gustafsson Foundation, Stockholm, Sweden (HSS), Astra Zeneca, Mölndal, Sweden (HSS/AS), The University Grants Commission, New Delhi, India (HSS/AS), Ministry of Science & Technology, Govt. of India (HSS/AS), Indian Medical Research Council, New Delhi, India (HSS/AS) and India-EU Co-operation Program (RP/AS/HSS) and IT-901/16 (JVL), Government of Basque Country and PPG 17/51 (JVL), JVL thanks to the support of the University of the Basque Country (UPV/EHU) PPG 17/51 and 14/08, the Basque Government (IT-901/16 and CS-2203) Basque Country, Spain; and Foundation for Nanoneuroscience and Nanoneuroprotection (FSNN), Romania. Technical and human support provided by Dr. Ricardo Andrade from SGIker (UPV/EHU) is gratefully acknowledged. Dr. Seaab Sahib is supported by Research Fellowship at the University of Arkansas, Fayetteville, AR, by Department of Community Health; Middle Technical University, Wassit, Iraq, and The Higher Committee for Education Development in Iraq, Baghdad, Iraq. We thank Suraj Sharma, Blekinge Inst. Technology, Karlskrona, Sweden and Dr. Saja Alshafeay, University of Arkansas Fayetteville, AR, USA, for computer and graphic support. The U.S. Government is authorized to reproduce and distribute reprints for Government purpose notwithstanding any copyright notation thereon. The views and conclusions contained herein are those of the authors and should not be interpreted as necessarily representing the official policies or endorsements, either expressed or implied, of the Air Force Office of Scientific Research or the U.S. Government.

Conflict of interest

There is no conflict of interest between any entity and organization mentioned here.

References

Afshary, K., Chamanara, M., Talari, B., Rezaei, P., Nassireslami, E., 2020. Therapeutic effects of minocycline pretreatment in the locomotor and sensory complications of spinal cord injury in an animal model. J. Mol. Neurosci. 70 (7), 1064–1072. https://doi.org/10.1007/s12031-020-01509-8. (Epub 2020 Mar 6).

Alm, P., Sharma, H.S., Sjöquist, P.O., 2000. Westman J. A new antioxidant compound H-290/51 attenuates nitric oxide synthase and heme oxygenase expression following hyperthermic brain injury. An experimental study using immunohistochemistry in the rat. Amino Acids 19 (1), 383–394.

Askvig, J.M., Watt, J.A., 2015. The MAPK and PI3K pathways mediate CNTF-induced neuronal survival and process outgrowth in hypothalamic organotypic cultures. J. Cell Commun. Signal. 9 (3), 217–231. https://doi.org/10.1007/s12079-015-0268-8. (Epub 2015 Feb 20).

Ayaz, S.B., Ikram, M., Matee, S., Ahmad, K., 2014. Spinal cord injury secondary to front rolls as part of military physical training: a case report. NeuroRehabilitation 34 (3), 473–477. https://doi.org/10.3233/NRE-141064.

Azadi, S., Johnson, L.E., Paquet-Durand, F., Perez, M.T., Zhang, Y., Ekström, P.A., van Veen, T., 2007. CNTF+BDNF treatment and neuroprotective pathways in the rd1 mouse retina. Brain Res. 1129 (1), 116–129. https://doi.org/10.1016/j.brainres.2006.10.031. (Epub 2006 Dec 6).

Cao, Q., He, Q., Wang, Y., Cheng, X., Howard, R.M., Zhang, Y., WH, D.V., Shields, C.B., Magnuson, D.S., Xu, X.M., Kim, D.H., Whittemore, S.R., 2010. Transplantation of ciliary neurotrophic factor-expressing adult oligodendrocyte precursor cells promotes remyelination and functional recovery after spinal cord injury. J. Neurosci. 30 (8), 2989–3001. https://doi.org/10.1523/JNEUROSCI.3174-09.2010.

Chen, S., Yi, M., Zhou, G., Pu, Y., Hu, Y., Han, M., Jin, H., 2019. Abdominal aortic transplantation of bone marrow mesenchymal stem cells regulates the expression of ciliary neurotrophic factor and inflammatory cytokines in a rat model of spinal cord ischemia-reperfusion injury. Med. Sci. Monit. 25, 1960–1969. https://doi.org/10.12659/MSM.912697.

Chen, X.B., Yuan, H., Wang, F.J., Tan, Z.X., Liu, H., Chen, N., 2015. Protective role of selenium-enriched supplement on spinal cord injury through the up-regulation of CNTF and CNTF-Ralpha. Eur. Rev. Med. Pharmacol. Sci. 19 (22), 4434–4442.

Christie, S.D., Comeau, B., Myers, T., Sadi, D., Purdy, M., Mendez, I., 2008. Duration of lipid peroxidation after acute spinal cord injury in rats and the effect of methylprednisolone. Neurosurg. Focus. 25 (5). https://doi.org/10.3171/FOC.2008.25.11.E5, E5.

Conti, A., Miscusi, M., Cardali, S., Germanò, A., Suzuki, H., Cuzzocrea, S., Tomasello, F., 2007. Nitric oxide in the injured spinal cord: synthases cross-talk, oxidative stress and inflammation. Brain Res. Rev. 54 (1), 205–218. https://doi.org/10.1016/j.brainresrev.2007.01.013.

de Rivero Vaccari, J.P., 2019. Carbon monoxide releasing molecule-3 inhibits inflammasome activation: a potential therapy for spinal cord injury. EBioMedicine 40, 17–18. https://doi.org/10.1016/j.ebiom.2019.01.020. Epub 2019 Jan 14.

Elliott, K.A., Jasper, H., 1949. Measurement of experimentally induced brain swelling and shrinkage. Am. J. Physiol. 157 (1), 122–129. https://doi.org/10.1152/ajplegacy.1949.157.1.122.

Enck, P., Greving, I., Klosterhalfen, S., Wietek, B., 2006. Upper and lower gastrointestinal motor and sensory dysfunction after human spinal cord injury. Prog. Brain Res. 152, 373–384. https://doi.org/10.1016/S0079-6123(05)52025-9.

Fritz, H.A., Lysack, C., Luborsky, M.R., Messinger, S.D., 2015. Long-term community reintegration: concepts, outcomes and dilemmas in the case of a military service member with a spinal cord injury. Disabil. Rehabil. 37 (16), 1501–1507. https://doi.org/10.3109/09638288.2014.967415 (Epub 2014 Oct 1).

Furlan, J.C., Kurban, D., Craven, B.C., 2020. Traumatic spinal cord injury in military personnel versus civilians: a propensity score-matched cohort study. BMJ Mil. Health 166 (E), e57–e62. https://doi.org/10.1136/jramc-2019-001197. (Epub 2019 May 31).

Gao, W., Li, J., 2017. Targeted siRNA delivery reduces nitric oxide mediated cell death after spinal cord injury. J. Nanobiotechnol. 15 (1), 38. https://doi.org/10.1186/s12951-017-0272-7.

Gordh, T., Sharma, H.S., Azizi, M., Alm, P., Westman, J., 2000. Spinal nerve lesion induces upregulation of constitutive isoform of heme oxygenase in the spinal cord. An immunohistochemical investigation in the rat. Amino Acids 19 (1), 373–381.

Grabher, P., Callaghan, M.F., Ashburner, J., Weiskopf, N., Thompson, A.J., Curt, A., Freund, P., 2015. Tracking sensory system atrophy and outcome prediction in spinal cord injury. Ann. Neurol. 78 (5), 751–761. https://doi.org/10.1002/ana.24508. (Epub 2015 Sep 18).

Grulova, I., Slovinska, L., Nagyova, M., Cizek, M., Cizkova, D., 2013. The effect of hypothermia on sensory-motor function and tissue sparing after spinal cord injury. Spine J. 13 (12), 1881–1891. https://doi.org/10.1016/j.spinee.2013.06.073. (Epub 2013 Sep 5).

Hall, E.D., Wang, J.A., Bosken, J.M., Singh, I.N., 2016. Lipid peroxidation in brain or spinal cord mitochondria after injury. J. Bioenerg. Biomembr. 48 (2), 169–174. https://doi.org/10.1007/s10863-015-9600-5.

Han, X., Yang, N., Xu, Y., Zhu, J., Chen, Z., Liu, Z., Dang, G., Song, C., 2011. Simvastatin treatment improves functional recovery after experimental spinal cord injury by upregulating the expression of BDNF and GDNF. Neurosci. Lett. 487 (3), 255–259. Epub 2010 Sep 19.

Hanigan, W.C., Sloffer, C., 2004. Nelson's wound: treatment of spinal cord injury in 19th and early 20th century military conflicts. Neurosurg. Focus. 16 (1). https://doi.org/10.3171/foc.2004.16.1.5, E4.

Haninec, P., Houst'ava, L., Stejskal, L., Dubový, P., 2003. Rescue of rat spinal motoneurons from avulsion-induced cell death by intrathecal administration of IGF-I and cerebrolysin. Ann. Anat. 185 (3), 233–238. https://doi.org/10.1016/S0940-9602(03)80030-4.

Hearn, J.H., Cross, A., 2020. Mindfulness for pain, depression, anxiety, and quality of life in people with spinal cord injury: a systematic review. BMC Neurol. 20 (1), 32. https://doi.org/10.1186/s12883-020-1619-5.

Hodgetts, S.I., Yoon, J.H., Fogliani, A., Akinpelu, E.A., Baron-Heeris, D., Houwers, I.G.J., Wheeler, L.P.G., Majda, B.T., Santhakumar, S., Lovett, S.J., Duce, E., Pollett, M.A., Wiseman, T.M., Fehily, B., Harvey, A.R., 2018. Cortical AAV-CNTF gene therapy combined with Intraspinal mesenchymal precursor cell transplantation promotes functional and morphological outcomes after spinal cord injury in adult rats. Neural Plast. 2018, 9828725. https://doi.org/10.1155/2018/9828725. (eCollection 2018).

Hu, Z., Deng, N., Liu, K., Zhou, N., Sun, Y., Zeng, W., 2020. CNTF-STAT3-IL-6 Axis mediates neuroinflammatory cascade across Schwann cell-neuron-microglia. Cell Rep. 31 (7), 107657. https://doi.org/10.1016/j.celrep.2020.107657.

Ip, N.Y., Yancopoulos, G.D., 1992. Ciliary neurotrophic factor and its receptor complex. Prog. Growth Factor Res. 4 (2), 139–155. https://doi.org/10.1016/0955-2235(92)90028-g.

Jia, C., Keasey, M.P., Lovins, C., Hagg, T., 2018. Inhibition of astrocyte FAK-JNK signaling promotes subventricular zone neurogenesis through CNTF. Glia 66 (11), 2456–2469. https://doi.org/10.1002/glia.23498.

Kallakuri, S., Purkait, H.S., Dalavayi, S., VandeVord, P., Cavanaugh, J.M., 2015. Blast overpressure induced axonal injury changes in rat brainstem and spinal cord. J. Neurosci. Rural Pract. 6 (4), 481–487. https://doi.org/10.4103/0976-3147.169767.

Leslie, D., 1946. Report of a fatal case of blast injury of the spinal cord. Med. J. Aust. 1, 188.

Li, L., Xiao, B., Mu, J., Zhang, Y., Zhang, C., Cao, H., Chen, R., Patra, H.K., Yang, B., Feng, S., Tabata, Y., Slater, N.K.H., Tang, J., Shen, Y., Gao, J., 2019. A MnO(2) nanoparticle-dotted hydrogel promotes spinal cord repair via regulating reactive oxygen species microenvironment and synergizing with mesenchymal stem cells. ACS Nano 13 (12), 14283–14293. https://doi.org/10.1021/acsnano.9b07598. (Epub 2019 Dec 2).

Liu, Z., Yao, X., Jiang, W., Li, W., Zhu, S., Liao, C., Zou, L., Ding, R., Chen, J., 2020. Advanced oxidation protein products induce microglia-mediated neuroinflammation via MAPKs-NF-kappaB signaling pathway and pyroptosis after secondary spinal cord injury. J. Neuroinflammation 17 (1), 90. https://doi.org/10.1186/s12974-020-01751-2.

Marsala, J., Orendácová, J., Lukácová, N., Vanický, I., 2007. Traumatic injury of the spinal cord and nitric oxide. Prog. Brain Res. 161, 171–183. https://doi.org/10.1016/S0079-6123(06)61011-X.

McDonald, S.D., Goldberg-Looney, L.D., Mickens, M.N., Ellwood, M.S., Mutchler, B.J., Perrin, P.B., 2018. Appraisals of DisAbility primary and secondary scale-short form

(ADAPSS-sf): psychometrics and association with mental health among U.S. military veterans with spinal cord injury. Rehabil. Psychol. 63 (3), 372–382. https://doi.org/10.1037/rep0000230 (Epub 2018 Jul 26).

Menon, P.K., Muresanu, D.F., Sharma, A., Mössler, H., Sharma, H.S., 2012. Cerebrolysin, a mixture of neurotrophic factors induces marked neuroprotection in spinal cord injury following intoxication of engineered nanoparticles from metals. CNS Neurol. Disord. Drug Targets 11 (1), 40–49. https://doi.org/10.2174/187152712799960781.

Nagahara, A.H., Tuszynski, M.H., 2011. Potential therapeutic uses of BDNF in neurological and psychiatric disorders. Nat. Rev. Drug Discov. 10 (3), 209–219. Review.

Noamani, A., Lemay, J.F., Musselman, K.E., Rouhani, H., 2020. Postural control strategy after incomplete spinal cord injury: effect of sensory inputs on trunk-leg movement coordination. J. Neuroeng. Rehabil. 17 (1), 141. https://doi.org/10.1186/s12984-020-00775-2.

Panahian, N., Maines, M.D., 2001. Site of injury-directed induction of heme oxygenase-1 and -2 in experimental spinal cord injury: differential functions in neuronal defense mechanisms? J. Neurochem. 76 (2), 539–554. https://doi.org/10.1046/j.1471-4159.2001.00023.x.

Papa, S., Pizzetti, F., Perale, G., Veglianese, P., Rossi, F., 2020. Regenerative medicine for spinal cord injury: focus on stem cells and biomaterials. Expert. Opin. Biol. Ther. 20 (10), 1203–1213. https://doi.org/10.1080/14712598.2020.1770725. (Epub 2020 Jun 4).

Sadowsky, C.L., Hammond, E.R., Strohl, A.B., Commean, P.K., Eby, S.A., Damiano, D.L., Wingert, J.R., Bae, K.T., McDonald 3rd, J.W., 2013. Lower extremity functional electrical stimulation cycling promotes physical and functional recovery in chronic spinal cord injury. J. Spinal Cord Med. 36 (6), 623–631. https://doi.org/10.1179/2045772313Y.0000000101 (Epub 2013 Mar 20).

Sahib, S., Sharma, A., Menon, P.K., Muresanu, D.F., Castellani, R.J., Nozari, A., Lafuente, J.V., Bryukhovetskiy, I., Tian, Z.R., Patnaik, R., Buzoianu, A.D., Wiklund, L., Sharma, H.S., 2020. Cerebrolysin enhances spinal cord conduction and reduces blood-spinal cord barrier breakdown, edema formation, immediate early gene expression and cord pathology after injury. Prog. Brain Res. 258, 397–438. https://doi.org/10.1016/bs.pbr.2020.09.012. (Epub 2020 Nov 12).

Sámano, C., Nistri, A., 2019. Mechanism of neuroprotection against experimental spinal cord injury by Riluzole or methylprednisolone. Neurochem. Res. 44 (1), 200–213. https://doi.org/10.1007/s11064-017-2459-6. (Epub 2017 Dec 30).

Schoenfeld, A.J., McCriskin, B., Hsiao, M., Burks, R., 2011. Incidence and epidemiology of spinal cord injury within a closed American population: the United States military (2000-2009). Spinal Cord 49 (8), 874–879. https://doi.org/10.1038/sc.2011.18. (Epub 2011 Mar 8).

Sharma, H.S., 2003. Neurotrophic factors attenuate microvascular permeability disturbances and axonal injury following trauma to the rat spinal cord. Acta Neurochir. Suppl. 86, 383–388.

Sharma, H.S., 2005a. Neuroprotective effects of neurotrophins and melanocortins in spinal cord injury: an experimental study in the rat using pharmacological and morphological approaches. Ann. N. Y. Acad. Sci. 1053, 407–421.

Sharma, H.S., 2005b. Pathophysiology of blood-spinal cord barrier in traumatic injury and repair. Curr. Pharm. Des. 11 (11), 1353–1389. Review.

Sharma, H.S., 2006. Post-traumatic application of brain-derived neurotrophic factor and glia-derived neurotrophic factor on the rat spinal cord enhances neuroprotection and improves motor function. Acta Neurochir. Suppl. 96, 329–334.

Sharma, H.S., 2007a. A select combination of neurotrophins enhances neuroprotection and functional recovery following spinal cord injury. Ann. N. Y. Acad. Sci. 1122, 95–111.

Sharma, H.S., 2007b. Neurotrophic factors in combination: a possible new therapeutic strategy to influence pathophysiology of spinal cord injury and repair mechanisms. Curr. Pharm. Des. 13 (18), 1841–1874. (Review).

Sharma, H.S., 2010. Selected combination of neurotrophins potentiate neuroprotection and functional recovery following spinal cord injury in the rat. Acta Neurochir. Suppl. 106, 295–300.

Sharma, H.S., 2011. Early microvascular reactions and blood-spinal cord barrier disruption are instrumental in pathophysiology of spinal cord injury and repair: novel therapeutic strategies including nanowired drug delivery to enhance neuroprotection. J. Neural Transm. 118 (1), 155–176. Epub 2010 Dec 16. (Review).

Sharma, H.S., Olsson, Y., 1990. Edema formation and cellular alterations following spinal cord injury in the rat and their modification with p-chlorophenylalanine. Acta Neuropathol. 79 (6), 604–610. https://doi.org/10.1007/BF00294237.

Sharma, H.S., Westman, J., 2003. Depletion of endogenous serotonin synthesis with p-CPA attenuates upregulation of constitutive isoform of heme oxygenase-2 expression, edema formation and cell injury following a focal trauma to the rat spinal cord. Acta Neurochir. Suppl. 86, 389–394.

Sharma, H.S., Nyberg, F., Olsson, Y., Dey, P.K., 1990a. Alteration of substance P after trauma to the spinal cord: an experimental study in the rat. Neuroscience 38 (1), 205–212. https://doi.org/10.1016/0306-4522(90)90386-i.

Sharma, H.S., Olsson, Y., Dey, P.K., 1990b. Early accumulation of serotonin in rat spinal cord subjected to traumatic injury. Relation to edema and blood flow changes. Neuroscience 36 (3), 725–730. https://doi.org/10.1016/0306-4522(90)90014-u.

Sharma, H.S., Olsson, Y., Nyberg, F., Dey, P.K., 1993. Prostaglandins modulate alterations of microvascular permeability, blood flow, edema and serotonin levels following spinal cord injury: an experimental study in the rat. Neuroscience 57 (2), 443–449. https://doi.org/10.1016/0306-4522(93)90076-r.

Sharma, H.S., Westman, J., Olsson, Y., Alm, P., 1996. Involvement of nitric oxide in acute spinal cord injury: an immunocytochemical study using light and electron microscopy in the rat. Neurosci. Res. 24 (4), 373–384. https://doi.org/10.1016/0168-0102(95)01015-7.

Sharma, H.S., Nyberg, F., Westman, J., Alm, P., Gordh, T., Lindholm, D., 1998. Brain derived neurotrophic factor and insulin like growth factor-1 attenuate upregulation of nitric oxide synthase and cell injury following trauma to the spinal cord. An immunohistochemical study in the rat. Amino Acids 14 (1–3), 121–129.

Sharma, H.S., Alm, P., Sjöquist, P.O., Westman, J., 2000a. A new antioxidant compound H-290/51 attenuates upregulation of constitutive isoform of heme oxygenase (HO-2) following trauma to the rat spinal cord. Acta Neurochir. Suppl. 76, 153–157.

Sharma, H.S., Nyberg, F., Gordh, T., Alm, P., Westman, J., 2000b. Neurotrophic factors influence upregulation of constitutive isoform of heme oxygenase and cellular stress response in the spinal cord following trauma. An experimental study using immunohistochemistry in the rat. Amino Acids 19 (1), 351–361.

Sharma, H.S., Westman, J., Gordh, T., Alm, P., 2000c. Topical application of brain derived neurotrophic factor influences upregulation of constitutive isoform of heme oxygenase in the spinal cord following trauma an experimental study using immunohistochemistry in the rat. Acta Neurochir. Suppl. 76, 365–369.

Sharma, H.S., Drieu, K., Westman, J., 2003. Antioxidant compounds EGB-761 and BN-52021 attenuate brain edema formation and hemeoxygenase expression following hyperthermic brain injury in the rat. Acta Neurochir. Suppl. 86, 313–319.

Sharma, H.S., Zimmermann-Meinzingen, S., Johanson, C.E., 2010. Cerebrolysin reduces blood-cerebrospinal fluid barrier permeability change, brain pathology, and functional deficits following traumatic brain injury in the rat. Ann. N. Y. Acad. Sci. 1199, 125–137. https://doi.org/10.1111/j.1749-6632.2009.05329.x.

Sharma, H.S., Menon, P.K., Lafuente, J.V., Aguilar, Z.P., Wang, Y.A., Muresanu, D.F., Mössler, H., Patnaik, R., Sharma, A., 2014. The role of functionalized magnetic iron oxide nanoparticles in the central nervous system injury and repair: new potentials for neuroprotection with Cerebrolysin therapy. J. Nanosci. Nanotechnol. 14 (1), 577–595. https://doi.org/10.1166/jnn.2014.9213.

Sharma, H.S., Patnaik, R., Muresanu, D.F., Lafuente, J.V., Ozkizilcik, A., Tian, Z.R., Nozari, A., Sharma, A., 2017. Histaminergic receptors modulate spinal cord injury-induced neuronal nitric oxide synthase upregulation and cord pathology: new roles of nanowired drug delivery for neuroprotection. Int. Rev. Neurobiol. 137, 65–98. https://doi.org/10.1016/bs.irn.2017.09.001. (Epub 2017 Nov 3).

Sharma, H.S., Feng, L., Muresanu, D.F., Castellani, R.J., Sharma, A., 2019. Neuroprotective effects of a potent bradykinin B2 receptor antagonist HOE-140 on microvascular permeability, blood flow disturbances, edema formation, cell injury and nitric oxide synthase upregulation following trauma to the spinal cord. Int. Rev. Neurobiol. 146, 103–152. https://doi.org/10.1016/bs.irn.2019.06.008. (Epub 2019 Jul 9).

Stankoff, B., Aigrot, M.S., Noël, F., Wattilliaux, A., Zalc, B., Lubetzki, C., 2002. Ciliary neurotrophic factor (CNTF) enhances myelin formation: a novel role for CNTF and CNTF-related molecules. J. Neurosci. 22 (21), 9221–9227. https://doi.org/10.1523/JNEUROSCI.22-21-09221.2002.

Vannemreddy, P., Ray, A.K., Patnaik, R., Patnaik, S., Mohanty, S., Sharma, H.S., 2006. Zinc protoporphyrin IX attenuates closed head injury-induced edema formation, blood-brain barrier disruption, and serotonin levels in the rat. Acta Neurochir. Suppl. 96, 151–156.

Winkler, T., Sharma, H.S., Stålberg, E., Badgaiyan, R.D., 2000. Neurotrophic factors attenuate alterations in spinal cord evoked potentials and edema formation following trauma to the rat spinal cord. Acta Neurochir. Suppl. 76, 291–296.

Wolman, L., 1967. Blast injury of the spinal cord. Paraplegia 5 (2), 83–88. https://doi.org/10.1038/sc.1967.8.

Wyndaele, J.J., 2011. More knowledge of worldwide incidence and epidemiology of spinal cord injury: data from the United States military. Spinal Cord 49 (8), 857. https://doi.org/10.1038/sc.2011.85. (eCollection 2017).

Xiong, L.L., Liu, F., Lu, B.T., Zhao, W.L., Dong, X.J., Liu, J., Zhang, R.P., Zhang, P., Wang, T.H., 2017. Bone marrow mesenchymal stem-cell transplantation promotes functional improvement associated with CNTF-STAT3 activation after hemi-sectioned spinal cord injury in tree shrews. Front. Cell. Neurosci. 11, 172. https://doi.org/10.3389/fncel.2017.00172. (Epub 2019 Jan 3).

Zheng, G., Zhan, Y., Wang, H., Luo, Z., Zheng, F., Zhou, Y., Wu, Y., Wang, S., Wu, Y., Xiang, G., Xu, C., Xu, H., Tian, N., Zhang, X., 2019. Carbon monoxide releasing molecule-3 alleviates neuron death after spinal cord injury via inflammasome regulation. EBioMedicine 40, 643–654. https://doi.org/10.1016/j.ebiom.2018.12.059.

CHAPTER

Diagnosis experience and literature review of patients with cervical, thoracic and lumbar multi-segment spinal stenosis: A case report

11

Chao He[a,†], Xu Longbiao[a,†], Ming Zhao[a], Lin Wang[b,*], and Hari Shanker Sharma[c,*]

[a]*Department of Neurosurgery, Zhuji People' Hospital of Zhejiang Province, Zhuji Affiliated Hospital of Shaoxing University, Zhuji, Zhejiang, China*

[b]*Department of Neurosurgery, Second Affiliated Hospital of Zhejiang University School of Medicine, Hangzhou, Zhejiang, China*

[c]*International Experimental Central Nervous System Injury & Repair (IECNSIR), Department of Surgical Sciences, Anesthesiology & Intensive Care Medicine, Uppsala University Hospital, Uppsala University, Uppsala, Sweden*

**Corresponding authors: Tel.: +86-13646829750, Fax: +86-057187068001 (Lin Wang); Tel.: +46-70-20-11-801; Fax: +46-18-24-38-99 (Hari Shanker Sharma), e-mail address: dr_wang@zju.edu.cn; sharma@surgsci.uu.se*

Abstract

Background: The incidence of cervical, thoracic and lumbar spinal canal stenosis is low. It is difficult to identify the main focus and responsible segment, and it is also difficult to select the sequence of staging surgery. We report a patient with triple stenosis.

Case presentation: In this paper, we introduced a 61-year-old female patient with cervical, thoracic and lumbar spinal canal stenosis who had previously undergone "lumbar discectomy" in the outer hospital. The postoperative effect was not good and the symptoms were poor. The diagnosis was "cervical spinal stenosis and lumbar spine surgery." The staged spinal canal decompression operation and Duhuo Jisheng Decoction (DHJSD) treatment were conducted in our hospital. After 3 months of follow-up, the functional and imaging results were satisfactory.

[†]He Chao and Xu Longbiao have equally contributed to this work as co-first authors.

Progress in Brain Research, Volume 266, ISSN 0079-6123, https://doi.org/10.1016/bs.pbr.2021.06.019
Copyright © 2021 Elsevier B.V. All rights reserved.

Conclusion: We should pay enough attention to the patients with spinal degenerative diseases who need surgery, and must pursue the unity of medical history, signs and images. In case of difficult patients, more comprehensive examination is required, and the main focus and responsible segment are determined through comprehensive analysis. The more important diseases that may exist cannot be covered up by focal lesion manifestations, so as to avoid unnecessary surgical trauma for patients. In addition, surgery combined with Chinese herbal medicine DHJSD therapy may be an effective treatment for this kind of disease.

Keywords

Cervical vertebral canal stenosis, Thoracic vertebral canal stenosis, Lumbar vertebral canal stenosis, Diagnostic strategy

1 Background

Cervical, thoracic and lumbar spinal canal stenosis may occur due to developmental, metabolic or degenerative reasons, but it is mostly found in segments with more spinal activity, such as cervical or lumbar (Li et al., 2015). The incidence rate of multi-segment triple spinal stenosis of cervical, thoracic and lumbar vertebrae is low. At present, relevant literature reports on diagnosis and treatment are relatively rare. Scholars Park and others analyzed the clinical data of 460 patients with lumbar stenosis, of which only 18 patients were complicated with triple stenosis. Japanese scholars found through autopsy of 1072 patients with lumbar stenosis that lumbar stenosis and thoracic stenosis are not caused by tandem stenosis, and the incidence rate of triple stenosis is relatively low (Bajwa et al., 2012; Luo et al., 2018; Kim et al., 2010; Park et al., 2015). In the following cases, we presented a case of triple spinal stenosis patient and describe our diagnosis and treatment experience in this disease.

2 Case presentation

The patient, a 61-year-old woman, was admitted to the hospital for "one year of weakness of both lower limbs and half a year of aggravation accompanied by walking instability." One year ago, the patient suffered from fatigue of both lower limbs without inducement, accompanied by numbness below the plane of both knees, and could not walk for a long distance. During the following 6 months, the symptoms gradually worsened, the both upper limbs felt normal, the movement was unimpeded, there was no palpitation, chest tightness, dizziness, nausea and other discomfort, which was not paid attention to at that time. In the following 6 months, the symptoms gradually worsened, with the feeling of stepping on cotton and unstable walking. The patient was diagnosed as "lumbar disc herniation + lumbar spinal canal stenosis" in the outer hospital and underwent lumbar discectomy. The above symptoms did not ease after operation. In the past 6 months, the patients felt that the appeal symptoms were further aggravated and accompanied by walking difficulties. In order

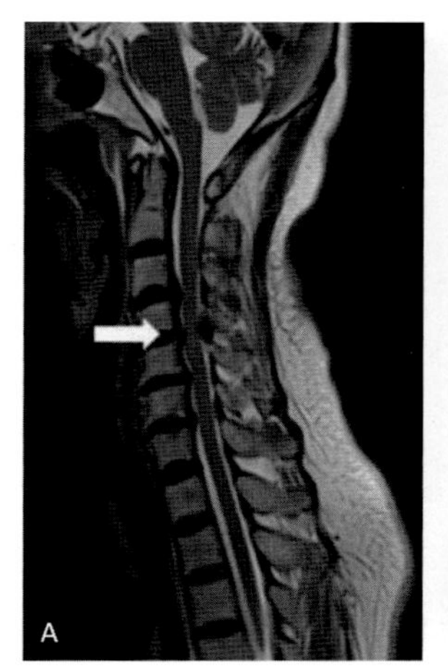

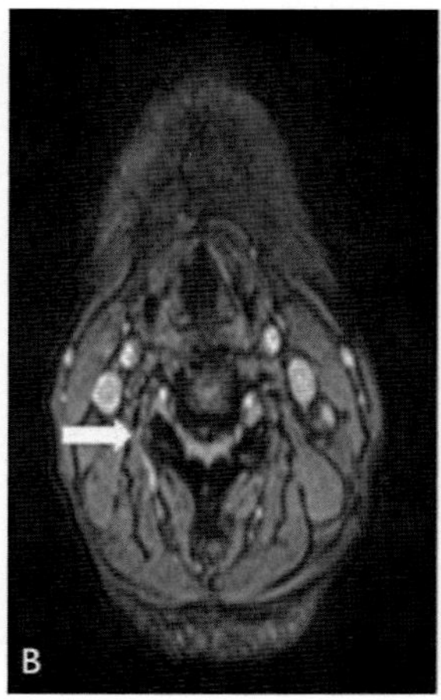

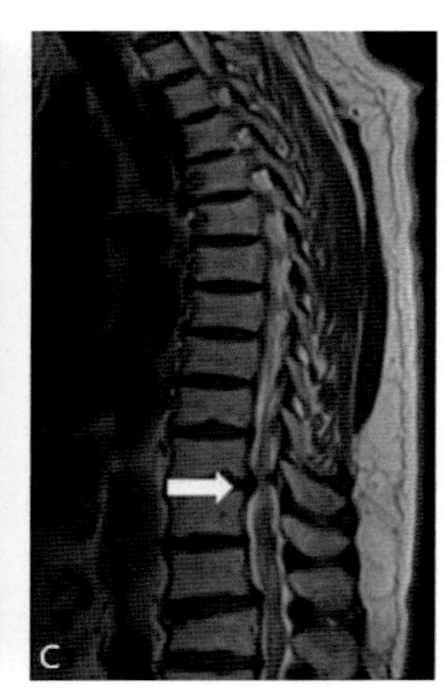

FIG. 1

(A and B) Cervical 4/5 spinal canal stenosis; (C) thoracic 10 spinal canal stenosis.

to make further diagnosis and treatment, they came to our hospital. The outpatient department was admitted to our department as "cervical spinal stenosis and lumbar postoperative surgery."

The patient underwent "left knee arthroplasty" 18 months ago in the Orthopedics Department of the Second Affiliated Hospital of Zhejiang University School of Medicine, and "left femoral internal fixation" 14 months ago due to left femoral fracture (Fig. 1).

Admission for physical examination: The general condition is good, and cranial nerve examination is the same. Surgical scars can be seen on the waist, left knee and thigh. Neck resistance (−), head muscle strength V, double upper limb muscle strength V, normal muscle tension, Hoffmann sign (+), right lower limb muscle strength IV, left lower limb muscle strength III, increased muscle tension, bilateral superficial hypoesthesia below knee joint, bilateral ankle clonus (+), Babinski sign (+). Imaging examination: Cervical magnetic resonance (June 18th, 2019) showed that cervical 3-cervical 7 intervertebral disc herniation, cervical 4/5 ligamentum flavum thickening and calcification, cervical canal stenosis, cervical spinal cord compression (Fig. 2A and B); Thoracic vertebral magnetic resonance (June 23rd, 2019) revealed that thoracic 10/11 and thoracic 11/12 intervertebral disc herniation, hypertrophy and ossification of ligamentum flavum, and corresponding spinal canal stenosis; Spinal cord edema at thoracic level 10–11 (Fig. 2C).

After admission, relevant examinations should be improved, and preoperative departments should discuss: Patients with cervical and thoracic vertebra stenosis, spinal cord compression, and clear surgical indications, one-stage surgical decompression treatment can be considered. Considering that the patient has undergone many operations, she is afraid of the operation and has great doubts about the effect of the operation. The patient has positive vertebral bundle sign and unstable walking of both lower limbs due to fatigue. Decompression of cervical spinal stenosis is more

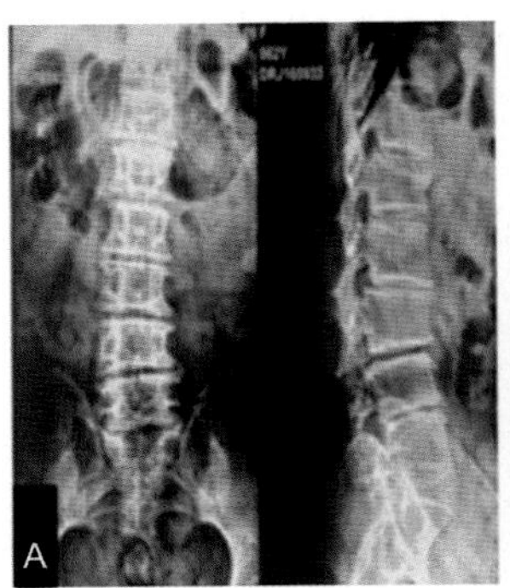
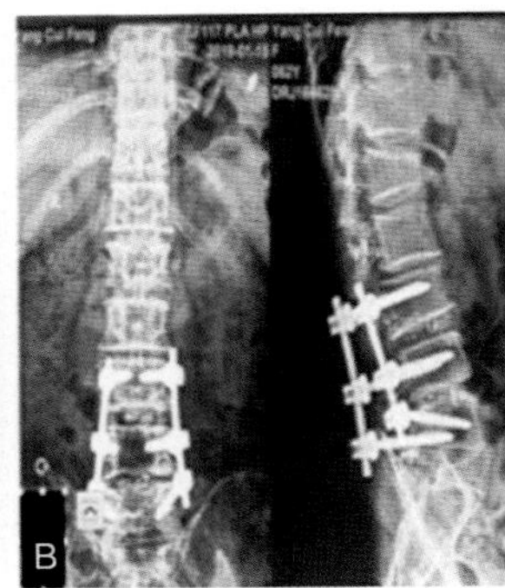
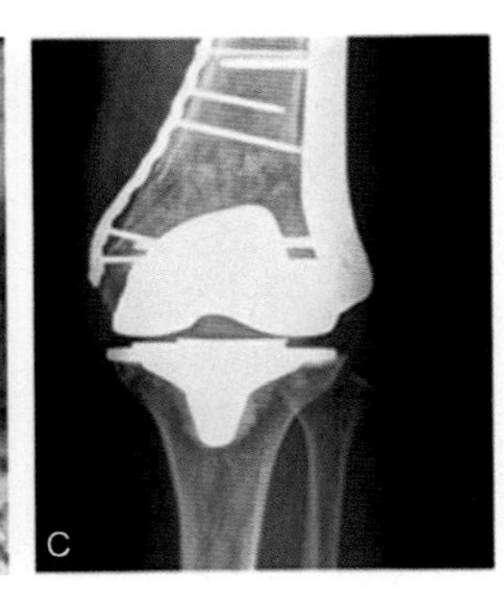

FIG. 2

(A and B) Preoperative and postoperative for lumbar disc herniation fusion and internal fixation; (C) Left knee arthroplasty and femoral internal fixation.

helpful to relieve the symptoms. Therefore, decompression of cervical spinal stenosis can be considered first to achieve results, and decompression of thoracic spinal stenosis can be considered after patients enhance confidence. After full communication with the patient's family, it is required to perform decompression in stages as planned.

The patient underwent cervical 4/5 spinal canal decompression + partial resection of ligamentum flavum under general anesthesia, and the operation went smoothly. The general condition of patients after operation is good. There is no obvious numbness and pain in both upper limbs. The numbness of both lower limbs is obviously relieved compared with that before operation. The muscle strength of the right lower limb is improved compared with that before operation. The left lower limb still feels heavy and weak. The muscle strength of both upper limbs is Grade V, the muscle strength of right lower limbs is Grade IV, and the muscle strength of left lower limbs is Grade III. Two weeks later, the thoracic 10/11 spinal canal decompression was performed under general anesthesia, and the operation went smoothly. After operation, there was no obvious numbness and pain in both upper limbs of the patient. The numbness of both lower limbs was significantly relieved compared with that before operation. The muscle strength of the right lower limb was improved compared with that before operation. The left lower limb still felt heavy and the sense of movement was improved. After surgery, this patient was treated with DHJSD, with the dosage of 100 mL per day for 3 months.

The patient was followed up for 3 months after operation, and the functional and imaging effects were satisfactory (Fig. 3). The standing balance was more stable than before, and the patient could walk for a long distance with the aid of auxiliary devices. Specialist physical examination: clear mind, good spirit, clear speech, answering relevant questions, and cooperating with body activities. The frontal striation and bilateral nasolabial sulcus are symmetrical, and the mouth angle is not skewed. The double pupils are equal in size and circle, with a diameter of about

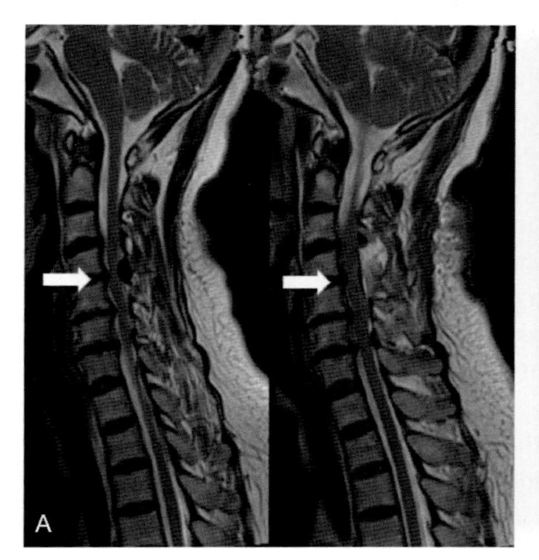

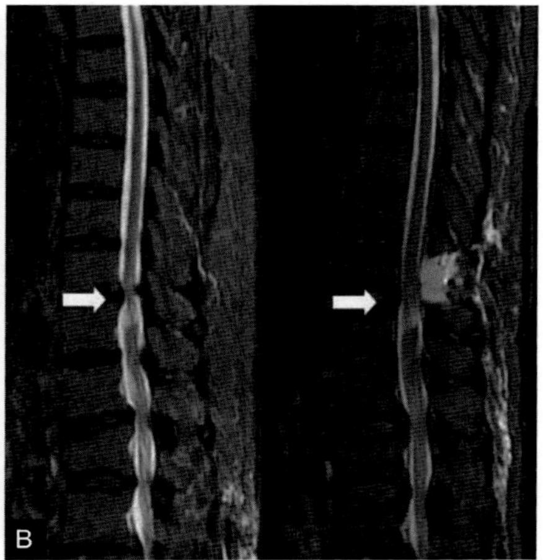

FIG. 3

(A) Preoperative and postoperative for cervical 4/5 spinal canal stenosis; (B) Preoperative and postoperative for thoracic 10 spinal canal stenosis.

3 mm. They are sensitive to light reflection, with the extension tongue in the middle, neck resistance (−) and head muscle strength of Grade V. The muscle tension of both upper limbs was normal, Hoffmann sign was negative, muscle strength was Grade V, tendon reflex (++). Muscle tension of both lower limbs increased, bilateral iliopsoas muscle strength Grade IV-, quadriceps femoris muscle strength Grade IV, tibialis anterior muscle and gastrocnemius muscle strength Grade V. Bilateral patellar clonus (−) and bilateral ankle clonus (+). The superficial sensation below the knee joint on both sides decreased slightly, the positional sensation was normal, and Babinski sign was weakly positive.

3 Discussion

Most patients with cervical, thoracic and lumbar multi-segment spinal stenosis have a long course of disease and often have experienced various treatment experiences. It is difficult to clearly express the development and changes of the disease (Guo et al., 2021; Cheng et al., 2020; Zang et al., 2019; Fu et al., 2019). This patient has undergone two surgical treatments for his symptoms before admission, and none of them has achieved the expected results or even worsened. The decision-making process of diagnosis and treatment is worthy of our reflection and discussion.

Such patients all have upper motor neuron symptoms and lower motor neuron symptoms, and the symptoms are interwoven with each other, which is easy to miss diagnosis and misdiagnosis (Yang et al., 2019; Huang et al., 2018; Naderi and Mertol, 2002; Yamada et al., 2018). In the process of diagnosis and treatment, the medical history should be inquired in detail to guide patients to provide valuable information for disease diagnosis. It is particularly important to carry out careful physical examination, which can identify the main focus and responsible segment by paying attention to the unique positive signs of cervical, lumbar and thoracic vertebral diseases and making analysis and judgment. In addition, it is necessary to perfect all kinds of auxiliary examinations, take X-rays of hyperextension and hyperflexion, CT plain scan and three-dimensional reconstruction, MRI of relevant brain and spine, nerve electrophysiology examination, etc. According to the patient's symptoms, positive signs and imaging data, the diagnosis can only be obtained by emphasizing the consistency of the combination of the three. Simple imaging signs without symptoms and signs cannot diagnose the disease (Li et al., 2019; Patel and Perloff, 2018).

The sequence of staging surgery is another major difficulty in this disease. This patient suffered from walking instability and weakness, foot stepping on cotton sensation, and intermittent claudication of spinal cord origin. The signs were bilateral muscle tension increased, tendon reflex active, Babinski sign and Hoffmann sign positive, ankle clonus positive, which were the typical symptoms of cervical spinal cord upper motor neuron compression. Therefore, we chose to treat the main focus and responsible segment of cervical spinal canal first. Currently, priority is given to segments, Wu et al. gave priority to cervical spine surgery for 222 patients with cervical spine complicated with lumbar spinal stenosis. Comparing the postoperative J0A score with Nurick score, it is concluded that the first cervical spine surgery can significantly improve the postoperative JOA score and ODI index and reduce the second-stage operation rate (Li et al., 2017), and Luo et al. pointed out that if lumbar spinal stenosis is treated first in patients with cervical spine as the main focus, the symptoms related to cervical spinal stenosis will deteriorate rapidly (Luo et al., 2019; Matsumoto et al., 2010). Therefore, for patients with "cervical, thoracic and lumbar triple stenosis," it is very necessary to perform full spinal MRI examination before spinal canal decompression surgery, especially before lumbar decompression patients' stenosis surgery, when positive signs that cannot be fully explained by lumbar lesions are found in physical examination, so as to exclude the possibility of multi-segment stenosis and other lesions such as space occupying lesions or vascular lesions (Fushimi et al., 2013; Huang and Al Zoubi, 2018). At present, limited by policies such as health insurance, most hospitals do not have the examination item of full spine magnetic resonance, nor can they do multi-site MR examination at one time. Moreover, the current outpatient scheduling system makes patients often accepted by different doctors at different outpatient times, and the diagnosis and treatment ideas lack consistency. This is also an important factor that causes clinical missed diagnosis or misdiagnosis.

At present, Surgical treatment plays an important role in treating spinal stenosis, which can reduce the clinical symptoms, while there are problems such as infection, complications, and high postoperative recurrence rate (Dave et al., 2019; Xiaolong et al., 2019). Except for surgical treatment, nonsurgical therapy, including traction, manipulation, acupuncture, physiotherapy and medicines, especially the Chinese herbal medicine showed good effects (Zhang et al., 2019; Ding et al., 2019; Ma et al., 2019; Li et al., 2018; Tang et al., 2018; Wei et al., 2013; Zhuomao et al., 2018). Chinese herbal medicine has been used in clinical practice for a long time and Duhuo Jisheng Decoction (DHJSD) is one of the representative prescriptions (Lin and Chen, 2007; Xiaocheng et al., 2018). The ancient classic prescriptions represented by DHJSD was recorded in the book "Bei Ji Qian Jin Yao fang" written by Sun Simiao of Tang Dynasty. It consists of 15 commonly used Chinese herbs. According to the theory of TCM, it has the effects of eliminating rheumatism, analgesics, nourishing liver and kidney, nourishing Qi and blood, and channeling meridians (Caibin et al., 2013; Xinhua et al., 2017). Basic research shows that it has the effects of antiinflammatory, analgesia, immune regulation, cartilage promotion, and fibrous ring repair (Yongmei and Jia, 2019). DHJSD has been widely used in the treatment of lumbar disc herniation (Sun et al., 2020; Yuanyuan, 2017), and other orthopedic diseases, such as knee arthritis and osteoporosis (Shangzhi and Juntao, 2018; Xu and Hong, 2018). A systematic review and meta-analysis for DHJSD in treatment of lumbar disc herniation also showed good effectiveness and safety (Wenming et al., 2016). In our case, after surgery, this patient was treated with DHJSD, with the dosage of 100 mL per day for 3 months. With the follow-up, the patient's symptoms improved significantly. It is proved that surgery combined with Chinese herbal medicine DHJSD therapy is an effective treatment for this patient.

4 Conclusion

We should pay enough attention to patients with spinal degenerative diseases who need surgery, and must pursue the unity of medical history, signs and images. In case of difficult patients, more comprehensive examination is required, and the main focus and responsible segment are determined through comprehensive analysis. The more important diseases that may exist cannot be covered up by focal lesion manifestations, so as to avoid unnecessary surgical trauma for patients.

Funding

This work was supported by Projects of Zhejiang Medical and Health Science and Technology (No. 2016KYB315, No. 2015ZA209, No. 2019ZD058, No. 2020PY018), Shaoxing Science and Technology Bureau (No. 2018C30150, No. 2018C30146), Zhejiang Nature Science Technology (GF18H09008).

References

Bajwa, N.S., Toy, J.O., Ahn, N.U., 2012. Is lumbar stenosis associated with thoracic stenosis? A study of 1,072 human cadaveric specimens. Spine J. 12 (12), 1142–1146.

Caibin, P., Yang, F., Xianxiang, L., 2013. Discussion on the connotation and extension of Duhuo Jisheng decoction. Arthritis Rheum. 2, 39–41.

Cheng, Z., et al., 2020. High-frequency spinal cord stimulation for treating pain in the lower limbs accompanied by bilateral para-anesthesia: a case report. J. Neurorestoratol. 8, 132–137. https://doi.org/10.26599/JNR.2020.9040012.

Dave, B.R., Degulmadi, D., Krishnan, A., et al., 2019. Risk factors and surgical treatment for recurrent lumbar disc prolapse: a review of the literature. Asian Spine J. 10, 1–9.

Ding, W., et al., 2019. Hand function reconstruction in patients with chronic incomplete lower cervical spinal cord injury by nerve segment insert grafting: a preliminary clinical report. J. Neurorestoratol. 7, 129–135. https://doi.org/10.26599/JNR.2019.9040013.

Fu, B., et al., 2019. Treatment of single-segment lumbar tuberculosis with minimally invasive posterior internal fixation combined with anterior small incision debridement. Brain Sci. Adv. 5, 203–212. https://doi.org/10.26599/BSA.2019.9050017.

Fushimi, K., Miyamoto, K., Hioki, A., et al., 2013. Neurological deterioration due to missed thoracic spinal stenosis after decompressive lumbar surgery: a report of six cases of tandem thoracic and lumbar spinal stenosis. Bone Joint J. 95 (10), 1388–1391.

Guo, X., et al., 2021. Clinical guidelines for neurorestorative therapies in spinal cord injury (2021 China version). J. Neurorestoratol. 9, 31–49. https://doi.org/10.26599/JNR.2021.9040003. http://jnr.tsinghuajournals.com/10.26599/JNR.2021.9040003. http://jnr.tsinghuajournals.com/10.26599/JNR.2021.9040003.

Huang, H., Al Zoubi, Z.M., 2018. A brief introduction to the special issue on clinical treatment of spinal cord injury. J. Neurorestoratol. 6 (1), 134–135.

Huang, H., Sharma, H.S., Chen, L., Otom, A., Al Zoubi, Z.M., Saberi, H., et al., 2018. Review of clinical neurorestorative strategies for spinal cord injury: exploring history and latest progresses. J. Neurorestoratol. 6 (1), 171–178.

Kim, B.S., Kim, J., Koh, H.S., et al., 2010. Asymptomatic cervical or thoracic lesions in elderly patients who have undergone decompressive lumbar surgery for stenosis. Asian Spine J. 4 (2), 65.

Li, W., Guo, S., Sun, Z., et al., 2015. Multilevel thoracic ossification of ligamentum flavum coexisted with/without lumbar spinal stenosis: staged surgical strategy and clinical outcomes. BMC Musculoskelet. Disord. 16 (1), 206.

Li, H., Chen, Z., Li, X., et al., 2017. Prioritized cervical or lumbar surgery for coexisting cervical and lumbar stenosis: prognostic analysis of 222 case. Int. J. Surg. 44, 344–349.

Li, Z., Hao, Y., Liu, M., et al., 2018. Clinical characteristics and risk factors of recurrent lumbar disk herniation: a retrospective analysis of three hundred and twenty-one cases. Spine 43, 1463–1469.

Li, J., He, L., Zhang, Y., 2019. Application of multishot diffusion tensor imaging in spinal cord tumors. Brain Sci. Adv. 5 (1), 59–64.

Lin, X.J., Chen, C.Y., 2007. Advances on study of treatment of lumbar disk herniation by Chinese medicinal herbs. Zhongguo Zhong Yao Za Zhi 32, 186–191.

Luo, F., et al., 2018. Genetically encoded neural activity indicators. Brain Sci. Adv. 4 (1), 1–15.

Luo, C.A., Kaliya-Perumal, A.K., Lu, M.L., et al., 2019. Staged surgery for tandem cervical and lumbar spinal stenosis: which should be treated first? Eur. Spine J. 28 (1), 61–68.

Ma, C., et al., 2019. Progress in research into spinal cord injury repair: tissue engineering scaffolds and cell transdifferentiation. J. Neurorestoratol. 7, 196–206. https://doi.org/10.26599/JNR.2019.9040024.

Matsumoto, M., Okada, E., Ichihara, D., et al., 2010. Age-related changes of thoracic and cervical intervertebral discs in asymptomatic subjects. Spine 35 (14), 1359–1364.

Naderi, S., Mertol, T., 2002. Simultaneous cervical and lumbar surgery for combined symptomatic cervical and lumbar spinal stenoses. Clin. Spine Surg. 15 (3), 229–232.

Park, M.S., Moon, S.H., Kim, T.H., et al., 2015. Asymptomatic stenosis in the cervical and thoracic spines of patients with symptomatic lumbar stenosis. Global Spine J. 5 (05), 366–371.

Patel, E.A., Perloff, M.D., 2018. Radicular pain syndromes: cervical, lumbar, and spinal stenosis. In: Seminars in Neurology. vol. 38. Thieme Medical Publishers, pp. 634–639. 06.

Shangzhi, L., Juntao, W., 2018. Clinical observation of Duhuo Shengtang decoction in treating postmenopausal women with osteoporosis. Zhongguo Zhong Yao Za Zhi 34, 181–183.

Sun, K., et al., 2020. A systematic review and meta-analysis for Chinese herbal medicine Duhuo Jisheng decoction in treatment of lumbar disc herniation. A protocol for a systematic review. Medicine 99 (9), e19310. https://doi.org/10.1097/MD.0000000000019310.

Tang, S., Mo, Z., Zhang, R., 2018. Acupuncture for lumbar disc herniation: a systematic review and meta-analysis. Acupunct. Med. 36, 62–70.

Wei, Z., Wei, G., Ping, Z., et al., 2013. Therapeutic effects of Chinese osteopathy in patients with lumbar disc herniation. Am. J. Chin. Med. 41, 983–994.

Wenming, Z., Shangquan, W., Ranxing, Z., et al., 2016. Evidence of Chinese herbal medicine Duhuo Jisheng decoction for knee osteoarthritis: a systematic review of randomised clinical trials. BMJ Open 6, e008973.

Xiaocheng, Z., Xu, J., Xiaozhong, Q., 2018. Advances in the conservative treatment of lumbar intervertebral disc herniation. China Med. Herald 15, 36–39.

Xiaolong, C., Uphar, C., Samuel, L., et al., 2019. Complication rates of different discectomy techniques for the treatment of lumbar disc herniation: a network meta-analysis. Eur. Spine J. 28, 2588–2601.

Xinhua, C., Yihua, F., Yu, Z., 2017. Clinical research progress on herbal oral administration forlumbar disc herniation. J. Tianjin Univ. Tradit. Chin. Med. 36, 237–240.

Xu, M., Hong, J., 2018. The clinical efficacy and prognosis of Duhuo Jisheng decoction in the treatment of cold dampness type lumbar intervertebral disc herniation. Shaanxi Zhongyi 39, 157–159.

Yamada, T., Yoshii, T., Yamamoto, N., et al., 2018. Surgical outcomes for lumbar spinal canal stenosis with coexisting cervical stenosis (tandem spinal stenosis): a retrospective analysis of 565 cases. J. Orthop. Surg. Res. 13 (1), 60.

Yang, X., et al., 2019. Instrumentation-related complications of lumbar degenerative disc diseases treated by minimally invasive transforaminal lumbar interbody fusion. Brain Sci. Adv. 5, 213–219. https://doi.org/10.26599/BSA.2019.9050019.

Yongmei, L., Jia, Q., 2019. Research progress of Duhuo Jisheng decoction in treating lumbar disc herniation. Yunnan J. Tradit. Chin. Med. Mater. Medica 40, 79–81.

Yuanyuan, H., 2017. The curative effect and mechanism of Duhuo Jisheng decoction on lumbar intervertebral disc protrusion of liver-kidney deficiency type. J. Clin. Res. Practice 2, 119–120.

Zang, J., et al., 2019. The treatment of neurotrophic foot and ankle deformity of spinal bifida: 248 cases in single center. J. Neurorestoratol. 7, 153–160. https://doi.org/10.26599/JNR.2019.9040016.

Zhang, Z., et al., 2019. The cell repair research of spinal cord injury: a review of cell transplantation to treat spinal cord injury. J. Neurorestoratol. 7, 55–62. https://doi.org/10.26599/JNR.2019.9040011.

Zhuomao, M., Renwen, Z., Jinfeng, C., et al., 2018. Comparison between oblique pulling spinal manipulation and other treatments for lumbar disc herniation: a systematic review and meta-analysis. J. Manip. Physiol. Ther. 41, 771–779.

Printed in the United States
by Baker & Taylor Publisher Services